SPECTRA OF DISCRETE STRUCTURES

Addressing the active and challenging field of spectral theory, this book develops the general theory of spectra of discrete structures, on graphs, simplicial complexes, and hypergraphs. In fact, hypergraphs have long been neglected in mathematical research, but because of the discovery of Laplace operators that can probe their structure, and their manifold applications from chemical reaction networks to social interactions, they have now become one of the most active areas of interdisciplinary research. The authors' analysis of spectra of discrete structures embeds intuitive and easily visualized examples, which are often quite subtle, within a general mathematical framework. They highlight novel research on Cheeger-type inequalities that connect spectral estimates with the geometry, more precisely the cohesion, of the underlying structure. Establishing mathematical foundations and demonstrating applications, this book will be of interest to graduate students and researchers in mathematics working on the spectral theory of operators on discrete structures.

This title is also available as open access on Cambridge Core.

Jürgen Jost is a founding Director and Scientific Member at the Max Planck Institute for Mathematics in the Sciences. He is also a PI at the federal AI project ScaDS.AI (Dresden/Leipzig), a guest member of the Max Planck Institute for Human Cognitive and Brain Sciences and an external faculty member of the Santa Fe Institute for the Sciences of Complexity. He is a recipient of the Gottfried-Wilhelm-Leibniz award of the Deutsche Forschungsgemeinschaft (DFG,1993) and is also a member of the German National Academy of Sciences Leopoldina, the Academy of Sciences and Literature at Mainz, and the Saxon Academy of Sciences and Humanities at Leipzig.

Raffaella Mulas is a tenured Assistant Professor at the Vrije Universiteit (VU) Amsterdam, where her research is supported by a Veni grant from the Dutch Research Council (NWO) and by a VU Startpremie grant. Previously, she was a Group Leader at the Max Planck Institute for Mathematics in the Sciences, where her research group was supported by a Minerva Fast Track Fellowship from the Max Planck Society, and a postdoc at the Alan Turing Institute in London.

Dong Zhang is an Assistant Professor at Peking University and was previously a postdoctoral researcher at the Max Planck Institute for Mathematics in the Sciences. His research area is discrete analysis, especially nonlinear spectral graph theory and discrete-to-continuous extensions.

CAMBRIDGE STUDIES IN ADVANCED MATHEMATICS

All the titles listed below can be obtained from good booksellers or from Cambridge University Press.
For a complete series listing, visit www.cambridge.org/mathematics.

Already Published
181 A. Agrachev, D. Barilari & U. Boscain *A Comprehensive Introduction to Sub-Riemannian Geometry*
182 N. Nikolski *Toeplitz Matrices and Operators*
183 A. Yekutieli *Derived Categories*
184 C. Demeter *Fourier Restriction, Decoupling and Applications*
185 D. Barnes & C. Roitzhcim *Foundations of Stable Homotopy Theory*
186 V. Vasyunin & A. Volberg *The Bellman Function Technique in Harmonic Analysis*
187 M. Geck & G. Malle *The Character Theory of Finite Groups of Lie Type*
188 B. Richter *Category Theory for Homotopy Theory*
189 R. Willett & G. Yu *Higher Index Theory*
190 A. Bobrowski *Generators of Markov Chains*
191 D. Cao, S. Peng & S. Yan *Singularly Perturbed Methods for Nonlinear Elliptic Problems*
192 E. Kowalski *An Introduction to Probabilistic Number Theory*
193 V. Gorin *Lectures on Random Lozenge Tilings*
194 E. Riehl & D. Verity *Elements of ∞-Category Theory*
195 H. Krause *Homological Theory of Representations*
196 F. Durand & D. Perrin *Dimension Groups and Dynamical Systems*
197 A. Sheffer *Polynomial Methods and Incidence Theory*
198 T. Dobson, A. Malnič & D. Marušič *Symmetry in Graphs*
199 K. S. Kedlaya *p-adic Differential Equations*
200 R. L. Frank, A. Laptev & T. Weidl *Schrödinger Operators: Eigenvalues and Lieb–Thirring Inequalities*
201 J. van Neerven *Functional Analysis*
202 A. Schmeding *An Introduction to Infinite-Dimensional Differential Geometry*
203 F. Cabello Sánchez & J. M. F. Castillo *Homological Methods in Banach Space Theory*
204 G. P. Paternain, M. Salo & G. Uhlmann *Geometric Inverse Problems*
205 V. Platonov, A. Rapinchuk & I. Rapinchuk *Algebraic Groups and Number Theory, I (2nd Edition)*
206 D. Huybrechts *The Geometry of Cubic Hypersurfaces*
207 F. Maggi *Optimal Mass Transport on Euclidean Spaces*
208 R. P. Stanley *Enumerative Combinatorics, II (2nd Edition)*
209 M. Kawakita *Complex Algebraic Threefolds*
210 D. Anderson & W. Fulton *Equivariant Cohomology in Algebraic Geometry*
211 G. Pineda Villavicencio *Polytopes and Graphs*
212 R. Pemantle, M. C. Wilson & S. Melczer *Analytic Combinatorics in Several Variables (2nd Edition)*
213 A. Yadin *Harmonic Functions and Random Walks on Groups*
214 Y. Kawamata *Algebraic Varieties: Minimal Models and Finite Generation*
215 J. Gillespie *Abelian Model Category Theory*
216 L. Anderson *Oriented Matroids*
217 Y. Motohashi *Essays in Classical Number Theory*
218 H. L. Montgomery & R. C. Vaughan *Multiplicative Number Theory II*
219 E. Rijke *Introduction to Homotopy Type Theory*
220 N. Berestycki & E. Powell *Gaussian Free Field and Liouville Quantum Gravity*
221 J. Jost, R. Mulas & D. Zhang *Spectra of Discrete Structures*

Spectra of Discrete Structures

JÜRGEN JOST
Max Planck Institute for Mathematics in the Sciences

RAFFAELLA MULAS
Vrije Universiteit Amsterdam

DONG ZHANG
Peking University

CAMBRIDGE
UNIVERSITY PRESS

Shaftesbury Road, Cambridge CB2 8EA, United Kingdom

One Liberty Plaza, 20th Floor, New York, NY 10006, USA

477 Williamstown Road, Port Melbourne, VIC 3207, Australia

314–321, 3rd Floor, Plot 3, Splendor Forum, Jasola District Centre,
New Delhi – 110025, India

Cambridge University Press is part of Cambridge University Press & Assessment,
a department of the University of Cambridge.

We share the University's mission to contribute to society through the pursuit of
education, learning and research at the highest international levels of excellence.

www.cambridge.org
Information on this title: www.cambridge.org/9781009641845
DOI: 10.1017/9781009641821

First published 2026

A catalogue record for this publication is available from the British Library

Library of Congress Cataloging-in-Publication Data
Names: Jost, Jürgen, 1956– author | Mulas, Raffaella, 1992– author | Zhang,
Dong (Of Beijing da xue) author
Title: Spectra of discrete structures / Jürgen Jost, Raffaella Mulas, Dong Zhang.
Description: Cambridge ; New York, NY : Cambridge University Press, 2026. |
Series: Cambridge studies in advanced mathematics ; 221 | Includes
bibliographical references and index.
Identifiers: LCCN 2025036413 (print) | LCCN 2025036414 (ebook) | ISBN
9781009641845 hardback | ISBN 9781009641821 ebook
Subjects: LCSH: Discrete mathematics | Spectral theory (Mathematics)
Classification: LCC QA297.4 .J67 2026 (print) | LCC QA297.4 (ebook)
LC record available at https://lccn.loc.gov/2025036413
LC ebook record available at https://lccn.loc.gov/2025036414

ISBN 978-1-009-64184-5 Hardback

Contents

Preface   *page* ix
Acknowledgments   xiv
Organization of the Book   xv

PART I   BASICS: FOUNDATIONAL MATERIAL,
ELEMENTARY ASPECTS, AND EXAMPLES

1   Introduction   3
1.1   Linear Partial Differential Equations, the Classical
      Laplace Operator, and Eigenvalue Problems   3
1.2   Laplace-Type Operators on Graphs   9
1.3   Other Sources and Directions of the Theory   12

2   The Abstract Setting   14
2.1   A Category Theoretical Perspective   14
      2.1.1   The Static Picture   14
      2.1.2   The Dynamical Picture   15
      2.1.3   Probing a (Dynamical) Structure   17
2.2   Relations and Norms   19
2.3   Complexes and Orientations   21
2.4   (Oriented) Hypergraphs and their Higher Order Analogs   23
      2.4.1   Generalizations   23
      2.4.2   Hypergraphs and Hypersimplical Complexes   25
      2.4.3   Oriented Hypergraphs and Oriented
              Hypersimplical Complexes   26
2.5   Scalar Products and Laplace Operators   28
      2.5.1   Simplicial Complexes   28
      2.5.2   Relative Complexes   32

v

2.5.3   Laplace Operators as Averaging Operators   34
2.5.4   Oriented Hypergraphs   36
2.6   Rayleigh Quotients and their Generalizations   37
2.6.1   Rayleigh Quotients   37
2.6.2   Critical Points and p-Laplacians   40
2.7   Symmetries   44
2.8   Homomorphisms   47
2.9   General Setting   48

**3   The Classical Case: Eigenvalues on Euclidean Domains and
      Riemannian Manifolds**   50
3.1   The Laplacian on Euclidean Domains   51
3.2   The Laplace–Beltrami Operator and the Hodge Laplacian
      on Riemannian Manifolds   59
3.3   The p-Laplacian   68

4   First Properties of the Spectrum   76
4.1   The Laplacian and Its Spectrum on Graphs   76
4.1.1   General Aspects   76
4.1.2   The Algebraic Graph Laplacian   80
4.1.3   The Normalized Graph Laplacian   85
4.2   Some Elementary Estimates   109
4.3   Cutting a Graph Optimally, or the Pólya–Cheeger
      Constant and Eigenvalue Estimates   113
4.4   Dividing a Graph Optimally into Opposite Classes, or
      the Dual Cheeger Constant and Estimates for the Largest
      Eigenvalue   123
4.5   Neighborhood Graphs   128
4.6   Signed Graphs   138
4.7   The General Case: Directed Graphs with Arbitrary Weights 150
4.8   Simplicial Complexes   155
4.9   Oriented Hypergraphs   161

**5   Eigenvalues and Eigenfunctions of the Laplace Operator via
      Floer Theory**   172
5.1   Rayleigh Quotients, Eigenfunctions, and Eigenvalues   172
5.2   Gradient Flow   175

PART II   EIGENVALUES AND EIGENFUNCTIONS ON
SIMPLICIAL COMPLEXES AND HYPERGRAPHS

6   Lovász Extensions   181
6.1   Extensions and Eigenvalue Problems   181
6.2   Definitions and Basic Properties of Lovász Extensions   183
6.3   Eigenvalue Problem for a Pair of Lovász Extensions   188
6.4   Lovász Extension and Cheeger Inequalities   198

7   Discrete p-Laplacians   205
7.1   Graph p-Laplacians ($p > 1$)   205
7.2   The Graph 1-Laplacian   213
7.3   The Graph ∞-Laplacian   224
7.4 p-Laplacians on Oriented Hypergraphs   226

8   Cheeger Inequalities   232
8.1   Cheeger Inequalities on Graphs   232
8.2   Cheeger Inequalities on Signed Graphs   233
8.3   Cheeger Inequalities on Simplicial Complexes   237
8.4   Cheeger Inequalities on Hypergraphs   255

9   Nodal Domains   269
9.1   Nodal Domain Theorems on Graphs and Matrices   270
9.2   Nonlinear Nodal Domain Theorems on Graphs   276
9.3   Nodal Domain Theorems on Hypergraphs   287

PART III   ADDITIONAL TOPICS: INTERLACING,
TENSORS, NONBACKTRACKING LAPLACIANS,
AND APPLICATIONS

10   Interlacing and Spectral Classes   297
10.1 Interlacing   297
      10.1.1   Deletion of Subcomplexes   299
      10.1.2   Simplicial Maps   305
      10.1.3   Coverings   311
      10.1.4   Contraction   314
10.2 Spectral Classes   319
      10.2.1   Asymptotic Classes of Graphs   319
      10.2.2   Spectral Asymptotics   322

11   Spectral Theory of Weighted Hypergraphs via Tensors   327
11.1 Tensors   327
11.2 Hypergraph Tensors   331

11.3 First Spectral Properties   333
11.4 Θ-Duplicate Vertices   341
11.5 The Hyperflower   343
11.6 Spectral Symmetries   348

12   The Nonbacktracking Laplacian   351
12.1 Basic Definitions   352
12.2 The Nonbacktracking Graph   356
12.3 First Spectral Properties of the Nonbacktracking Laplacian   357
12.4 Spectral Gap from 1   360
12.5 Cycles   363
12.6 Isospectrality   370

13   Applications   373
13.1 Social Sciences   373
        13.1.1   Eigenvector Centrality   374
        13.1.2   Cohesion of Networks   375
13.2 Biological Networks   376
13.3 Computer Science   378
13.4 Nonlinear Dynamics   384

Remarks on Notation   388
References   389
Index   403

Preface

Invariants associated with mathematical objects or data sets should be informative and interpretable, easy to compute, and stable in the presence of small perturbations. In particular, they should allow us to quickly distinguish qualitatively different structures. Certain invariants that satisfy these requirements can be obtained from the spectra, that is, the collection of eigenvalues, of linear operators for functions defined on that structure. And arguably the most useful ones are Laplace-type operators. Such a Laplace-type operator can be defined when a structure consists of elements that may stand in relations permitting the definition of a neighborhood relation. Then, one compares the value of a function at an element to the average of the function values at its neighbors. Such Laplace operators were first investigated for smooth structures, such as domains in Euclidean space or Riemannian manifolds. There were some analytical issues to be overcome, which among other things, stimulated the development of functional analysis and the theory of elliptic partial differential equations.

In this monograph, we look at structures that are discrete and finite, and therefore, there are no such analytical difficulties. Nevertheless, both historically and in our own research, the smooth theory could inspire much of the discrete theory. The neighborhood structures that we are considering could be those of a graph, a hypergraph, or a simplicial complex. In the case of a graph, we only have relations between pairs of elements. When relations can involve larger sets of elements, we obtain a hypergraph. And when any nontrivial subset of related elements also supports a relation, we get a simplicial complex. Although a simplicial complex can be considered as a special case of a hypergraph, the theories that we develop and explore for the two types of structures are quite different. An important insight underlying our theory is that a hypergraph should not be viewed as a deficient simplicial complex, but rather as a structure in its own right that can be enhanced with an orientation.

In graph theory, one usually considers the algebraic Laplacian which has many useful properties. While our theory is general, when the analysis becomes more specific, we shall mostly consider the normalized Laplacian instead for several reasons. The normalized Laplacian satisfies a conservation law and is therefore naturally related to diffusion processes and random walks. Also, the analogy with the smooth case seems stronger than that with the algebraic Laplacian. And finally, it seems better suited for the more abstract aspects of the theory that cover not only graphs but also hypergraphs and simplicial complexes.

So, what can the spectrum of a Laplace operator tell us about the geometry of the underlying structure? Perhaps most importantly, the smallest nonvanishing eigenvalue estimates how difficult it is to divide the structure, by severing as few relations as possible, into two large components, known as Cheeger cut (Cheeger had established such a relation in the Riemannian case). This has long been known for graphs, but we will derive new results in this direction for simplicial complexes and hypergraphs. Also, we will systematically investigate the nodal sets of the corresponding eigenfunction, although the nodal decomposition need not necessarily provide an optimal Cheeger cut. In order to understand this and other phenomena more deeply, we will also systematically study nonlinear versions of the Laplace operator, for a parameter $1 \leq p \leq \infty$. Here, $p = 2$ corresponds to the standard Laplacian, while the extreme cases $p = 1$ and $p = \infty$ are the most difficult, but also most insightful. An important tool that we shall systematically utilize will be the Lovász extension that turns a discrete optimization problem into a continuous one.

The Laplacian spectrum does not determine the underlying structure completely. For instance, there are isospectral graphs, that is, non-isomorphic graphs with the same Laplacian spectrum. Nevertheless, many classes of (hyper)graphs can be characterized by their spectral properties, and we can also look at the spectral asymptotics for families of (hyper)graphs. Moreover, small changes, such as additions or deletions of vertices or edges, have only a small influence on the spectrum, and this can be made precise by interlacing inequalities. Our corresponding analysis will depend on the systematic use of so-called Rayleigh quotients. These are functions defined on a sphere in some Euclidean space whose critical values yield the eigenvalues and whose critical points provide the corresponding eigenfunctions. This naturally leads to Morse theoretical conceptualizations, and it provides us with the opportunity to illustrate the geometry of Floer theory.

For the hypergraph Laplacian, there are various possibilities. On the one hand, we can study a previously introduced linear operator that uses orientations, which led us to the concept of a chemical hypergraph where a hyperedge is conceived as representing a chemical reaction that connects two sets of vertices: the educts and the products. On the other hand, we can also define a tensorial Laplace operator whose eigenvalue equations are polynomial and no longer linear as in the graph case. Both approaches yield mathematical interesting results that is useful for the quantitative and qualitative analysis of hypergraphs.

Our original interest in the spectral theory of discrete structures was partly motivated by questions arising from the analysis of empirical data in fields such as biology, neuroscience, chemistry, or the social sciences. Many such data arise as networks and can therefore be modeled as graphs, simplicial complexes, or hypergraphs. We have gene regulation, protein interaction, and neural or friendship networks, for example, which can be modeled as graphs. In topological data analysis, simplicial complexes are constructed from discrete metric data. Chemical reaction systems or coauthorship networks between scientists are naturally considered as hypergraphs. Spectra of linear operators naturally led to useful invariants for investigating, characterizing, or classifying such structures. Spectral methods have therefore found, and continue to find, many applications in network analysis across a wide range of empirical domains. The systematic study of network properties was first undertaken by social scientists. Since we did not want to start a mathematical monograph with a discussion of social science literature, these applications will be discussed only in Chapter 13. Of course, we need to be selective there and will highlight only a few key examples.

This motivation also had consequences for our research design. Empirical data are neither fully regular nor completely random. Therefore, on the one hand, we have not explored in depth the theory of graphs with specific regular structures. For instance, the important theory of expander graphs uses the strong regularity assumption that all vertices in a graph have the same number of neighbors. This is also the case for the Cayley graphs, which represent the structure of a discrete group geometrically. More generally, algebraic graph theory is a well-researched field, but this is not the thrust of our research. On the other hand, we have also not delved into random graph theory. This theory started with the work of Erdös and Rényi, who connected a given set of vertices in a purely random manner and then investigated how the resulting structure almost surely depended on the density of the random connections. This has found many applications in mathematics and statistical physics, for instance, in the direction of Wigner's semicircle law for spectra of random matrices. More generally, methods from random graph theory are powerful for deriving existence results for graphs with particular structural properties. Even though, according to some such arguments, almost all graphs may exhibit a particular property, the proofs are not explicit, and it may still be a challenge to exhibit a concrete graph with the property in question. Also, in network analysis and statistical physics, several other schemes for the construction of random graphs with specific features, in particular those exhibiting power law behavior, have been proposed, and again, this is an active field of research that is not covered in our book.

Thus, we investigate the spectral properties of general graphs, simplicial complexes, and hypergraphs, with a theory that is as systematic and comprehensive as possible. However, we have found that complete generality is not feasible, in the sense that hypergraphs should better not be considered as generalized or somehow deficient simplicial complexes. To recall, in a hypergraph, sets of vertices are related or connected by hyperedges. For a simplicial complex, it is required that whenever such a vertex set is related, then all its subsets stand in relation as well. Simplicial complexes are therefore well suited for the schemes of algebraic topology, more precisely (co)homology theory with its boundary relations. Thus, the (co)homology theory of simplicial complexes is clear and canonical. While people have also proposed such theories for hypergraphs, none can be considered clear or canonical. In that situation, one of the key insights underlying this book, as indicated previously, is that Laplace operators of hypergraphs should be constructed in a fundamentally different way with no analog for simplicial complexes. In fact, we impose and utilize an additional structure on hypergraphs, that of an orientation of hyperedges, and this eventually drives the theory.

As already mentioned, we have to relate the properties of spectra to the geometry of the underlying objects. For this book, the most important of these relations are the Cheeger-type inequalities. In such inequalities, a bound on an eigenvalue indicates how easy or difficult it is to separate a structure into disjoint substructures by cutting as few relations as possible. Other such inequalities estimate how close a structure is to or how different it is from a bipartite structure, that is, one where we have two classes of elements, with relations only between elements of opposite classes.

In fact, a general theory of Cheeger-type inequalities constitutes the core of this book. A Cheeger-type inequality relates an eigenvalue of a Laplace operator with a global quantity of the underlying structure, expressing how difficult it is to cut that structure into two large pieces. The larger the part of the structure, for instance, an edge set for a graph, needed for the cut, the larger that constant, and according to the inequality, that is bounded from above and below by an eigenvalue of the Laplacian. There are two conceptual issues. First, the Cheeger-type constant is a discrete quantity, whereas the Laplacian and its eigenvalues are continuous objects, in the sense that they succumb to the tools of analysis. This issue will be systematically resolved via Lovász-type extensions that turn a discrete optimization problem into a continuous one. The second issue is that the Cheeger-type constant is an L^1-quantity, whereas the Laplacian naturally operates on L^2-spaces. That issue will be resolved by looking at L^p-Laplacians. The case $p \to 1$ is the most difficult, but also the most useful, as it turns the inequality into an equality. Equipped with such techniques, we can also derive novel Cheeger-type inequalities for simplicial complexes and hypergraphs.

We shall supply many examples, particularly those that exhibit certain extremal properties of the spectra that we investigate.

Our interest is with *finite* structures. Recently, two monographs [171, 174] appeared that address the spectral theory of infinite graphs. They are complementary to ours. Another book [178] also addresses relations between the spectrum and the geometry of the underlying graph, but with a focus on the trace formula and inverse problems, that is, topics not covered in our book. In fact, Kurasov considers graphs as metric spaces, which leads to infinite spectra. This is very different from our approach that works with finite structures.

Acknowledgments

Some parts of this monograph are based on, or represent, a joint work with Fatihcan Atay, Anirban Banerjee, Frank Bauer, Tobias Böhle, Michael Casey, Kung-Ching Chang, Marzieh Eidi, Francesco Galuppi, Danijela Horak, Bobo Hua, Christian Kuehn, Wilmer Leal, Shiping Liu, Florentin Münch, Areejit Samal, Emil Saucan, Peter Stadler, Leo Torres, Lorenzo Venturello, and Giulio Zucal.

Jürgen Jost was partially supported by grant number 1514 from the German-Israeli Foundation (GIF). Raffaella Mulas was supported by the Dutch Research Council (NWO) through grant number VI.Veni.232.002. Dong Zhang was supported by grants from the National Natural Science Foundation of China (No. 12401443). Funding from Max Plank Digital Library made it possible for this book to be published open access, making the digital version freely available for anyone to read and reuse under a Creative Commons license.

Organization of the Book

This book consists of three parts. Part I gives an introduction, provides basic and foundational material, and presents the more elementary case of graph Laplacians, with many examples.

Part II is the core of this book. It systematically develops the theory of eigenvalues and eigenfunctions. A main aim is the derivation of Cheeger-type inequalities that relate the smallest positive eigenvalue of a Laplacian with the difficulty of decomposing the underlying structure. This difficulty is expressed by a constant named after Cheeger, who first considered the analogous problem on Riemannian manifolds. Our systematic approach needs to overcome two difficulties. The first difficulty is that this Cheeger-type constant is a discrete quantity, whereas the Laplacian and its eigenvalues are continuous ones, in the sense of critical points of an operator on a function space. This is addressed by the scheme of Lovász extension that translates a discrete optimization problem into a continuous one. The second difficulty is that these Cheeger-type constants are naturally L^1-quantities, while the Laplacian operates on L^2-spaces. This will be overcome by deforming the ordinary Laplacian into an L^1-operator. The spectral theory of the latter operator, however, is more difficult than that of the ordinary Laplacian.

Part III rounds off this book with selected additional topics and a discussion of applications of spectral theory.

Here, the individual chapters are explained in detail. The introductory Chapter 1 sets the historical stage for our project. Chapters 2–4 are independent of each other, and a reader can therefore begin where he or she likes. Chapter 2 develops the abstract perspective that will guide the later chapters, and it should therefore be the right starting point for those who like the systematic approach. Chapter 3 can serve as a motivation or introduction for those trained in classical analysis and/or Riemannian geometry, as it describes the eigenvalue problems for Laplacians in those contexts. This theory was developed earlier than the discrete case with which we are concerned in this book, and in fact it motivated some of the developments in the latter case. Finally, Chapter 4 presents elementary examples and basic constructions in detail. Readers who prefer to see concrete examples initailly may therefore choose to start with this chapter.

Since these three Chapters 2–4 are designed to be independent of each other, there will be some overlap between them, that is, certain fundamental constructions are presented, and some important results are proved in each of them.

Chapter 5 shows the construction of Floer theory, which can naturally capture the geometry of the eigenfunctions and eigenvalues in function spaces. This chapter can be read directly after Chapter 2, and perhaps even independently. The subsequent chapters do not depend on this chapter.

This concludes Part I, and we then move to the Part II. Chapter 6 develops constructions relating discrete and continuous optimization problems that will be applied in Chapters 7 and 8. The first of those, Chapter 7, enlarges the linear schemes dominating much of the rest of the book by more general nonlinear ones, considering the p-Laplacians for $1 \leq p \leq \infty$, and not only the case $p = 2$ underlying the linear theory. These will be useful in Chapter 8, where we solve the general problem of deriving Cheeger-type inequalities for simplicial complexes and hypergraphs.

Chapter 9 derives bounds on the number of nodal domains of eigenfunctions, that is, connected regions with a constant sign of such an eigenfunction. To put this into perspective, typically the lowest eigenvalue is 0, with a constant eigenfunction, that is, one that never changes sign. An eigenfunction for the next eigenvalue has only two such nodal domains, one where it is positive and another where it is negative throughout. We can similarly control the number of nodal domains of higher eigenfunctions.

This concludes Part II. In the Part III, a few additional topics will covered. The Chapters 10–13 are independent of each other and of those in Part II.

In Chapter 10, we control how much certain local operations affect the spectrum of a simplicial complex, and this also leads to defining asymptotic classes of graphs or other objects in terms of their spectral properties. This should be compared with notions of graph limits developed in other contexts.

Chapter 11 develops yet another perspective on the spectral theory of higher order objects, that of eigenvalues of tensors.

The non-backtracking Laplacian treated in Chapter 12 exhibits some novel features that can, for instance, be exploited for more efficiently searching a graph.

And finally, as already indicated, Chapter 13 discusses various applications, in the social and biological sciences, in computer science, for general network

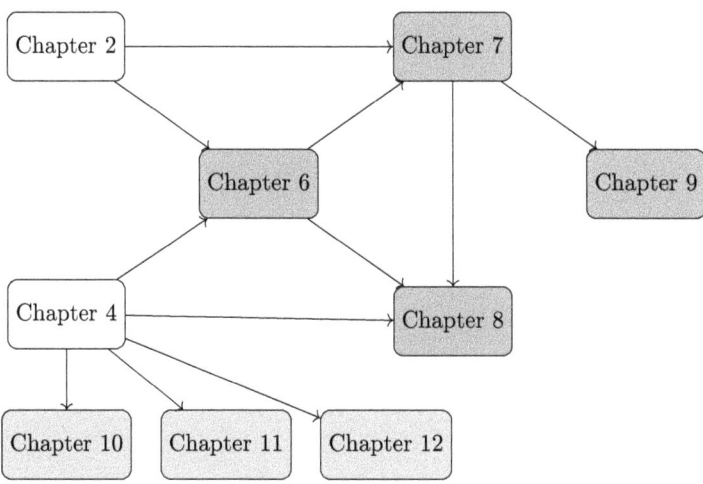

Figure 0.1 Interconnections among the chapters of this book.

analysis, and for emerging features in networks of coupled nonlinear oscillators, a research in statistical physics and nonlinear dynamics that is important for understanding pattern formation.

In summary, this book collects spectral properties for certain discrete structures, and Figure 0.1 illustrates interconnections among the chapters.

The chapters will start with a short summary in italic.

PART I

Basics: Foundational Material, Elementary
Aspects, and Examples

1

Introduction

This chapter provides a historical introduction. Spectra first appeared in the study of linear differential operators, such as the Laplace, the heat, and the wave operator, which originated in the eighteenth century. A Laplace operator on networks was first introduced by Kirchhoff in the nineteenth century.

1.1 Linear Partial Differential Equations, the Classical Laplace Operator, and Eigenvalue Problems

In recent decades, many empirical structures have been described as networks that is, constituted by (possibly weighted and/or directed) relations between basic constituents or elements. In fact, network analysis is a well-established subject within the general theory of complex systems, and quite a number of monographs have been devoted to it, besides an enormous number of research papers in several different disciplines. Mathematically, such a network is described as a (possibly weighted and/or directed) graph, or perhaps as a more general structure such as a simplicial complex or a hypergraph. The resulting graphs, however, typically are less regular or possess less structure than those traditionally studied in graph theory. Also, they can be quite large, if not sometimes gigantic. Since computational approaches to problems such as subgraph isomorphism tend to be difficult and often NP hard or even complete [65, 235, 242], efficient techniques are needed to study these empirical structures. One such approach includes the introduction of a some linear operator defined in terms of the network structure that operates on functions defined on the vertices. As a linear operator, it possesses a spectrum consisting of finitely many eigenvalues, as the structure, while possibly large, is finite. And when the operator is symmetric, these eigenvalues are real. The computation of the eigenvalues of symmetric linear operators is computationally relatively easy, and there exist fast algorithms that can produce the spectrum, or some relevant parts of it, for quite large networks. The question is what this spectrum can tell us about the underlying network. One quickly realizes that there are so-called graphs, that is, non-isomorphic graphs that nevertheless have the same spectrum, and therefore the spectrum cannot recover the graph completely. Nevertheless, as we shall explore in this monograph, many important qualitative properties of the underlying graph can be inferred from its spectrum. And perhaps we may not even be interested in all details, as empirical networks usually possess a certain amount of randomness, resulting from the stochastic process that brought them into being, such as biological evolution. While, as mentioned, such empirical networks tend to be less regular, more diffuse, partly random, and in any case qualitatively different from those preferred by many graph theorists, there exists an important mathematical paradigm that has become fundamental for mathematical approaches to network analysis. This refers to the 1959 paper [91] by Paul Erdős and Alfréd Rényi that introduced the concept of a random graph [91] and derived and explained several of its characteristic properties. The Book [32] is a systematic presentation of the theory of random graphs. More recently, other types of random networks were introduced that seem to match some of the features better than that are typically found in empirical networks. Since here we present a general approach to the spectral analysis of such structures, these details are not so important for our purposes.

The spectral theory of linear operators is a wide subject, with diverse sources from mathematics and physics and with ramifications into many different, perhaps almost all, mathematical fields. Even though the setting that we adopt in this book is more confined and the basic story has nothing to do with network analysis, it might be useful to provide a wider historical perspective.

The story might start with the introduction of the wave operator, also called the d'Alembertian, by Jean-Baptiste Le Rond d'Alembert (1717–83) in 1747. We can follow the account of the story given by Bernhard Riemann (1826–66) in his habilitation thesis (see pp. 227–271 in [236]) (profound historical information can also be found in [80]). D'Alembert was the first mathematician to formulate a partial differential equation. He postulated that for a vibrating string, modeled as a one-dimensional line of length ℓ, its extension $y(x,t)$ at time t, that is, its deviation from rest, on the interval $x \in [0,\ell]$ satisfies the following differential equation:

$$\frac{\partial^2 y}{\partial t^2} = \alpha^2 \frac{\partial^2 y}{\partial x^2}, \tag{1.1.1}$$

where α is independent of t, and also of x, if the string is uniformly thick. D'Alembert found that the general solution is given by

$$f(x + \alpha t) + g(x - \alpha t) \tag{1.1.2}$$

by introducing the variables $\xi = x + \alpha t$ and $\eta = x - \alpha t$ and rewriting (1.1.1) as

$$4\frac{\partial^2 y}{\partial \xi \partial \eta} = 0. \tag{1.1.3}$$

This led to a dispute with Leonhard Euler (1707–83) about which types of functions f and g should be admitted in (1.1.2), with d'Alembert insisting that the functions should be analytic. Later, Daniel Bernoulli (1700–82) entered the debate and proposed to consider solutions of the form

$$y = \sum a_k \sin \frac{k\pi x}{\ell} \cos \frac{k\pi \alpha(t - \beta_k)}{\ell}. \tag{1.1.4}$$

(It is assumed that the string is fixed at its endpoints, that is, $y(0,t) = y(\ell,t) = 0$.)

But Euler argued at the time that a sum of analytic functions could not represent general functions of the form (1.1.2), and therefore Bernoulli's approach was more restricted than his own. Then, also the young Joseph-Louis Lagrange (1736–1813) entered the debate, but a solution had to wait for almost 50 years until Jean-Baptiste Joseph Fourier (1768–1830) proposed his general theory of series expansions in terms of trigonometric functions. A mathematically rigorous proof of such an expansion for a general class of functions was subsequently provided by Peter Gustav Lejeune Dirichlet (1805–59). For the general solution, Riemann developed the modern theory of integration, which was continued by Henri Lebesgue (1875–1941). The Lebesgue integral enabled mathematicians to define the spaces $L^p(\Omega)$ of functions whose p-th power is integrable on some domain Ω. Importantly, these spaces are complete with respect to the corresponding norm. This made the development of modern functional analysis possible, with the fundamental concept of a Banach space. In particular, we have the Hilbert space $L^2(\Omega)$, which is indispensable for the mathematical formulation of quantum mechanics.

In a somewhat different direction, the analysis of the set of points where a Fourier series might not converge ultimately led Georg Cantor (1845–1918) to the creation of modern set theory.

But what does all that have to do with spectra? The answer is easy. When we try to compute a solution of (1.1.1) by separation of variables, that is, make the ansatz

$$y(x,t) = h(x)\phi(t), \tag{1.1.5}$$

the equation becomes, using now subscripts for partial derivatives,

$$h(x)\phi_{tt}(t) = \alpha^2 \phi(t) h_{xx}(x). \tag{1.1.6}$$

Therefore,

$$\frac{h_{xx}(x)}{h(x)} = \frac{1}{\alpha^2} \frac{\phi_{tt}(t)}{\phi(t)} \tag{1.1.7}$$

both have to be constant. When we substitute the value of this constant as $-\lambda$, we obtain the following eigenvalue equation:

$$- h_{xx}(x) = \lambda h(x). \tag{1.1.8}$$

On the interval $[0, \ell]$ with the boundary conditions $h(0) = 0 = h(\ell)$, (1.1.8) is satisfied by the following trigonometric functions:

$$h(x) = \sin \frac{n\pi x}{\ell}, \tag{1.1.9}$$

for the values

$$\lambda = \left(\frac{n\pi}{\ell}\right)^2. \tag{1.1.10}$$

Thus, in modern terminology, the functions (1.1.9) are the eigenfunctions of the operator $-h_{xx}$, the values (1.1.10) are the eigenvalues, and the Fourier expansion justifying Bernoulli's solution (1.1.4) is simply an expansion in terms of the eigenfunctions of the operator $-h_{xx}$. The eigenvalues (1.1.10) then represent the possible vibration modes of the string.

When we have vibrations of objects described by more than one spatial dimension, we obtain the wave equation as follows:

$$\frac{\partial^2 y}{\partial t^2} = -\alpha^2 \Delta y(x, t), \tag{1.1.11}$$

with the operator

$$- \Delta h(x) = \sum_{i=1}^{n} \frac{\partial^2 h}{\partial (x^i)^2}(x), \tag{1.1.12}$$

for $x = (x^1, \ldots, x^n)$. The minus sign in (1.1.12) is historically incorrect, but in line with the conventions adopted in this book. The operator $(-)\Delta$ is called the *Laplace operator*, or more shortly, the *Laplacian*, although it was first introduced and studied by Lagrange [129]. It played a fundamental role in Laplace's *Celestial Mechanics*.

For a solution of (1.1.11), we may again propose a separation of variables, as follows:

$$y(x, t) = h(x)\phi(t), \tag{1.1.13}$$

and the equation becomes

$$h(x)\phi_{tt}(t) = -\alpha^2 \phi(t)\Delta h(x). \tag{1.1.14}$$

Therefore,

$$\frac{-\Delta h(x)}{h(x)} = \frac{1}{\alpha^2}\frac{\phi_{tt}(t)}{\phi(t)} = -\lambda \tag{1.1.15}$$

for a constant λ. We obtain the following eigenvalue equation:

$$\Delta_{xx}(x) = \lambda h(x), \tag{1.1.16}$$

for the Laplace operator, which, for given boundary conditions on a bounded domain, again has nontrivial solutions only for a discrete set of eigenvalues $\lambda \geq 0$. As before, a general solution of (1.1.11) can be expanded in terms of the eigenfunctions of Δ. The eigenvalues and eigenfunctions depend on the geometry of the domain Ω, and therefore in general can no longer be written in terms of trigonometric functions (unless Ω is of a rectangular form $[0, \ell_1] \times \ldots \times [0, \ell_n]$), but the principle of an expansion of a solution of a linear differential equation in terms of eigenfunctions remains the same.

Fourier had considered the heat equation:

$$\frac{\partial y}{\partial t} = -\Delta y(x,t), \tag{1.1.17}$$

but the principle of an expansion in terms of eigenfunctions of $-\Delta$ applies here as well.

In this book, we shall investigate this principle not for differential operators on Euclidean (or more general) domains, but rather for difference operators on discrete structures. But before we get into that, we should discuss some other historical strands.

When Bernhard Riemann had introduced the concepts of Riemannian geometry in his habilitation address [237, 238], Eugenio Beltrami (1835–1900) realized that in Riemannian geometry, there is a natural analog of the Laplace operator. This operator is now called the Laplace–Beltrami operator. Like the Laplace operator, it operates on functions, now defined on Riemannian manifolds. A further important generalization emerged when operating with differential forms. Élie Cartan (1869–1951) had used the exterior algebra created by Hermann Graßmann (1809–77) to develop a calculus of differential forms. When ω is a p-form, that is, in local coordinates, a linear combination of expressions of the type $\eta(x)dx^{i_1} \wedge \ldots \wedge dx^{i_p}$, one can define its differential $d_p\omega$, which for this expression is the $(p+1)$-form,

$$d_p\eta(x)dx^{i_1} \wedge \ldots \wedge dx^{i_p} = \sum_j \frac{\partial \eta(x)}{\partial x^j}dx^j \wedge dx^{i_1} \wedge \ldots \wedge dx^{i_p}. \tag{1.1.18}$$

Since the Riemannian metric induces a natural scalar product on q-forms, one has an adjoint operator $d_p^\star$ mapping a $(p + 1)$-form to a p-form, and one considers the operator,

$$\Delta_p = d_{p-1} d_{p-1}^\star + d_p d_p^\star. \tag{1.1.19}$$

This operator is called the *Hodge Laplacian*, but the theory was developed more or less simultaneously and independently by Hermann Weyl (1885–1955), Georges de Rham (1903–90), William Hodge (1903–75), and Kunihiko Kodaira (1915–97). For $p = 0$ (where $d_{-1} = 0$), it becomes the Laplace–Beltrami operator, which, when the metric is Euclidean, is the Laplace operator. Beno Eckmann (1917–2008) then introduced a combinatorial operator that is analogous to Δ_p for simplicial complexes. The eigenvalues of all these operators, the Laplace operator on a domain in Euclidean space, the Laplace–Beltrami operator on a Riemannian manifold, the Hodge Laplacian on differential forms, or the Eckmann Laplacian on simplicial complexes, all encode important geometric information about the underlying spaces.

But before we come to that, the reader may have wondered about the origin of the terms *eigenvalue* and *eigenfunction*.

While Gottfried Wilhelm Leibniz (1646–1716) had already introduced the determinant of a matrix, and linear algebra in Cartesian spaces had been developed by Arthur Cayley (1825–91) and Graßmann, but was generally accepted only much later, it was the theory of integral equations that ultimately led to general spectral theory. When David Hilbert (1862–1943) learned about the work of Erik Ivar Fredholm (1866–1927) on general integral equations, he realized that for symmetric equations, a much richer theory could be developed. This led to the analysis of compact linear operators in Hilbert spaces in terms of their spectra, that is, the collection of their eigenvalues, and more generally, to bounded or even unbounded such operators. It turned out that this provided the appropriate mathematical framework for developing the quantum mechanics discovered by Werner Heisenberg (1901–76) and Erwin Schrödinger (1887–1961), at the hands of John von Neumann (1903–57). There, operators represent quantum mechanical observables, and their eigenvalues are the possible outcomes of observations. Also, the Schrödinger equation is naturally studied on spaces of complex-valued functions in the Hilbert space L^2. And this is true more generally for linear partial differential equations, such as the Laplace equation $\Delta y = 0$, the heat equation (1.1.17), and the wave equation (1.1.11), where Fourier expansions converge in Hilbert spaces. And the terms *eigenvalue* and *eigenfunction*, to finally come to that, are derived from Hilbert's German expressions *Eigenwert* and *Eigenfunktion* in

[134]. (The German *eigen* means *proper, particular, characteristic, and inherent.*) A curious anecdote highlights the humorous misunderstanding that they were named after the biophysical chemist and Nobel laureate Manfred Eigen (1927–2019).

1.2 Laplace-Type Operators on Graphs

The theory of linear partial differential equations, Fourier expansions, or linear operators in Hilbert spaces had to overcome many analytical obstacles for their successful development and applications. In particular, it took a long time before mathematicians had developed the appropriate versions of convergence and could successfully apply them in those theories. In this book, however, we treat a situation that faces none of these analytical difficulties. We consider Laplace-type operators on finite discrete structures, such as graphs, simplicial complexes, or hypergraphs. Strangely, the corresponding theory was developed much later than its analytical analog just described, and it really got off the ground only when it became inspired by the latter.

Nevertheless, there does exist an origin from the nineteenth century. In 1847, Gustav Kirchhoff (1824–87) introduced a discrete version of the Laplace operator, for the study of electrical networks [140, 173]. He defined the operator of a graph with adjacency matrix A and diagonal degree matrix D as

$$K := D - A. \tag{1.2.1}$$

Here, for a graph with N vertices, the adjacency matrix has entry $a_{vv'} = 1$ when the vertices $v \neq v'$ are connected by an edge, and $a_{vv'} = 0$ else, and the degree of vertex v is $\deg_v = \sum'_v a_{vv'}$. (This is for an unweighted graph; in a weighted graph, we can consider $a_{vv'}$ as the weight of the edge from v to v'.) The spectrum of this operator has been widely studied in algebraic graph theory (see, for instance, [29, 38, 72]).

While this operator can be seen as a discrete version of the Laplace operator, a version that is a better analog in many respects is the Chung Laplacian, as follows:

$$\mathcal{L} := \mathrm{Id} - D^{-1/2} A D^{-1/2}, \tag{1.2.2}$$

where Id is the $N \times N$ identity matrix. This operator was first systematically studied by Fan Chung [61]. For instance, $\mathcal{L}$ is the operator that generates random walks on the graph, in the sense of [68] (see [146]). Putting

$$\Delta := D^{-1/2} \mathcal{L} D^{1/2}, \tag{1.2.3}$$

the matrices Δ and $\mathcal{L}$ become similar, which implies that they have the same spectrum, that is, the same eigenvalues counted with multiplicity. While K in (1.2.1) and the operator (1.2.2) are obviously symmetric operators, the operator (1.2.3) is also symmetric with respect to some scalar product (see [145]), as will become clear below, see (4.1.32).

As already indicated several times, the general question is what the spectrum of some Laplace operator, be it a continuous or a discrete one, can tell us about the underlying structure. For example, what can we learn from the spectrum of the Laplace operator operating on functions that vanish on the boundary of some bounded domain in $\mathbb{R}^n$ about the geometry of that domain? Or what can we learn from the spectrum of the Laplace–Beltrami operator on a compact Riemannian manifold about the geometry of that manifold? Or what can the spectrum of (1.2.1) or (1.2.2) tell us about the structure of the underlying finite graph?

In fact, in 1882, the physicist Arthur Schuster (1851–1934)[47] put forth a question: *Can one determine the shape of an object by listening to its vibrations?* Mark Kac [164] famously reformulated it as: *Can one hear the shape of a drum?* But recalling the previous discussion of the wave operator and the expansion of its solutions by Bernoulli and Fourier, the question was modified as: *Can one reconstruct the shape of a mathematical drum from the eigenvalues of its Laplace operator?*, where *mathematical drum* stands for a Euclidean domain. A negative answer was given in 1992 by Carolyn Gordon, David L. Webb, and Scott Wolpert in a paper titled "One cannot hear the shape of a drum" [114]. In other words, one cannot reconstruct the exact shape of an object from the eigenvalues of the Laplacian. Nevertheless, one can infer important information. In 1911, Weyl [272] proved an asymptotic formula for the eigenvalues of a bounded domain in terms of its dimension and volume, and Weyl's reasoning directly extends to Riemannian manifolds. In fact, in 1953, Minakshisundaram [209] showed that for a compact Riemannian manifold without boundary, one can hear the dimension, the volume, and the total scalar curvature (see, for instance, [47]). The situation for the discrete Laplacians on graphs is analogous.

The results of Weyl use the asymptotics of the spectrum, that is, the large eigenvalues. But even a single small eigenvalue, in fact the first positive eigenvalue, can reveal important geometric features. Again, this started in the context of continuous problems. While the general idea of relating isoperimetric and Sobolev inequalities emerged already in [94, 204], the decisive step was taken by Cheeger [56] in Riemannian geometry. For a compact, connected n-dimensional manifold M, he defined

$$h(M) := \inf_D \frac{\text{vol}_{n-1}(\delta D)}{\text{vol}_n(D)},$$

where $D \subset M$ is a smooth n-submanifold with boundary δD and $0 < \text{vol}_n(D) \le \text{vol}(M)/2$. Cheeger proved that the first nonvanishing eigenvalue $\lambda_{\min}(M)$ of the Laplace–Beltrami operator satisfies

$$\lambda_{\min}(M) \ge \frac{1}{4}h^2(M).$$

Peter Buser [41] later showed that for each compact manifold, there exists Riemannian metrics for which the inequality becomes sharp. In [42], he also proved that if the Ricci curvature of a compact unbordered Riemannian n-manifold M is bounded below by $-(n-1)a^2$, for some $a \ge 0$, then

$$\lambda_{\min}(M) \le 2a(n-1)h + 10h^2.$$

Therefore, $h(M)$ can be used to estimate $\lambda_{\min}(M)$ and vice versa.

It was then discovered that an analogous relation holds for the Laplacian of a graph. In fact, what is now called the *Cheeger constant* of a simple graph $G = (V, E)$ with vertex set V and edge set E had already been introduced in 1951 by George Pólya and Gábor Szegö [230], who called it the *isoperimetric constant* and defined it as

$$h(G) := \min_{\emptyset \ne S \subsetneq V} \frac{|E(S, \bar{S})|}{\min\{\text{vol}(S), \text{vol}(\bar{S})\}},$$

where $E(S, \bar{S})$ denotes the set of edges between S and its complement $\bar{S} := V \setminus S$, while the *volume* of S, denoted $\text{vol}(S)$, is the sum of the vertex degrees in S. Finding a set S realizing the Cheeger constant, also called a *Cheeger cut*, means finding a small *edge cut* $E(S, \bar{S})$ such that, if removed from G, it divides the graph into two disconnected components that have roughly equal volume. Therefore, h measures how different G is from a disconnected graph, and it is largest for the complete graph.

Jozef Dodziuk [82] and Noga Alon with Vitali Milman [4] derived analogous estimates for the graph Cheeger constant and for the first nonvanishing eigenvalue of the Kirchhoff Laplacian (1.2.1) associated to a connected graph. Also, Fan Chung [61] proved the Cheeger inequalities for the *symmetric Chung Laplacian* (1.2.2) of a graph. Chung's estimate reads as

$$\frac{1}{2}h(G)^2 \le \lambda_2(G) \le 2h(G), \tag{1.2.4}$$

where $\lambda_2(G)$ is the first nonzero eigenvalue of the connected graph G (the first eigenvalue is always $\lambda_1(G) = 0$, for a constant eigenfunction). Thus, as in the Riemannian case, $h(G)$ can be used to estimate $\lambda_2(G)$ and vice versa.

Moreover, the eigenvectors corresponding to $\lambda_2(G)$ can be used in order to approximate the Cheeger cut, as follows. An eigenvector for $\mathcal{L}(G)$ can be seen as a function $f\colon V \to \mathbb{R}$ and, if f is an eigenfunction with eigenvalue $\lambda_2(G)$, then f must achieve both positive and negative values, and the edges between the sets

$$\{v \in V\colon f(v) \geq 0\} \quad \text{and} \quad \{v \in V\colon f(v) < 0\}$$

approximate the Cheeger cut. Since solving the Cheeger cut problem is NP-hard [256], while the eigenvalues and the eigenvectors of $\mathcal{L}(G)$ can be found quickly, spectral clustering based on these results is a very common tool and has many applications (see, for instance, [49, 53, 253, 270, 271]).

1.3 Other Sources and Directions of the Theory

Like most other mathematical subjects, the spectral theory of discrete structures has multiple origins and expands into different research directions, besides those that we have cited for the history and that we develop in this book. Thus, it seems appropriate to sketch those that are perhaps the most important ones.

An important source was the paper [199] by Grigorii Aleksandrovich Margulis where he found an explicit construction of an infinite family of graphs, which are what he called *concentrators* and which are, with a shift of emphasis, now called *expanders*. Mark Pinsker [229] had already shown the existence of such graphs by a probabilistic argument. Concentrators are graphs where there exist many disjoint paths from any set of k inputs to k outputs. Expanders are graphs for which every subset of a vertex set has a large neighborhood. This means that many edges have to lead from that subset to its complement. Expanders can be used to construct superconcentrators [102]. See also the discussion in [61]. In either case, the graph should be sparse, that is, not have many edges. Such families of graphs are important in computer science, for instance, for efficient network communication. Other applications include error-correcting codes, cryptography, or complexity theory. We shall discuss this in Section 13.3. In that theory, one is mainly interested in d-regular graphs, that is, in graphs where all vertices have the same number d of neighbors. It turns out that if one wishes to make the expansion property quantitative, a good measure is a lower bound for the smallest nonvanishing eigenvalue λ_2 of the Laplace operator. The lower bound in (1.2.4) here explains why this is so. As mentioned earlier, the subject will be taken up in Section 13.3, but here we refer to [136, 176, 192]. In fact, this is a well-researched field with a rather extensive literature.

Since the d-regularity of the graphs makes the graphs more regular and uniform, one can also profitably invoke many other mathematical areas and apply their tools. In fact, Margulis' original construction was of a number theoretic nature, and subsequently, again by number theoretical ideas, other classes of expander graphs have been found, such as the Ramanujan graphs. For a number and group theoretical survey, see [193].

An important topic is that of *random graphs*. The basic construction was introduced by Paul Erdös (1913–96) and Alfred Rényi (1921–70) [91]. Given a set of N vertices, and some $p \in (0, 1)$, every pair of vertices is connected by an edge with probability p, independently of other pairs. The typical properties of such random graphs scale like $\frac{p}{N}$. They have been intensively studied (see, for instance, [32]). Also, their spectra have many interesting features (see, for instance, [136]). Random constructions can also be used for proving the existence of graphs with certain specified properties, but such constructions do not provide explicit examples.

Another direction that is complementary to what we cover in this book is the spectral geometry of infinite graphs. This topic also involves important analytical aspects, such as the heat flow on the underlying graph, and it is advantageously treated from the general perspective of Dirichlet forms. Here, we refer to the recent comprehensive monograph [171].

2

The Abstract Setting

We are interested in discrete structures like graphs, hypergraphs, and simplicial complexes, and we want to study them via spectral properties of Laplace type operators defined on them. These operators perform some kind of local averaging. The eigenvalues and eigenfunctions of such operators can be obtained as critical values and points of Rayleigh quotients. Symmetry groups of the underlying structures have representations on eigenspaces.

Let us start abstractly.

2.1 A Category Theoretical Perspective

A category consists of *objects* $A, B, C, \ldots$ and *morphisms* $f : A \to B, g : B \to C, \ldots$ between them. (Because of their graphical representation, morphisms are also called *arrows*.) Morphisms can be composed if the codomain of one is the domain of the others, for example, in the case at hand, we can form $g \circ f : A \to C$. The composition is associative, and every object carries (at least) the identity morphism, which is neutral for compositions. A *functor* from one category to another maps objects to objects and morphisms between objects to morphisms between their images.

For an introduction to category theory, see, for instance, [147] or any other relevant textbook.

2.1.1 The Static Picture

We start with basic category $\mathbf{I}$ with just two objects $0, 1$ and, besides their identity morphisms, only a single morphism

$$1 \longrightarrow 0$$

and a *functor*

$$H : \mathbf{I} \to \mathbf{Sets}$$

into the category of sets. Thus, we have two sets R and U and a map

$$h: R \to U. \tag{2.1.1}$$

We shall usually assume that the sets R and U are finite.

For our purposes, we shall take

$$U = \mathcal{P}(V),$$

the set of subsets of a set V. We note that $\mathcal{P}(V)$ defines a simplicial complex (see Definition 2.3.1), its Čech complex. The vertices of this complex are the elements of $\mathcal{P}(V)$, that is, the subsets of V. Such subsets $V_0, V_1, \ldots, V_k$ define a k-simplex whenever

$$V_0 \cap \ldots \cap V_k \neq \emptyset.$$

Turning to the other set involved, R, we shall interpret an $r \in R$ as a *relation*. Thus, to each relation $r \in R$, we associate a set of elements of V. Without loss of generality, we can assume that $V = \bigcup_{r \in R} h(r)$.

This is a basic *hypergraph*. Thus, a hypergraph is a cover of a vertex set V. If every $h(r) \subset V$ has precisely two elements, we have an ordinary *graph*.

This is not yet enough structure for us, however. So, we turn to Section 2.1.2.

2.1.2 The Dynamical Picture

We consider the basic category **J** with two objects and two nontrivial arrows

$$1 \rightrightarrows 0$$

and a *functor*

$$H: \mathbf{J} \to \mathbf{Sets}.$$

Thus, we have two sets R and U and two maps

$$h_i, h_t: R \to U. \tag{2.1.2}$$

This is a *directed* graph, where R is the edge set and U is the vertex set, and the two arrows h_i, h_t map an edge to its initial and terminal vertex, respectively.

For our purposes, again

$$U = \mathcal{P}(V)$$

is the set of subsets of a set V.

This gives us more structure, and we can interpret a relation $r \in R$ as a *hyperedge*. That is, to r, we associate two sets of elements of V, the input

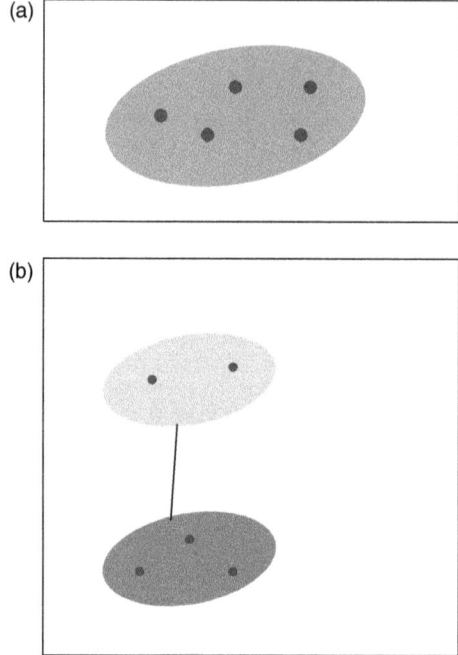

Figure 2.1 An ordinary (a) and an oriented (b) hypergraph.

and the output of the hyperedge. We then have a *directed hypergraph*. Thus, a directed hypergraph consists of two families of subsets of a vertex set V.

We get an induced map

$$h_i(r) \mapsto h_t(r)$$

that maps an input of a hyperedge to its output. This, however, in general does **not** induce a simplicial map of the Čech complexes $\Sigma_i \to \Sigma_t$. Therefore, we should construct other invariants than those coming from these Čech complexes. We shall usually ignore the direction, that is, also consider the reverse direction and speak of *orientations*. In [153], we called this a *chemical hypergraph*.

In short, in an *oriented hypergraph*, one hyperedge can carry two orientations, corresponding to two directions.

In a *directed* hypergraph, a hyperedge goes in only one direction.

The difference between an ordinary and an oriented hypergraph is illustrated Figure 2.1. The hyperedge between the top and the bottom vertex set does not have a direction, so this is not a directed hypergraph.

Figure 2.2 shows a hypergraph with overlapping hyperedges.

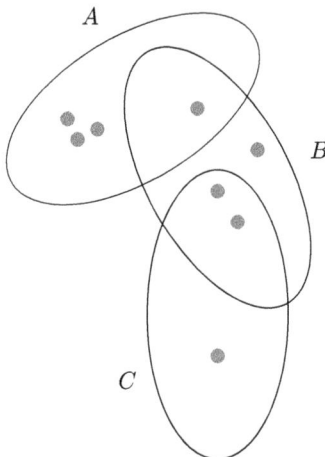

Figure 2.2 A hypergraph with overlapping hyperedges.

The hyperedges A and B have one vertex in common; the hyperedges B and C have two. There is no vertex that is in both A and C, and hence, in particular, none simultaneously contained in all three.

Another example, the hyperflower, introduced in Definition 4.9.9 and depicted in Figure 4.2, will play an important role in Chapter 4.9 and Section 11.5.

2.1.3 Probing a (Dynamical) Structure

We associate the Hilbert space $L^2(U)$ of functions to target U in (2.1.1) or (2.1.2). Giving a weight w_u to each $u \in U$, we define a scalar product for functions on U

$$(f,g)_U := \sum_u w_u f(u)g(u)$$

and let R operate on $L^2(U)$.

We can then consider *differences:* For a function $f : U \to \mathbb{R}$ and $h_i, h_t : R \to U$, we put

$$\delta f(r) := \sum_{u \in h_i(r)} f(u) - \sum_{v \in h_t(r)} f(v).$$

When we exchange $h_i(r)$ and $h_t(r)$, we get a $-$ sign. Therefore, we should consider oriented relations r. We denote the change of orientation, that is, exchanging $h_i(r)$ and $h_t(r)$, by a $-$ sign

$$r \mapsto -r.$$

δ satisfies an important property, *closedness:* If

$$h_t(r_1) = h_i(r_2) \quad \text{and } h_t(r_2) = h_i(r_3) \quad \text{and } h_t(r_3) = h_i(r_1)$$

then

$$\delta f(r_1) + \delta f(r_2) + \delta f(r_3) = 0.$$

Thus, δ annihilates cycles.

As the next step, we also define a scalar product on functions on R

$$(\gamma, \omega)_R := \sum w_r \gamma(r) \omega(r)$$

to obtain the adjoint δ^*

$$(f, \delta^* \gamma)_U = (\delta f, \gamma)_R.$$

Since δ is antisymmetric, that is, $\delta f(-r) = -\delta f(r)$, δ^* vanishes on symmetric γ. Thus, if $\gamma(-r) = \gamma(r)$, $\delta^*(\gamma) = 0$. Hence, we may assume antisymmetry,

$$\gamma(-r) = -\gamma(r).$$

(If we wanted to include symmetric γ, we should choose a non-antisymmetric δ.)

We interpret

$$\delta f(r) := \sum_{u \in h_i(r)} f(u) - \sum_{v \in h_t(r)} f(v)$$

as a *boundary operator*. δ and its adjoint δ^*

$$(f, \delta^* \gamma)_U = (\delta f, \gamma)_R$$

allow us to define the *Laplacians*

$$\Delta_U := \delta^* \delta : L^2(U) \to L^2(U)$$
$$\Delta_R := \delta \delta^* : L^2(R) \to L^2(R)$$

Since

$$(\Delta_U f, g)_U = (\delta f, \delta g)_R = (f, \Delta_U g)_U$$
$$(\Delta_R \gamma, \omega)_R = (\delta^* \gamma, \delta^* \omega)_U = (\gamma, \Delta_R \omega)_R$$

Δ_U and Δ_R are both symmetric and non-negative and therefore have real, non-negative eigenvalues. Also, their nonzero eigenvalues coincide.

Let us summarize the steps that yield these operators [153].

1. We give weights to the hyperedges. In the simplest case, each hyperedge has weight 1, and therefore each vertex v acquires weight hypdeg $v :=$ $\sum_{h:\, v \in h}(|h|-1)$, $|h|$ being the number of vertices contained in the hyperedge h;
2. define a scalar product for functions defined on hyperedges and a scalar product for functions defined on vertices, based on the weights;
3. define the boundary operator for functions defined on vertex sets, mapping them to functions defined on hyperedges;
4. find the adjoint operator based on the scalar products and
5. define the Laplace operators as the two different compositions of the boundary operator and its adjoint.

2.2 Relations and Norms

We now develop an analytical perspective, as opposed to the categorical one mentioned earlier. Now, the formal setting will be more concrete than before, but the analytical one will be more general. As before, we consider a set V. In the sequel, it will be usually finite, but for the moment, we only need it to be a compact space. (Most of what we are going to say also applies to infinite or noncompact spaces, but we do not want to address the analytical issues involved with that.) On this set, we consider observables, that is, functions $f : V \to \mathbb{R}$.

And as before, we furthermore assume that the elements of V can stand in relations. The observables should help us explore those relations. For instance, when V is finite (in which case, we write $|V|$ for the number of its elements), the simplest such relations are binary neighborhood relations, $v \sim v'$, and whenever such a relation holds, we may consider the difference $\delta f(v, v') :=$ $f(v') - f(v)$. The relations could be weighted, with weights $w_{vv'} \in \mathbb{R}$, offering us the notational convenience to put $w_{vv'} = 0$ when v and v' are not related. This convention will often be used in the sequel, usually without being explicitly declared. We call V *connected* if for any two $v, v' \in V$, we can find $v_0 = v, v_1, \ldots, v_m = v'$ with $w_{v_{i-1}v_i} \neq 0$ for $i = 1, \ldots, m$. We sometimes also write $w(v, v')$ instead of $w_{vv'}$.

Weights induce us to consider $\delta f(v, v') := w_{vv'}(f(v') - f(v))$. When the relations are symmetric, that is, $w_{vv'} = w_{v'v}$, and non-negative, i.e. $w_{vv'} \geq 0$ for all v, v', things become simpler. (When the weights are non-symmetric, one might have to consider observables $f : V \to \mathbb{C}$.) When V is a differentiable manifold (possibly with boundary), as the δ-operator, we may consider the differential $df(v)$ and apply it to tangent vectors.

We could also have relations not between individual elements but between subsets of V. When $V_1, V_2 \subset V$ are related, we may consider differences of the form $\sum_{v \in V_2} f(v) - \sum_{v' \in V_1} f(v')$, and again, this can be refined by weights if desired. We can also work with higher order differentials as on simplicial complexes or on differentiable manifolds. In particular, in that case, we might have weights $w(v_1, \ldots, v_k)$ between k vertices, for all values of k that are relevant in the context in question.

To become quantitative, we also need norms $\|.\|$, both on the set of observables on V and on their differentials on the set E of relations. We may then consider quotients

$$\frac{\|\delta f\|_E}{\|f\|_V}, \tag{2.2.1}$$

assuming, of course, that $\|f\|_V \neq 0$. Here, we have used the fact that both the observables and their differentials constitute vector spaces. Since norms are homogeneous, that is, $\|\lambda g\| = |\lambda| \|g\|$ when $\lambda \in \mathbb{R}$, we may normalize $\|f\|_V = 1$, that is, consider

$$\|\delta f\|_E \text{ on the unit sphere } \{\|f\|_V = 1\}. \tag{2.2.2}$$

It turns out to be advantageous to generalize this further, by not taking the norms as such, but some functions of them. The simplest possibility consists in raising them to power p and consider

$$\frac{\|\delta f\|_E^p}{\|f\|_V^p}, \tag{2.2.3}$$

for some choice of $1 \leq p \leq \infty$. The choice $p = 2$ is the most convenient one, and we have used in (2.2.3) to define the Laplace operators. It occurs naturally when the norms come from scalar products. In that case, the resulting eigenvalue problems become *linear*, which is of course a great mathematical advantage. Nevertheless, as we shall see, other choices of p, and in particular $p = 1$, that is, the original formulation (2.2.1), while more difficult to analyze mathematically, can provide important insight.

The basic idea then consists in looking for the critical values and the critical points of this functional (2.2.3) on the unit sphere, that is, with the normalization $\{\|f\|_V = 1\}$. These critical values will constitute the spectra that we are interested in.

We can of course also take more general monotonically increasing functions F, G and consider

$$\frac{F(\|\delta f\|_E)}{G(\|f\|_V)}. \tag{2.2.4}$$

Again, suitable choices of F and G will let us obtain insight about the underlying structures.

2.3 Complexes and Orientations

We explained in Section 2.2 that we want to consider differences

$$\delta f(v, v') = f(v') - f(v) \qquad (2.3.1)$$

of observations at points v, v' that are related or more general versions thereof. Now, such a difference (2.3.1) has a sign that changes if we interchange v and v'. While this does not matter if we take norms, as in (2.2.1), (2.2.3), and (2.2.4), for an analysis of the underlying structure of relations, this seems relevant and important.

Thus, the relations are naturally *directed*, that is, of the form $v' \to v$, and we may speak of v' as the *tail* and v as the *head* of the relation underlying (2.3.1).

In the *symmetric* case, we could then simply consider both differences $df(v, v') = f(v') - f(v)$ and $df(v', v) = f(v) - f(v')$, which would then give us a factor of 2 in (2.2.1) but would otherwise not change much.

Nevertheless, even in this case, it turns out that we should not neglect signs and consider oriented relations of the form $v' \to v$. This becomes clear when we also consider higher order relations. In the simplest case, such higher order relations can be expressed geometrically in terms of a *simplicial complex*.

Definition 2.3.1. Let V be a (usually finite) set of vertices. A *simplicial complex* Σ on V is a subset of its power set, $\Sigma \subset \mathcal{P}(V)$, that is, closed under taking subsets, that is, for a *simplex* $C \in \Sigma$, any of its subsets $C' \subset C$ is also a simplex in Σ. The empty set is also considered as a simplex.

A simplex C with $q+1$ vertices is called a q-*simplex* (where we consider q as its dimension), its subsimplices are called its *faces*, and its $(q-1)$-dimensional faces are called its *facets*. The dimension of a simplicial complex is the largest dimension among its simplices. A one-dimensional simplicial complex is a *graph*.

Let Σ_q be the collection of the q-simplices of Σ. We let $C_q = C_q(\Sigma)$ be the Abelian group with coefficients in $\mathbb{R}$ generated by the elements of Σ_q. We also write $C^q = C^q(\Sigma)$ for the linear functions from C_q to $\mathbb{R}$.

We usually assume that a simplicial complex is *connected*. This means that for any two of its nonempty simplices s, s', there exists a chain of simplices $s_0 = s, s_1, \ldots, s_m = s'$ such that any two adjacent simplices in this chain have at least one vertex in common.

We usually and naturally assume that all elements of the vertex set V participate in the simplicial complex Σ. Thus, Σ_0 is the vertex set V. We also note that we assume that all the vertices constituting a simplex are different from each other.

We shall also need *orientations* of simplices.

Definition 2.3.2. An *orientation* of a q-simplex is an ordering of its vertices up to even permutation. An odd permutation of the vertices changes an oriented q-simplex σ_q into the oppositely oriented simplex $-\sigma_q$.

We require that $f \in C^q$ satisfies

$$f(-\sigma_q) = -f(\sigma_q), \tag{2.3.2}$$

for every oriented q-simplex.

For instance, for $q = 1$, we have

$$f(v_1, v_0) = -f(v_0, v_1).$$

Orientations introduce signs, and we can cancel terms with opposite signs. This will lead to the simplifications that (co)homology provides.

Definition 2.3.3. Let $f \in C^{q-1}$. We then define its *coboundary* $\delta f : C^q \to \mathbb{R}$ as

$$\delta f(v_0, v_1, \ldots, v_q) = \sum_{i=0}^{q} (-1)^i f(v_0, \ldots, \hat{v}_i, \ldots, v_q), \tag{2.3.3}$$

where a $\hat{\ }$ over a vertex means that it is omitted.

When we want to emphasize the dimension, we write

$$\delta^q : C^q \to C^{q+1}. \tag{2.3.4}$$

Equation (2.3.3) applies to any q-simplex with vertices $v_0, \ldots, v_q$. On the right-hand side, we sum over its facets. By the defining property of a simplicial complex, they are all contained in Σ_{q-1}, and so, f is defined for all of them. We then extend (2.3.3) by linearity to C^q.

The basic property of this coboundary operator, which provides the foundation for cohomology theory and which is easily checked, is

Proposition 2.3.1.

$$\delta^{q+1} \circ \delta^q = 0 \quad \text{for } q = 0, 1, \ldots \tag{2.3.5}$$

or written more shortly

$$\delta \circ \delta = 0. \tag{2.3.6}$$

To see the mechanism, let us consider the case $q = 0$ and a simplex with vertices v_0, v_1, v_2. Then,

$$\delta^1 \circ \delta^0 f(v_0, v_1, v_2) = \delta^0 f(v_1, v_2) - \delta^0 f(v_0, v_2) + \delta^0 f(v_0, v_1) \qquad (2.3.7)$$
$$= f(v_2) - f(v_1) - (f(v_2) - f(v_0)) + f(v_1) - f(v_0)$$
$$= 0.$$

Assuming (2.3.2), we can rewrite (2.3.7) as

$$\delta f(v_1, v_2) + \delta f(v_2, v_0) + \delta f(v_0, v_1) = f(v_2) - f(v_1) + f(v_0)$$
$$- f(v_2) + f(v_1) - f(v_0) = 0,$$

which makes the algebraic structure clearer.

Definition 2.3.4. The q-th *cohomology group* of the simplicial complex Σ is

$$H^q(\Sigma) := \ker \delta^q / \operatorname{im} \delta^{q-1}. \qquad (2.3.8)$$

The dimension $b_q(\Sigma)$ of $H^q(\Sigma)$ is called the q-th *Betti number* of Σ.

Remark 2.3.1. We can also let $C_q(\Sigma, \mathbb{F})$ be the linear space with coefficients in an Abelian group $\mathbb{F}$, generated by the elements of Σ_q, and let $C^q(\Sigma, \mathbb{F})$ be the linear functions from $C_q(\Sigma, \mathbb{F})$ to $\mathbb{F}$, and then we can define the cohomology group $H^q(\Sigma, \mathbb{F})$ in the same way. It is usual to take $\mathbb{F}$ to be a commutative ring (e.g., the integer ring $\mathbb{Z}$) or even a field (e.g., the field $\mathbb{C}$ of the complex numbers or the finite field $\mathbb{Z}_p := \mathbb{Z}/p\mathbb{Z}$). As an interesting example, we refer to [254] for Cheeger constants defined on a simplicial complex which use the cohomology over the finite field $\mathbb{Z}_2$. In this book, we mainly work with real coefficients.

2.4 (Oriented) Hypergraphs and their Higher Order Analogs

2.4.1 Generalizations

The concept of a simplicial complex obviously generalizes that of a graph. Further generalizations are possible and meaningful and will be treated in this book. In fact, we have already introduced (oriented) hypergraphs in Section 2.1. Before discussing that structure in more detail, to motivate what follows, let us consider two examples where such structures naturally arise in applications.

1. Social relations and interactions: Social network theory considers binary relations or interactions between persons, and they are thus naturally modeled by a graph. However, often, more than two persons constitute a unit

of interaction. For instance, when we consider relations between scientists, we can consider research groups or coauthorships of scientific papers as the basic relations. In particular, when we consider groups of coauthors, they often overlap, as a scientist can write different papers with different coauthors. Moreover, when we look at research groups, we also naturally encounter higher order units, such as research institutes or university departments.

2. Networks of chemical reactions. Here, usually, a set of inputs, ingredients, or educts is transformed by a reaction into a set of outputs or products, which can then, in turn, become inputs for further reactions. The sets of inputs and outputs connected by a reaction need not necessarily be disjoint, as there could be catalyzers that participate in the reaction (in fact, may enable it) but are not changed by it.

These two examples suggest somewhat different generalizations of graphs or simplicial complexes, as we had already discussed in Section 4.9.

1. Instead of edges connecting two vertices, constituting a graph, we can look at *hyperedges* connecting arbitrary numbers of vertices. We can then look at higher order relations connecting hyperedges, and if we want, we can iterate this.

2. Instead of vertices connected by edges, we can look at *hypervertices*, that is, collections of vertices, connected by edges. In contrast to the first generalization above, where a single collection of vertices can be connected by a hyperedge, here we would naturally have two collections of vertices that are connected by an edge. This will naturally lead us to the concept of an oriented hypergraph (already introduced from a different perspective in Section 4.9). Since we could also consider graphs with self-loops, that is, where an edge is allowed to connect a vertex with itself, we should also admit the possibility that these two collections of vertices are not necessarily disjoint.

    We may then also consider relations involving more than two hypervertices.

Construction 1 simply generalizes edges (two-element sets) to hyperedges (multielement sets), while construction 2 just extends vertices (single-element sets) to hypervertices (multielement sets). Essentially, however, the difference is not so much in the content but in the conceptualization. Construction 2 naturally gives oriented hypergraphs. It has the disadvantage that in many cases, the resulting graph, with hypervertices as its vertices, is disconnected. On the other hand, one can perhaps more naturally encode restrictions on connections between hypervertices. Here is an example: Take as hypervertices the members of the powerset of some finite set S, that is, all the nonempty subsets of S. Then, connect any two of them by an edge whenever their intersection is empty.

We shall formally define those two types of generalizations in the next two sections. In brief, the point of view for the first type will be to consider hyperedges instead of edges, where a hyperedge may connect more than two vertices. For the second type, we consider hypervertices instead of vertices. A hypervertex is a collection of vertices. An edge then connects two hypervertices.

2.4.2 Hypergraphs and Hypersimplical Complexes

Here, we explore the first type of generalization suggested in the previous section.

Definition 2.4.1. A *hypergraph* is a pair $\Gamma = (V, E)$, where V is a finite set of nodes or vertices, and E is a multiset of elements $e \subseteq V$ called *hyperedges*. Given $v \in V$, we define its *degree*, denoted $\deg v$, as the number of hyperedges containing v.

Since E is a multiset, different hyperedges can contain exactly the same nodes; however, we label the hyperedges so that we can treat E as a set, and we write $E = \{e_1, \ldots, e_M\}$.

The notion of hypergraph can be generalized by iterating this principle.

Definition 2.4.2. An *ℓ-th order hypergraph* is a sequence $\Upsilon = (\Upsilon_1, \ldots, \Upsilon_\ell)$ such that, for each $q = 1, \ldots, \ell - 1$, $(\Upsilon_q, \Upsilon_{q+1})$ is a hypergraph.

Remark 2.4.1. Higher order hypergraphs can have interesting empirical interpretations. For instance, taking up an example from the previous section, we can consider a third-order hypergraph that represent researchers, as follows:

- The elements of Υ_1 are researchers;
- The elements of Υ_2 are research groups; and
- The elements of Υ_3 are research institutes.

We can also consider the hyperedges of some hypergraph as the basic elements of a higher structure:

Definition 2.4.3. Given a hypergraph $\Gamma = (V, E)$, a *hypersimplicial complex* on Γ is a simplicial complex on E.

We now consider hypergraphs with the additional structure that each vertex–hyperedge incidence is given a coefficient.

Definition 2.4.4. Let C be a set such that $0 \in C$. A *hypergraph with coefficients in* C is a triple $\Gamma = (V, E, \psi)$ where $\Gamma = (V, E)$ is a hypergraph and $\varphi \colon V \times E \to C$ such that

$$\varphi(v, e) \neq 0 \iff v \in e$$

is the *incidence function*. If $V = \{v_1, \ldots, v_n\}$ and $E = \{e_1, \ldots, e_m\}$, the *incidence matrix* of Γ is the $n \times m$ matrix $\mathcal{I} = (\mathcal{I}_{ij})$ whose entries are

$$\mathcal{I}_{ij} := \varphi(v_i, e_j).$$

We may adopt the implicit convention that $\varphi(v, e) = 0$ means that $v \notin e$.

Definition 2.4.5. A *hypergraph with real coefficients* is a hypergraph with coefficients in $\mathbb{R}$. In this case, the *degree* of a vertex $v \in V$ is defined as

$$\deg v := \sum_{e \in E} \varphi(v, e)^2.$$

A *weighted hypergraph* is a hypergraph with real coefficients such that $w(e) := \sqrt{\varphi(v, e)} > 0$ does not depend on v, for each $e \in E$ and $v \in e$.

In this case, in particular,

$$\deg v = \sum_{e \in E \colon v \in e} w(e).$$

Definition 2.4.6. A *complex unit hypergraph* is a hypergraph with coefficients in the complex unit circle.

We adopt the following definitions and notations for any hypergraph Γ with vertex set V and hyperedge set E, with or without coefficients.

For the maximal (minimal) degree of the vertices in Γ, we use the symbol $\overline{d} = \overline{d}(\Gamma)$ ($\underline{d} = \underline{d}(\Gamma)$). Given $e \in E$, we denote its *cardinality*, that is, the number of vertices contained in e, by $|e|$. For the maximal (minimal) cardinality of the hyperedges in Γ, we use the symbol $\overline{c} = \overline{c}(\Gamma)$ ($\underline{c} = \underline{c}(\Gamma)$). We say that Γ is *d-regular* if $\deg v = d$ for each $v \in V$; it is *c-uniform* if $|e| = c$ for each $e \in E$. We say that Γ is *connected* if, for every pair of vertices $v, v' \in V$, there exists a *path* that connects v and v', that is, there exist $\hat{v}_1, \ldots, \hat{v}_k \in V$ and $\hat{e}_1, \ldots, \hat{e}_{k-1} \in E$ such that $\hat{v}_1 = v$, $\hat{v}_k = v'$, and $\{\hat{v}_i, \hat{v}_{i+1}\} \subseteq \hat{e}_i$ for each $i = 1, \ldots, k - 1$.

2.4.3 Oriented Hypergraphs and Oriented Hypersimplical Complexes

We now formalize the second type of generalization suggested in Section 2.4.1.

Definition 2.4.7. An *oriented hypergraph* is a hypergraph $\Gamma = (V, E, \varphi)$ with coefficients in $\{-1, 0, +1\}$. If $\varphi(v, e) = 1$ (respectively, $\varphi(v, e) = -1$), v is said to be an *output* (respectively, *input*) for e. If $v \neq v'$ and

$$\varphi(v, e) = \varphi(v', e) \neq 0 \quad (\text{respectively}, \varphi(v, e) = -\varphi(v, e) \neq 0),$$

the vertices v and v' are *co-oriented* (respectively, *antioriented*) in e.

By Definition 2.4.5, the degree of a vertex in an oriented hypergraph is

$$\deg v = \sum_{e \in E} \varphi(v, e)^2 = |\{e \in E : v \in e\}|.$$

On an oriented hypergraph, any process (such as a random walk or a diffusion process) is allowed to go in either direction, that is, from the inputs to the outputs or from the outputs to the inputs. Thus, calling sets of vertices in- or outputs are just labels. When we want to distinguish between them, we consider

Definition 2.4.8. A *directed hypergraph* is an oriented hypergraph with the additional assumption that hyperedges are *directed*, from the inputs toward the outputs.

On a directed hypergraph, processes can only follow the directions. Therefore, a random walker or a diffusion process on a directed hypergraph can only go from the inputs toward the outputs.

Remark 2.4.2. Our definition of directed hypergraph is slightly more general than that in [104].

We now come to the definition of hypersimplicial complex.

Definition 2.4.9. Let V again be a finite set. An *oriented hypersimplicial complex* is a simplicial complex on some $\mathcal{A} \subset \mathcal{P}(V)$, the power set of V. The elements of $\mathcal{A}$ are called *hypervertices*, to distinguish them from the vertices, the elements of V. All notions then are defined analogously to Definition 2.3.1, with the prefix *hyper* and the qualification *oriented* (to distinguish it from the structures introduced in Section 2.4.2).

In particular, a one-dimensional oriented hypersimplicial complex is an *oriented hypergraph*.

The additional structural refinement that an oriented hypersimplicial complex offers when compared to an ordinary simplicial complex is that different zero-dimensional hypersimplices, that is, hypervertices, can share some elements of V, that is, vertices. We refer to the example in Section 2.4.1 where hypervertices are sets of educts or products of chemical reactions.

Thus, when we have a function $g: V \to \mathbb{R}$, we can define a cumulative function via

$$f(V_0) = \sum_{v \in V_0} g(v) \quad \text{for } V_0 \in \mathcal{A}. \tag{2.4.1}$$

Again, for functions f defined on $\mathcal{A}$ and extended linearly as before to linear combinations, we can form the differences

$$\delta f(V_0, V_1) = \sum_{v' \subset V_1} f(v') - \sum_{v \subset V_0} f(v), \tag{2.4.2}$$

for $V_0, V_1 \in \mathcal{A}$. Equation (2.3.3) can then be extended analogously.

2.5 Scalar Products and Laplace Operators

2.5.1 Simplicial Complexes

We consider a simplicial complex Σ. We recall that, concerning the orientations, according to (2.3.2), we have for any $\phi \in C^q$, where C^q are the linear functions on chains of q-dimensional (hyper)simplices,

$$\phi(-\sigma_q) = -\phi(\sigma_q), \tag{2.5.1}$$

that is, changing the orientation yields a minus sign.

We then choose positive definite inner products $(\cdot, \cdot)_q$ on the C^q.

Definition 2.5.1. The adjoint $(\delta^q)^*: C^{q+1} \to C^q$ of the coboundary operator δ^q is defined by

$$(\delta^q f_1, f_2)_{q+1} = (f_1, (\delta^q)^* f_2)_q,$$

for $f_1 \in C^q$ and $f_2 \in C^{q+1}$.

We then have the arrows

$$C^{q-1} \underset{\delta^{q-1*}}{\overset{\delta^{q-1}}{\rightleftarrows}} C^q \underset{\delta^{q*}}{\overset{\delta^q}{\rightleftarrows}} C^{q+1}. \tag{2.5.2}$$

Lemma 2.5.1.

$$(\operatorname{im} \delta^q)^\perp = \ker(\delta^q)^*, \tag{2.5.3}$$

$$(\operatorname{im} (\delta^q)^*)^\perp = \ker \delta^q. \tag{2.5.4}$$

Proof If $\delta^q f = 0$, then

$$((\delta^q)^* g, f) = (g, \delta^q f) = 0 \quad \text{for all } g \in C^{q+1},$$

and conversely, if this vanishes for all $g \in C^{q+1}$, then $\delta^q f = 0$, because the scalar product is positive definite, showing (2.5.4). The proof of (2.5.3) is of course analogous. □

Lemma 2.5.2.

$$\text{im } \delta^{q-1} \cap \text{im } (\delta^q)^* = \{0\}. \tag{2.5.5}$$

Proof If $f = \delta^{q-1} h = (\delta^q)^* g$, then

$$(f, f) = (\delta^{q-1} h, (\delta^q)^* g) = (\delta^q \delta^{q-1} h, g) = 0$$

by (2.3.5), hence $f = 0$. □

Corollary 2.5.1.

$$C^q = \text{im } \delta^{q-1} \oplus \text{im } (\delta^q)^* \oplus (\ker \delta^q \cap \ker(\delta^{q-1})^*, \tag{2.5.6}$$

□

where $\oplus$ indicates orthogonality with respect to the scalar product $(\cdot, \cdot)$. Equation (2.5.6) is a discrete analogue of the Hodge decomposition in Riemannian geometry (see [150]).

Proof By the proof of Lemma 2.5.2, im δ^{q-1} and im $(\delta^q)^*$ are orthogonal, and by Lemma 2.5.1, any element that is orthogonal to both of them has to be contained in both $\ker \delta^q$ and $\ker(\delta^{q-1})^*$. □

We now define the following three operators on C^q:

Definition 2.5.2. (i) The *q-dimensional up Laplace operator* or simply *q-up Laplace operator* of the (hyper)simplicial complex Σ is

$$L^q{}_{up} := (\delta^q)^* \delta^q.$$

(ii) The *q-dimensional down Laplace operator* or *q-down Laplace operator* is

$$L^q{}_{down} := \delta^{q-1} (\delta^{q-1})^*.$$

(iii) The *q-dimensional Laplace operator* or *q-Laplace operator* is

$$L^q := L^q{}_{up} + L^q{}_{down} = (\delta^q)^* \delta^q + \delta^{q-1} (\delta^{q-1})^*.$$

Note that we usually omit the argument Σ, that is, write L^q instead of $L^q(\Sigma)$.

For $q = 0$, that is, when we look at the operators on the vertices of a simplicial complex, we have $L^0{}_{down} = 0$, and hence

$$L^0 = L^0{}_{up}. \tag{2.5.7}$$

Similarly, for $q = \dim \Sigma$, the up-Laplacian vanishes, and

$$L^{\dim \Sigma} = L^{\dim \Sigma}{}_{down}. \qquad (2.5.8)$$

By construction, we have

Lemma 2.5.3. *The operators* L_{up}^q, L_{down}^q, *and* L^q *are self-adjoint.*

$\square$

Also,

Lemma 2.5.4. *The operators* $L = L_{up}^q, L_{down}^q, L^q$ *are non-negative, that is, they satisfy*

$$(Lf, f) \geq 0 \text{ for all } f \in C^q. \qquad (2.5.9)$$

Proof We have

$$(L_{up}^q f, f) = ((\delta^q)^* \delta^q f, f) = (\delta^q f, \delta^q f) \geq 0 \qquad (2.5.10)$$

$$(L_{down}^q f, f) = (\delta^{q-1}(\delta^{q-1})^* f, f) = ((\delta^{q-1})^* f, (\delta^{q-1})^* f) \geq 0 \quad (2.5.11)$$

$$(L^q f, f) = (\delta^q f, \delta^q f) + ((\delta^{q-1})^* f, (\delta^{q-1})^* f) \geq 0. \qquad (2.5.12)$$

$\square$

In particular, from (2.5.10) to (2.5.12),

Corollary 2.5.2.

$$L_{up}^q f = 0 \text{ if and only if } \delta^q f = 0 \qquad (2.5.13)$$

$$L_{down}^q f = 0 \text{ if and only if } (\delta^{q-1})^* f = 0 \qquad (2.5.14)$$

$$L^q f = 0 \text{ if and only if } \delta^q f = 0 \text{ and } (\delta^{q-1})^* f = 0. \qquad (2.5.15)$$

Since the operators L_{up}^q, L_{down}^q and L^q are self-adjoint, non-negative operators on finite-dimensional Hilbert spaces, we have

Theorem 2.5.1. *The eigenvalues of the operators* $L_{up}^q(\Sigma)$, $L_{down}^q(\Sigma)$, *and* $L^q(\Sigma)$ *are real and non-negative.*

Corollary 2.5.2 characterizes the eigenvalue 0. The other eigenvalues then are positive.

Definition 2.5.3. The collection of the eigenvalues (counted with multiplicity) of a self-adjoint operator on a finite dimensional Hilbert space is called its *spectrum*.

We now look at scalar products on the spaces of cochains, as needed for the definition of the Laplace operators. Here, we only consider positive inner products, and when we shall speak about a scalar product in the sequel, we shall always assume that it is positive definite.

Each oriented simplex σ generates a cochain, consisting of the real multiples of the function f_σ with $f_\sigma(\sigma) = 1$, $f_\sigma(\tau) = 0$ for all oriented simplices $\tau \neq \pm\sigma$ (and of course $f_\sigma(-\sigma) = -1$). We assume that the cochains generated by different simplices are orthogonal to each other. This restricts the possible scalar products. A scalar product with this property can be obtained from a *weight function* w that associates to every simplex σ a positive real number.

Definition 2.5.4. A *weighted simplicial complex* (Σ, w) is a simplicial complex with a non-negative weight $w(\sigma)$ for every simplex.

A positive inner product on the space C^q can be written in terms of a positive weight function w as

$$(f, g)_q = \sum_{\sigma \in \Sigma_q} w(\sigma) f(\sigma) g(\sigma), \qquad (2.5.16)$$

where Σ_q is the space of q-simplices. We shall, however, also systematically consider cases where some of the weights may vanish. In that case, the corresponding product (2.5.16) will of course no longer be positive.

In the sequel, we shall write $\mathrm{sgn}(\sigma, \sigma') = \pm 1$ for two orientations of a simplex when those orientations coincide/differ. We also write $\partial\sigma$ for the cellular boundary of a simplex, that is, for the collection of its facets.

By simple linear algebra, the q-up Laplace operator is then given by

$$(L_{up}^q f)([\sigma]) = \sum_{\substack{\rho \in \Sigma_{q+1}: \\ \sigma \in \partial\rho}} \frac{w(\rho)}{w(\sigma)} f([\sigma])$$

$$+ \sum_{\substack{\sigma' \in \Sigma_q : \sigma \neq \sigma', \\ \sigma, \sigma' \in \partial\rho}} \frac{w(\rho)}{w(\sigma)} \mathrm{sgn}([\sigma], \partial[\rho]) \, \mathrm{sgn}([\sigma'], \partial[\rho]) f([\sigma']),$$

$$(2.5.17)$$

and the q-down Laplace operator is

$$(L_{down}^q f)([\sigma]) = \sum_{\tau \in \partial\sigma} \frac{w(\sigma)}{w(\tau)} f([\sigma])$$

$$+ \sum_{\sigma' : \sigma \cap \sigma' = \tau} \frac{w(\sigma')}{w(\tau)} \mathrm{sgn}([\tau], \partial[\sigma]) \, \mathrm{sgn}([\tau], \partial[\sigma']) f([\sigma']).$$

$$(2.5.18)$$

For our purposes, however, we need some relation between the weights in different dimensions.

Definition 2.5.5. *The degree of a q-simplex σ of Σ is*

$$\deg \sigma := \sum_{\rho \in \Sigma_{q+1}:\, \sigma \in \partial \rho} w(\rho). \tag{2.5.19}$$

Definition 2.5.6. If the weight function w on Σ satisfies

$$w(\sigma) = \deg \sigma, \tag{2.5.20}$$

for every $\sigma \in \Sigma_q$, we call the Laplace operator defined on the cochain complex of Σ the *weighted normalized combinatorial Laplace operator*. If, in addition, the weights of the facets of Σ are equal to 1, then the Laplace operator is called the *normalized combinatorial Laplace operator*.

When (2.5.20) holds, (2.5.17) simplifies, and the normalized combinatorial up-Laplacian is given by

$$(L_{up}^q f)([\sigma]) = f([\sigma])$$
$$+ \frac{1}{\deg \sigma} \sum_{\substack{\sigma' \in \Sigma_q:\, \sigma' \neq \sigma, \\ \sigma, \sigma' \in \partial \rho}} w(\rho) \, \mathrm{sgn}([\sigma], \partial[\rho]) \, \mathrm{sgn}([\sigma'], \partial[\rho]) f([\sigma']). \tag{2.5.21}$$

2.5.2 Relative Complexes

Let Σ be a simplicial complex and Σ', a subcomplex. Then, the q-th cochain group of the pair (Σ, Σ') is

$$C^q(\Sigma, \Sigma') := \{f \in C^q(\Sigma) | f([F]) = 0, \text{ for every } F \in \Sigma_q(\Sigma')\},$$

and the coboundary operator $\delta^q : C^q(\Sigma) \to C^{q+1}(\Sigma)$ induces a homomorphism of relative groups $\delta^q_{rel} : C^q(\Sigma, \Sigma') \to C^{q+1}(\Sigma, \Sigma')$ by

$$\delta^q_{rel}(f)([\bar{F}]) = f(\partial_q[\bar{F}]), \quad \text{for } f \in C^q(\Sigma, \Sigma'). \tag{2.5.22}$$

Definition 2.5.7. The q-th *cohomology group* of the pair (Σ, Σ_0) is

$$H^q(\Sigma, \Sigma_0) := \ker \delta^q_{rel} / \operatorname{im} \delta^{q-1}_{rel}. \tag{2.5.23}$$

(For relative cohomology, see for instance [128] or [203].)

When we have inner products on the $C^q(\Sigma, \Sigma')$, we can define the corresponding Laplace operators.

Definition 2.5.8. The *Laplace operators* of a pair (Σ, Σ') are

$$L_{up}^q := L_{up}^q(\Sigma, \Sigma') := (\delta_{rel}^q)^* \delta_{rel}^q, \quad L_{down}^q := L_{down}^q(\Sigma, \Sigma') := \delta_{rel}^{q-1}(\delta_{rel}^{q-1})^*.$$
(2.5.24)

The *Laplace operator* is

$$L^q := L^q(\Sigma, \Sigma') := L_{up}^q + L_{down}^q.$$
(2.5.25)

Again, and obviously, these operators depend on the inner products. As before, we shall consider such products that are defined through weights on the simplices. And we shall allow for some of the weights to vanish.

We shall not only consider positive-valued weight functions but also permit values equal to 0. Thus, we consider weight functions

$$w : \bigcup_i \Sigma_q \to \mathbb{R}^+ \cup \{0\}.$$
(2.5.26)

When some of the weights are 0, we say that (Σ, w) is a *degenerate weighted simplicial complex*. The preceding constructions carry over to degenerate simplicial complexes, with the difference that $(\delta^q)^*$ is a formal adjoint, defined as the standard adjoint on its nondegenerate counterpart and extended by zero on simplices of weight zero, that is,

$$(\delta^q)^* \bar{f}([F]) = \begin{cases} \sum_{\substack{\bar{F} \in \Sigma_{q+1} : \\ F \in \partial \bar{F}}} \frac{w(\bar{F})}{w(F)} \operatorname{sgn}([F], \partial[\bar{F}]) f([F]) & \text{for } w(F) \neq 0, \\ 0 & \text{for } w(F) = 0. \end{cases}$$
(2.5.27)

On degenerate simplicial complexes, the appropriate version of the discrete Hodge Theorem says that the kernel of the q-Laplace operator is isomorphic to the direct sum of the q-th homology group of a complex and the vector space generated by q-simplices of weight zero.

A graph with loops constitutes a special case of a weighted graph. For the graph Laplace operator, a loop effects only the degree of the corresponding vertex. The same is true for simplicial complexes. We can define *simplicial complexes with loops* in analogy to graphs with loops:

Definition 2.5.9. Let $[\Sigma]$ be an oriented simplicial complex. If we allow oriented faces in which a vertex appears more than once, to be elements of $[\Sigma]$, then we say that $[v_0, \ldots, v_i, v_{i+1} \ldots, v_i, v_{j+1}, v_{q+1}]$ is a *loop* on the simplex

$$[v_0, \ldots, v_i, v_{i+1} \ldots, \hat{v}_i, v_{j+1}, v_{q+1}].$$

The boundary of a loop is zero. Thus, the Laplace operator defined on a simplicial complex $[\Sigma]$ with loops coincides with the one defined on Σ. For the degree of a face, we have to add the weight of the loop to its total degree.

Figure 2.3 In (b), we see the proper difference of graph (a) and its subgraph Σ'.

Therefore, simplicial complexes with loops are already contained in the definition of weighted complexes, and we need not treat them separately. We include
simplicial complexes with loops here because they will arise naturally when
we consider simplicial maps, which do not preserve dimensionality.

Definition 2.5.10. A weighted simplicial complex $(\Sigma', w_{\Sigma'})$ is a *subcomplex* of
(Σ, w_Σ) if Σ' is a subcomplex of Σ and $w_{\Sigma'}(F) \leq w_\Sigma(F)$ for every simplex F
of Σ'.

Definition 2.5.11. A *proper difference* of the weighted simplicial complexes
(Σ, w_Σ) and $(\Sigma', w_{\Sigma'})$ is a degenerate weighted simplicial complex (Θ, w_Θ),
such that $\Theta \equiv \Sigma$, $w_\Theta := w_\Sigma - w_{\Sigma'}$, and $\Theta' = \{F \in \Sigma \mid w_\Theta(F) > 0\}$ is a
simplicial complex.

As an example, we can take a subgraph of an unweighted graph. We delete
all the edges of the subgraph. The remaining edges have weight 1, those that
are deleted get weight 0. The degrees, that is, the weights of the vertices, are
decreased by how many edges they lose.

In Figure 2.3, Σ is the triangle in graph (a). Every vertex has degree, hence
weight 2. Σ' is the bottom edge with its two end points, which have weight 1
in Σ', and the proper difference Θ is graph (b), where the two lower vertices
now have weight $2 - 1 = 1$ (and the bottom edge has weight 0 and is hence not
drawn).

2.5.3 Laplace Operators as Averaging Operators

We consider the case of pairwise relations, that is, in the terminology of Section
2.5.1, the case $q = 0$, vertices connected by edges. In particular, $L^0_{down} = 0$,
and so for $q = 0$, (2.5.21) becomes

$$L^0 f(v) = f(v) - \frac{1}{\deg v} \sum_{\rho=[v,v']} w(\rho) f(v'), \qquad (2.5.28)$$

with deg $v = \sum_{\rho=[v,v']} w(\rho)$ as the normalizing factor. The operator L^0 is the Laplacian of the graph formed by the vertices and edges of a simplicial complex. In (2.5.28), we compare the value of f at v with the average of the values of f at all other points connected to v by some edge. This is the important *averaging property* of the zero-dimensional Laplacian.

This averaging property can also be taken as the essence of a general definition of Laplace operators. Let Ω be a domain in $\mathbb{R}^n$ or some more abstract space, and $w: \Omega \times \Omega \to \mathbb{R}$, a non-negative, symmetric function. The quantity $w(x,y)$ can be interpreted as some kind of edge weight between the points x, y for any pair $(x,y) \in \Omega \times \Omega$. When x, y are vertices in a graph, the integrals become sums, and we are back in the case of (2.5.28). We define the average $\bar{w}: \Omega \to \mathbb{R}$ of w by

$$\bar{w}(x) = \int_\Omega w(x,y)dy$$

and assume that $\bar{w}$ is positive almost everywhere. On a graph, while w is an edge function, $\bar{w}$ would be a vertex function, $\bar{w}(x)$ being the degree of the vertex x with edge weights $w(x,y)$. We first use $\bar{w}(x)$ and $w(x,y)$ to define the L^2-norms for functions $u: \Omega \to \mathbb{R}$ and vector fields p, that is, $p: \Omega \times \Omega \to \mathbb{R}$,

$$(u_1,u_2)_{L^2} := \int_\Omega u_1(x)u_2(x)\bar{w}(x)dx$$

$$(p_1,p_2)_{L^2} := \int_{\Omega \times \Omega} p_1(x,y)p_2(x,y)w(x,y)dxdy$$

and the corresponding norms $|u|$ and $|p|$.

The discrete derivative of a function $u: \Omega \to \mathbb{R}$ is defined by

$$Du(x,y) = u(y) - u(x). \tag{2.5.29}$$

Even though Du does not depend on w, it is in some sense analogous to a gradient, as we shall see (2.5.32). Its pointwise norm then is

$$|Du|(x) = \left(\frac{1}{\bar{w}(x)} \int_\Omega (u(y) - u(x))^2 w(x,y)dy\right)^{\frac{1}{2}}. \tag{2.5.30}$$

The *divergence* of a vector field $p: \Omega \times \Omega \to \mathbb{R}$ is defined by

$$\operatorname{div} p(x) := \frac{1}{\bar{w}(x)} \int_\Omega (p(x,y) - p(y,x))w(x,y)dy. \tag{2.5.31}$$

Note that, in contrast to Du for a function u, the divergence of a vector field depends on the weight w. For $u: \Omega \to \mathbb{R}$ and $p: \Omega \times \Omega \to \mathbb{R}$, we then have the integration by parts formula

$$(Du,p)_{L^2} = -(u,\operatorname{div} p)_{L^2}. \tag{2.5.32}$$

With the vector field Du and the divergence operator div, we can define a Laplacian for functions

$$\Delta u(x) := -\text{div}\,(Du) = u(x) - \frac{1}{\bar{w}(x)} \int_\Omega u(y)w(x,y)dy\,, \tag{2.5.33}$$

which in the case of a graph coincides with the Laplacian (2.5.28). Again, the principle is to compare the value of a function at a point with the average of the values on some neighborhood. For such a principle on Riemannian manifolds and more general metric spaces, see [143, 144].

2.5.4 Oriented Hypergraphs

We consider an oriented hypergraph Γ with vertex set V and hyperedge set E. For a vertex v, as in Section 2.4.3, we let

$$\deg v := |\text{hyperedges containing } v| \tag{2.5.34}$$

and, again, we assume that $\deg v > 0$ for all $v \in V$. As before, this *degree* will serve as a normalizing factor. Of course, one could choose a different such factor, but our choice here is in agreement with the conventions of Section 2.5.1. We then obtain the corresponding *scalar products*.

Definition 2.5.12. For $f, g : V \to \mathbb{R}$, let

$$(f,g)_V := \sum_{v \in V} \deg v \cdot f(v) \cdot g(v). \tag{2.5.35}$$

For $\omega, \gamma : E \to \mathbb{R}$, let

$$(\omega, \gamma)_E := \sum_{e \in E} \omega(e) \cdot \gamma(e). \tag{2.5.36}$$

The *boundary operator* then maps functions on vertices to functions on hyperedges that change their sign upon a change of orientation, as always.

Definition 2.5.13. For $f : V \to \mathbb{R}$ and $e \in E$, let

$$\delta f(e) := \sum_{v^j \text{ output of } e} f(v^j) - \sum_{v_i \text{ input of } e} f(v_i). \tag{2.5.37}$$

Thus, the convention is that inputs will have a $-$ sign, while outputs will have a $+$ sign.

We then have

Lemma 2.5.5. *The adjoint of the operator δ with respect to the scalar products (2.5.35), (2.5.36) is*

$$\delta^*(\gamma)(v) = \frac{\sum_{h_{out} : v \text{ output}} \gamma(h_{out}) - \sum_{h_{in} : v \text{ input}} \gamma(h_{in})}{\deg v}. \tag{2.5.38}$$

We then obtain our *Laplace operators*

Lemma 2.5.6. *The Laplace operator for functions $f: V \to \mathbb{R}$ on the vertex set V of an oriented hypergraph is given by*

$$L^0 f(v) := \delta^* \delta = \frac{\sum_{h_{in}:\, v\ input} \left(\sum_{v'\ input\ of\ h_{in}} f(v') - \sum_{z'\ output\ of\ h_{in}} f(z') \right)}{\deg v}$$
$$- \frac{\sum_{h_{out}:\, v\ output} \left(\sum_{\hat{v}\ input\ of\ h_{out}} f(\hat{v}) - \sum_{\hat{z}\ output\ of\ h_{out}} f(\hat{z}) \right)}{\deg v}.$$

(2.5.39)

The Laplacian for functions $\gamma: E \to \mathbb{R}$ on the hyperedge set E, with $\gamma(h^+) = -\gamma(h^-)$ under a change of orientation, is given by

$$L^1 \gamma(h) := \delta \delta^* = \sum_{v_i\ input\ of\ h} \frac{\sum_{h_{in}:\, v_i\ input} \gamma(h_{in}) - \sum_{h_{out}:\, v_i\ output} \gamma(h_{out})}{\deg v_i}$$
$$- \sum_{v^j\ output\ of\ h} \frac{\sum_{h'_{in}:\, v^j\ input} \gamma(h'_{in}) - \sum_{h'_{out}:\, v^j\ output} \gamma(h'_{out})}{\deg v^j}.$$

(2.5.40)

2.6 Rayleigh Quotients and their Generalizations

2.6.1 Rayleigh Quotients

The eigenvalues of our Laplace operators admit a variational characterization described in the following general theorem.

Theorem 2.6.1 (Courant-Fischer-Weyl min-max principle)**.** *Let H be an N-dimensional vector space with a positive definite scalar product $(\cdot, \cdot)$. Let $\mathcal{H}_k$ be the family of all k-dimensional subspaces of H. Let $A: H \to H$ be a self-adjoint linear operator. Then the eigenvalues $\lambda_1 \le \ldots \le \lambda_N$ of A can be obtained by*

$$\lambda_k = \min_{H_k \in \mathcal{H}_k} \max_{g(\neq 0) \in H_k} \frac{(Ag, g)}{(g, g)} = \max_{H_{N-k+1} \in \mathcal{H}_{N-k+1}} \min_{g(\neq 0) \in H_{N-k+1}} \frac{(Ag, g)}{(g, g)}. \quad (2.6.1)$$

The vectors g_k realizing such a min-max or max-min then are corresponding eigenvectors, and the min-max spaces H_k are spanned by the eigenvectors for the eigenvalues $\lambda_1, \ldots, \lambda_k$, and analogously, the max-min spaces H_{N-k+1} are spanned by the eigenvectors for the eigenvalues $\lambda_k, \ldots, \lambda_N$.

Thus, we also have

$$\lambda_k = \min_{g \in H, (g,g_j)=0 \text{ for } j=1,...,k-1} \frac{(Ag,g)}{(g,g)} = \max_{g \in H, (g,g_\ell)=0 \text{ for } \ell=k+1,...,N} \frac{(Ag,g)}{(g,g)}. \tag{2.6.2}$$

In particular,

$$\lambda_1 = \min_{g \in H} \frac{(Ag,g)}{(g,g)}, \qquad \lambda_N = \max_{g \in H} \frac{(Ag,g)}{(g,g)}. \tag{2.6.3}$$

Definition 2.6.1.

$$R_A(g) := \frac{(Ag,g)}{(g,g)} \tag{2.6.4}$$

is the *Rayleigh quotient* (or the *Rayleigh–Ritz quotient*) of g.

We shall usually implicitly assume that $g \neq 0$ in (2.6.4), without always explicitly mentioning it; else, we use the convention that $R_A(g) = 0$ if $g = 0$.

With this convention, we have the following obvious but useful factorization result.

Lemma 2.6.1. *Let* $A, B: H \to H$ *be self-adjoint linear operators, with* $A = a^\star a$, $B = b^\star b$. *Then*

$$R_A(g) = \frac{(ag,ag)}{(g,g)}, \tag{2.6.5}$$

$$R_{AB}(g) = \frac{(abg,abg)}{(bg,bg)} \frac{(bg,bg)}{(g,g)} = R_{AB}(bg)R_B(g). \tag{2.6.6}$$

$\square$

In fact, without loss of generality, we may assume $(g,g) = 1$ in (2.6.1). Then, we get rid of the denominators, and the variational problems reduce to problems on the unit sphere in H (or more precisely, since g and $-g$ always yield the same value, on the corresponding projective space). That perspective will also be adopted in Chapter 5.

Proof We want to find an orthonormal basis of H consisting of eigenfunctions of A,

$$u_k, \ k = 1,...,N.$$

This is achieved as follows. We iteratively define, with $H_1 := H$,

$$H_k := \{g \in H : (g,u_i) = 0 \text{ for } i \leq k-1\}, \tag{2.6.7}$$

we put

$$\lambda_k := \inf_{g \in H_k \setminus \{0\}} \frac{(Ag,g)}{(g,g)}, \tag{2.6.8}$$

and we claim that those are the eigenvalues, which thus can be obtained as these infima. Since $H_k \subset H_{k-1}$, we have

$$\lambda_k \geq \lambda_{k-1}. \tag{2.6.9}$$

We may find a function u_k that realizes the infimum in (2.6.8), that is,

$$\lambda_k = \frac{(Au_k, u_k)}{(u_k, u_k)}. \tag{2.6.10}$$

Since then for every $\phi \in H_k, t \in \mathbb{R}$,

$$\frac{(A(u_k + t\phi), u_k + t\phi)}{(u_k + t\phi, u_k + t\phi)} \geq \lambda_k, \tag{2.6.11}$$

the derivative of that expression with respect to t vanishes at $t = 0$, and we obtain, using (2.6.10) and the self-adjointness of A,

$$0 = (Au_k, \phi) - \lambda_k(u_k, \phi) \tag{2.6.12}$$

for all $\phi \in H_k$; in fact, this even holds for all $\phi \in H$, and not only for those in the subspace H_k, since for $i \leq k - 1$

$$(u_k, u_i) = 0 \tag{2.6.13}$$

and

$$(Au_k, u_i) = (u_k, Au_i) = \lambda_i(u_k, u_i) = 0, \tag{2.6.14}$$

since $u_k \in H_k$. Thus, if we also recall (2.6.12),

$$(Au_k, \phi) = \lambda_k(u_k, \phi) \tag{2.6.15}$$

for all $\phi \in H$ whence

$$Au_k = \lambda_k u_k. \tag{2.6.16}$$

Since, as noted (2.6.13), we may require

$$(u_k, u_k) = 1 \tag{2.6.17}$$

for $k = 1, \ldots, N$, and since u_k is mutually orthogonal by construction, we have constructed an orthonormal basis of H consisting of eigenfunctions of A.

Thus, we may expand any $g \in H$ as

$$g = \sum_k (g, u_k) u_k. \tag{2.6.18}$$

We then also have

$$(g, g) = \sum_k (g, u_k)^2, \tag{2.6.19}$$

since u_k satisfies

$$(u_j, u_k) = \delta_{jk}, \tag{2.6.20}$$

which is the condition for being an orthonormal basis. Finally, using (2.6.19) and (4.1.46), we obtain

$$(Ag, g) = \sum_k \lambda_k (g, u_k)^2. \tag{2.6.21}$$

We have (2.6.8),

$$\lambda_k = \min \left\{ \frac{(Ag, g)}{(g, g)} : g \neq 0, (g, u_j) = 0 \text{ for } j = 1, \ldots, k-1 \right\}. \tag{2.6.22}$$

Dually,

$$\lambda_k = \max \left\{ \frac{(Ag, g)}{(g, g)} : g \neq 0 \text{ linear combination of } u_j \text{ with } j \leq k \right\}. \tag{2.6.23}$$

That maximum is realized when g is a multiple of the k-th eigenfunction, and so is the minimum in (2.6.22).

When L is a $k + 1$-dimensional subspace, there is some g in L satisfying

$$(g, u_j) = 0 \text{ for } j = 0, \ldots, k - 1. \tag{2.6.24}$$

Equations (2.6.19) and (2.6.21) yield

$$\frac{(Ag, g)}{(g, g)} = \frac{\sum_{j \geq k} \lambda_j (g, u_j)^2}{\sum_{j \geq k} (g, u_j)^2} \geq \lambda_k. \tag{2.6.25}$$

This gives us

$$\max_{g \in L \setminus \{0\}} \frac{(Ag, g)}{(g, g)} \geq \lambda_k. \tag{2.6.26}$$

This yields the second equality in (2.6.2), and the first follows by dualizing the argument. □

2.6.2 Critical Points and p-Laplacians

We take another look at the quotient (2.6.5), that is,

$$R(g) = \frac{(ag, ag)}{(g, g)} \tag{2.6.27}$$

for the self-adjoint linear operator $A = a^\star a \colon H \to H$. In fact, $a \colon H \to H_1$ may map H into another Hilbert space H_1, and the product (ag, ag) will then be taken in H_1.

By homogeneity, instead of looking for critical points of $R_A(g)$ via a mini-maxing scheme as in Section 2.6.1, by homogeneity, we look for critical points of (ag, ag) under the condition $(g, g) = 1$. Equivalently, we can introduce a Lagrange multiplier λ and search for critical points of

$$(ag, ag) - \lambda((g, g) - 1). \tag{2.6.28}$$

And a critical g_λ then has to satisfy $(g_\lambda, g_\lambda) = 1$ and

$$\frac{d}{dt}_{|t=0} ((a(g_\lambda + t\varphi), a(g_\lambda + t\varphi)) - \lambda(g_\lambda + t\varphi, g_\lambda + t\varphi)) = 0 \tag{2.6.29}$$

for all $\varphi \in H$, and therefore

$$(a\varphi, ag_\lambda) - \lambda(\varphi, g_\lambda) = 0, \tag{2.6.30}$$

and hence, the eigenvalue equation

$$Ag_\lambda = \lambda g_\lambda, \tag{2.6.31}$$

that is, the Lagrange multiplier becomes an eigenvalue. Conversely, when g_λ solves (2.6.31), it is a critical point of (2.6.27) with critical value λ.

Equation (2.6.31) is, of course, a linear equation, but we now want to generalize this in a nonlinear fashion. When we write $\|g\| = (g, g)^{1/2}$, the Rayleigh quotient (2.6.27) becomes

$$R(g) = \frac{\|ag\|^2}{\|g\|^2},$$

and we can then naturally consider

$$R_p(g) = \frac{\|ag\|^p}{\|g\|^p} \tag{2.6.32}$$

for some exponent p, which we first restrict to $1 < p < \infty$, even though subsequently the cases $p = 1$ and $p = \infty$ will be of particular interest.

Analogously to (2.6.28), we introduce a Lagrange multiplier λ to require $\|g\|^p = 1$ and search for critical points of

$$\|ag\|^p - \lambda(\|g\|^p - 1). \tag{2.6.33}$$

Analogously to (2.6.30), critical points now have to satisfy

$$(a\varphi, (ag_\lambda, ag_\lambda)^{\frac{p}{2}-1} ag_\lambda) - \lambda(\varphi, (g_\lambda, g_\lambda)^{\frac{p}{2}-1} g_\lambda) = 0,$$

and hence the nonlinear eigenvalue equation

$$a^\star((ag_\lambda, ag_\lambda)^{\frac{p}{2}-1} ag_\lambda) = \lambda(g_\lambda, g_\lambda)^{\frac{p}{2}-1} g_\lambda, \tag{2.6.34}$$

or when written in terms of norms,

$$A_p g_\lambda := a^\star(\|ag_\lambda\|^{p-2} ag_\lambda) = \lambda \|g_\lambda\|^{p-2} g_\lambda. \qquad (2.6.35)$$

In the spirit of the minimaxing scheme of Section 2.6.1, there is also a variational characterization of these nonlinear eigenvalues. We discuss here only the case where the Hilbert spaces H, H_1 are finite dimensional, of dimension n, so that we can identify them with $\mathbb{R}^n$ with its Euclidean structure (although in Section 3.3, the infinite dimensional case will occur). The appropriate concept is (see for instance [152, 277])

Definition 2.6.2. The *Krasnoselskii* $\mathbb{Z}_2$ *genus* of a centrally symmetric set S in $\mathbb{R}^n$ is

$$\text{genus}(S) :=$$
$$\begin{cases} \min\{k \in \mathbb{Z}^+ : \exists \text{ odd continuous } h \colon S \setminus \{0\} \to \mathbb{S}^{k-1}\} & \text{if } S \setminus \{0\} \neq \emptyset, \\ 0 & \text{if } S \setminus \{0\} = \emptyset. \end{cases}$$

For $k \geq 1$, we let $\text{Gen}_k := \{S \subset \mathbb{R}^n : S \text{ centrally symmetric with genus } (S) \geq k\}$.

Here, $\mathbb{Z}_2$ is the symmetry group for the reflection $j(x) = -x$ of $\mathbb{R}^n$, and *centrally symmetric* simply means invariance under j. The typical example of such a centrally symmetric set is the sphere $\mathbb{S}^{n-1} \subset \mathbb{R}^n$.

Lemma 2.6.2. *The genus of the sphere $\mathbb{S}^{n-1} \subset \mathbb{R}^n$ is n.*

Proof The identity map $\mathbb{S}^{n-1} \to \mathbb{S}^{n-1}$ is odd and continuous. Therefore, the genus is at most n. But it follows from the Borsuk-Ulam Theorem [37] that there exists no such odd continuous map from $\mathbb{S}^{n-1}$ to $\mathbb{S}^{k-1}$ for $k < n$. $\qquad\square$

Lemma 2.6.3. *1. The genus of $\mathbb{R}^n \setminus \mathbb{R}^m$ is $n - m$ $(n \geq m)$.*
2. If $S_1 \subset S_2$, then $\text{genus}(S_1) \leq \text{genus}(S_2)$
3. If $\text{genus}(S) \geq k$ for $S \subset \mathbb{R}^n$, then $S \cap \mathbb{R}^m \neq \emptyset$ for $m \geq n - k$

Proof For $x = (x^1, \ldots, x^n) \in \mathbb{R}^n \setminus \mathbb{R}^m$, we have $(x^{m+1})^2 + \ldots (x^n)^2 \neq 0$, where we let $\mathbb{R}^m = \{(y^1, \ldots, y^n) \in \mathbb{R}^n : y^{m+1} = \cdots = y^n = 0\}$, and so the map

$$(x^1, \ldots, x^n) \mapsto \left(\frac{x^{m+1}}{(x^{m+1})^2 + \ldots (x^n)^2}, \ldots, \frac{x^n}{(x^{m+1})^2 + \ldots (x^n)^2} \right) \in \mathbb{R}^n \setminus \mathbb{R}^m$$

is continuous and odd on $\mathbb{R}^n \setminus \mathbb{R}^m$. Hence, the genus is $\leq n - m$. The opposite inequality follows from the argument of Lemma 2.6.2 (but will not be needed in the sequel). This proves the first statement. The second one is a direct consequence of the definition because a continuous map restricted to a subset of its domain remains continuous. The third statement is a consequence of the two others. □

Thus, the idea is to perform minimaxing over families that (in the sense of Definition 2.6.2) have the topology of spheres. According to the general theory of Lusternik and Schnirelman, presented for instance in [152, 277], for $k = 1, \ldots, n$, the constants

$$\lambda_k(A_p) := \inf_{S \in \text{Gen}_k} \sup_{f \in S \setminus \{0\}} R_p(f) \tag{2.6.36}$$

then are eigenvalues of A_p. They satisfy

$$\lambda_1 \leq \ldots \leq \lambda_n$$

and, if $\lambda = \lambda_{k+1} = \ldots = \lambda_{k+l}$ for $0 \leq k < k + l \leq n$, then

$$\text{genus}(\{\text{eigenfunctions corresponding to } \lambda\}) \geq l.$$

In the preceding, we have required $p > 1$, because if we took $p = 1$, (2.6.34) would become

$$a^\star \frac{ag_\lambda}{\|ag_\lambda\|} = \lambda \frac{g_\lambda}{\|g_\lambda\|}, \tag{2.6.37}$$

and these quotients become undefined at points where $g_\lambda = 0$. (In our applications, the Hilbert space H and H_1 are function spaces, and the functions may then have zeroes without being identically 0.)

In order to handle this issue, we define the operator A_p for $p = 1$ as a set-valued map. The Rayleigh quotients, as their name indicates, are quotients of functionals

$$\frac{F(f)}{G(f)},$$

and when both F and G are differentiable, a critical point g satisfies the eigenvalue equation

$$\nabla F(g) - \frac{F(g)}{G(g)} \nabla G(g) = 0$$

for the eigenvalue $\lambda = \frac{F(g)}{G(g)}$ as we have explored earlier. In the nondifferentiable case, as in [130], we use the generalized gradient of Clarke [63], defining

Definition 2.6.3. The *generalized gradient* of a functional $R \colon H \to \mathbb{R}$ at $g \in H$ is

$$\partial R(g) := \{ g^* \in H \colon \lim_{f \to g, t \to 0} \sup \frac{R(f + t\phi) - R(f)}{t} \geq (g^*, \phi) \text{ for all } \phi \in H \}. \tag{2.6.38}$$

$g \in H$ is a *critical point* of R if

$$0 \in \partial R(g). \tag{2.6.39}$$

[130] then shows

Lemma 2.6.4. *Let F, G be convex, Lipschitz continuous, even and positively p-homogeneous for $p \geq 1$ (i.e., $F(\mu f) = \mu^p F(f)$ for all $\mu \geq 0$, and the same for G). Then, a necessary condition that f is a critical point of $\frac{F(f)}{G(f)}$ is that*

$$0 \in \partial F(f) - \lambda \partial G(f), \tag{2.6.40}$$

where $\lambda = \frac{F(f)}{G(f)}$. When G is continuously differentiable at f, this is also sufficient.

In fact, in the present situation, instead of the generalized gradient ∂, we can also and will use the *subgradient*, see for instance [88, 239].

Definition 2.6.4. $R \colon \mathbb{R}^n \to \mathbb{R} \cup \{\pm\infty\}$ is called *subdifferentiable* at g if there exists some $g^* \in H$ such that

$$R(h) - R(g) \geq (h - g, g^*), \tag{2.6.41}$$

and the set of all g^* with this property is called the *subgradient* $\nabla R(g)$.

(When we ignore the lim sup in (2.6.38) and put $f = g, t\phi = h - g$, then (2.6.38) reduces to (2.6.41).)

Thus, the subgradient of a subdifferentiable R at g consists of the slopes of all affine linear functions that are not larger than R and agree with R at g. Our eigenvalue equation then becomes

$$0 \in \nabla F(g_\lambda) - \lambda \nabla G(g_\lambda). \tag{2.6.42}$$

Thus, we take (2.6.40) or (2.6.42) as our eigenvalue equation for the operator A_1 when $\frac{F(f)}{G(f)}$ is the Rayleigh quotient $R_1(f)$. This condition will be spelled out in more detail in Sections 3.3 and 7.2. We should, however, point out that the setting there will be slightly different. We shall consider spaces of functions $f \colon \Sigma \to \mathbb{R}$, and instead of working with the p-th powers of the L^2-norm, we shall consider the L^p-norms. But the abstract setting and the problems to be overcome will remain the same.

2.7 Symmetries

Let V be some set with relations $\omega(v_1, \ldots, v_k)$ for certain values of k as in Section 2.2. We abbreviate that structure as (V, ω).

Definition 2.7.1. An *automorphism* of (V,ω) is a bijection $\sigma\colon V \to V$ with the property that

$$\omega(\sigma(v_1),\ldots,\sigma(v_k)) = \omega(v_1,\ldots,v_k) \qquad (2.7.1)$$

for all relations. For such an automorphism σ and $f\colon V \to \mathbb{R}$, we put

$$\sigma^* f(v) = f(\sigma(v)).$$

Remark 2.7.1. This concept also applies to hypersimplicial complexes, with obvious modifications.

Obviously, the automorphism of (V,ω) forms a group, its *automorphism group*. We then have the obvious, but important

Lemma 2.7.1. *If σ is an automorphism of (V,ω), then for all Laplacians*

$$L(\sigma^*\eta)(v) = \sigma^*(L\eta)(v) \qquad (2.7.2)$$

for all $v \in V$ and all η to which it can be applied, for instance to $f\colon V \to \mathbb{R}$ when the relations hold for $k = 2$, that is, for pairs of vertices.

Proof This holds because L is a linear operator. □

Thus, Laplacians commute with automorphisms σ. In particular, the spectrum is not affected by automorphisms, and eigenspaces stay invariant. We conclude

Corollary 2.7.1. *Laplacian eigenspaces constitute representations of the automorphism group of (V,ω).* □

We can use Lemma 2.7.1 to decompose the spectrum of L. For simplicity, we consider the case $k = 2$, and hence, the space of functions $f\colon V \to \mathbb{R}$ and an automorphism τ of (V,ω) with

$$\tau^2 = \mathrm{id}. \qquad (2.7.3)$$

Then, τ has two possible eigenvalues, ± 1, on the space of functions $f\colon V \to \mathbb{R}$, and L leaves those two eigenspaces $L_\pm$ invariant. Also, $V = V_0 + V_1$ where $\tau(v_0) = v_0$ precisely if $v_0 \in V_0$. That is, V_0 is the set of those vertices that are fixed by τ. Moreover, we can write $V_1 = V' \cup V''$ where V', V'' are disjoint and $\tau(V') = V''$. Since also $\tau(V'') = V'$ because of (2.7.3), V' and V'' play symmetric roles.

Without loss of generality, we assume that V' (and hence also V'') is connected, as otherwise we can rearrange the decomposition of V_1 and/or write τ as the composition of several such automorphisms.

Lemma 2.7.2. *There is a $|V'|$-dimensional space of functions $f : V \to \mathbb{R}$ consisting of eigenfunctions of L that vanish on V_0 and that are antisymmetric on V' and V'', that is, $f(v'') = -f(v')$ if $v'' = \tau(v') \in V''$ for $v' \in V'$, and a remaining $(|V'|+|V_0|)$-dimensional space of eigenfunctions that are symmetric on V' and V'', that is, $f(v'') = f(v')$.*

Proof The first class of functions are those that are eigenfunctions of τ for the eigenvalue -1, where the second class has eigenvalue 1. By Lemma 2.7.1, these are unions of eigenspaces of L. Since $|V''| = |V'|$ and $V = V_0 \cup V' \cup V''$, this generates the space of all functions on V. $\square$

Definition 2.7.2. An *induced substructure* or a *motif* $(\hat{V}, \omega)$ consists of some nonempty vertex set $\hat{V} \subset V$ and the same relations between its elements as in V.

Let $\hat{V}$ be such a motif. We then have the induced Laplacian

$$L_{V,\hat{V}} f(v) = f(v) - \frac{1}{\deg_V v} \sum_{v'} \omega_{vv'} f(v'), \qquad (2.7.4)$$

where $\deg_V v$ denotes the degree of v in V.

Definition 2.7.3. The motif V' with vertex set V' is a *duplicated motif* if V' and V'' are disconnected.

Lemma 2.7.3. *Let V' be a duplicated motif in V, and let $v_0 \in V_0$ be a neighbor of some $v' \in V'$, that is, $\omega_{v',v_0} \neq 0$. Then, v_0 is also a neighbor of $v'' = \tau(v') \in V''$.*

Proof Since $v_0 \in V_0$ is fixed by the automorphism τ, and since $\omega_{v',v_0} = \omega_{\tau(v'),v_0}$, the claim follows. $\square$

Lemma 2.7.4. *Let V' be a duplicated motif in V. Then, we find a basis of eigenfunctions of the Laplacian L of V of functions f satisfying either*

1.

$$L_{V,V'} f(v) = \begin{cases} f(v) & \text{for } v \in V' \text{ (and also for } v \in V'') \\ -f(\tau(v)) & \text{for } v \in V'' \\ 0 & \text{for } v \in V_0 \end{cases} \qquad (2.7.5)$$

2. or

$$f(\tau(v)) = f(v) \text{ for } v \in V'. \qquad (2.7.6)$$

The latter eigenfunctions are those of V^τ obtained as the quotient of V by τ, with the induced weights.

Proof If v_0 in V_0 and $f(v_0) = 0$ and if f is antisymmetric, then also $Lf(v_0) = 0$ because the contributions from its neighbors v' and $v'' = \tau(v')$ (see Lemma 2.7.3) cancel each other in $Lf(v_0)$. The result then follows from Lemma 2.7.2 because under our assumptions, a neighbor w of $v \in V'$ is contained either in V' or in V_0, in which case, for an antisymmetric f, $f(z) = 0$, and therefore, we can restrict the computation in (2.7.5) to the induced Laplacian, that is, consider only the vertices in V'. □

For the case of hypergraphs with real coefficients, the role of symmetries was studied in [154], and we have used that reference for our exposition.

2.8 Homomorphisms

As in Section 2.7, let U, V be sets with relations $\omega_U(u_1, \ldots, u_k)$ and $\omega_V(v_1, \ldots, v_k)$, respectively, for certain values of k as in Section 2.2.

Definition 2.8.1. A *homomorphism* of $h: (U, \omega_U) \rightarrow (V, \omega_V)$ is a map $h: U \rightarrow V$ with the property that

$$\omega_V(h(u_1), \ldots, h(u_k)) = \omega_U(u_1, \ldots, u_k) \tag{2.8.1}$$

for all relations. For such a homomorphism h and $f: V \rightarrow \mathbb{R}$, we put

$$h^* f(u) = f(h(u)).$$

When $\hat{\Gamma}$ and Γ are (unweighted) graphs,[1] a homomorphism $h: \hat{\Gamma} \rightarrow \Gamma$ thus maps the vertices of $\hat{\Gamma}$ to vertices of Γ and preserves edges, that is, $h(v_1) \sim h(v_2)$ if and only if $v_1 \sim v_2$. A special case is

Definition 2.8.2. Let $\hat{\Gamma}$ and Γ be graphs. A *covering* is a homomorphism $\pi: \hat{\Gamma} \rightarrow \Gamma$ that maps the edges emanating from each vertex v of $\hat{\Gamma}$ bijectively onto the edges emanating from its image $\pi(v)$ in Γ.

We also say that the *stars* of vertices are mapped bijectively. Importantly, we do not require that π induces a bijection between the vertex sets. When Γ is connected, then each vertex of Γ must have the same number of preimages under h, and if that number is k, we speak of a k-covering. See [6] and the references therein for more detailed discussions of coverings. We shall analyze 2-coverings in Section 4.6.

---

[1] The definitions should be clear but will be explicitly formulated in Section 4.1.

2.9 General Setting

We shall work in the context of a general discrete structure. Let $\mathcal{M} :=$ $(M_1,\ldots,M_\ell)$ be a sequence of multisets. For each $q = 1,\ldots,\ell$, consider a Hilbert space $C(M_q)$ equipped with a scalar product $(\cdot,\cdot)_q$, and consider two operators

$$C(M_q) \overset{\delta^q}{\underset{\beta^q}{\rightleftarrows}} C(M_{q+1}).$$

We call $(\mathcal{M},(\delta^q)_{q\geq 1},(\beta^q)_{q\geq 1})$ a *discrete structure* because it contains some groups of objects and the interaction relations between them.

Most of the discrete objects considered in this book can be recovered by this general structure, including (directed, oriented, weighted) hypergraphs, simplicial complexes, and Lefschetz complexes. Moreover, we shall use this to formalize new objects we are interested in, such as hypersimplicial complexes.

We consider **Laplace operators** of the following type.

(i) The *q-up Laplace operator* of $\mathcal{M}$ is

$$L_{up}^q := \beta^q \delta^q,$$

(ii) The *q-down Laplace operator* of $\mathcal{M}$ is

$$L_{down}^q := \delta^{q-1}\beta^{q-1}, \text{and}$$

(iii) The *q-Laplace operator* of $\mathcal{M}$ is

$$L^q := L^q{}_{up} + L^q{}_{down}.$$

The **linear case** is the most important one. Suppose that both β^q and δ^q are linear. Then, the nonzero eigenvalues of L^q_{down} and L_{up}^{q-1} coincide. Also, since $(L_{up}^q)^* = (\delta^q)^*(\beta^q)^*$ has the same eigenvalues with L_{up}^q, in principle, we only need to study L_{up}^q.

Example 2.9.1. Recall, from Section 2.4.3, that a directed hypergraph is an oriented graph $\Gamma = (V,E,\varphi)$ with the additional assumption that hyperedges are directed, from the inputs toward the outputs. In this case, each hyperedge e can be seen as a directed edge from its input set $\text{in}(e)$ toward its output set $\text{out}(e)$. Hence, E can be seen as the edge set of a directed graph whose vertex set is some $\mathcal{A} \subset \mathcal{P}(V)$. Consider

$$C(V) \xrightarrow{\delta^0} C(\mathcal{A}) \xrightarrow{\delta^1} C(E),$$

where $\delta^0 f(A) := \sum_{v\in A} f(v)$ and $\delta^1 g(e) := g(\text{out}(e)) - g(\text{in}(e))$. By letting $\delta := \delta^1\delta^0$, we have $\delta f(e) = \sum_{v'\in\text{out}(e)} f(v') - \sum_{v\in\text{in}(e)} f(v)$. Let $\beta\colon C(E) \to$

$C(V)$ be defined by $\beta q(v) := \sum_{e\,:\,v\in\text{out}(e)} q(e)$. Then, the Laplacian can be defined as $L = \beta\delta$.

In many instances, we shall also consider **non-linear cases**. We can also take β^q and δ^q as nonlinear operators, as shown by the next two examples.

Example 2.9.2. Let $\Gamma = (V, E)$ be a k-uniform hypergraph with nodes $v_1, \ldots, v_N$. Let $\delta f(e) := \prod_{v\in e} f(v)$, and let $\beta g(v) := \sum_{e\in E\,:\,v\in e} g(e)$. Then, the nonlinear operator $\beta\delta$ is related to the tensors associated with Γ that we shall treat in Chapter 11. In fact, $\beta\delta$ is the k-th order N-dimensional tensor T such that $T_{i_1,\ldots,i_k} = 1$ if $\{v_{i_1}, \ldots, v_{i_k}\} \in E$, and $T_{i_1,\ldots,i_k} = 0$ otherwise. Also, the corresponding *H-eigenvalue problem* can be formulated as $\beta\delta f = \lambda f^k$.

Example 2.9.3. Let $\Gamma = (V, E, \varphi)$ be a directed hypergraph. Let

$$\delta f(e) := \max_{v\in\text{out}(e)} f(v) - \min_{v'\in\text{in}(e)} f(v'),$$

and consider a set-valued map β such that $\beta\delta f$ coincides with the subgradient of $\|\delta f\|_p^p := \sum_{e\in E} |\delta f(e)|^p$. Then, $\beta\delta$ coincides with the Lovász p-Laplacian for hypergraphs that we shall treat in Chapters 6 and 8.

In many instances, we shall also consider **linear and adjoint cases**. In this book, the most important case is where δ^q is linear and $\beta^q = (\delta^q)^*$. Hence, as suggested by (2.2.4), we shall consider the critical points of functionals in the form of

$$f \mapsto \frac{F(\delta f)}{G(f)},$$

where $F\colon C(M_{q+1}) \to \mathbb{R}$ and $G\colon C(M_q) \to \mathbb{R}$ are locally Lipschitz functions. Thus, we shall study the general eigenvalue problem

$$0 \in \nabla_f F(\delta f) - \lambda\nabla_f G(f),$$

where ∇_f indicates the Clarke derivative at f (see Definition 2.6.3). Since

$$\nabla_f F(\delta f) = \delta^*\nabla F(\delta f) = \delta^* \circ \nabla F \circ \delta(f),$$

it is convenient to set $\beta = \delta^*$ (or $\beta = \delta^* \circ \nabla F$). That is, the additional assumption $\beta = \delta^*$ helps us to study the eigenvalue problems we are interested in by variational methods.

3

The Classical Case: Eigenvalues on Euclidean Domains and Riemannian Manifolds

This chapter provides an overview of the spectral theory of the Laplacian on functions and differential forms. It also introduces the p-Laplacian for functions. We emphasize variational aspects of eigenvalues and Cheeger's inequality.

Eigenvalue problems for Laplace operators were first systematically developed in the continuous case, a long time before they found attention in graph theory. The classical problem is the Dirichlet problem for the Laplace operator on a domain in Euclidean space. The general theory was presented in [69] and applied in [70] not only to the Dirichlet problem for the Laplacian but also to a range of other problems arising in the calculus of variations and the theory of partial differential equations.

The topic was then taken up in Riemannian geometry, and eigenvalue problems for the Laplace–Beltrami operator on closed Riemannian manifolds or on domains in such manifolds were intensively studied, see, for instance, the monographs [28, 55, 181] and the key contribution [182]. In particular, in that context, Cheeger [56] discovered his famous inequality, relating the second (i.e., the first nonvanishing) eigenvalue of the Laplace–Beltrami operator on a closed Riemannian manifold to the volume of hypersurfaces cutting that manifold into two large parts.

This theory then later inspired the analogous theory of eigenvalues of graph Laplacians, starting with [4, 61, 82], although eigenvalues of the algebraic graph Laplacian had already been studied earlier from a different perspective.

In this chapter, we therefore summarize the theory in the Euclidean and Riemannian case, as this theory is well understood and motivates much of what we shall present in this monograph in the discrete case.

In fact, the Euclidean and Riemannian theory is more difficult in the sense that it requires the application of compactness schemes and the solution of convergence issues, which are not present in the discrete case. The mathematical difference between the finite discrete and the smooth setting is that in the latter, differential operators such as the Laplacian are unbounded operators with only dense domains in appropriate Hilbert or Banach spaces, whereas the discrete operators we shall consider are bounded (see for instance [142] for a discussion). For this reason, we shall only indicate some of the corresponding proofs, without providing all details. These can be readily found in the literature.

There is an important difference between the setting of this chapter and that of subsequent chapters where we shall only treat finite discrete structures. In the smooth case discussed here, differential operators on Euclidean domains or manifolds are unbounded operators with only dense domains in Hilbert space. In contrast, in the discrete setting, normalized discrete Laplacians are bounded operators. This will avoid many analytical difficulties.

3.1 The Laplacian on Euclidean Domains

In this section, we use the reference [149]. On $\mathbb{R}^n$, we have the *Laplace operator*

$$\Delta f(x) := -\sum_{i=1}^{n} \frac{\partial^2 f}{\partial (x^i)^2}(x) \tag{3.1.1}$$

for $x = (x^1, \ldots, x^n)$. (While in Chapter 6, we shall use the notation $\mathbf{x}$ for vectors, here we denote a vector by a simple x, in order to follow the conventions of Riemannian geometry, which we shall discuss in the next section.) With the standard symbols of vector calculus, (3.1.1) can also be written as

$$\Delta f(x) = -\text{div}\,\nabla f(x). \tag{3.1.2}$$

While most analysts prefer not to use the minus sign in (3.1.1) and (3.1.2), we use that sign in line with our general convention that makes the Laplacian a positive operator. In the definition, one could take f as a C^2 (twice continuously differentiable)-function, but it turns out that for the analysis of the eigenvalue problem, it is natural to consider a wider class of functions.

We let $\Omega \subset \mathbb{R}^n$ be a domain, that is, an open bounded set. In order to exclude trivial cases, we always assume that Ω is nonempty. The basic function space is $L^2(\Omega)$, the space of measurable functions that are square integrable. It is a Hilbert space with respect to the scalar product

$$(f, g)_0 := \int_\Omega f(x)g(x)dx \tag{3.1.3}$$

and we have the norm

$$\|f\|_0 := \|f\|_{L^2(\Omega)} = (f, f)_0^{1/2}. \tag{3.1.4}$$

The *Sobolev space* $H^{1,2}(\Omega)$ is defined as the completion of $C^\infty(\Omega)$ with respect to the scalar product

$$(f,g)_1 := \int_\Omega f(x)g(x)dx + \int_\Omega Df(x)\cdot Dg(x)dx, \qquad (3.1.5)$$

where $Df = \left(\frac{\partial f}{\partial x^1},\ldots,\frac{\partial f}{\partial x^n}\right)$ is the vector of partial derivatives of f, and $\cdot$ denotes the Euclidean scalar product. We then also have the corresponding Sobolev norm

$$\|f\|_1 := \|f\|_{H^{1,2}(\Omega)} = (f,f)_1^{1/2}. \qquad (3.1.6)$$

The space $H_0^{1,2}(\Omega) \subset H^{1,2}(\Omega)$ is defined as the completion of $C_0^\infty(\Omega)$ with respect to (3.1.5), where the latter space consists of those C^∞-functions that vanish outside some relatively compact $\Omega' \subset \Omega$. [1] Functions $f \in H^{1,2}(\Omega)$ have so-called weak derivatives $Df = (D_1 f,\ldots,D_n f)$ that satisfy $D_i f \in L^2(\Omega)$ and the integration by parts formula

$$\int D_i f \phi dx = -\int f \frac{\partial \phi}{\partial x^i} dx \quad \text{for all } \phi \in C_0^\infty(\Omega). \qquad (3.1.7)$$

When $f \in C^1(\Omega)$, then $D_i f$ is the partial derivative $\frac{\partial f}{\partial x^i}$.

We want to study the *eigenfunctions* of Δ in $H_0^{1,2}(\Omega)$, thus $f \in H_0^{1,2}(\Omega)$ with

$$\Delta f(x) = \lambda f(x) \quad \text{for all } x \in \Omega, \qquad (3.1.8)$$

where Δf again has to be defined in the weak sense by an integration by parts formula. We shall see, however, that all solutions of (3.1.8) will be smooth, and so, in the end, regularity is not an issue. When we find such an $f \neq 0$, then λ is an *eigenvalue* of Δ. These eigenvalues are non-negative real numbers. In fact, the Laplace operator is symmetric and non-negative, because for ϕ, ψ in $C_0^\infty(\Omega)$, we have

$$(\Delta\phi,\psi)_0 = (D\phi,D\psi)_0 = (\phi,\Delta\psi)_0.$$

In order to study the eigenvalues, we need some important calculus results. As a preparation, we recall some basic results about Hilbert spaces.

Lemma 3.1.1. *Any bounded sequence in a Hilbert space* $(H,\langle\cdot,\cdot\rangle_H)$ *contains a weakly convergent subsequence, that is, a sequence* $(h_n)_{n\in\mathbb{N}}$ *for which* $\langle h_n,\ell\rangle_H$ *converges for any* $\ell \in H$. *Conversely, weakly convergent sequences are bounded.*

---

[1] The subscripts 0 or 1 in the scalar products and norms indicate the number of derivatives that enter. Thus, in the same manner, one can also define higher order Sobolev spaces. The subscript 0 in $H_0^{1,2}(\Omega)$ indicates that we are looking at functions that vanish on the boundary of Ω.

Moreover, the norm $\|.\|_H$ *is lower semi-continuous with respect to weak convergence, that is, the weak limit h satisfies*

$$\|h\|_H \leq \liminf_{n\to\infty} \|h_n\|_H. \tag{3.1.9}$$

Weak convergence is to be distinguished from strong convergence, that is, convergence with respect to the norm $\|.\|_H$, which would mean $\|h_n - h\|_H \to 0$.

The first result about the Sobolev space $H_0^{1,2}(\Omega)$ that we shall need is the *Poincaré Inequality*:

Theorem 3.1.1. *Any* $f \in H_0^{1,2}(\Omega)$ *satisfies*

$$\|f\|_{L^2(\Omega)} \leq c\,\mathrm{Vol}(\Omega)^{1/n}\|Df\|_{L^2(\Omega)}, \tag{3.1.10}$$

with some universal constant c.

We shall understand in a moment that the best value of $c\,\mathrm{Vol}(\Omega)^{1/n}$ that can be achieved is the square root of the first eigenvalue λ_1.

The most important tool for the scheme to follow, however, is *Rellich's compactness Theorem*. Both $H_0^{1,2}(\Omega)$ and $L^2(\Omega)$ are Hilbert spaces, and the former in fact is a subspace of the latter because any function whose $H^{1,2}$-norm is bounded is also bounded in L^2. However, the $H^{1,2}$-norm is much stronger than the L^2-norm, and we have

Theorem 3.1.2 (Rellich). *The embedding of* $H_0^{1,2}(\Omega)$ *into* $L^2(\Omega)$ *is compact. That means that any sequence with bounded* $H^{1,2}(\Omega)$*-norm (3.1.6) contains a subsequence that converges with respect to the* $L^2(\Omega)$*-norm (3.1.4). Equivalently, by Lemma 3.1.1, a sequence that converges weakly in* $H^{1,2}(\Omega)$ *converges strongly in* $L^2(\Omega)$*, that is, with respect to the* $L^2(\Omega)$*-norm.*

We can now start our scheme and define

$$\lambda_1 := \inf_{f\in H\setminus\{0\}} \frac{(Df, Df)}{(f,f)}. \tag{3.1.11}$$

By Theorem 3.1.1,

$$\lambda_1 > 0. \tag{3.1.12}$$

(This is in contrast to the situation we encounter elsewhere, where $\lambda_1 = 0$, and we only have $\lambda_2 > 0$.) We let $(f_n)_{n\in\mathbb{N}}$ be a minimizing sequence for (3.1.12), so that

$$\lim_{n\to\infty} \frac{(Df_n, Df_n)}{(f_n, f_n)} = \lambda_1. \tag{3.1.13}$$

We may assume that

$$\|f_n\| = 1 \quad \text{for all } n, \tag{3.1.14}$$

and by (3.1.13), then also

$$\|Df_n\| \le K \quad \text{for all } n. \tag{3.1.15}$$

Since bounded sequences in Hilbert spaces contain weakly convergent subsequences, we may assume that $(f_n)_{n \in \mathbb{N}}$ converges weakly in the Hilbert space $H := H_0^{1,2}(\Omega)$ to some $u_1 \in H$, and by the Rellich compactness Theorem 3.1.2, $(f_n)_{n \in \mathbb{N}}$ then also converges strongly in $L^2(\Omega)$ to u_1; by (3.1.14), it follows that

$$\|u_1\| = 1.$$

Furthermore, it follows, because of lower semi-continuity of $\|Df\|_{L^2(\Omega)}$ for weak convergence in H (Lemma 3.1.1; by Theorem 3.1.1, $\|Df\|_{L^2(\Omega)}$ defines a norm in H), and the definition of λ_1 is that

$$\lambda_1 \le (Du_1, Du_1) \le \lim_{n \to \infty} (Df_n, Df_n) = \lambda_1,$$

so

$$\frac{(Du_1, Du_1)}{(u_1, u_1)} = \lambda_1.$$

We can now implement the scheme of the *Rayleigh quotients*. We assume that

$$(\lambda_1, u_1), \dots, (\lambda_{m-1}, u_{m-1})$$

have already been determined iteratively, with $\lambda_1 \le \lambda_2 \le \dots \le \lambda_{m-1}$,

$$\Delta u_i(x) = \lambda_i u_i(x) \quad \text{in } \Omega,$$

and

$$(u_i, u_j) = \delta_{ij} \quad \text{for } i, j = 1, \dots, m-1. \tag{3.1.16}$$

We set

$$H_m := \{f \in H : (f, u_i) = 0 \text{ for } i = 1, \dots, m-1\}$$

and

$$\lambda_m := \inf_{f \in H_m \setminus \{0\}} \frac{(Df, Df)}{(f, f)}.$$

We note that H_m is a Hilbert space itself because as the orthogonal complement of a finite dimensional subspace, it is closed. In fact, if $(f_n)_{n \in \mathbb{N}} \subset H_m$ converges to f, then, as $(f_n, u_i) = 0$ for all $n \in \mathbb{N}$, $(f, u_i) = 0$ for $i = 1, \dots, m$, so $f \in H_m$. Also, since $H_m \subset H_{m-1}$, we have

$$\lambda_m \ge \lambda_{m-1}, \tag{3.1.17}$$

As before, we then find a $u_m \in H_m$ with $\|u_m\| = 1$ and

$$\lambda_m = (Du_m, Du_m) = \frac{(Du_m, Du_m)}{(u_m, u_m)}. \tag{3.1.18}$$

Then

$$\Delta u_m = \lambda_m u_m \quad \text{in } \Omega. \tag{3.1.19}$$

This is followed by a Lagrange multiplier argument as in the general scheme, but here there are also some regularity issues. We first consider $\phi \in H_m$. The following expression is differentiable in t and has a minimum at $t = 0$,

$$\frac{(D(u_m + t\phi), D(u_m + t\phi))}{(u_m + t\phi, u_m + t\phi)}.$$

Therefore,

$$\begin{aligned}
0 &= \frac{d}{dt} \frac{(D(u_m + t\phi), D(u_m + t\phi))}{(u_m + t\phi, u_m + t\phi)}\Big|_{t=0} \\
&= 2\left(\frac{(Du_m, D\phi)}{(u_m, u_m)} - \frac{(Du_m, Du_m)}{(u_m, u_m)}\frac{(u_m, \phi)}{(u_m, u_m)}\right) \\
&= 2\left((Du_m, D\phi) - \lambda_m(u_m, \phi)\right).
\end{aligned}$$

$u_m \in H_m$, $(u_m, f) = 0$ for all f in the orthogonal complement of H_m. Since that complement is spanned by the u_i for $i = 1, \ldots, m - 1$,

$$(u_m, u_i) = 0$$

and then also

$$(Du_m, Du_i) = (Du_i, Du_m) = \lambda_i(u_i, u_m) = 0.$$

Altogether,

$$(Du_m, D\phi) - \lambda_m(u_m, \phi) = 0 \tag{3.1.20}$$

for all $\phi \in H$. Writing the scalar product out, u_m is a solution of

$$\int_\Omega Du_m(x) D\phi(x) dx - \lambda_m \int_\Omega u_m(x)\phi(x) dx = 0,$$

for all $\phi \in H_0^{1,2}(\Omega)$. Thus, u_m is a weak solution of an elliptic partial differential equation, and by elliptic regularity theory (see for instance [149]), $u_m \in C^\infty(\Omega)$ and

$$\Delta u_m(x) = \lambda_m u_m(x) \quad \text{for all } x \in \Omega. \tag{3.1.21}$$

We next observe that the sequence of the eigenvalues λ_m is unbounded.

Lemma 3.1.2. $\lim_{m \to \infty} \lambda_m = \infty$.

Proof If not, (3.1.18) would imply

$$\|Du_m\| \le K \quad \text{for all } m \in \mathbb{N}.$$

By the Rellich compactness Theorem 3.1.2, a subsequence of $(u_m)_{m \in \mathbb{N}}$ would converge to some limit $u \in L^2(\Omega)$, but this is not possible since the u_m are mutually orthonormal, $(u_\ell, u_m) = \delta_{\ell m} = 1$ for $\ell = m$ and $= 0$ for $\ell \neq m$. $\square$

We now arrive at the main result.

Theorem 3.1.3. *For any (nonempty) open and bounded $\Omega \subset \mathbb{R}^n$, the eigenvalue problem*

$$\Delta u = \lambda u = 0$$

has countably many eigenvalues $\lambda = \lambda_m > 0$ with pairwise orthonormal vectors $u_m \in H_0^{1,2}(\Omega)$ and

$$(Du_m, Du_\ell) = \lambda_m \delta_{m\ell}. \tag{3.1.22}$$

Moreover,

$$\lambda_m \to \infty \quad \text{for } m \to \infty.$$

The u_m constitutes a complete orthonormal basis of $L^2(\Omega)$. Therefore, for $f \in L^2(\Omega)$, we have the expansion

$$f = \sum_{i=1}^{\infty} (f, u_i) u_i, \tag{3.1.23}$$

and if $f \in H_0^{1,2}(\Omega)$, also

$$(Df, Df) = \sum_{i=1}^{\infty} \lambda_i (f, u_i)^2. \tag{3.1.24}$$

Proof (3.1.22) follows from (3.1.20) and the fact the that u_m are orthonormal. For showing the expansions (3.1.23) and (3.1.24), we put for $f \in H$

$$\alpha_i := (f, u_i) \quad \text{for } i \in \mathbb{N}$$

and

$$f_m := \sum_{i=1}^{m} \alpha_i u_i, \quad \phi_m := f - f_m.$$

Then,

$$(\phi_m, u_i) = 0 \quad \text{for } i = 1, \dots, m \tag{3.1.25}$$

and by definition of λ_{m+1},

$$(D\phi_m, D\phi_m) \geq \lambda_{m+1}(\phi_m, \phi_m). \tag{3.1.26}$$

By (3.1.20) and (3.1.25), we also have

$$(D\phi_m, Du_i) = 0 \quad \text{for } i = 1,\ldots,m. \tag{3.1.27}$$

From (3.1.25), we obtain

$$(\phi_m, \phi_m) = (f, f) - (f_m, f_m), \tag{3.1.28}$$

and from (3.1.27)

$$(D\phi_m, D\phi_m) = (Df, Df) - (Df_m, Df_m). \tag{3.1.29}$$

From (3.1.26) and (3.1.29),

$$(\phi_m, \phi_m) \leq \frac{1}{\lambda_{m+1}}(Df, Df).$$

By Lemma 3.1.2, the right-hand side goes to 0, and therefore the sequence ϕ_m converges to 0 in $L^2(\Omega)$. Hence,

$$f = \lim_{m \to \infty} f_m = \sum_{i=1}^{\infty} (f, u_i)u_i \quad \text{in } L^2(\Omega),$$

which shows (3.1.23). Next,

$$Df_m = \sum_{i=1}^{m} \alpha_i Du_i,$$

so by (3.1.22),

$$\begin{aligned}
(Df_m, Df_m) &= \sum_{i=1}^{m} \alpha_i^2 (Du_i, Du_i) \\
&= \sum_{i=1}^{m} \lambda_i \alpha_i^2.
\end{aligned} \tag{3.1.30}$$

Since from (3.1.29), $(Df_m, Df_m) \leq (Df, Df)$ and all $\lambda_i > 0$,

$$\sum_{i=1}^{\infty} \lambda_i \alpha_i^2$$

converges.

Now for $m \leq n$,

$$(D\phi_m - D\phi_n, D\phi_m - D\phi_n) = (Df_n - Df_m, Df_n - Df_m) = \sum_{i=m+1}^{n} \lambda_i \alpha_i^2.$$

Therefore, $(D\phi_m)$ is also a Cauchy sequence in $L^2(\Omega)$. Thus, the convergence of ϕ_m to 0 takes place not only in $L^2(\Omega)$ but even in $H^{1,2}(\Omega)$.

But then, (3.1.29) yields

$$(Df, Df) = \lim_{m \to \infty} (Df_m, Df_m) = \sum_{i=1}^{\infty} \lambda_i \alpha_i^2 \quad \text{(cf. (3.1.30))}.$$

It remains to verify completeness, that is, our scheme yields all the eigenvalues of Δ on Ω and all the eigenvectors are linear combinations of u_i. By the eigenvalue equations, the eigenvectors corresponding to different eigenvalues are L^2-orthogonal. Therefore, also an eigenvector for an eigenvalue $\lambda \neq \lambda_m$ for all m would be orthogonal to all the u_m, contradicting (3.1.23). This shows completeness.

$\square$

The eigenvalues of a domain $\Omega \subset \mathbb{R}^n$ satisfy many inequalities in terms of the geometry of Ω. For instance, we have *Weyl's estimate* [273]. If $N(\lambda)$ is the number of eigenvalues $\leq \lambda$, then for $\lambda \to \infty$,

$$N(\lambda) \sim \frac{\omega_n}{(2\pi)^n} \text{Vol}(\Omega) \lambda^{n/2}, \tag{3.1.31}$$

where ω_n is the volume of the unit ball in $\mathbb{R}^n$, the remainder term is of order lower than $n/2$, and it has been analyzed in terms of the geometry of Ω (see for instance [55]). Thus,

$$(\lambda_j)^{n/2} \sim \frac{(2\pi)^n}{\omega_n \text{Vol}(\Omega)} j. \tag{3.1.32}$$

These estimates generalize the simple case of an interval $(0, R)$ where the eigenfunctions $u_j(x) = \sin(\frac{j}{R}\pi x)$ for $x \in [0, R]$ correspond to the eigenvalues $\lambda_j = (\pi \frac{j}{R})^2$. And the rectangle $(0, R_1) \times \ldots \times (0, R_n) \subset \mathbb{R}^n$ then has the eigenvalues $\pi^2 \left(\frac{j_1^2}{R_1^2} + \ldots + \frac{j_n^2}{R_n^2} \right)$ for positive integers $j_1, \ldots, j_n$. Details can be found in [69].

More generally, we can also consider the functional

$$\inf_{f - \phi \in H \setminus \{0\}} \frac{(Df, Df)}{(f - \phi, f - \phi)} \tag{3.1.33}$$

for some function $\phi \in L^2(\Omega)$.

The eigenvalue equation would then be

$$\Delta u = \lambda(u - \phi). \tag{3.1.34}$$

Again, the eigenvalues would be ≥ 0.

Such problems occur in many contexts, for instance in image denoising, see for instance [12, 241]. There, one wants to denoise a function $\phi \colon \Omega \to \mathbb{R}$ by smoothing it, and to achieve that, one minimizes a functional of the form

$$\int_\Omega |Du|^2 dx + \rho \int_\Omega |u - \phi|^2 dx. \tag{3.1.35}$$

Here, ϕ is the given input that stands for the noisy image, and one wishes to recover the original image as well as possible. The smoothness term $\int_\Omega |Du|^2 dx$ is supposed to regularize ϕ and thereby suppress the noise, as that is not smooth. The fidelity term $\int_\Omega |u - \phi|$ controls the deviation of the denoised version u from ϕ. The smoothness and the fidelity term are balanced by the parameter $\rho > 0$. Formally, a minimizer u has to satisfy an equation of the form

$$\Delta u + \rho(u - \phi|) = 0, \tag{3.1.36}$$

but this is not an eigenvalue equation of the form (3.1.34), because $\rho > 0$.

3.2 The Laplace–Beltrami Operator and the Hodge Laplacian on Riemannian Manifolds

In this section, we use the reference [150], but we do not spell out all details, as this section is probably most useful for those readers who already have some background in Riemannian geometry. Nevertheless, we shall explain all important conceptual points, to make the analogy with the discrete case as clear as possible.

Let M be a compact and connected Riemannian manifold of dimension n. Again, M is assumed to be nonempty. Compactness is assumed to avoid certain convergence issues that would require a discussion not related to our topic. Nevertheless, much of the theory can be and has been extended to noncompact (but complete) manifolds, see for instance [87]. Usually, some control at infinity is required. Although the analysis of the previous section can be combined with that of the present one to treat Riemannian manifolds with boundary, here, for simplicity, we shall only discuss closed manifolds.

In particular, M is a differentiable manifold. As a manifold, it can be covered by local coordinate charts, with coordinates $x = (x^1, \ldots, x^n)$ taken from some domain $\Omega \subset \mathbb{R}^n$. As usual, our notation will not make a distinction between a point $x \in M$ and its representation in local coordinates. We shall see, however, that all vectorial quantities change their expressions under coordinate changes. Here, tensor calculus enters, giving precise computational rules for such transformations.

We shall assume that all expressions that depend on $x \in M$ do so smoothly.

As a differentiable manifold, M then also carries vector fields, that is, objects that in local coordinates are expressed as

$$V(x) = v^i(x) \frac{\partial}{\partial x^i}. \tag{3.2.1}$$

Here and in the remainder of this section, we use the *Einstein summation convention* that omits a summation sign for products carrying the same index in a lower and in an upper position. That is, for instance

$$a^i b_i := \sum_{i=1}^{n} a^i b_i \quad \text{or} \quad \alpha_{jk} \gamma^j := \sum_{j=1}^{n} \alpha_{jk} \gamma^j. \tag{3.2.2}$$

Importantly, when we change coordinates from x to $y = y(x)$, then we need to transform the tangent vectors $\frac{\partial}{\partial x^i}$. The rule is

$$\frac{\partial}{\partial x^i} = \frac{\partial y^j}{\partial x^i} \frac{\partial}{\partial y^j} \quad \text{for } i = 1, \ldots n, \tag{3.2.3}$$

or conversely, since coordinate changes are diffeomorphisms,

$$\frac{\partial}{\partial y^j} = \frac{\partial x^i}{\partial y^j} \frac{\partial}{\partial x^i} \quad \text{for } j = 1, \ldots n, \tag{3.2.4}$$

where, to repeat, the convention (3.2.2) is applied, that is, in (3.2.3), we sum over i, and in (3.2.4), over j.

The vector $\frac{\partial}{\partial x^i}$ constitutes a basis for the tangent vectors at each point. A vector field as in (3.2.1) then is such a linear combination with coefficients that depend on the point x, for the moment, smoothly.

In addition to tangent vectors, we have cotangent vectors, and 1-forms as the corresponding fields. These are of the form

$$\omega(x) = \omega_j(x) dx^j, \tag{3.2.5}$$

where the dx^j are dual to the $\frac{\partial}{\partial x^i}$, that is, satisfy

$$dx^j \left(\frac{\partial}{\partial x^i} \right) = \delta_i^j := \begin{cases} 1 & \text{for } i = j \\ 0 & \text{for } i \neq j. \end{cases} \tag{3.2.6}$$

The dx^j transform opposite to the $\frac{\partial}{\partial x^i}$, so as to preserve (3.2.6), that is, to also have $dy^m \left(\frac{\partial}{\partial y^\ell} \right) = \delta_\ell^m$. Thus, the transformation rule is

$$dy^j = \frac{\partial y^j}{\partial x^i} dx^i. \tag{3.2.7}$$

We also need exterior p-forms or p-forms for short. These are obtained by taking exterior products of the dx^i, and they are thus of the form

$$\omega = \omega_{i_1 \ldots i_p} dx^{i_1} \wedge \ldots \wedge dx^{i_p}, \quad \text{with } 1 \leq i_1 < \ldots < i_p \leq n. \tag{3.2.8}$$

The key property of the exterior product is its antisymmetry, with the basic relation

$$dx^i \wedge dx^j = -dx^j \wedge dx^i, \tag{3.2.9}$$

and therefore in particular

$$dx^i \wedge dx^i = 0. \tag{3.2.10}$$

Therefore, we may assume that the coefficients $\omega_{i_1 \ldots i_p}$ in (3.2.8) are antisymmetric in the sense that they change their sign whenever two indices are interchanged. (This is also the reason why we can require $1 \leq i_1 < \ldots < i_p \leq n = \dim M$ in (3.2.8).) For instance, a 2-form is given as

$$\omega = \omega_{ij} dx^i \wedge dx^j = -\omega_{ij} dx^j \wedge dx^i, \tag{3.2.11}$$

which yields the antisymmetry

$$\omega_{ji} = -\omega_{ij}. \tag{3.2.12}$$

In particular, because of (3.2.10), we may assume that $\omega_{ii} = 0$ for all i.

A p-form can be applied to p tangent vectors, on the basis of the characteristic property of the exterior product $\wedge$. For $p = 2$, we have

$$dx^i \wedge dx^j \left(\frac{\partial}{\partial x^k}, \frac{\partial}{\partial x^\ell} \right) = \delta^i_k \delta^j_\ell - \delta^j_k \delta^i_\ell.$$

Thus, for $V = v^k \frac{\partial}{\partial x^k}, W = w^\ell \frac{\partial}{\partial x^\ell}$, and $\omega = \omega_{ij} dx^i \wedge dx^j$,

$$\omega(V, W) = \omega_{ij}(v^i w^j - v^j w^i).$$

In particular,

$$\omega(W, V) = -\omega(V, W).$$

We can capture this antisymmetry by introducing *orientations*. An orientation $[V_1, \ldots, V_p]$ of p vectors $V_1, \ldots, V_p$ is an ordering up to even permutation. An odd permutation causes a minus sign. Thus, an orientation is an element of $\{\pm 1\}$. If we let $[V_1, \ldots, V_k, \ldots, V_\ell, \ldots, V_p]$ be an ordered collection of p vectors, then

$$[V_1, \ldots, V_\ell, \ldots, V_k, \ldots, V_p] = -[V_1, \ldots, V_k, \ldots, V_\ell, \ldots, V_p], \tag{3.2.13}$$

where we have interchanged V_k and V_ℓ and left all other V_j in their original position. And then the application of a p-form ω also yields a minus sign when we change the orientation, that is,

$$\omega(V_1,\ldots,V_\ell,\ldots,V_k,\ldots,V_p) = -\omega(V_1,\ldots,V_k,\ldots,V_\ell,\ldots,V_p). \quad (3.2.14)$$

The space of d-forms on a d-dimensional vector space X is one-dimensional. We may take some basis $V_1,\ldots,V_d$ of X. Then, $V_1 \wedge \ldots \wedge V_d$ is a basis for the d-forms, and we may declare some such basis as positive (i.e., positively oriented) and then also call any other basis $W_1 \wedge \ldots \wedge W_d$ if $W_1,\ldots,W_d$ is obtained from $V_1,\ldots,V_d$ by a base change with positive determinant. And it is negative when the base change has a negative determinant.

We denote the space of p-forms on M by $\Omega^p(M)$. We then have the fundamental

Definition 3.2.1. The *exterior derivative $d := d^p : \Omega^p(M) \to \Omega^{p+1}(M)$ ($p = 0, 1,\ldots,\dim M$)* is defined by

$$d(\eta(x)dx^{i_1} \wedge \ldots \wedge dx^{i_p}) = \frac{\partial \eta(x)}{\partial x^j} dx^j \wedge dx^{i_1} \wedge \ldots \wedge dx^{i_p}, \quad (3.2.15)$$

and it is then extended by linearity to all of $\Omega^p(M)$.

We easily check

Lemma 3.2.1. *The exterior derivative d is independent of the choice of coordinates. If $\omega \in \Omega^p(M), \vartheta \in \Omega^q(M)$, then $d(\omega \wedge \vartheta) = d\omega \wedge \vartheta + (-1)^p \omega \wedge d\vartheta$.*

The key property of the exterior derivative is

Theorem 3.2.1.

$$d \circ d = 0, \quad (3.2.16)$$

or more explicitly

$$d^{p+1} \circ d^p = 0 \quad \text{for } p = 0, 1,\ldots,\dim M.$$

Proof Since d is a linear operator, it suffices to check (3.2.16) on forms of the type

$$\omega(x) = f(x)dx^{i_1} \wedge \ldots \wedge dx^{i_p}.$$

In that case

$$d \circ d(\omega(x)) = d\left(\frac{\partial f}{\partial x^j} dx^j \wedge dx^{i_1} \wedge \ldots \wedge dx^{i_p}\right)$$

$$= \frac{\partial^2 f}{\partial x^j \partial x^k} dx^k \wedge dx^j \wedge dx^{i_1} \wedge \ldots \wedge dx^{i_p}$$

$$= 0,$$

because of the antisymmetry (3.2.9), since $\frac{\partial^2 f}{\partial x^j \partial x^k} = \frac{\partial^2 f}{\partial x^k \partial x^j}$ is symmetric. □

So far, we have only used the differentiable but not the Riemannian structure of M. But in order to define an adjoint of d and then a Laplace operator, we need scalar products. And scalar products between tangent vectors or 1-forms come from a Riemannian metric. Such a *Riemannian metric* is a positive definite and symmetric 2-tensor that is written in local coordinates as $(g_{ij})_{i,j=1,\ldots,d}$. Again, these quantities depend on the point $x \in M$, that is, $g_{ij} = g_{ij}(x)$, and the following expressions then are meant to hold at every $x \in M$.

In more detail, the metric tensor is written as

$$g_{ij} dx^i \otimes dx^j, \tag{3.2.17}$$

which can then be applied via the rule (3.2.5) to two tangent vectors to yield the scalar product defined by[2]

$$\langle \frac{\partial}{\partial x^i}, \frac{\partial}{\partial x^j} \rangle = g_{ij}. \tag{3.2.18}$$

Euclidean space is a Riemannian manifold (noncompact, of course), and in the standard coordinates, its metric is given by

$$g_{ij} = \delta_{ij} = \begin{cases} 1 & \text{if } i = j \\ 0 & \text{else.} \end{cases} \tag{3.2.19}$$

In other coordinates, the metric may look different, but a Riemannian metric is Euclidean if and only if it can be brought into the form (3.2.19) by a coordinate transformation.

In order to keep the duality between tangent vectors and 1-forms, we also put

$$\langle dx^i, dx^j \rangle = g^{ij}, \tag{3.2.20}$$

where (g^{ij}) denotes the *inverse* metric of (g_{ij}).

In fact, it is a general principle that a scalar product on some vector space then also induces products on all other vector spaces that are algebraically constructed from it. For instance, the relation between (3.2.18) and (3.2.20) simply

---

[2] Here, we use a $\cdot$ for the scalar product. Usually, in Riemannian geometry, $\langle \cdot, \cdot \rangle$ is used for the pointwise Riemannian product, and $(\cdot, \cdot)$ for the L^2-product (see [150]).

comes from the fact that the space of 1-forms is dual to the space of tangent vectors. Thus, by this principle, we also obtain an induced product on the space of p-forms, via

$$\langle dx^{i_1} \wedge \ldots \wedge dx^{i_p}, dx^{j_1} \wedge \ldots \wedge dx^{j_p} \rangle = g^{i_1 j_1} \cdots g^{i_p j_p}. \qquad (3.2.21)$$

Moreover, p- and q-forms with $p \neq q$ are orthogonal.

More generally, if $\omega^{i_1}, \ldots \omega^{i_p}, \eta^{j_1}, \ldots \eta^{j_p}$ are 1-forms, then

$$\langle \omega^{i_1} \wedge \ldots \wedge \omega^{i_p}, \eta^{j_1} \wedge \ldots \wedge \eta^{j_p} \rangle = \det(\omega^i \cdot \eta^j). \qquad (3.2.22)$$

This implies, in particular, that the *volume form*

$$d\mathrm{vol} := \sqrt{\det(g_{ij})} dx^1 \wedge \ldots \wedge dx^n \qquad (3.2.23)$$

transforms properly and does not depend on the choice of coordinates, at least when all coordinate transformations have a positive determinant. The manifold M is called *orientable* when it can be covered by coordinate charts for which all coordinate transformations have such a positive determinant on a class of basis vector fields of the tangent spaces. We shall therefore assume now that M is orientable.

We can then integrate functions f on M via

$$\int_M f d\mathrm{vol}. \qquad (3.2.24)$$

And we can then use our pointwise products $\langle \cdot, \cdot \rangle$ obtained from the Riemannian metric g_{ij} to define L^2-products by integration. For instance, for p-forms ω, η,

$$(\omega, \eta) := \int_M \langle \omega, \eta \rangle d\mathrm{vol}. \qquad (3.2.25)$$

And this then allows us to define the adjoint $d^*: \Omega^{p+1} \to \Omega^p$ of $d: \Omega^p \to \Omega^{p+1}$ via

$$(d^* \eta, \omega) = (\eta, d\omega) \quad \text{for } \eta \in \Omega^{p+1}, \omega \in \Omega^p. \qquad (3.2.26)$$

We should point out that d^* in (3.2.26) is only the formal adjoint of d because we did not specify the domains of definition of these operators. While that can be readily done, this is not our concern here.

Definition 3.2.2. The *Laplace(–Beltrami) operator* on $\Omega^p(M)$ is

$$\Delta = dd^* + d^* d: \Omega^p(M) \to \Omega^p(M).$$

$\omega \in \Omega^p(M)$ is called *harmonic* if

$$\Delta\omega = 0.$$

More precisely, one should write

$$d_p : \Omega^p(M) \to \Omega^{p+1}(M)$$
$$d_p^* : \Omega^{p+1}(M) \to \Omega^p(M).$$

Then,

$$\Delta_p = d_{p-1}d_{p-1}^* + d_p^* d_p : \Omega^p(M) \to \Omega^p(M). \qquad (3.2.27)$$

Nevertheless, we shall usually omit the index p.

The definition also directly implies

Corollary 3.2.1. Δ *is (formally) self-adjoint, that is,*

$$(\Delta\alpha,\beta) = (\alpha,\Delta\beta) \quad \text{for } \alpha,\beta \in \Omega^p(M).$$

$\square$

Lemma 3.2.2.

$$(\Delta\omega,\omega) = (dd^*\omega,\omega) + (d^*d\omega,\omega) = (d^*\omega,d^*\omega) + (d\omega,d\omega) \geq 0. \quad (3.2.28)$$

In particular, Δ *is non-negative, and*

$$\Delta\omega = 0 \text{ if and only if } d\omega = 0 \text{ and } d^*\omega = 0. \qquad (3.2.29)$$

Proof (3.2.28) follows from the definitions. Since both terms on the right-hand side of (3.2.28) are non-negative and vanish only if $d\omega = 0 = d^*\omega$, (3.2.29) also follows. $\square$

For $p = 0$, $d_0^* = 0$, and so, $\Delta = d^*d$. In fact, one computes (see [150]) that for $p = 0$, Δ is the *Laplace–Beltrami operator* given in local coordinates by

$$\Delta f = -\frac{1}{\sqrt{\det(g_{k\ell})}} \frac{\partial}{\partial x^j} \left(\sqrt{\det(g_{k\ell})} g^{ij} \frac{\partial f}{\partial x^i} \right). \qquad (3.2.30)$$

When the Riemannian metric is Euclidean, the Laplace–Beltrami operator (3.2.30) reduces to the Laplace operator defined in (3.1.1). This is seen from the characterization (3.2.19) of the Euclidean metric.

Returning to a compact manifold, for $p = 0$, $\Delta f = 0$ is equivalent to $df = 0$. A 0-form f is simply a function, and so we have

Corollary 3.2.2. *On a compact Riemannian manifold, every harmonic function is constant.* $\square$

Thus, the considerations in this section generalize those of Section 3.1 in two ways. On one hand, the Laplace–Beltrami operator generalizes the Euclidean Laplace operator for functions. On the other hand, we have also defined a Laplace operator, also called the *Hodge Laplacian*, for *p*-forms. There, we see some phenomena that are not present for functions.

The eigenvalues of the Laplace–Beltrami operator of a Riemannian manifold M encode its geometric features, and conversely, these eigenvalues can be estimated in terms of geometric quantities. The classical estimate of Weyl [273] also holds on Riemannian manifolds, relating the asymptotics of the spectrum to the volume $\mathrm{Vol}(M) = \int_M d\mathrm{vol}$ of M (see (3.2.23)). If $N(\lambda)$ is the number of nonzero eigenvalues $\leq \lambda$, then for $\lambda \to \infty$,

$$N(\lambda) \sim \frac{\omega_n}{(2\pi)^n} \mathrm{Vol}(M)\lambda^{n/2}, \qquad (3.2.31)$$

where ω_n is the volume of the unit ball in $\mathbb{R}^n$, while the remainder term is of order lower than $n/2$, and it has been analyzed in terms of the geometry of M (see for instance [55]). Thus,

$$(\lambda_j)^{n/2} \sim \frac{(2\pi)^n}{\omega_n \mathrm{Vol}(M)}(j-1). \qquad (3.2.32)$$

These estimates generalize the simple case of a circle of length $2\pi r$ where the eigenfunctions $u_j(x) = \cos(\frac{j-1}{r}x)$ for $x \in [0, 2\pi r]$ correspond to the eigenvalues $\lambda_j = (\frac{j-1}{r})^2$.

Later studies of eigenvalues on Riemannian manifolds often concentrated on the first nontrivial eigenvalue, λ_2. Lichnerowicz [183] found that this eigenvalue can be controlled from below in terms of a local quantity, the Ricci curvature of M, when that quantity happens to be positive throughout M. Li and Yau [182] later developed a systematic approach for controlling eigenvalues of Riemannian manifolds, which triggered much further research. See, for example, [150] for a survey of results and references.

The first eigenvalue of the Laplace–Beltrami operator is again $\lambda_1 = 0$, with a constant eigenfunction. The eigenfunctions for the other eigenvalues are orthogonal to the constants, that is, satisfy

$$\int_M u\, d\mathrm{vol} = 0. \qquad (3.2.33)$$

The Cheeger estimate controls the second eigenvalue in terms of the global quantity, the *Cheeger constant*

$$h(M) := \inf \frac{\mathrm{Vol}_{n-1}(S)}{\min(\mathrm{Vol}_n(M_1), \mathrm{Vol}_n(M_2))} \qquad (3.2.34)$$

where the infimum is taken over all $(n-1)$-dimensional submanifolds S of M that divide M into two pieces M_1, M_2, and Vol_{n-1} is its volume. This quantity becomes small when it is easy to cut M into two large pieces. This means that

M can be divided into two pieces M_1 and M_2 of large volume by cutting M along a hypersurface S of small volume. Cheeger's estimate [56] is

$$\lambda_2 \geq \frac{h(M)^2}{4}. \tag{3.2.35}$$

Let us sketch the argument, suppressing as usual the analytical difficulties. Let u_2 be an eigenfunction u_2 with eigenvalue λ_2. Since $\int_M u_2 = 0$ by (3.2.33), M consists of two pieces M_1, M_2 with $u_{2,|M_1} < 0, u_{2,|M_2} > 0$; we may assume that M_1 is the one of smaller volume. Then,

$$\Delta u_2 = \lambda_2 u_2 \text{ in } M_1, \qquad u_2 = 0 \text{ on } \partial M_1. \tag{3.2.36}$$

We multiply (3.2.36) by u_2 and integrate by parts over M_1.[3] We obtain

$$\lambda_2 = \frac{\int_{M_1} \langle du_2, du_2 \rangle}{\int_{M_1} (u_2)^2}. \tag{3.2.37}$$

By the Cauchy–Schwarz inequality, the function $\phi := (u_2)^2$ satisfies, since $d(u_2)^2 = 2u_2 du_2$,

$$\left(\int_{M_1} \|d\phi\| \right)^2 \leq 4 \int_{M_1} \langle du_2, du_2 \rangle \int_{M_1} (u_2)^2. \tag{3.2.38}$$

This implies

$$\lambda_2 \geq \frac{1}{4} \left(\frac{\int_{M_1} \|d\phi\|}{\int_{M_1} \phi} \right)^2. \tag{3.2.39}$$

We shall now apply Federer's coarea formula ([93]; see, for example, [152] for an easy proof), which says that for every open $\Omega \subset M$ and every smooth $\phi: \Omega \to \mathbb{R}$, we have

$$\int_\Omega \|d\phi\| = \int_{-\infty}^{\infty} \text{Vol}_{n-1}(\phi^{-1}(t) \cap \Omega) dt. \tag{3.2.40}$$

Since for $t > 0$, $\phi^{-1}(t)$ is a closed hypersurface in the interior of M_1 as ϕ vanishes on ∂M_1, (3.2.40) yields

$$\int_{M_1} \|d\phi\| = \int_{-\infty}^{\infty} \text{Vol}_{n-1}(\phi^{-1}(t) \cap M_1) dt$$

---

[3] For the analytical difficulties resulting from the fact that we do not yet know how smooth the separating hypersurface $\partial M_1 = \partial M_2$ is, see for instance [93]. Actually, this hypersurface is smooth except possibly for a singular set of high codimension. In fact, it turns out that it has constant mean curvature, which in the smooth case can be seen by a Lagrange multiplier argument.

$$\geq h(M) \int_0^\infty \mathrm{Vol}_d(\{\phi \geq t\} \cap M_1)dt \qquad (3.2.41)$$

$$= h(M) \int_{M_1} \phi. \qquad (3.2.42)$$

Since the volume of M_1 was assumed to be smaller than that of M_2 and $\{\phi \geq t\}$ is a subset of M_1, its volume is the smaller one of the two parts into which $\phi^{-1}(t)$ dissects M. (3.2.42), combined with (3.2.39), yields (3.2.35).

When M is disconnected (which, however, we exclude by our general assumptions), then $\lambda_2 = 0$, because then we do not need any hypersurface to separate M into two pieces. When M is connected, then $\lambda_2 > 0$. When λ_2 becomes small, then M is almost disconnected, as we can cut it along a hypersurface with small volume into two large pieces.

3.3 The p-Laplacian

In this section, we shall develop the scheme set up in Section 2.6.2 in an infinite dimensional context. In order to present the material in this section in the analytical context to which it naturally belongs, we shall use the exposition of [157].

We consider again a bounded domain $\Omega \subset \mathbb{R}^n$. Since integration by parts will play an important role, we assume that it has a boundary $\partial\Omega$ that is piecewise Lipschitz. While substantially more general cases can be covered, here we want to avoid such technical issues. The theory also naturally extends to domains in Riemannian manifold, analogously to what was presented in Section 3.2, but it seems that not all details can be found in the literature, however.

The case treated in Section 3.1 corresponds to $p = 2$. In fact, the index p indicates the power to which functions are integrable and the corresponding norms. These norms define Banach spaces. The case $p = 2$ is special because then the Banach space is a Hilbert space.

Let first $1 < p < \infty$. In the same way that $L^p(\Omega)$ generalizes $L^2(\Omega)$, the Sobolev space $H^{1,p}(\Omega)$ generalizes $H^{1,2}(\Omega)$. Details can be found in [149], for instance. For $u \in H^{1,p}(\Omega)$, we consider the functional

$$I_p(u) = \int_\Omega |\nabla u|^p dx. \qquad (3.3.1)$$

Its Euler–Lagrange operator is the p-Laplacian

$$\Delta_p u = -\mathrm{div}\left(|\nabla u|^{p-2}\nabla u\right). \qquad (3.3.2)$$

For $p = 2$, this is the Laplace operator (3.1.2).

Generalizing the setting of Section 3.1, the eigenvalue problem concerns critical points of the Rayleigh quotient

$$\frac{\int_\Omega |\nabla u|^p\, dx}{\int_\Omega |u|^p\, dx}, \tag{3.3.3}$$

requiring as always $u \not\equiv 0$. Equivalently, we may look for critical points of I_p under the constraint

$$\int_\Omega |u|^p\, dx = 1, \tag{3.3.4}$$

Again, we need a boundary condition, which we choose as the Dirichlet condition $u \in H_0^{1,p}(\Omega)$, which is the proper formulation of

$$u \equiv 0 \text{ on } \partial\Omega. \tag{3.3.5}$$

On a compact Riemannian manifold M with boundary ∂M, we can do the same when we integrate in (3.3.1), (3.3.4) with respect to the Riemannian volume measure, and let ∇ and div denote the Riemannian gradient and divergence operators. When M is closed, we do not need a boundary condition, of course.

The equation for eigenfunctions and eigenvalues now is

$$\Delta_p u = \lambda |u|^{p-2} u. \tag{3.3.6}$$

For $1 < p < \infty$, the functionals in (3.3.1) and (3.3.4) are strictly convex, and the spectral theory is similar to that for $p = 2$. [268] analyzes the situation on a Riemannian manifold. We do not spell out the details here.

The case $p = 1$ is more difficult since the functionals are no longer strictly convex. (3.3.6) then formally becomes

$$-\operatorname{div}\left(\frac{\nabla u}{|\nabla u|}\right) = \lambda\frac{u}{|u|}. \tag{3.3.7}$$

In (3.3.6) for $p > 1$, we may put the right-hand side $= 0$ at points where $u = 0$, but this is no longer possible in (3.3.7). Some references are [51, 166, 168, 169, 185, 196, 207, 208, 224, 243], by Kawohl, Schuricht and their students and collaborators, and by Chang who systematically studied this eigenvalue problem.

We need a so-called weak formulation, that is, require that the equations holds after integration by parts. For the details, we introduce a substitute z of $\frac{\nabla u}{|\nabla u|}$ and a substitute s of $\frac{u}{|u|}$, turning (3.3.7) into

$$-\operatorname{div} z = \lambda s, \tag{3.3.8}$$

where $s \in L^\infty(\Omega)$ satisfies

$$s(x) \in \operatorname{Sgn}(u(x)) \tag{3.3.9}$$

with

$$\operatorname{Sgn}(t) := \begin{cases} \{1\} & \text{if } t > 0, \\ [-1,1] & \text{if } t = 0, \\ \{-1\} & \text{if } t < 0, \end{cases}$$

and the vector field $z \in L^\infty(\Omega, \mathbb{R}^n)$ satisfies

$$\|z\|_\infty = 1, \quad \operatorname{div} z \in L^n(\Omega), \quad -\int_\Omega u \operatorname{div} z \, dx = I(u), \tag{3.3.10}$$

where

$$I(u) = \int_\Omega |Du| dx + \int_{\partial\Omega} |u^{\partial\Omega}| d\mathcal{H}^{n-1}. \tag{3.3.11}$$

We have to explain (3.3.11), and this needs some background. While for $p > 1$, the natural space to work in is $H^{1,p}(\Omega)$, for $p = 1$, it is no longer $H^{1,1}(\Omega)$, but rather $BV(\Omega)$. This is the space of functions of bounded variation (for a short introduction, see for instance [152]). It consists of all functions in $L^1(\Omega)$ for which

$$|Du|(\Omega) = \sup \left\{ \int_\Omega u \operatorname{div} g \, dx : g = (g^1, \dots g^n) \right.$$
$$\left. \in C_0^\infty(\Omega, \mathbb{R}^n), |g(x)| \le 1 \text{ for all } x \in \Omega \right\} < \infty. \tag{3.3.12}$$

Here, $|Du| dx$ is a measure, and $|Du|(\Omega)$ is its *total variation*. Thus, this quantity is again defined through integration by parts against smooth functions, but the important point here is that we only integrate against the divergence of test functions. This makes the space larger and for the present purposes more convenient, than the Sobolev space $H^{1,1}(\Omega)$. In particular, functions that jump across hypersurfaces are in $BV(\Omega)$, and we can reconstruct the size of such a hypersurface by evaluating the norm of that jump. We shall now explain that.

When $u \in C^1(\Omega)$, we have

$$\int_\Omega u \operatorname{div} g \, dx = -\int_\Omega \sum_i g^i \frac{\partial u}{\partial x^i} dx,$$

and this explains that BV-functions permit such an integration by parts in a weak sense. More precisely, for a BV-function u, its distributional gradient is represented by a finite $\mathbb{R}^n$ valued signed measure $|Du| dx$, and we can write

$$\int_\Omega u \operatorname{div} g \, dx = -\int_\Omega g |Du| dx \text{ for } g \in C_0^\infty(\Omega, \mathbb{R}^n). \tag{3.3.13}$$

Also, $u \in BV(\Omega)$ has a well-defined trace $u^{\partial\Omega} \in L^1(\partial\Omega)$, and (3.3.13) generalizes to

$$\int_\Omega u \operatorname{div} h \, dx = -\int_\Omega h |Du| dx$$
$$+ \int_{\partial\Omega} u^{\partial\Omega}(hv) d\mathcal{H}^{n-1} \text{ for } h \in C^1(\Omega, \mathbb{R}^n) \cap C(\overline{\Omega}, \mathbb{R}^n)$$

$$(3.3.14)$$

where v is the outer unit normal of $\partial\Omega$, and $\mathcal{H}^{n-1}$ is $(n-1)$-dimensional Hausdorff measure.

BV-functions can be discontinuous along hypersurfaces. By definition, a Borel set $E \subset \Omega$ has finite perimeter, given by the following quantity, if its characteristic function χ_E satisfies

$$|D\chi_E|(\Omega)\left(= \sup\left\{\int_E \operatorname{div} g : g \in C_0^\infty(\Omega, \mathbb{R}^n), |g| \le 1\right\}\right) < \infty.$$

If the boundary of E is a compact Lipschitz hypersurface, then the perimeter of E coincides with the Hausdorff measure $\mathcal{H}^{n-1}(\partial E)$. More generally, if $E \subset \Omega$, we have

$$|D\chi_E| := |D\chi_E|(\mathbb{R}^n) = |D\chi_E|(\Omega) + \mathcal{H}^{n-1}(\partial E \cap \partial\Omega). \quad (3.3.15)$$

With these concepts, we can return to the eigenvalue problem for the 1-Laplacian. (3.3.8), however, is not yet satisfactory because, in general, it has too many solutions. It becomes rather arbitrary on sets of positive measure where u vanishes, see [208]. Therefore, one seeks solutions that are the critical points of a variational principle. Since the functional is not smooth, one needs an appropriate notion of a critical point. The vanishing of the weak slope of [75] seems to be the right one. Inner variations provide another necessary criterion [208]. Viscosity solutions provide another criterion which, however, is still not stringent enough [169].

Here, however, we do not go into the general theory, but rather look only at the first eigenvalue because that is related to the *Cheeger constant* of Ω. That constant is defined as

$$h_1(\Omega) := \inf_{E \subset \overline{\Omega}} \frac{|D\chi_E|}{|E|} \quad (3.3.16)$$

where $|E|$ is the Lebesgue measure of E. A set realizing the infimum in (3.3.16) is called a Cheeger set, and every bounded Lipschitz domain Ω possesses at least one Cheeger set. For such a Cheeger set $E \subset \Omega$, $\partial E \cap \Omega$ is smooth except possibly for a singular set of Hausdorff dimension at most $n-8$ and of constant mean curvature $\frac{1}{n-1} h_1(\Omega)$ at all regular points. The Cheeger set of a domain need not be unique.

Importantly, $h_1(\Omega)$ equals the first eigenvalue of the 1-Laplacian. In fact,

$$h_1(\Omega) = \inf_{u \in BV(\Omega), u \neq 0} \frac{\int_\Omega |Du|dx + \int_{\partial\Omega} |u^{\partial\Omega}|d\mathcal{H}^{n-1}}{\int_\Omega |u|dx} =: \lambda_{1,1}(\Omega) \quad (3.3.17)$$

is the smallest $\lambda \neq 0$ for which there is a nontrivial solution u of (3.3.7), and such a u is of the form χ_E for a Cheeger set, up to a multiplicative factor, of course. This eigenvalue can also be obtained as the limit of the first eigenvalues of the p-Laplacians for $p \to 1$. If $\lambda_{1,p}(\Omega)$ denotes the smallest nonzero eigenvalue of (3.3.6), then

$$\lim_{p \to 1^+} \lambda_{1,p}(\Omega) = \lambda_{1,1}(\Omega). \quad (3.3.18)$$

We also have the lower bound

$$\lambda_{1,p}(\Omega) \geq \left(\frac{h_1(\Omega)}{p}\right)^p \quad (3.3.19)$$

generalizing the original Cheeger bound for $p = 2$.

More generally, for any family of eigenvalues $\lambda_{k,p}(\Omega)$ of (3.3.6), $\lim_{p \to 1^+} \lambda_{k,p}(\Omega)$ is an eigenvalue of (3.3.7). The converse is not true; however, (3.3.7) may have more solutions than can be obtained as limits of solutions of (3.3.6). This again is a testimony to the subtleties of the eigenvalue problem for the 1-Laplacian.

The functional $|Du|$ appears also in image denoising, in the total variation (TV) models introduced in [240]. There, as in (3.1.35), one wants to denoise a function $\phi: \Omega \to \mathbb{R}$ by smoothing it, and in the TV models, one minimizes a functional of the form

$$\int_\Omega |Du|dx + \rho \int_\Omega |u - \phi|dx. \quad (3.3.20)$$

Again, ϕ is the noisy image. The smoothness term is now $\int_\Omega |Du|dx$, and the fidelity term is $\int_\Omega |u - \phi|$, and the two are balanced by the parameter $\rho > 0$. Formally, a minimizer u has to satisfy an equation of the form

$$\text{div}\left(\frac{Du}{|Du|}\right) = \rho\frac{u - \phi}{|u - \phi|}, \quad (3.3.21)$$

which is similar to (3.3.7).

In practice, the L^1-formulation turns out to be better than the L^2-formulation (3.1.35), although the latter may look more natural and is mathematically. The intuitive reason is that the L^1-norm emphasizes solutions with fewer degrees of freedom, as in the theory of compressed sensing ([99]).

Still, when the model is applied to actual data, the performance is not so good. People have therefore modified (3.3.20) to what is called a nonlocal model in image processing [109]. In [141], such a model was derived from geometric considerations, and this may also provide some insight into the relation with the discrete models considered in this monograph. This model utilizes the structures defined in (2.5.29)–(2.5.33).

x, y stand for patches in the image on Ω, a domain in $\mathbb{R}^n$, or some more abstract space, and for $w: \Omega \times \Omega \to \mathbb{R}$, a non-negative, symmetric function, $w(x, y)$ is the edge weight between $(x, y) \in \Omega \times \Omega$. Here x, y stand for patches in the image, and in our setting, they could also be vertices in a graph (in which case the integrals would become sums). We can use the concepts developed in Section 2.5.3 and define the average $\bar{w}: \Omega \to \mathbb{R}$ of w by

$$\bar{w}(x) = \int_\Omega w(x, y) dy$$

and assume that $\bar{w}$ is positive almost everywhere. $\bar{w}(x)$ and $w(x, y)$ are used to define the L^2-norms for functions $u: \Omega \to \mathbb{R}$ and vector fields p, that is, $p: \Omega \times \Omega \to \mathbb{R}$,

$$(u_1, u_2)_{L^2} := \int_\Omega u_1(x) u_2(x) \bar{w}(x) dx$$

$$(p_1, p_2)_{L^2} := \int_{\Omega \times \Omega} p_1(x, y) p_2(x, y) w(x, y) dx dy$$

and the corresponding norms $|u|$ and $|p|$.

The discrete derivative of an image $u: \Omega \to \mathbb{R}$ is

$$Du(x, y) = u(y) - u(x).$$

We can then introduce the nonlocal TV (or BV) functional of [141] as

$$
\begin{aligned}
TV_w(u) \quad &:= \int_\Omega |Du| \bar{w}(x) dx \\
&= \int_\Omega \left(\int_\Omega (u(y) - u(x))^2 w(x, y) dy \right)^{\frac{1}{2}} \sqrt{\bar{w}(x)} dx.
\end{aligned}
\tag{3.3.22}
$$

The nonlocal TV model then is

$$
\begin{aligned}
ROF_w(u) \quad &= TV_\omega(u) + \mu \int_\Omega |u - f| \bar{w}(x) dx. \\
&= \int_\Omega \left(\int_\Omega (u(y) - u(x))^2 w(x, y) dy \right)^{\frac{1}{2}} \sqrt{\bar{w}(x)} dx \\
&\quad + \rho \int_\Omega \int_\Omega |u(x) - \phi(x)| w(x, y) dx dy.
\end{aligned}
\tag{3.3.23}
$$

In the other direction, going to the limit $p \to \infty$ obviously is not entirely trivial. The Rayleigh quotient (3.3.3) becomes (after normalizing the exponent)

$$\frac{\|\nabla u\|_{\infty, \Omega}}{\|u\|_{\infty, \Omega}}.
\tag{3.3.24}$$

Formally, $\|\nabla u\|_{\infty, \Omega} = \lim_{p \to \infty} \|\nabla u\|_{p, \Omega}$. But in contrast to the p-norms, the ∞-norm does not evaluate a function at all of its arguments, but only at those where extrema are reached. Therefore, one may suspect that a naive definition of critical points of (3.3.24) may not fully specify a solution.

Nevertheless, the limiting operator of Δ_p was identified in [7] as

$$\Delta_\infty u = -\sum_{i,j=1}^n \frac{\partial u}{\partial x^i} \frac{\partial u}{\partial x^j} \frac{\partial^2 u}{\partial x^i \partial x^j}. \tag{3.3.25}$$

In general, however, solutions of the variational problems associated with $p = \infty$, like (3.3.24), need not be smooth everywhere, and so, one needs to invoke the concept of viscosity solutions of (3.3.25) (see for instance [163]).

In particular, the eigenvalue problem for the first eigenvalue λ_1 is

$$\min(|\nabla u(x)| - \lambda_1 u(x), \Delta_\infty u(x)) = 0 \tag{3.3.26}$$

for a positive solution u, and the eigenvalue is explicitly determined as

$$\lambda_1 = \frac{1}{\max_{x \in \Omega} \mathrm{dist}(x, \partial\Omega)}, \tag{3.3.27}$$

and conversely, any positive solution of (3.3.26) minimizes the Rayleigh quotient (3.3.10), and

$$\lambda_1 = \lim_{p \to \infty} (\lambda_1(p))^{1/p}, \tag{3.3.28}$$

where $\lambda_1(p)$ is of course the first eigenvalue of Δ_p on Ω. On the unit ball, the corresponding solution is $u_\infty(x) = 1 - |x|$. This is not smooth at the origin, and in fact, the value of λ_1 is determined at that point only, as everywhere else $\Delta_\infty u(x) = 0$, by (3.3.25).

The higher eigenfunction problem is more complicated to formulate, as the higher eigenfunctions change sign, see [162]. When the function $u(x) > 0$, one requires (3.3.26), and where $u(x) < 0$, one requires $\max(-|\nabla u(x)| - \lambda_1 u(x), \Delta_\infty u(x)) = 0$. At points where $u(x) = 0$, we should have $\Delta_\infty u(x) = 0$. In the simple case of the unit interval $(0, 1) \subset \mathbb{R}$, all eigenfunctions and eigenvalues can be determined explicitly. We start with the function

$$v(x) := \begin{cases} x & \text{for } -\frac{1}{2} \le x \le \frac{1}{2} \\ 1 - x & \text{for } \frac{1}{2} < x \le 1 \\ -1 - x & \text{for } -1 \le x < -\frac{1}{2} \end{cases}, \tag{3.3.29}$$

which is defined on the interval $[-1, 1]$ and put

$$u_k(x) := v(kx) \text{ for } x \in [0, 1] \text{ and } k = 1, 2, \ldots \tag{3.3.30}$$

It turns out that there also exist other versions of the p- and the ∞-Laplacian that generalize the standard Euclidean Laplacian. See [89, 167, 227]. In particular, we may consider

$$\Delta_p^G u = \frac{1}{p}|\nabla u|^{2-p}\Delta_p u = -\frac{1}{p}|\nabla u|^{2-p}\operatorname{div}(|\nabla u|^{p-2}\nabla u) \qquad (3.3.31)$$

for $1 \le p < \infty$ (again, the $-$ sign does not agree with the conventions in the literature, but follows those of the present paper), and

$$\Delta_\infty^G u = |\nabla u|^{-2}\Delta_\infty u. \qquad (3.3.32)$$

Since $\Delta_1^G = -|\nabla u|\operatorname{div}(|\nabla u|^{-1}\nabla u)$, one then has

$$\Delta_p^G u = \frac{1}{p}\Delta_1^G + \frac{p-1}{p}\Delta_\infty^G \text{ for } 1 < p < \infty. \qquad (3.3.33)$$

Also, note that Δ_p^G defined in (3.3.31) is 1-homogeneous (i.e., $\Delta_p^G(cf) = c\Delta_p^G(f)$ for any constant c), while the original operator Δ_p defined in (3.3.2) is $(p-1)$-homogeneous (i.e., $\Delta_p(cf) = \operatorname{sign}(c)|c|^{p-1}\Delta_p(f)$ for any nonzero constant c). Therefore, both the eigenvalue problems $\Delta_p^G = \lambda u$ and $\Delta_p u = \lambda|u|^{p-2}u$ are scale-preserving, that is, if v is an eigenfunction, so is cv for any nonzero constant c (see for instance [27]). However, if we consider the modified eigenvalue problem $\Delta_p = \lambda u$ involving the original p-Laplacian, the scale-preserving property holds only for $p = 2$.

4

First Properties of the Spectrum

This chapter discusses the elementary properties of Laplace operators on graphs and hypergraphs. Many interesting examples will illustrate how special eigenvalues emerge. We also introduce discrete Pólya–Cheeger constants and their dual versions and provide the initial steps relating spectral clustering, spectra of neighborhood graphs, signed Laplacians, and spectra of simplicial complexes and hypergraphs.

4.1 The Laplacian and Its Spectrum on Graphs

In this section, we shall analyze Laplacians on *graphs*. From the general perspective, this means that we consider the case $q = 0$, because then simplices of dimension > 1 play no role, and only the vertices and the edges enter. But when vertices are connected by edges, we have a graph. This section can also serve to explicate the general theory in the simplest situation, and therefore, we try to make the reasoning here self-contained, instead of drawing upon that general theory.

The case of main interest will be the normalized combinatorial Laplacian, but we shall also discuss the algebraic Laplacian that is commonly used in graph theory.

4.1.1 General Aspects

We assume in this section, unless explicitly stated otherwise, that **the graph** Γ **is finite and connected**. Let it have N *vertices*, sometimes also called *nodes* for the sake of variety, and let V be the vertex set. For a subset W of V, we let $\overline{W}$ be its complement $V \backslash W$ in V.

$[v, v']$ denotes an oriented edge with vertices $v, v' \in V$. In this section, the orientation will play no role, and so, we write (v, v') or (vv') for the underlying unoriented edge.

Definition 4.1.1. We say that two vertices v, v' in a graph are *neighbors*, in symbols

$$v \sim v',$$

if they are connected by an edge. We shall also denote that edge by (v, v') or (vv'), according to the conventions just established.

We also say that the graph has a *self-loop* at v if $v \sim v$.

The *degree* $\deg v$ of a vertex v is the number of its neighbors (including itself when v has a self-loop).

Remark 4.1.1. For simplicial complexes, we do not allow self-loops. Therefore, when we make general statements that also apply to simplicial complexes of dimension > 1, we require that there are no self-loops. Hence, in this section, if not explicitly stated otherwise, **the graph Γ has no self-loops**.

We consider the space $L^2(\Gamma)$ of functions $f: V \to \mathbb{R}$ on the vertex set V of a graph equipped with a scalar product

$$(f, g) = \sum_v b(v) \, f(v) g(v) \tag{4.1.1}$$

for some positive function b on the vertex set. We may consider $b(v)$ as the weight of the vertex v. We also let the edges carry weights. We write the weight of the unoriented edge $e = (vv')$ in a weighted graph as $b(e) = w_{vv'}$. And since weights are assigned to unoriented simplices, we have $w_{v'v} = w_{vv'}$. If we wish that the vertex and the edge weights are consistent, we should assume $b(v) = \sum_{v' \sim v} w_{vv'}$, but for the moment, we do not require that. But as for the vertex weights, we require the edge weights $b(e)$ to be positive. This then is the setting of this section.

We have from (2.5.7) and (2.5.17)

Definition 4.1.2. The Laplace operator associated with the scalar product b of (4.1.1) is given by

$$L^0(f)(v) = \frac{1}{b(v)} \sum_{e=(vv')} b(e)(f(v) - f(v')). \tag{4.1.2}$$

The Laplacian (4.1.2) thus is the weighted sum of the weighted differences between the value of a function at a vertex v and at its neighbors. This is also in accordance with the general scheme developed in Section 2.5.

Lemma 4.1.1. *Let Γ be a connected graph.*

1. *L^0 is self-adjoint with respect to the scalar product (4.1.1), that is,*

$$(f, L^0 g) = (L^0 f, g) \tag{4.1.3}$$

for all $f, g \in L^2(\Gamma)$.

2. L^0 *is non-negative:*

$$(L^0 f, f) \geq 0 \qquad (4.1.4)$$

for all f.

3. $L^0 f = 0$ *precisely when f is constant.*

Proof (4.1.3) follows from the definitions of the scalar product and the Laplacian. To see this in detail, we compute

$$
\begin{aligned}
(f, L^0 g) &= \sum_{v,v',v \sim v'} b(e)(f(v) - f(v'))g(v) \\
&= - \sum_{v,v',v \sim v'} b(e)(f(v) - f(v'))g(v') \\
&= \frac{1}{2} b(e) \sum_{v,v',v \sim v'} (f(v) - f(v'))(g(v) - g(v')) \\
&= b(e) \sum_e (f(v) - f(v'))(g(v) - g(v')), \qquad (4.1.5)
\end{aligned}
$$

and by symmetry, this then also equals $(L^0 f, g)$. And (4.1.5) also implies (4.1.4).

When $L^0 f = 0$, there can neither be a vertex v with $f(v) \geq f(v')$ for all $v' \sim v$ with strict inequality for at least one such v', that is, a nontrivial local maximum, nor a nontrivial local minimum, as this would contradict the fact that, by (4.1.2), $L^0 f(v) = 0$ means that the value $f(v)$ is the weighted average of the values at the neighbors of v. Since Γ is connected, f then has to be a constant (when Γ is not connected, a solution of $L^0 f = 0$ is constant on every connected component of Γ). $\qquad \square$

Remark 4.1.2. The argument in the preceding proof is a special case of the maximum principle.

The preceding properties have consequences for the eigenvalues of L^0. We write them as λ_k so that the eigenvalue equation becomes

$$L^0 u_k = \lambda_k u_k, \qquad (4.1.6)$$

with u_k being a corresponding *eigenfunction*. We order the eigenvalues as

$$\lambda_1 = 0 < \lambda_2 \leq \ldots \leq \lambda_N.$$

That $\lambda_2 > 0$ follows from the last item of Corollary 4.1.1, which is a consequence of Lemma 4.1.1.

Corollary 4.1.1. *1. By 1, the eigenvalues are real.*

2. By 2, they are non-negative.

3. By 3, the smallest eigenvalue is $\lambda_1 = 0$. Since we assume that Γ is connected, this eigenvalue is simple, that is,

$$\lambda_k > 0 \tag{4.1.7}$$

for $k > 1$.

□

Formula (4.1.5) suggests to consider the difference operator 2.3.1. For neighbors $v \sim v'$,

$$\delta f(v, v') := f(v') - f(v). \tag{4.1.8}$$

The operator δ can be considered as a map from functions on the vertices of Γ to functions on the edges of Γ. In order to make the latter space also an L^2-space, we introduce the product

$$(\delta f, \delta g) := \sum_{e=(vv')} b(e)(f(v') - f(v))(g(v') - g(v)), \tag{4.1.9}$$

where we sum over edges and not over vertices. If we did the latter, we would need to put in a factor of $1/2$ because each edge would then be counted twice, as in the derivation of (4.1.5).

We now recall the Rayleigh quotients from Section 2.6.1. We may find an orthonormal basis of $L^2(\Gamma)$ consisting of eigenfunctions of L^0,

$$u_k, \ k = 1, \dots, N$$

for the eigenvalues $\lambda_k, \ k = 1, \dots, N$.

We may expand any function f on Γ as

$$f(v) = \sum_k (f, u_k) u_k(v) \tag{4.1.10}$$

and

$$(f, f) = \sum_k (f, u_k)^2 \tag{4.1.11}$$

and

$$(\delta f, \delta f) = \sum_k \lambda_k (f, u_k)^2. \tag{4.1.12}$$

We now recall Theorem 2.6.1.

Lemma 4.1.2. *Let $\mathcal{H}^k$ be the collection of all k-dimensional linear subspaces of H. We have*

$$\lambda_k = \max_{H_{k-1} \in \mathcal{H}^{k-1}} \min \left\{ \frac{(\delta u, \delta u)}{(u, u)} : u \neq 0, (u, f) = 0 \text{ for all } f \in H_{k-1} \right\} \quad (4.1.13)$$

and dually

$$\lambda_k = \min_{H_k \in \mathcal{H}^k} \max \left\{ \frac{(\delta u, \delta u)}{(u, u)} : u \in H_k \setminus \{0\} \right\}. \quad (4.1.14)$$

In (4.1.13), we consider the minimal Rayleigh quotient under $k - 1$ constraints, and we maximize that with respect to the constraints. In (4.1.14), we consider the maximal Rayleigh quotient for k degrees of freedom, and we minimize that with respect to those degrees of freedom.

An obvious consequence is

Corollary 4.1.2. *1. An eigenfunction u for an eigenvalue λ satisfies*

$$\lambda = \frac{(\delta u, \delta u)}{(u, u)}. \quad (4.1.15)$$

2. The second smallest eigenvalue of L^0 is given by

$$\lambda_2 = \min \left\{ \frac{(\delta u, \delta u)}{(u, u)} : \sum_v b(v)u(v) = 0 \right\} \quad (4.1.16)$$

(because

$$\sum_v b(v)u(v) = 0 \quad (4.1.17)$$

is equivalent to $(u, u_1) = 0$ where u_1 is a constant, that is, an eigenfunction for λ_1).

3. The largest eigenvalue of L^0 is given by

$$\lambda_N = \max \left\{ \frac{(\delta u, \delta u)}{(u, u)} : u \in L^2(\Gamma) \setminus \{0\} \right\}. \quad (4.1.18)$$

$\square$

4.1.2 The Algebraic Graph Laplacian

In this section, we consider the algebraic graph Laplacian. This is the Laplace operator that is usually treated in texts on graph theory, for instance, in [38, 113]. The role of the eigenvalue λ_2 of this Laplacian for connectivity properties was recognized by Fiedler [95, 96]. Here, we do not go deeply into this subject but only record some elementary properties of the algebraic graph Laplacian.

Definition 4.1.3. The *algebraic graph Laplacian*, also called the *Kirchhoff Laplacian*, is the Laplace operator (4.1.2) for the weights $b(v) = 1 = b(e)$ for all vertices and edges. Thus, the underlying scalar product is

$$(f,g) = \sum_v f(v)g(v), \tag{4.1.19}$$

and the operator is

$$\Lambda(f)(v) = \sum_{v' \sim v} (f(v) - f(v')) = \deg v \, f(v) - \sum_{v' \sim v} f(v'). \tag{4.1.20}$$

In order to formulate this in slightly more abstract terms, we define the *adjacency matrix A* on $V \times V$ by

$$A(v,v') := \begin{cases} 1 & \text{if } v' \sim v \\ 0 & \text{else} \end{cases} \tag{4.1.21}$$

and the diagonal *degree matrix D* by

$$D(v,v') := \begin{cases} \deg v & \text{if } v = v' \\ 0 & \text{else,} \end{cases} \tag{4.1.22}$$

we can write

$$\Lambda = D - A. \tag{4.1.23}$$

The adjacency matrix has some useful properties. We interpret $A(v,v')$ as the number of paths of length 1 between v and v'. There is such a path precisely if $A(v,v') = 1$, and none if it is 0. Therefore, if we take the power A^k of the adjacency matrix, then

Lemma 4.1.3. $A^k(v,v')$ *is the number of paths of length k between v and v'. So, $A^k(v,v)$ is the number of paths of length k that start and end at v. In particular, $A^2(v,v)$ is the degree of v.*

□

Lemma 4.1.4. *Let Γ be a graph without self-loops. The eigenvalues of Λ satisfy*

$$\sum_k \lambda_k = \sum_v \deg v. \tag{4.1.24}$$

(a)   (b)   (c)

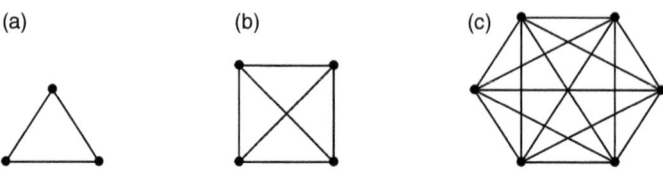

Figure 4.1 The complete graphs K_3 (a), K_4 (b) and K_6 (c).

If Γ is – as always – connected, eigenfunctions u_k for eigenvalues λ_k for $k \geq 2$ satisfy

$$\sum_v u_k(v) = 0. \tag{4.1.25}$$

Proof The sum of the eigenvalues of a matrix equals its trace. The adjacency matrix A is trace-free, and so, in (4.1.24), we only have the trace of D.

(4.1.25) is (4.1.17) for the algebraic Laplacian, since here all $b(v) = 1$.

□

Naturally, we should consider some examples.

Definition 4.1.4. The *complete graph K_N* is the graph on N vertices with $v \sim v'$ for any two different vertices. We also let K_N^0 be the graph on N vertices with $v \sim v'$ for any two not necessarily different vertices.

Thus, every vertex of K_N^0 has a self-loop. The vertices of K_N all have degree $N - 1$, while those of K_N^0 have degree N.

Figure 4.1 shows some complete graphs.

We can easily determine the spectra of K_N and K_N^0.

Lemma 4.1.5. *The spectra of K_N and K_N^0 both consist of the eigenvalue $\lambda_1 = 0$ and the eigenvalue $\lambda_2 = \ldots = \lambda_N = N$.*

Proof For a complete graph K_N, all the degrees $\deg v$ have the value $N - 1$ and are therefore equal. Hence, also all the nontrivial eigenvalues are equal. They are

$$\lambda_2 = \ldots = \lambda_N = N, \tag{4.1.26}$$

since

$$\Lambda u(v) = (N - 1)u(v) - \sum_{v' \neq v} u(v') = Nu(v) \tag{4.1.27}$$

for any u with values $u(v_1) = 1, u(v_2) = -1$ for two distinct vertices and $u(v) = -0$ on all other vertices.

$$\lambda_2 = \ldots = \lambda_N = N + 1 \tag{4.1.28}$$

For K_N^0, we have the same eigenfunctions and eigenvalues, that is, the non-trivial one is N. In fact, the algebraic Laplacian is the same. If J is the $N \times N$-matrix with all entries $= 1$, then this is the adjacency matrix $A(K_N^0)$, while the adjacency matrix $A(K_N)$ is $J - \mathrm{Id}$. And since the degree matrix $D(K_N^0)$ of K_N^0 is $N\mathrm{Id}$, while $D(K_N) = (N-1)\mathrm{Id}$, the two Laplacians agree, $D(K_N^0) - A(K_N^0) = D(K_N) - A(K_N)$.

□

Thus, the spectrum of the algebraic graph Laplacian does not distinguish these two graphs.

Of course, the spectrum of K_N checks with (4.1.24), since for K_N, both sides are equal to $(N-1)N$.

From the preceding, we record

$$\Lambda(K_N) = N\mathrm{Id} - J. \tag{4.1.29}$$

This has a corollary.

Corollary 4.1.3. *For a graph Γ with N vertices, let $\overline{\Gamma}$ be the complementary graph, that is, the graph with the same vertex set in which two vertices are connected by an edge precisely if they are not connected in Γ. Then, its eigenvalues are given by*

$$\lambda_k(\overline{\Gamma}) = N - \lambda_{N-k+2}(\Gamma) \quad for\ k = 2, \ldots, N. \tag{4.1.30}$$

Proof The Laplacians satisfy

$$\Lambda(\overline{\Gamma}) + \Lambda(\Gamma) = \Lambda(K_N) = N\mathrm{Id} - J \tag{4.1.31}$$

by (4.1.29). Let u_k be an eigenfunction of Γ for λ_k, $k \geq 2$. Since we may assume (4.1.25), we have $Ju_k = 0$. Hence,

$$Nu_k = (N\mathrm{Id} - J)u_k = \Lambda(\overline{\Gamma})u_k + \Lambda(\Gamma)u_k = \Lambda(\overline{\Gamma})u_k + \lambda_k u_k$$

from which the result follows.

□

The next example is the *complete bipartite* graph $K_{m,n}$, which will be treated in more detail in Section 4.1.3. It consists of two classes of vertices, V_1 with m and V_2 with n members. Every vertex in V_1 is connected with every vertex in V_2, but no vertices in the same class are connected.

Figure 4.2 shows $K_{3,5}$ and $K_{1,6}$. The latter is an example of a *star graph*.

Like all graphs, $K_{m,n}$ has the eigenvalue 0 with a constant eigenfunction. The function that is equal to n on all vertices in V_1 and equal to $-m$ on all vertices of V_2 satisfies (4.1.25) and is an eigenfunction for the eigenvalue $m+n$.

(a)   (b)

Figure 4.2 The complete bipartite graphs $K_{3,5}$ (a) and $K_{1,6}$ (b).

(a)   (b)

Figure 4.3 Two graphs ((a) and (b)) that are isospectral for the algebraic graph Laplacian.

Let $v, v' \in V_1$. We put $f(v) = 1, f(v') = -1$, and $f(w) = 0$ for all other vertices. Then, f is an eigenfunction with eigenvalue n, and there are $m - 1$ linearly independent such functions, and hence, we get the eigenvalue n with multiplicity $m-1$. Analogously, we get the eigenvalue m with multiplicity $n-1$. Thus, we have altogether found all the $m + n$ eigenvalues of the graph $K_{m,n}$ with its $m + n$ vertices. In fact, the functions f will also be eigenfunctions for the normalized graph Laplacian; see Corollary 4.1.11, but the corresponding eigenvalue will be 1.

The complement graph of $K_{m,n}$ is the disjoint union $K_m + K_n$ of the complete graphs K_m and K_n. The eigenvalues of $K_{m,n}$ could therefore also have been derived from those of K_m and K_n by Corollary 4.1.3.

Figure 4.3 shows another example of two graphs, taken from [269], with the same spectrum of the algebraic Laplacian. These graphs both have the spectrum $0, 3 - \sqrt{5}, 2, 3, 3, 3 + \sqrt{5}$. In fact, the second graph is bipartite, that is, the vertices can be grouped into two classes without connections inside either class (see Definition 4.1.5). The first one is not bipartite, however, and so the spectrum of the algebraic graph Laplacian cannot determine bipartiteness. This is different for the spectrum of the normalized graph Laplacian, as we shall see in Lemma 4.1.12.

4.1.3 The Normalized Graph Laplacian

We now move to the *normalized graph Laplacian*, the Laplace operator on graphs that is of most interest to us.

Let us put this operator into the general perspective developed in Section 2.5.1. In general, the simplices of a simplicial complex have weights. But since the dimension of a graph as a simplicial complex is 1, the edges are the highest-dimensional simplices, and therefore, according to our convention, as formalized in Definition 2.5.6, their weights are 1. And the weight of a vertex then is its degree. This defines an *unweighted graph*, and in this section, if not explicitly stated otherwise, **the graph Γ is unweighted**.

We now equip the space $L^2(\Gamma)$ of functions $f : V \to \mathbb{R}$ on the vertex set V of a graph with the scalar product

$$(f,g) = \sum_v \deg v \, f(v) g(v). \tag{4.1.32}$$

This then is the setting of this section.

Remark 4.1.3. The constructions of this section naturally extend to *weighted graphs*. We write the weight of the unoriented edge (vv') in a weighted graph as $w_{vv'}$. And since weights are assigned to unoriented simplices, we have $w_{v'v} = w_{vv'}$. The degree of a vertex v in a weighted graph then is $\deg v = \sum_{v' \sim v} w_{vv'}$. And we can then use this in the scalar product (4.1.32) and also in the subsequent formula for the Laplacian $\Delta = L^0$, see (4.1.35).

We have from (2.5.7) and (2.5.17)

Lemma 4.1.6. *The normalized combinatorial Laplace operator is given by*

$$L^0(f)(v) = f(v) - \frac{1}{\deg v} \sum_{v' \sim v} f(v') \tag{4.1.33}$$

We shall simply speak of L^0 of (4.1.33) as the *Laplacian* of the graph in question, and we shall usually write

$$\Delta := L^0. \tag{4.1.34}$$

As explained in Section 2.5.3, the Laplacian (4.1.33) evaluates the difference between the value of a function at a vertex v and the average of the values at its neighbors.

Proof According to the general scheme developed in Section 2.5, we have

$$L^0(f)(v) = f(v) - \frac{1}{\deg v} \sum_{v' \sim v} w_{vv'} f(v') = \frac{1}{\deg v} \sum_{v' \sim v} w_{vv'} (f(v) - f(v')),$$

$$\tag{4.1.35}$$

a special case of (4.1.2). And, as explained, since the dimension of a graph as a simplicial complex is 1, the edges are the highest-dimensional simplices, and therefore, according to Definition 2.5.6, their weights are 1, and this then leads to (4.1.35).   □

Our eventual aim will be to analyze the spectrum of the operator L^0, that is, the collection of its eigenvalues. When we speak of the eigenvalues of a graph in the sequel, we shall always mean those of L^0. Let us formulate the problem in slightly more abstract terms. With the *adjacency matrix A* (4.1.21) and the diagonal *degree matrix D* (4.1.22), we can write

$$\Delta = \text{Id} - D^{-1} A. \tag{4.1.36}$$

Remark 4.1.4. This should be compared with (4.1.23).

The scheme also works in the weighted case, putting $A(v, v') := w_{vv'}$.

Remark 4.1.5. Concerning our notation and conventions, we point out:

• While some of the subsequent material is taken from [146], we should alert the reader that our conventions here are different from those of [146]. In fact, our Δ is the negative of that in that reference. The reason lies in the systematic development of Section 2.5 of which Δ is a special case. There is some analogous notational discrepancy for the Laplace operator used in analysis and geometry, also usually denoted by Δ. Analysts usually define the Laplacian of a function f on $\mathbb{R}^n$ as $\Delta f = \sum_{i=1}^{n} \frac{\partial^2 f}{\partial x_i^2}$. That operator is a negative operator because integration by parts yields (for compactly supported functions, say) $\int f \Delta f = - \int |df|^2$. The geometers therefore prefer to introduce a minus sign into the preceding definition in order to turn the Laplacian into a positive operator, actually for reasons that are completely analogous to those of Section 2.5. And in fact, Eckmann's Theorem 4.8.1 is analogous to the Hodge Theorem in Riemannian geometry that represents de Rham cohomology classes by harmonic differential forms (see [150]).

Another notational difference between the present text and [146] is that here we denote the smallest eigenvalue by λ_1 (and, therefore, the largest one is λ_N, with N being the number of vertices), whereas there, it had been denoted by λ_0. The reason was that the smallest eigenvalue is always 0, and so, there is also some advantage in that notation.

Most of the conventions, however, agree with those of [217], from which some of the material in this chapter is taken.

• In graph theory, often a different Laplacian is considered, the algebraic graph Laplacian (4.1.20) that we have briefly discussed in Section 4.1.2. While that operator is also quite useful in graph theory, we prefer the operator

(4.1.33) because it is derived from systematic constructions on simplicial complexes, and also, because it is naturally related to random walks and diffusion processes on graphs.

We now list the important properties of Δ. The first three have already been stated in Lemma 4.1.1 and Corollary 4.1.1. As before, we denote the eigenvalues by λ_k so that the eigenvalue equation becomes

$$\Delta u_k = \lambda_k u_k, \tag{4.1.37}$$

with u_k being a corresponding *eigenfunction*, and we order the eigenvalues as

$$\lambda_1 = 0 < \lambda_2 \leq \dots \leq \lambda_N.$$

Lemma 4.1.7. *Let Γ be a connected graph.*

1. *Δ is self-adjoint with respect to the scalar product $(\cdot, \cdot)$, that is,*

$$(f, \Delta g) = (\Delta f, g) \tag{4.1.38}$$

for all $f, g \in L^2(\Gamma)$. Hence, all eigenvalues are real
2. *Δ is non-negative:*

$$(\Delta f, f) \geq 0 \tag{4.1.39}$$

for all f. Hence, all eigenvalues are non-negative.
3. *$\Delta f = 0$ precisely when f is constant. Hence, the smallest eigenvalue is $\lambda_1 = 0$. Since we assume that Γ is connected, this eigenvalue is simple, that is,*

$$\lambda_k > 0 \tag{4.1.40}$$

for $k > 1$.
4. *When Γ has no self-loops, then the trace of Δ is N. Hence*

$$\sum_{k=1}^{N} \lambda_k = N. \tag{4.1.41}$$

Proof The first three have already been derived in Lemma 4.1.1 and Corollary 4.1.1. Concerning the last one, since for loopless graphs, the adjacency matrix A has only 0s on the diagonal, the last claim follows from (4.1.36). $\square$

By (4.1.33), λ is an eigenvalue with eigenfunction f if

$$f(v) - \frac{1}{\deg v} \sum_{v' \sim v} f(v') = \lambda f(v), \tag{4.1.42}$$

or equivalently

$$(1 - \lambda)f(v) = \frac{1}{\deg v} \sum_{v' \sim v} f(v'). \qquad (4.1.43)$$

We next consider the difference operator (2.3.1) that was also displayed in (4.1.8). For neighbors $v \sim v'$,

$$\delta f(v,v') := f(v') - f(v). \qquad (4.1.44)$$

The operator δ can be considered as a map from functions on the vertices of Γ to functions on the edges of Γ. In order to make the latter space also an L^2-space, we consider the product

$$(\delta f, \delta g) := \sum_{e=(vv')} (f(v') - f(v))(g(v') - g(v)), \qquad (4.1.45)$$

which is of course the special case of (4.1.9) for edge weights $b(e) = 1$. Note that we are summing here over edges and not over vertices. If we did the latter, we would need to put in a factor of $1/2$ because each edge would then be counted twice. We also point out that in contrast to the product of (4.1.32), $(f,g) = \sum_v \deg v\, f(v)g(v)$, we do not include weights here. The reason is that, here, the sum should be considered as a sum of edges and not one over vertices, and since we are considering unweighted graphs at this point, the edges do not carry any natural weights. This is in accordance with the general principle of Definition 2.5.6.

We also note that (4.1.45) does not depend on the orientations of the edges (and therefore, our convention to count edges only once in the undirected case is appropriate).

The product (4.1.45) encodes more information about the graph than the product (4.1.32). The latter only depends on the weights (as given by the vertex degrees) but not on the connection structure of the graph. There exist many structurally quite diverse graphs with the same degree sequence, and given a graph, one can rewire it by a cross exchange of edges without changing the degrees of the nodes. Namely, given vertices $v_1 \sim v_1'$ and $v_2 \sim v_2'$, but without edges between v_1 and v_2, or between v_1' and v_2', we create a new graph by deleting the edges between v_1 and v_1' and between v_2 and v_2' and inserting new edges between v_1 and v_2 and between v_1' and v_2'. That operation preserves the degrees of all vertices and therefore also the product (4.1.32) for any functions f,g on the graph. (4.1.45), in contrast, is affected because the edge set is changed.

From (4.1.5), we recall

$$(\delta f, \delta g) = (f, \Delta g). \qquad (4.1.46)$$

Thus, our product (4.1.45) is naturally related to the Laplacian Δ.

As in Section 4.1.1, we can use the Rayleigh quotients from Section 2.6.1 and Theorem 2.6.1 and Lemma 4.1.2 to characterize the eigenvalues.

Corollary 4.1.4. *1. Any eigenfunction u for some eigenvalue λ satisfies*

$$\lambda = \frac{(\Delta u, u)}{(u,u)} = \frac{(\delta u, \delta u)}{(u,u)}. \tag{4.1.47}$$

2. Since the eigenfunctions can be taken orthogonal to each other and since $(u, u_1) = 0$ (where u_1 is a constant, that is, an eigenfunction for λ_1) means

$$\sum_v \deg v\, u(v) = 0, \tag{4.1.48}$$

the second smallest eigenvalue of Δ is

$$\lambda_2 = \min \left\{ \frac{(\delta u, \delta u)}{(u,u)} : \sum_v \deg v\, u(v) = 0 \right\}. \tag{4.1.49}$$

3. The largest eigenvalue of Δ is

$$\lambda_N = \max \left\{ \frac{(\delta u, \delta u)}{(u,u)} : u \in L^2(\Gamma) \setminus \{0\} \right\}. \tag{4.1.50}$$

4. The eigenvalues satisfy

$$0 \le \lambda \le 2. \tag{4.1.51}$$

Proof The first three items have already been stated in Corollary 4.1.2, and the lower bound in (4.1.51) is also obvious, for instance, from (4.1.47). It remains to show the upper bound.

By the 3rd statement of Corollary 4.1.4, the largest eigenvalue is

$$\lambda_N = \max_{u \neq 0} \frac{\sum_{e=(v_0,v_1)} (u(v_0) - u(v_1))(u(v_0) - u(v_1))}{\sum_v \deg v\, u(v)^2}. \tag{4.1.52}$$

The largest possible value of this quantity is 2, and this value is realized when

$$u(v_1) = -u(v_0) \tag{4.1.53}$$

for every edge.   □

Recalling (4.1.45), we can rewrite (4.1.47) to obtain

Corollary 4.1.5. *Any eigenfunction u for some eigenvalue $\lambda \neq 0$ satisfies*

$$\lambda = \frac{\sum_{e=(v_0,v_1)} (u(v_0) - u(v_1))^2}{\sum_v \deg v\, u(v)^2}. \tag{4.1.54}$$

□

Let us also derive another characterization.

Corollary 4.1.6. *Any eigenfunction u for some eigenvalue $\lambda \neq 0$ satisfies*

$$2 - \lambda = \frac{\sum_v \frac{1}{\deg v} \sum_{v',v'' : v' \sim v, v'' \sim v} (u(v') - u(v''))^2}{\sum_{v,v' : v \sim v'} (u(v) - u(v'))^2} \tag{4.1.55}$$

Proof We first derive some general identities for a function u on the vertex set of Γ.

$$\sum_v \frac{1}{\deg v} \sum_{v',v'' : v' \sim v, v'' \sim v} (u(v') - u(v''))^2$$

$$= \sum_v \left(\sum_{v'' : v \sim v''} \frac{1}{\deg v} \sum_{v' : v' \sim v} (u(v'') - u(v'))^2 \right)$$

$$= \sum_v \left(\sum_{v'' : v \sim v''} \frac{1}{\deg v} \sum_{v' : v' \sim v} u(v')^2 - \frac{2}{\deg v} \sum_{v' : v' \sim v} u(v'')u(v') + u(v'')^2 \right)$$

$$= \sum_v \left(2 \sum_{v' : v' \sim v} u(v')^2 - \frac{2}{\deg v} \left(\sum_{v' : v' \sim v} u(v') \right)^2 \right)$$

$$= 2 \sum_v \sum_{v' : v' \sim v} u(v')^2 - \sum_v 2 \deg v \left(\frac{1}{\deg v} \sum_{v' : v' \sim v} u(v') \right)^2.$$

We now observe that we can replace u by $u - u(v)$ in the first and hence also in all subsequent lines. This yields

$$\sum_v \frac{1}{\deg v} \sum_{v',v'' : v' \sim v, v'' \sim v} (u(v') - u(v''))^2$$

$$= 2 \sum_v \sum_{v' : v' \sim v} (u(v') - u(v))^2 - \sum_v 2 \deg v \left(\frac{1}{\deg v} \sum_{v' : v' \sim v} (u(v') - u(v)) \right)^2$$

$$= 2 \sum_v \sum_{v' : v' \sim v} (u(v') - u(v))^2 - \sum_v 2 \deg v (\Delta u(v))^2.$$

When u now is an eigenfunction, $\Delta u = \lambda u$ for some eigenvalue λ, then, by (4.1.47), $\sum_v \sum_{v' : v' \sim v} (u(v') - u(v))^2 = 2\lambda \sum_v \deg v \, u(v)^2$, and we obtain

$$\sum_v \frac{1}{\deg v} \sum_{v',v'' : v' \sim v, v'' \sim v} (u(v') - u(v''))^2 = 2\lambda(2 - \lambda) \sum_v \deg v \, u(v)^2. \tag{4.1.56}$$

Using (4.1.47) again, we can also reformulate this as

$$2 - \lambda = \frac{\sum_v \frac{1}{\deg v} \sum_{v',v'': v'\sim v, v''\sim v}(u(v') - u(v''))^2}{\sum_v \sum_{v': v'\sim v}(u(v') - u(v))^2}. \tag{4.1.57}$$

$\square$

Let us also go into the details of the computation of the normalized combinatorial Laplacian L^1 on a graph. We consider an oriented edge $e = [v_0, v_1]$. Let $f: V \to \mathbb{R}$ be a function on the vertex set V of our graph. Then (4.1.44) becomes

$$\delta f(e) = f(v_1) - f(v_0), \tag{4.1.58}$$

and so, for a function $\omega: E \to \mathbb{R}$ on the edges of our graph, which, according to (2.5.1), satisfies $\omega[v_1, v_0] = -\omega[v_0, v_1]$,

$$\sum_v \deg v \, f(v)\delta^*\omega(v) = (f, \delta^*\omega)$$

$$= (\delta f, \omega)$$

$$= \sum_e \delta f(e)\omega(e)$$

$$= \sum_{e=[v_0,v_1]} (f(v_1) - f(v_0))\omega(e)$$

$$= \sum_{v_0,v_1} \deg v_1 \, f(v_1)\omega(v_0,v_1) - \sum_{v_0,v_1} \deg v_0 \, f(v_0)\omega(v_0,v_1)$$

and hence

$$\delta^*\omega(v) = \frac{1}{\deg v} \sum_{v\in e}(-1)^{i(e)}\omega(e) \tag{4.1.59}$$

with

$$i(e) = \begin{cases} 1 & \text{if } e = (v, v') \\ 0 & \text{if } e = (v', v). \end{cases}$$

Moreover, for $e = [v_0, v_1]$ as before,

$$L^1\omega(e) = \delta\delta^*\omega(e) = \delta f(e), \text{ with } f(v) = \frac{1}{\deg v} \sum_{v\in e}(-1)^{i(e)}\omega(e),$$

$$= \frac{1}{\deg v_0} \sum_{v_0\in g=[v_0,v']} \omega(g) - \frac{1}{\deg v_1} \sum_{v_1\in f=[v_1,v']} \omega(f). \tag{4.1.60}$$

From (4.8.6), we obtain

Corollary 4.1.7. *The operators Δ in (4.1.33) and L^1 in (4.1.60) on a graph have the same nonzero eigenvalues.*   □

Moreover, by Theorem 4.8.1, the multiplicities of the eigenvalue 0 of Δ and L^1 on a graph Γ with vertex set V and edge set E equal the Betti numbers b_0 and b_1. Thus, also

$$\chi(\Gamma) = |V| - |E| = b_0 - b_1 = \dim \ker \Delta - \dim \ker L^1. \qquad (4.1.61)$$

In particular, when the graph is connected, $b_0 = 1$, that is, the eigenvalue 0 of Δ is simple, and by (4.1.61), the multiplicity of the eigenvalue 0 of L^1 then is $|E| - |V| + 1$. A connected graph with $|E| = |V| - 1$ is called a *tree*, and we conclude

Corollary 4.1.8. *A connected graph $\Gamma = (V, E)$ is a tree if and only if the kernel of L^1 is trivial.*   □

More generally, we can express the vanishing of the Laplacian of a function ω at an edge in terms of diffusion through that edge as governed by ω. In fact, from (4.1.60), we have that for an edge $e = [v_0, v_1]$, $L^1\omega(e) = \delta\delta^*\omega(e) = 0$ precisely if what comes into the initial vertex v_0 and leaves at the terminal vertex v_1 is balanced by what exits at v_0 and enters at v_1. We note that in particular, the edge $e = [v_0, v_1]$ enters into (4.1.60) with the contribution $\left(\frac{1}{\deg v_0} + \frac{1}{\deg v_1}\right)\omega(e)$, and from this subtraction, it is subtracted what enters into v_0 from other edges and what leaves at v_1 into other edges, while the opposite contributions, leaving at v_0 or entering at v_1, have a positive sign.

In particular, on a *chain*, that is, a set of different vertices $\{v_0, \ldots, v_m\}$ with edges $e_1 = [v_0, v_1], \ldots, e_m = [v_{m-1}, v_m]$, the only solution of $L^1\omega = 0$ is $\omega \equiv 0$, but on a *cycle*, that is, when we close the chain by an edge $e_0 = [v_m, v_0]$, $\omega \equiv constant$ is also a solution.

We now apply these principles to study the eigenvalue problem for some other simple examples. We first consider the complete graphs K_N and K_N^0, see Definition 4.1.4.

As in Lemma 4.1.5, we can easily determine the spectra of K_N and K_N^0.

Lemma 4.1.8. *The spectrum of K_N consists of the eigenvalue $\lambda_1 = 0$ and the eigenvalue $\lambda_2 = \ldots = \lambda_N = \frac{N}{N-1}$.*
K_N^0 has $\lambda_1 = 0$ and $\lambda_2 = \ldots = \lambda_N = 1$.

Proof For a complete graph K_N, all the degrees $\deg v$ have the value $N - 1$ and are therefore equal. Hence, also all the nontrivial eigenvalues are equal. We have

$$\lambda_2 = \dots = \lambda_N = \frac{N}{N-1} \qquad (4.1.62)$$

since

$$\Delta u(v) = u(v) - \frac{1}{N-1}\sum_{v' \neq v} u(v') = \frac{N}{N-1}u(v) \qquad (4.1.63)$$

for any u that is orthogonal to the constants, that is,

$$\frac{1}{N}\sum_{v' \in V} \deg v'\, u(v') = \frac{N-1}{N}\sum_{v' \in V} u(v') = 0. \qquad (4.1.64)$$

Similarly, on K_N^0, we have

$$\lambda_2 = \dots = \lambda_N = 1 \qquad (4.1.65)$$

since $\Delta u = u$ whenever u is orthogonal to the constants.

□

In fact, we could have deduced this result directly from Lemma 4.1.5 because of the following:

Lemma 4.1.9. *Let Γ be a graph with constant degree, that is, $\deg v = k$ for some k and all vertices v. Then, the eigenvalues of the algebraic Laplacian Λ are k times those of the normalized Laplacian Δ.*

Proof In that case, the two Laplacians are related by

$$\Lambda = k\Delta. \qquad (4.1.66)$$

and therefore also the eigenvalues differ by that factor.   □

The example of complete graphs shows that neither the smallest nonzero nor the largest eigenvalue of a graph needs to be simple.

In fact, the spectral properties found in Lemma 4.1.8 characterize the complete graphs.

Lemma 4.1.10. *A graph with N vertices without self-loops is complete if and only if its spectrum consists of 0 as a simple eigenvalue and $\frac{N}{N-1}$ with multiplicity $N-1$.*

More precisely, for a graph with N vertices without self-loops that is not complete, we have

$$\lambda_2 \leq 1 \text{ and } \lambda_N > \frac{N}{N-1}. \qquad (4.1.67)$$

Proof We have already verified that the spectrum of K_N has those values. For the converse direction, by Corollary 4.1.4,

$$\lambda_2 = \min_{g\,:\,\sum_v \deg v \, g(v)=0} \frac{(\Delta g, g)}{(g, g)}. \tag{4.1.68}$$

When Γ is not complete, we can find two vertices v_1, v_2 that are not connected by an edge and take g with $g(v_1), g(v_2) \neq 0$, but

$$\deg v_1 \, g(v_1) + \deg v_2 \, g(v_2) = 0 \text{ and } g(v) = 0 \text{ for all other } v. \tag{4.1.69}$$

Inserting this into the Rayleigh quotient (4.1.68) makes that expression 1, and the minimum therefore is ≤ 1. The second inequality of (4.1.67) follows by taking two connected vertices v_1 and v_2 and g as before and verifying that the Rayleigh quotient then becomes $> \frac{N}{N-1}$. □

From Lemma 4.1.10, we see that there are graphs, such as the complete graphs K_N, that are completely determined by their spectrum. Thus, there are no other graphs that are isospectral with K_N. And in fact, there do not even exist graphs whose spectrum is very close to that of K_N. This follows from Theorem 4.1.1, which was found in [73].

Theorem 4.1.1. *For a noncomplete graph with N vertices,*

$$\lambda_N \geq \frac{N+1}{N-1}. \tag{4.1.70}$$

Proof The proof is taken from [155]. We may assume that the graph is connected. In fact, for a nonconnected graph, the smallest connected component has at most $N/2$ vertices. Thus, its largest eigenvalue, and hence also the largest eigenvalue λ_N of the entire graph, satisfies

$$\lambda_N \geq \frac{N/2}{N/2 - 1} = \frac{N}{N-2} > \frac{N+1}{N-1},$$

which proves the theorem for nonconnected graphs.

Thus, we may indeed assume that the graph is connected. Since it is not complete, there exists a vertex v with $\deg v \leq N-2$. As the graph is connected, we can find some other vertex w that is a neighbor of some neighbor of v but not a neighbor of v itself. Then, $\deg v' \leq N-2$ as well, since v is not a neighbor of w. For any collection $v_1, v_2, \ldots, v_m$ of vertices, let $S(v_1, \ldots, v_m)$ be the set of their joint neighbors, and let $s(v_1, \ldots, v_m)$ be its cardinality. Let $S = S(v, v')$ be the nonempty set of vertices that are neighbors of both v and v', and let s be the cardinality of S, that is, the number of joint neighbors of v and w. We shall construct a function f with $(f, \Delta f) \geq \frac{N+1}{N-1}(f, f)$. Then, the result follows from (4.1.50).

We put

$$f(u) := \begin{cases} -1 & : u \in S, \\ \frac{N-1}{2}\frac{s}{\deg v} & : u = v, \\ \frac{N-1}{2}\frac{s}{\deg v'} & : u = v', \\ 0 & : \text{else.} \end{cases}$$

f satisfies

$$\deg v\, \Delta f(v) = \deg v f(v) + s = \frac{N+1}{2}s$$

and thus, $\Delta f(v) = \frac{N+1}{N-1} f(v)$. Likewise, $\Delta f(v') = \frac{N+1}{N-1} f(v')$. We now claim that $-\Delta f(u) \geq \frac{N+1}{N-1}$ for all $u \in S$. We observe $s \geq \max(1, \deg v + \deg v' + 2 - N)$ and calculate

$$-\Delta f(u) = \frac{\deg u - s(u,v,v') + f(v) + f(v')}{\deg u} \geq 1 + \frac{1 - s + f(v) + f(v')}{\deg u}.$$

$$(4.1.71)$$

As $f(v) + f(v') \geq s$, we can use $\deg u \leq N - 1$ and continue

$$\frac{1 - s + f(v) + f(v')}{\deg u} \geq \frac{1-s}{N-1} + \frac{s}{2\deg v} + \frac{s}{2\deg v'} \qquad (4.1.72)$$

$$= \frac{1}{N-1} + s\left(\frac{1}{2\deg v} + \frac{1}{2\deg v'} - \frac{1}{N-1}\right).$$

Since $\deg v \leq N - 2$ and $\deg v' \leq N - 2$, the term in brackets is positive, and thus,

$$s\left(\frac{1}{2\deg v} + \frac{1}{2\deg v'} - \frac{1}{N-1}\right)$$

$$\geq \max(1, \deg v + \deg v' + 2 - N)\left(\frac{1}{2\deg v} + \frac{1}{2\deg v'} - \frac{1}{N-1}\right).$$

$$(4.1.73)$$

We write $D := (\deg v + \deg v')/2$, then the harmonic–arithmetic mean gives

$$\frac{1}{2\deg v} + \frac{1}{2\deg v'} \geq \frac{1}{D}.$$

Thus,

$$s\left(\frac{1}{2\deg v} + \frac{1}{2\deg v'} - \frac{1}{N-1}\right) \geq \max(1, 2D + 2 - N)\left(\frac{1}{D} - \frac{1}{N-1}\right).$$

$$(4.1.74)$$

We aim to show that the latter term is at least $1/(N-1)$, which, by multiplying with $D(N-1)$ and subtracting D is equivalent to

$$\max(1, 2D + 2 - N)(N - 1 - D) - D \geq 0. \qquad (4.1.75)$$

If $D \leq (N-1)/2$, then the maximum equals 1 and the inequality follows immediately. If $D \geq (N-1)/2$, then $\max(1, 2D+2-N) = 2D+2-N$, and so the left-hand side becomes a concave quadratic polynomial in D with its zero points at $D = N-2$ and $D = (N-1)/2$. Thus, the inequality (4.1.75) holds for all D between those zeroes. But by assumption, D lies between the zero points, which proves the claim that $-\Delta f(u) \geq \frac{N+1}{N-1}$ for all $u \in S$. In particular, this shows that $f\Delta f \geq \frac{N+1}{N-1} f^2$. Summation over the vertex set proves the claim of the theorem for all connected graphs.

$\square$

Remark 4.1.6. Reference [73] could also determine the graphs for which (4.1.70) becomes an equality; see also [155] for a simpler proof. These graphs are the complete graphs with a single edge removed, or when N is odd, two complete graphs of size $(N + 1)/2$ joined at a single vertex. The simplest non-trivial example for the latter case, two copies of K_3 joined at a single vertex, will be examined in Figure 4.5. For all other graphs, the inequality (4.1.70) is strict, and one may derive more precise lower bounds for them; see [155].

We now come to another class of graphs where in contrast to complete graphs, connections between vertices are restricted. This will in particular lead to a behavior of the largest eigenvalue λ_N that is opposite to what we see on complete graphs.

Definition 4.1.5. A graph Γ is *bipartite* if its vertex set V can be decomposed as $V = V_+ \cup V_-$ with disjoint V_+, V_- such that there are no edges connecting vertices in the same components. That is, every edge has one of its endpoints in V_+ and the other in V_-.

For instance, every tree is bipartite. Starting with any vertex v_0 of a tree, one puts v_0 into V_+, all neighbors of v_0 into V_-, the neighbors of those neighbors into V_+ again, and so on. Likewise, cycles, that is, graphs with vertices $v_1, \ldots, v_N$ and edges $(v_1, v_2), \ldots (v_{N-1}, v_N), (v_N, v_1)$ for even N, are bipartite, with V_+ containing the vertices with an even index. Cycles with an odd N are not bipartite, however. In fact, we have

Lemma 4.1.11. *A graph is bipartite if and only if it does not contain cycles of odd length.*

Proof Without loss of generality, we may assume that our graph Γ is connected, that is, any two vertices can be joined by a path. (Otherwise, we would apply the subsequent reasoning to every connected component of Γ.) Take any vertex v_0 in Γ. Let V_+ be the class of all vertices v that have an even distance from v_0, that is, a shortest path connecting v_0 has an even number of edges. Likewise, let V_- consist of those vertices that have an odd distance to v_0. If there is no edge connecting any two vertices in V_+, or any two vertices in V_-, the graph is bipartite. Assume now that there is an e edge connecting $v_1, v_2 \in V_+$, for instance. We then create a cycle of odd length by taking the path from V_0 to v_1, which has an even length, the edge from v_1 to v_2, which has length 1, and then the path of even length from v_2 to v_0. □

The cycle constructed in the preceding proof may traverse some edges back and forth, that is, it need not consist of disjoint edges. Any such cycle, however, has a subcycle with disjoint edges. We simply take away all edges traversed back and forth and take a component of the rest. The parity of the length, that is, whether it is even or odd, is not affected by this procedure.

Lemma 4.1.12. *A graph Γ is bipartite if and only if its largest eigenvalue is 2. And a corresponding eigenfunction is a multiple of the function that is $+1$ on one class and -1 on the other.*

Proof In (4.1.53) in the proof of Corollary 4.1.4, we have seen that the largest possible value of the Rayleigh quotient giving λ_N is realized for a function u that satisfies $u(v_1) = -u(v_0)$ whenever $v_1 \sim v_0$, that is, a function of the type claimed in the Lemma. In that case, $\lambda_N = 2$. And that is possible if and only if the graph is bipartite. □

From Lemma 4.1.12, we see that a single eigenvalue can already encode a distinctive characteristic feature of a graph, in the case at hand being bipartite. More generally,

Lemma 4.1.13. *A graph Γ is bipartite if and only if whenever λ is an eigenvalue, then $2 - \lambda$ is an eigenvalue as well.*

Proof Let λ be an eigenvalue with eigenfunction f on a bipartite graph with vertex set $V = V_+ \cup V_-$. Then,

$$\tilde{f} := \begin{cases} f(v) & \text{for } v \in V_+ \\ -f(v) & \text{for } v \in V_- \end{cases} \tag{4.1.76}$$

is an eigenfunction for the eigenvalue $2 - \lambda$. In fact, for the eigenfunction f, we have by (4.1.43)

$$(1 - \lambda)f(v) = \frac{1}{\deg v} \sum_{v' \sim v} f(v'), \qquad (4.1.77)$$

and hence

$$(1 - (2 - \lambda))f(v) = (\lambda - 1)f(v) = -\frac{1}{\deg v} \sum_{v' \sim v} f(v'), \qquad (4.1.78)$$

and on a bipartite graph; therefore, $\tilde{f}$ from (4.1.76) is an eigenfunction for the eigenvalue $2 - \lambda$.

Conversely, if whenever λ is an eigenvalue, then so is $2 - \lambda$, then since 0 is always an eigenvalue (with the constants as eigenfunctions), 2 has to be an eigenvalue. By Lemma 4.1.12, this implies that the graph is bipartite. □

The proof also yields

Corollary 4.1.9. *When λ is an eigenvalue of a bipartite graph with eigenfunction f, then according to (4.1.76), we obtain an eigenfunction for $2 - \lambda$ by changing the sign of f on one of the components. Unless f vanishes identically on one component, this function $\tilde{f}$ is linearly independent of f. Thus, only eigenfunctions for the eigenvalue 1 can vanish identically on one of the components.*

More generally, on a bipartite graph, we only need to know half of the eigenvalues and eigenfunctions to reconstruct the entire spectrum. □

This Corollary will be quite useful later when we compute the spectra of certain classes of bipartite graphs.

Corollary 4.1.10. *Let Γ be a connected graph. If 2 is an eigenvalue of Γ, then it is simple.*

Proof By Lemma 4.1.13, when 2 is an eigenvalue, the graph is bipartite, and furthermore, its multiplicity equals that of the eigenvalue 0. By Lemma 4.1.7, that eigenvalue is simple since Γ is connected. □

We have now seen that the eigenvalues 0 and 2, the minimal and maximal possible values by Lemma 4.1.12, play special roles. The eigenvalue 1 is also very interesting. According to (4.1.43), f is an eigenfunction for the eigenvalue 1 if and only if

$$\sum_{v' \sim v} f(v') = 0 \text{ for every vertex } v. \qquad (4.1.79)$$

This can, for instance, be achieved as follows. Suppose the graph contains two vertices v_1 and v_2 that are not connected by an edge but have the same neighbors, that is,

$$v \sim v_1 \text{ if and only if } v \sim v_2. \tag{4.1.80}$$

In such a situation, we can define

$$f(v) = \begin{cases} +1 & \text{for } v = v_1 \\ -1 & \text{for } v = v_2 \\ 0 & \text{else.} \end{cases} \tag{4.1.81}$$

This function f then satisfies (4.1.79) because $f(v') = 0$ for all neighbors of v_1 and for all neighbors of v_2, and for any v that has both v_1 and v_2 as its neighbors, the sum also vanishes, since $f(v_2) = -f(v_1)$. And since v_1 and v_2 have the same neighbors, any v is either a neighbor of both or of neither of them. Since we know that $\Delta = L^0$ and L^1 have the same nonzero eigenvalues, in the preceding situation, we should also generate an eigenvalue 1 of L^1. This is simply achieved by putting

$$\omega(e) = \begin{cases} +1 & \text{for } e = (v, v_1) \\ -1 & \text{for } e = (v, v_2) \text{ (or equivalently 1 for } e = (v_2, v)) \\ 0 & \text{else.} \end{cases} \tag{4.1.82}$$

Then, at any $v \neq v_1, v_2$, there is no contribution to $L^1\omega$ as given in (4.1.60) because what comes in equals what flows out. At v_1, for an edge $e = (v, v_1)$, we get the contribution $\deg v_1 \left(\frac{1}{\deg v_1} \omega(e) \right) = 1$, and analogously at v_2, and hence the ω of (4.1.82) is an eigenvector for the eigenvalue 1. Geometrically, v_2 is a source, and v_1 is a sink for that ω.

We also observe the remarkable fact that eigenfunctions can be very local, that is, be nonzero only on a small set of vertices in the present case of the eigenfunction f from (4.1.81) for Δ, only on two vertices, which, in fact, is the minimal possible number for the support of an eigenfunction.

The preceding can be turned into a constructive process the *duplication of vertices*. We start with some vertex v_1 in a graph Γ and construct a graph Γ' that has one additional vertex v_2 that is not connected with v_1 but has the same neighbors as it. Such a graph Γ' has the eigenvalue 1. Also, when Γ already possesses the eigenvalue 1 from a previous vertex duplication process, then this eigenvalue with its eigenfunction carries over to Γ'. To see this, consider an eigenfunction f_1 for the eigenvalue 1. It satisfies (4.1.79), and if we extend it to Γ' by putting $f_1(v_2) = 0$, this still holds for all vertices v of Γ. (4.1.79) also holds for v_2 because it holds for v_1, and v_1 and v_2 have the same neighbors.

Thus, let us start with the simplest nontrivial graph, consisting of two vertices v_+, v_-, connected by an edge $e = (v_+, v_-)$. This graph is denoted by $K_{1,1}$, for a reason that will become apparent in a moment. When we now iteratively duplicate v_+ $m - 1$ times and v_- $n - 1$ times, we obtain a bipartite graph with vertex set V_+ having m and V_- having n vertices. Also, every vertex in V_+ is connected with every vertex in V_-. This is the *complete bipartite* graph $K_{m,n}$. And, we already know its spectrum. As a bipartite graph, it has the eigenvalues 0 and 2 (which are simple by Lemma 4.1.13 as the graph is connected), and since $K_{m,n}$ is the result of $m + n - 2$ vertex duplications, it possesses the eigenvalue 1 with multiplicity $m + n - 2$. Since $K_{m,n}$ has $m + n$ vertices, we therefore have found all its eigenvalues. This shows

Corollary 4.1.11. *The spectrum of the complete bipartite graph $K_{m,n}$ consists of the eigenvalues 0 and 2 with multiplicity 1 and the eigenvalue 1 with multiplicity $m + n - 2$.*

This also leads to the following observation. Any two graphs $K_{m,n}$ and $K_{m',n'}$ with $m' + n' = m + n$ have the same spectrum, or as one says, they are *isospectral*. (For systematic constructions of isospectral graphs, see for instance, [112, 258].) Thus, graphs cannot always be distinguished by their spectra. The different complete bipartite graphs with the same number of vertices can, however, be distinguished by the multiplicity of the eigenvalue 0 of L^1 because they have different first Betti numbers; see (4.1.61). In fact, $K_{m,n}$ has mn edges, and since by (4.1.61), $m + n - mn = 1 - b_1$, the multiplicity of that eigenvalue is $b_1 = mn - m - n + 1$. (Note that of course $K_{m,n}$ and $K_{n,m}$ are isomorphic graphs, that is, there exists a bijection between the vertex sets that maps edges to edges; in fact, we simply have to interchange V_+ and V_- to effect that isomorphism.)

Of course, (4.1.79) has a simple algebraic explanation.

Lemma 4.1.14. *The multiplicity of the eigenvalue 1 equals the dimension of the kernel of the adjacency matrix (4.1.21).*

Proof Obvious, since by (4.1.21),

$$0 = \sum_{v'} a_{vv'} f(v') = \sum_{v' \sim v} f(v') \quad \text{for every } v$$

precisely if f is in the kernel of the adjacency matrix $(a_{vv'})$.   □

Thus, it is easy to construct further examples of graphs with the eigenvalue $\lambda = 1$, for example:

1. A *cycle graph* C_N consists of nodes $v_1, \ldots, v_N$ such that the node v_i is connected to the nodes v_{i-1} and v_{i+1} mod N (i.e., there is a connection

between v_1 and v_N). When $N = 4m$, we put $u(v_j) = 0$ when j is odd and $u(v_j) = (-1)^{j/2}$ for even j. Then, (4.1.79) holds, and this function is an eigenfunction for the eigenvalue 1. (For $m = 1$, we have in fact a node duplication but no longer for $m > 1$.)

2. The same construction works for the *wheel graph* W_N where we add to C_N a single central vertex v_0 that is connected to all other vertices. For $N = 4m$, we use the function u from the previous example, extended by $u(v_0) = 0$.

    (In fact, all eigenfunctions of W_N vanish at the central vertex v_0, except for the multiples of the function with $u(v_0) = -3$ and $u(v) = 1$ at all other vertices. This function belongs to the eigenvalue $4/3$. For even N, we also have the eigenvalue $5/3$, with an eigenfunction that vanishes again at v_0 and alternatingly assumes the values ± 1 at the other vertices.)

3. We connect C_3 and C_5 at a single node v_0. We put $u(v_0) = 1$, $u(v) = -1$ at the two other nodes from C_3, $u(v) = 1$ at the two neighbors of v_0 in C_5, and $u(v) = -1$ at the remaining two nodes in C_5. This u again satisfies (4.1.79).

4. A bipartite graph with an odd number of vertices must possess the eigenvalue 1 because by Lemma 4.1.13 or Corollary 4.1.9, all other eigenvalues occur in pairs, and so, 1 is needed to produce an odd number of eigenvalues.

Corollary 4.1.12. *Let Γ with N vertices have an induced subgraph Γ_0 with M vertices, that is, a graph containing M of the vertices of Γ and all edges between those vertices that are present in Γ, and assume that Γ_0 does not have the eigenvalue 1. Then, the multiplicity of the eigenvalue 1 of Γ is at most $N - M$.*

The *proof* follows directly from Lemma 4.1.14.

For instance, when Γ contains K_M as an induced subgraph (such an induced subgraph K_M is also called an *M-clique*), then the multiplicity of 1 is at most $N - M$, see [57].

Remark 4.1.7. We observe that a complete graph K_N^0 with self-loops and a complete bipartite graph $K_{m,n}$ with $m + n = N$ have almost the same spectrum. While K_N^0 has the eigenvalue 1 with multiplicity $N - 1$, $K_{m,n}$ has this eigenvalue with multiplicity $N - 2$. But as we shall understand in Section 4.5, the appropriate comparison is that between $K_{m,n}$ and the disjoint union of K_m^0 and K_n^0. Both have the eigenvalue 1 with multiplicity $N - 2$, and instead of the eigenvalue 2 of $K_{m,n}$, the latter graph has an additional eigenvalue 0 because it has two components.

One of the complete bipartite graphs deserves a special name.

Definition 4.1.6. The graph $K_{1,n}$ is called a *star graph*.

Thus, a star graph consists of one central vertex, a hub, that is connected to all other vertices, but none of those are connected with each other. A star graph is a tree. In fact, conversely, any tree is bipartite, as we have observed earlier.

The preceding can also be understood in terms of symmetries (see Section 2.7 for the abstract setting). When we have two vertices v_1, v_2 as in (4.1.80), that is, $v \sim v_1$ if and only if $v \sim v_2$ for all vertices (and so, in particular, $v_1 \nsim v_2$, as we are currently not allowing for self-loops), then we have a graph automorphism $\tau: V \to V$ (meaning that $v \sim v'$ if and only if $\tau(v) \sim \tau(v')$) given by

$$\tau(v_1) = v_2, \quad \tau(v_2) = v_1, \quad \tau(v) = v \text{ for all } v \neq v_1, v_2. \qquad (4.1.83)$$

The Laplacian commutes with τ, that is, with $\tau_* f(v) := f(\tau v)$,

$$\Delta(\tau_* f)(v) = \tau_*(\Delta f)(v). \qquad (4.1.84)$$

In particular, it leaves the eigenspaces of τ invariant. The eigenvalues of τ are 1 and -1, and the latter is spanned by the function f of (4.1.81). Therefore, this space is one-dimensional and mapped to itself by Δ, and from (4.1.84) we then see that its eigenvalue for Δ is 1. The eigenspace of τ for the eigenvalue 1 consists of functions with $f(v_1) = f(v_2) = 0$, and that space then is spanned by the other eigenfunctions of Δ.

For another example, let τ again be as in (4.1.83), but assume now that v_1 and $v_2 = \tau(v_1)$ are connected by an edge. Then, the eigenvalue equation for f as in (4.1.81) at v_1 becomes

$$f(v_1) - \frac{1}{\deg v_1} f(v_2) = \lambda f(v_1),$$

that is, since $f(v_2) = -f(v_1)$,

$$\lambda = 1 + \frac{1}{\deg v_1}. \qquad (4.1.85)$$

Let $e = (v_1, v_2)$ and $e' = (v'_1, v'_2)$ be duplicate edges, that is, $v \sim v_j$ if and only if $v \sim v'_j$ for $j = 1, 2$ and $v \neq v_1, v_2, v'_1, v'_2$ and $v_j \nsim v'_k$ for $j, k = 1, 2$. Then, we can determine two eigenvalues and eigenfunctions of Δ by solving

$$f(v_1) - \frac{1}{\deg v_1} f(v_2) = \lambda f(v_1)$$

$$f(v_2) - \frac{1}{\deg v_2} f(v_1) = \lambda f(v_2)$$

analogously for v'_1, v'_2

$$f(v) = 0 \quad \text{for all other } v.$$

Figure 4.4 An example of edge duplication.

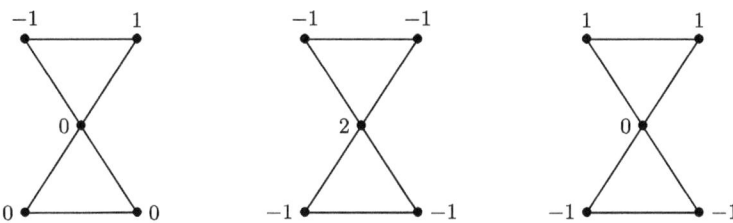

Figure 4.5 Eigenfunctions of the graph of Figure 4.4.

This yields

$$\lambda = 1 \pm \frac{1}{\sqrt{\deg v_1 \deg v_2}}, \tag{4.1.86}$$

see [19].

Let us determine the spectra of some further classes of graphs. We first look at the graph in Figure 4.4, which is obtained by duplicating an edge of a triangle. According to (4.1.86), this generates the eigenvalues $1/2$ and $3/2$. And according to (4.1.85), two eigenvalues $3/2$ are generated by the symmetries of switching the two vertices inside a petal, but the two classes are not disjoint, and so, this construction only yields two eigenvalues $3/2$ altogether. In fact, all eigenfunctions arising from these constructions vanish at the central vertex. But, a further eigenvalue $3/2$ arises from the symmetry exchanging the two triangles, with the central vertex being a fixed point, and this yields an eigenfunction that is nonzero at the center. Thus, we have determined all the eigenvalues of this graph. Let us indicate the corresponding eigenfunctions in Figure 4.5.

The first produces the eigenvalue $3/2$ with multiplicity 2, the second again yields the eigenvalue $3/2$, so that it has multiplicity 3 altogether, while the last one yields the eigenvalue $1/2$. And of course, we have the eigenvalue 0 with multiplicity 1. In passing, we observe that for this graph, the inequality (4.1.70) for the largest eigenvalue becomes an inequality. We can view this graph as two copies of K_3 joined at a single vertex.

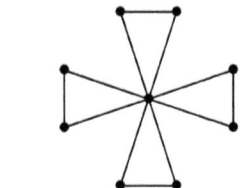

Figure 4.6 The petal graph P_4.

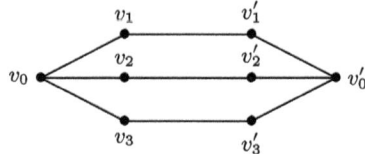

Figure 4.7 A book graph.

We can then construct the family of graphs P_ℓ, called the *petal graphs*, where ℓ triangles are joined at a single central vertex v_0. In Figure 4.4, we see the case $\ell = 2$, and in Figure 4.6, $\ell = 4$.

Iterating the above construction, or alternatively, applying (4.1.85) and (4.1.86) repeatedly, we obtain the eigenvalue $3/2$ with multiplicity $\ell + 1$, $1/2$ with multiplicity $\ell - 1$, and 0 with multiplicity 1. Since P_ℓ has $2\ell + 1$ vertices, this is the entire spectrum.

In particular, for petal graphs with more than two petals, like for complete graphs with more than two vertices, neither the smallest nonzero nor the largest eigenvalue is simple.

Petal graphs have an odd number of vertices. There also is a family with the same spectral properties but an even number of vertices. These are the *book graphs*. They are obtained by joining two star graphs. One star graph has a central vertex v_0 connected to m peripheral vertices $v_1, \ldots, v_m$, which are not connected to each other. The other star similarly has a central vertex v_0' connected to m peripheral vertices $v_1', \ldots, v_m'$. We then join them by connecting v_i to v_i' for $i = 1, \ldots, m$, as shown in Figure 4.7.

For the book graph on $N = 2m + 2$ nodes, 0 and 2 are eigenvalues with multiplicity 1, since Γ is connected and bipartite. Moreover, since the book graph is obtained by the m-fold duplication of the edge (v_1, v_1') in the line graph with vertices v_0, v_1, v_1', and v_0', by (4.1.86), and again, we have a symmetry, this time without a fixed point, obtained by exchanging the two stars, and so, similar to the petal graph, $\lambda = 1/2$ and $\lambda = 3/2$ are eigenvalues with multiplicity m each. In fact, the corresponding eigenfunctions can be constructed as follows:

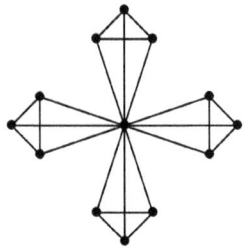

Figure 4.8 A windmill graph.

1. By letting $\sum_{i=1}^{m} f(v_i) = 0$, $f(v_0) = f(v_0') = 0$, $f(v_i') = \mp f(v_i)$, we obtain $m - 1$ linearly independent eigenfunctions;
2. By letting $f(v_i) = -f(v_i') = 1$ for $i = 1, \ldots, m$ and $f(v_0) = -f(v_0') = 2$, we obtain one more eigenfunction for $1/2$ and similarly by letting $g(v_i) = g(v_i') = -1$ for $i = 1, \ldots, m$ and $g(v_0) = g(v_0') = 2$, we obtain one more eigenfunction for $3/2$.

Hence, also in this case, we have a spectral gap of size 1 about 1.

In fact, the spectral gap at 1 here is the largest possible; we have

Theorem 4.1.2 ([158]). *For any connected graph Γ on $N \geq 3$ nodes,*

$$\min_i |1 - \lambda_i| \leq \frac{1}{2}. \qquad (4.1.87)$$

Moreover, equality is achieved if and only if Γ is either a petal graph (for N odd) or a book graph (for N even).

Remark 4.1.8. Petal graphs are also known as *friendship graphs*. The famous Friendship Theorem [90] states that the only finite graphs with the property that every two vertices have exactly one neighbor in common are precisely the petal graphs. In [62], it has been proved that, among connected graphs, the petal graphs are uniquely determined by the eigenvalues of the adjacency matrix. And, Theorem 4.1.2 implies that both petal graphs and book graphs are characterized by their normalized Laplacian spectra.

In fact, the star and the petal graphs belong to a larger family, that of the *windmill graphs* $Wd(n, \ell)$ obtained by taking ℓ copies of the complete graph K_{n-1} and then adding a single central vertex connected to all others. Equivalently, we take ℓ copies of K_n and join all of them at a single vertex. Thus, $Wd(n, \ell)$ has $\ell(n - 1) + 1$ vertices. The star graphs S_ℓ then are $Wd(2, \ell)$ and the petals P_ℓ are $Wd(3, \ell)$. Figure 4.8 shows $Wd(4, 4)$: Again, the spectra of the windmill graphs are easily determined. We have the eigenvalue 0, as always.

Figure 4.9 An eigenfunction on the bistar $S_{3,3}$.

The eigenvalue $\frac{n}{n-1}$ occurs with multiplicity $\ell(n-2)+1$ and $\frac{1}{n-1}$ with multiplicity $\ell - 1$. We get $\ell(n-2)$ for $\frac{n}{n-1}$ by taking an eigenfunction for that eigenvalue on one of the K_n that vanishes at the central vertex. The remaining eigenfunction for that eigenvalue is obtained by taking the same eigenfunction on all the K_n that has the same value b on all noncentral vertices. And if it then has the value a at the remaining vertex, we assign the value ℓa to the central vertex. And for the eigenvalue $\frac{1}{n-1}$, we take a function with the value $+1$ on all noncentral vertices of one wing, -1 on those of another wing, and 0 everywhere else, including the central vertex. While we have just constructed the eigenvalues and eigenfunctions explicitly, of course they can be derived from general symmetry principles.

We observe that all windmill graphs have only two nonvanishing eigenvalues, and the gap between them is 1.

We next consider the *symmetric bistar* $S_{q,q}$ where to each vertex of a central edge, q other vertices are connected. Figure 4.9 shows $S_{3,3}$ and indicates an eigenfunction.

The corresponding eigenfunction and eigenvalue λ need to satisfy

$$a - 1 = \lambda a,$$

$$1 + \frac{1}{q+1} - \frac{q}{q+1} a = \lambda,$$

whence

$$a_\pm = \frac{q}{2} \pm \sqrt{\frac{1}{4q^2} + \frac{q+1}{q}},$$

$$\lambda_\pm = \frac{a_\pm - 1}{a_\pm}.$$

Thus, $a_- = -1$, leading to the eigenvalue $\lambda = 2$, reflecting the bipartiteness of $S_{q,q}$, whereas $a_+ = \frac{q+1}{q}$, leading to $\lambda = \frac{1}{q+1}$, which for $q = 1$ is $= 1/2$, but which goes to 0 for $q \to \infty$.

In Figure 4.10, we seek an eigenfunction of the form
which leads to

$$-b - 1 = \lambda(-b),$$

$$1 - \frac{1}{q+1} + \frac{q}{q+1} b = \lambda.$$

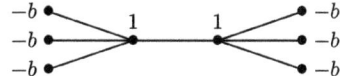

Figure 4.10 Another eigenfunction on $S_{3,3}$.

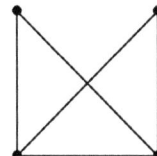

Figure 4.11 Duplicating a vertex of K_3.

The solution $b = -1$ leads to a constant eigenfunction for the eigenvalue 0, but $b = \frac{q+1}{q}$ gives the eigenvalue $\frac{2q+1}{q+1}$, which tends to 2 for $q \to \infty$. Since $S_{q,q}$ is obtained from $S_{1,1}$ by repeated node duplication, the remaining eigenvalue is 1 with multiplicity $2q - 2$, with eigenfunctions as above. Thus, we have obtained all eigenvalues and a basis of the eigenfunctions for $S_{q,q}$. Note that all the eigenfunctions are either symmetric or antisymmetric for some automorphism of $S_{q,q}$. This, in fact, can be turned into a general principle.

Duplicating a node produces the eigenvalue 1 but may change some of the other eigenvalues of the graph we started with. Thus, let us duplicate one of the nodes of K_3 to obtain the graph of Figure 4.11.

This has the eigenvalue 0, the eigenvalue 1 because the two upper nodes are duplicates, the eigenvalue 4/3 with an eigenfunction that is ± 1 on the two lower and 0 on the two upper nodes, and finally the eigenvalue 5/3, as the sum of the eigenvalues is 4 by (4.1.41). Another reason is that this graph is K_4 minus one edge (between the two upper vertices), which is a case for which Theorem 4.1.1 is sharp. The graph K_3 that we started with had the eigenvalue 3/2 with multiplicity 2.

When, however, as in Figure 4.12, we duplicate all three vertices of K_3 to obtain the graph we have the eigenvalue 1 with multiplicity 3 because of the three vertex duplications, and the eigenvalue 3/2 of K_3 with multiplicity 2, with one eigenfunction indicated in Figure 4.12, obtained from the corresponding eigenfunction on K_3. The same mechanism works to show more generally

Corollary 4.1.13. *Let Γ be obtained by duplicating all the vertices of some graph Γ_0. Then Γ retains all the eigenvalues of Γ_0 and in addition has the eigenvalue 1 with multiplicity given by the number of vertices of Γ_0.*   □

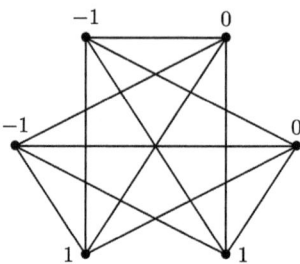

Figure 4.12 Duplicating all three vertices of K_3, and an eigenfunction.

The graph we have just constructed is the complete *tripartite graph* $K_{2,2,2}$. In a general tripartite graph, the vertex set V consists of three disjoint vertex sets V_1, V_2, and V_3 without any connections between members of the same V_i, for $i = 1, 2, 3$. The complete tripartite graph K_{n_1,n_2,n_3} has three such vertex sets with V_i containing n_i vertices, and for which any vertex in some V_i is connected to all vertices in the other two vertex sets $V_j, j \neq i$. It is obtained by repeated vertex duplications from $K_3 = K_{1,1,1}$, and the above argument shows that K_{n_1,n_2,n_3} has the eigenvalue 1 with multiplicity $N - 3$, where $N = n_1 + n_2 + n_3$. (By Lemma 4.1.14, this also follows from [57].) And it also follows from our considerations, recalling (4.1.41), that the sum of the other two nontrivial eigenvalues is 3, and hence, since none of them is 2, as the graph is not bipartite, they both are larger than 1. When $n_1 = n_2 = n_3$, they are both $3/2$, but for other values, they may be different, as our examples show. And again, complete tripartite graphs are characterized by their spectrum.

Of course, these notions and results permit an obvious generalization when we duplicate the vertices of K_k.

Corollary 4.1.14. *A complete k-partite graph $K_{n_1,\ldots,n_k}$ has the eigenvalue 1 with multiplicity $N - k$ (where $N = \sum n_j$). The sum of the remaining $k - 1$ nonvanishing eigenvalues is k, and when $n_1 = \ldots = n_k$, they are all equal to $\frac{k}{k-1}$.*   □

The preceding considerations are of course special cases of the general scheme established in Section 2.7.

We also recall that since Γ is connected, the trivial eigenvalue $\lambda_1 = 0$ is simple. If Γ had two components, then the next eigenvalue λ_2 would also become 0. A corresponding eigenfunction would be equal to a constant on each component, the two values chosen such (4.1.64) is satisfied; in particular, one of the two would be positive, the other one, negative. We therefore expect that for graphs with a pronounced community structure, that is, for ones that can be broken up into two large components by deleting only a few edges as discussed above, the eigenvalue λ_2 should be close to 0. Formally, this is easily seen from the variational characterization

$$\lambda_2 = \min \left\{ \frac{\sum_{e=(v,v')\in E} (u(v') - u(v))^2}{\sum_v \deg v \, u(v)^2} : \sum_v \deg v \, u(v) = 0 \right\} \qquad (4.1.88)$$

(see (2.6.8) and observe that $\sum_v \deg v \, u(v) = 0$ is equivalent to $(u, u_1) = 0$ as the eigenfunction u_1 is constant). Namely, if two large components of Γ are only connected by few edges, then one can make u constant on either side, with opposite signs so as to respect the normalization (4.1.64) with only a small contribution from the numerator.

More generally, when Γ consists of several clusters with only very few connections between them, one should find several eigenvalues close to 0.

The strategy for obtaining an eigenfunction for the second eigenvalue λ_2 is, according to (4.1.88), to do the same as one's neighbors. Because of the constraint $\sum_v \deg v \, u(v) = 0$, this is not globally possible, however. The second eigenfunction thus exhibits oscillations with the lowest possible frequency. Thus, if we take such a second eigenfunction u_2 and consider the connected components that remain after deleting all edges at whose endpoints u_2 has different signs, then there are precisely two such components, one on which u_2 is positive and one on which it is negative.

This issue will be taken up very systematically in Chapter 8, but in Section 4.3, we shall offer a preliminary elementary treatment.

4.2 Some Elementary Estimates

In Lemma 4.1.12, we had seen that the largest eigenvalue λ_N is 2 when the graph is bipartite and smaller when it is not. Therefore, we may naturally ask in which sense $2 - \lambda_N$ quantifies how different the graph is from being bipartite. In order to develop some intuition about this question, let us consider the following example. Let Γ_0 be a bipartite graph with N vertices. We consider a highest eigenfunction $\bar{u}$ that is $+1$ on one class and -1 on the other class of vertices, as described in Lemma 4.1.12. In particular by (4.1.47),

$$\frac{\frac{1}{2} \sum_{v_1, v_2 : v_1 \sim v_2} (\bar{u}(v_1) - \bar{u}(v_2))^2}{\sum_v \deg v \, \bar{u}(v)^2} = 2. \qquad (4.2.1)$$

By adding another vertex v_0 and connecting it to one of the vertices v_1 of Γ_0, we obtain a new bipartite graph Γ_1. We extend $\bar{u}$ by $\bar{u}(v_0) = 0$ to Γ_1. Thus, the numerator and the denominator of (4.2.1) are both increased by 1.

Let the volume $\sum_v \deg v$ of Γ_0 be sufficiently large. Then, on Γ_1 for any small $\varepsilon > 0$,

$$\frac{\frac{1}{2} \sum_{v_1, v_2 : \, v_1 \sim v_2} (\bar{u}(v_1) - \bar{u}(v_2))^2}{\sum_v \deg v \, \bar{u}(v)^2} > 2 - \varepsilon. \tag{4.2.2}$$

This is not affected when we construct a graph Γ by attaching another graph Γ_2 at v_0 and extend $\bar{u}$ by 0 to all of Γ_2. For instance, Γ_2 could be a complete graph K_M with M vertices, for any M. In particular, the difference $2 - \lambda_N$, which by our construction is smaller than ε, is not very sensitive to the global shape of Γ_2. This implies, for instance, that $2 - \lambda_N$ cannot control a global quantity like the *clustering coefficient* C, defined as

$$C := \frac{3 \times \text{ number of triangles}}{\text{number of connected triples of vertices}}, \tag{4.2.3}$$

where a *triangle* is a triple of mutually connected vertices. The clustering coefficient measures how many connections there exist between the neighbors of a node. C becomes maximal if Γ is a fully connected graph. In contrast, C vanishes when Γ is a bipartite graph. In that sense, the clustering coefficient expresses an averaged difference from a graph being bipartite.

Our construction of attaching a complete graph K_M to a bipartite graph Γ_0 through a connecting node produces a graph with C arbitrarily close to its maximal value 1 when M is sufficiently large. But since, by our construction, $2 - \lambda_N < \varepsilon$ is arbitrarily small, the largest eigenvalue does not control the clustering coefficient.

But we can use Corollary 4.1.6 to control $2 - \lambda_N$ by a local clustering measure. We let Δ be the collection of triangles in Γ, that is, triples (v_1, v_2, v_3) of vertices for which $v_i \sim v_j$ for $i \neq j = 1, 2, 3$. For an edge $e = (v, v')$, we then define

$$c(e) = c(v, v') = \sum_{v'' : \, (v, v', v'') \in \Delta} \frac{1}{\deg v''} \tag{4.2.4}$$

and put

$$c(\Gamma) := \min_{e \in E} c(e). \tag{4.2.5}$$

We then have

Theorem 4.2.1. *The largest eigenvalue λ_N of the graph Γ satisfies*

$$2 - \lambda_N \geq c(\Gamma). \tag{4.2.6}$$

Proof We use (4.1.55), that is,

$$2 - \lambda_N = \frac{\sum_z \frac{1}{\deg z} \sum_{z', z'' : \, (z, z') \in E, (z, z'') \in E} (u(z') - u(z''))^2}{\sum_{v, v' : \, (v, v') \in E} (u(v) - u(v'))^2}$$

where u is an eigenfunction for the eigenvalue λ_N. For every edge $(v, v') \in E$, the term $(u(v) - u(v'))^2$ in the denominator also appears in the numerator for every triangle $(v, v', z) \in \triangle$, with $z' = v$, $z'' = v'$. This yields the estimate (4.2.6).   □

For the graph K_3, that is, a single triangle, we have $c(e) = 1/2$ for every edge, and the highest eigenvalue is $3/2$, and so, (4.2.6) is sharp in this case.

A more refined estimate can be found in [24].

With the scheme developed in this section, the control in the other direction, that is, estimating the largest eigenvalue from below, does not quite work because of the following example. Consider a graph with many cycles of odd length, but all of length at least 5. Here, $c(e) = 0$ for every edge as there are no triangles, but $2 - \lambda_{N-1} \neq 0$ because the graph is not bipartite as bipartite graphs can only have cycles of even length.

However, we can control the largest eigenvalue from below in a different way, as we will see in Section 4.4.

In contrast to the largest eigenvalue λ_N, the smallest nonvanishing eigenvalue λ_2 cannot be controlled by the minimum of such a purely local constant. We also need to invoke some global quantities. We recall that the *volume* of Γ is defined as

$$\text{vol } \Gamma = \sum_v \deg v. \tag{4.2.7}$$

Definition 4.2.1. The *diameter* $d(\Gamma)$ of Γ is the maximal distance between two vertices of Γ. Here, the *distance* between two vertices is the minimal number of edges needed to connect them.

For instance, a complete graph has diameter 1, whereas a chain with $m + 1$ vertices has diameter m.

For an edge $e = (v, v')$, let

$$b(e) := \text{number of triangles containing } e, \tag{4.2.8}$$

and

$$b(\Gamma) := \min_{e \in E} b(e). \tag{4.2.9}$$

Then we have refining Lemma 1.9 in [61]

Theorem 4.2.2.

$$\lambda_2 \geq \frac{1 + \frac{1}{2}b(\Gamma)}{d(\Gamma)\text{vol } \Gamma}. \tag{4.2.10}$$

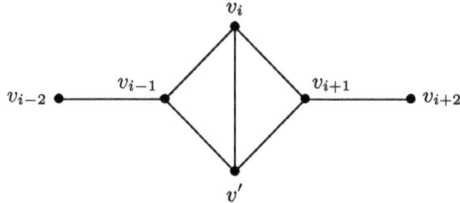

Figure 4.13 A graph used in the proof of Theorem 4.2.2.

Proof For $x_0, x_1, \ldots, x_m \in \mathbb{R}$, we have the inequality

$$\sum_{i=1}^{m}(x_{i-1} - x_i)^2 \geq \frac{1}{m}(x_0 - x_m)^2. \tag{4.2.11}$$

We let u be an eigenfunction for λ_2 and choose a vertex v_0 where $|u(v)|$ assumes its maximum. Since $\sum_v \deg v\, u(v) = 0$ for such an eigenfunction, there is some other vertex v_m with $u(v_0)u(v_m) < 0$, and let us assume that the distance between v_0 and v_m is m, with a path $v_0 \sim v_1 \ldots \sim v_m$ connecting them. We recall

$$\lambda_2 = \frac{\sum_{v \sim v'}(u(v) - u(v'))^2}{\sum_v \deg v\, u(v)^2}.$$

For an estimate from below, we only count some of the terms $(u(v) - u(v'))^2$ with $v \sim v'$. These include, first of all, the terms $(u(v_{i-1}) - u(v_i))^2$ for $i = 1, \ldots, m$. Next, we count the terms $(u(v_{i-1}) - u(v'))^2$ and $(u(v') - u(v_i))^2$ for triangles (v_{i-1}, v_i, v'). Here, we have to distinguish two cases.

1. v' is not contained in any triangle with another edge (v_{j-1}, v_j) from the path. We then use (4.2.11) to get

$$(u(v_{i-1}) - u(v'))^2 + ((u(v') - u(v_i))^2) \geq \frac{1}{2}(u(v_{i-1}) - u(v_i))^2.$$

2. v' is a vertex of a triangle for two adjacent edges, as indicated in Figure 4.13. We note that since $v_0, v_1, \ldots, v_m$ is a shortest path, it is not possible that more than two of its edges or two non-adjacent edges are contained in some triangles with the same third vertex. In the situation of Figure 4.13, we may assume

$$(u(v_{i-1})-u(v'))^2+(u(v')-u(v_{i+1}))^2) \geq (u(v_{i-1})-u(v_i))^2+(u(v_i)-u(v_{i+1}))^2),$$

since replacing v_i by v' also yields a shortest path.

We therefore get

$$\lambda_2 \geq \frac{(1 + \frac{1}{2}b(\Gamma)) \sum_{i=1,\ldots,m}((u(v_{i-1}) - u(v_i))^2}{\text{vol}\,\Gamma\, u^2(v_0)}$$

$$\geq \frac{\frac{1}{d(\Gamma)}(1 + \frac{1}{2}b(\Gamma))(u(v_0) - u(v_m))^2}{\text{vol}\,\Gamma\, u^2(v_0)}$$

$$\geq \frac{1 + \frac{1}{2}b(\Gamma)}{d(\Gamma)\text{vol}\,\Gamma},$$

where we have used (4.2.11) again in the second-to-last step. □

We note that, in contrast to (4.2.6), the estimate (4.2.10) is not sharp for K_3.

4.3 Cutting a Graph Optimally, or the Pólya–Cheeger Constant and Eigenvalue Estimates

We shall now take up the discussion at the end of Section 4.2 and introduce a quantity h that will control the first nontrivial eigenvalue λ_2 of a graph. This quantity was introduced by Pólya and Szego in graph theory in 1951, see [230]. An analogous quantity was defined in Riemannian geometry by Cheeger in 1970 [56], and his important insight was that this quantity could be used to estimate the first non-vanishing eigenvalue of the Laplace–Beltrami operator of the underlying Riemannian manifold. Dodziuk [82] and Alon-Milman [4] then derived an analogous estimate for the algebraic graph Laplacian, and Chung [61] carried this over to the normalized Laplacian that we are using here. The relevant constant on a graph therefore is usually also called the Cheeger constant.

Let us first recall the general setting from Section 3.2. We want to look at isoperimetric problems. In Euclidean space, for some (sufficiently regular) set S, one compares the volume $|S|$ of S with that of its boundary, $|\partial S|$. For instance, for some fixed volume $|S|$, one may want to minimize $|\partial S|$, or conversely, for fixed $|\partial S|$, one wants to maximize $|S|$. The solution in either case is given by a ball, and the boundary is a sphere. One can pose the same problem in a Riemannian manifold and show the existence of a solution. One can also try to divide a compact Riemannian manifold M into two large pieces S and $M \setminus S$ with small $|\partial S|(= |\partial(M \setminus S)|)$, that is, try to minimize

$$\frac{|\partial S|}{\min(|S|, |M \setminus S|)}. \tag{4.3.1}$$

Here, we want to examine analogous problems on graphs. For a subset $A \subset V$ of the vertex set of a graph Γ, we want to control the measure of its boundary $|\partial A|$. For any $B \subset V$, its measure or volume is, as we recall, $\mathrm{vol}\, B = \sum_{v \in B} \deg v$. But a question that arises here is how to define that boundary. In fact, there are two options, a vertex boundary and an edge boundary. For the vertex boundary, we consider pairs v, v' with $v \sim v'$, $v \in A$, $v' \notin A$, and we may then define the *vertex boundary* as either the set of all those v or as the set of all those v'. Thus, there are two possible choices, an external and an internal vertex boundary, and in general, their volumes may be quite different from each other. We shall encounter them at the end of Section 6.3. When we, instead, define the *edge boundary*, there is a canonical choice. The edge boundary ∂A simply consists of the set $E(A, V \setminus A)$ of all edges $e = (v, v')$ with $v \in A$, $v' \notin A$, and the measure of any subset $E_0 \subset E$ of the edge set of Γ is simply the number of edges it contains. (We are discussing here unweighted graphs; in the weighted case, one would take the sum of the weights of the edges in E_0.) The problem analogous to (4.3.1) then is to minimize

$$\frac{|\partial A|}{\min(|A|, |M \setminus A|)}. \tag{4.3.2}$$

This is called the *Cheeger problem*, in analogy with the problem (4.3.1), which is the problem that Cheeger himself [56] actually studied by relating it to the first nonzero Laplacian eigenvalue. (For the minimization problem (4.3.1) itself and its history, see for instance [110]. And the problem of minimizing (4.3.2) can already be found in [230].)

In this section, we shall treat the relation between (4.3.2) and the eigenvalue λ_2. Let us first prepare the setting.

Let $\phi \colon V \to \mathbb{R}_{\geq 0}$ be a function, with $M := \max \phi(v)^2$. We put, for $t \geq 0$,

$$\Phi_v(t) := \begin{cases} 1 & \text{if } \phi(v) \geq t \\ 0 & \text{else.} \end{cases} \tag{4.3.3}$$

Let $V_\phi^+(\tau) := \{v \in V : \phi(v) \geq \tau\}$. Then, for every $t \geq 0$,

$$\sum_{u \sim v} |\Phi_u(\sqrt{t}) - \Phi_v(\sqrt{t})| = |E(V_\phi^+(\sqrt{t}), \overline{V_\phi^+(\sqrt{t})})|, \tag{4.3.4}$$

where $\overline{S} = V \setminus S$ is the complement of a subset $S \subset V$ and where $|E(V_1, V_2)|$ denotes the number of edges with one endpoint in V_1 and the other in V_2, for $V_1, V_2 \subset V$. Thus,

$$\int_0^M \sum_{u \sim v} |\Phi_u(\sqrt{t}) - \Phi_v(\sqrt{t})| dt = \int_0^M \left| E\left(V_\phi^+(\sqrt{t}), \overline{V_\phi^+(\sqrt{t})}\right) \right| dt$$

$$\geq \inf_t \frac{|E(V_\phi^+(\sqrt{t})), \overline{V_\phi^+(\sqrt{t})}|}{\sum_{v:\,\phi(v)\geq\sqrt{t}} \deg(v)} \int_0^M \sum_{v:\,\phi(v)\geq\sqrt{\tau}} \deg(v)d\tau$$

$$= \inf_t \frac{|E(V_\phi^+(\sqrt{t})), \overline{V_\phi^+(\sqrt{t})}|}{\sum_{v:\,\phi(v)\geq\sqrt{t}} \deg(v)} \sum_v \deg(v)\phi(v)^2, \tag{4.3.5}$$

where the last equality follows from the fact that

$$\int \sum_{v:\,\phi(v)\geq\sqrt{\tau}} \deg(v)d\tau = \sum_v \deg(v) \int_{\tau:\,0\leq\tau\leq\phi(v)^2} 1d\tau = \sum_v \deg(v)\phi(v)^2.$$

We put, for $\emptyset \neq S \subsetneq V$,

$$\eta_1(S) := \frac{|E(S,\overline{S})|}{\operatorname{vol}(S)}. \tag{4.3.6}$$

For purposes that will become clear in Section 4.6, we shall first derive a trivial toy version of the estimate that we are after. We introduce the constant

$$h_1(\Gamma) := \min_S \eta_1(S). \tag{4.3.7}$$

Of course, $h_1 = 0$, but what will be relevant will be the structure of the argument to follow. From (4.3.5), we then obtain

$$\int_0^M \sum_{u\sim v} |\Phi_u(\sqrt{t}) - \Phi_v(\sqrt{t})|dt \geq h_1 \sum_v \deg(v)\phi(v)^2. \tag{4.3.8}$$

Now, fix $v \sim u$ and assume that $\phi(v) \leq \phi(u)$. Then, $\left|\Phi_u(\sqrt{t}) - \Phi_v(\sqrt{t})\right| = 1$ if and only if $\phi(v) < \sqrt{t} \leq \phi(u)$, while $\left|\Phi_u(\sqrt{t}) - \Phi_v(\sqrt{t})\right| = 0$ otherwise. Hence,

$$\int_0^M \left|\Phi_u(\sqrt{t}) - \Phi_v(\sqrt{t})\right| dt = \int_{t\in[0,M]:\,\phi(v)<\sqrt{t}\leq\phi(u)} 1dt$$
$$= \phi(u)^2 - \phi(v)^2$$
$$= (|\phi(u) - \phi(v)|)(|\phi(u) + \phi(v)|). \tag{4.3.9}$$

From (4.3.8) and (4.3.9), we obtain

$$h_1 \leq \frac{\sum_{u\sim v} |\phi(u) - \phi(v)|(|\phi(u)| + |\phi(v)|)}{\sum_v \deg(v)\phi(v)^2}$$
$$\leq \frac{\sqrt{\sum_{u\sim v} |\phi(u) - \phi(v)|^2} \sqrt{\sum_{u\sim v}(|\phi(u)| + |\phi(v)|)^2}}{\sum_v \deg(v)\phi(v)^2}$$
$$\leq \frac{\sqrt{\sum_{u\sim v} |\phi(u) - \phi(v)|^2} \sqrt{2\sum_u \deg(u)|\phi(u)|^2}}{\sum_v \deg(v)\phi(v)^2}.$$

Hence,

$$h_1 \leq \sqrt{\frac{2 \sum_{u \sim v} |\phi(u) - \phi(v)|^2}{\sum_v \deg(v)\phi(v)^2}},$$

for all $\phi: V \to \mathbb{R}_{\geq 0}$. It follows that, for all (not necessarily non-negative) functions $\phi: V \to \mathbb{R}$,

$$h_1 \leq \sqrt{\frac{2 \sum_{u \sim v} (|\phi(u)| - |\phi(v)|)^2}{\sum_v \deg(v)\phi(v)^2}} \leq \sqrt{\frac{2 \sum_{u \sim v} |\phi(u) - \phi(v)|^2}{\sum_v \deg(v)\phi(v)^2}}. \qquad (4.3.10)$$

Of course, we can take here a constant function ϕ and get $h_1 = 0$, which in fact can also be readily seen from its definition (4.3.7). The important thing, however, is that inside the root on the right-hand side, we have the Rayleigh quotient of the function ϕ, and subsequently, we shall insert eigenfunctions of Δ into such inequalities to control eigenvalues in terms of such global quantities as h_1, or conversely, estimate those quantities by eigenvalues.

With this argument in place, we now go for the real thing. We consider a subset S of the vertex set V and its complement $\bar{S} = V \setminus S$. For any subsets $V_1, V_2 \subset V$, $|E(V_1, V_2)|$ denotes the number of edges with one endpoint in V_1 and the other in V_2. Since the conventions here can be confusing, let us point out explicitly that each edge with $v_1 \in V_1$ and $v_2 \in V_2$ is counted only once. (This is different from the convention in [11].) In particular, $|E(V_1, V_1)|$ is the number of edges with both endpoints in V_1. Thus, for $S \subset V$, we have

$$\mathrm{vol}\, S = 2|E(S,S)| + |E(S,\bar{S})|. \qquad (4.3.11)$$

(We consider here again the unweighted case, but everything naturally extends to weighted graphs. On a weighted graph, of course, the edges carry their weights, and $|E(V_1, V_2)|$ then becomes the sum of the weights of the edges between V_1 and V_2.)

We then put

$$\eta(S) := \frac{|E(S,\bar{S})|}{\min(\mathrm{vol}(S), \mathrm{vol}(\bar{S}))} = \max(\eta_1(S), \eta_1(\bar{S})), \qquad (4.3.12)$$

with $\eta_1(S) = \frac{|E(S,\bar{S})|}{\mathrm{vol}(S)}$, and introduce the *(Pólya)–Cheeger constant*

$$h := \min_S \eta(S). \qquad (4.3.13)$$

Thus, we have the minimization problem of cutting V into two parts $S, \bar{S}$ by removing as few edges as possible, but such that both parts have a large volume, since the numerator in (4.3.12) is symmetric in $S, \bar{S}$. In particular, when Γ is not connected, then $h = 0$, because we can let S be one component, and then, there

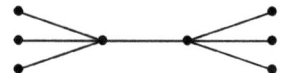

Figure 4.14 The bistar of Figure 4.9.

are no edges to its complement $\overline{S}$, that is, $|E(S, \overline{S})| = 0$. When Γ is connected, however, this is not possible, and we have $h > 0$ in that case.

In the example of the bistar $S_{q,q}$ of Figure 4.14, it is obvious that it is best to cut the central edge. This disconnects the graph into two pieces $S, \overline{S}$ of equal volume $2q + 1$ (in each piece, there are q vertices with a single neighbor, a central vertex, and that central vertex has $q + 1$ neighbors). Thus,

$$h = \frac{1}{2q + 1}. \tag{4.3.14}$$

Thus, the larger q, the smaller the Pólya–Cheeger constant.

The star graph $K_{1,n}$ becomes disconnected by cutting any edge, thereby cutting off a single peripheral vertex. Then, $h = 1$. This does not depend on n. In fact, the star graph cannot be cut into two large connected components at all. Whichever way we cut it, there is at most one large component. The rest will just be a collection of isolated vertices. Similarly for the petal graph where again the second largest component after a cut contains at most two vertices.

We next consider the complete graph K_N and cut it into two components. The smaller then has $n \leq N/2$ vertices and volume $n(N - 1)$. We need to cut $n(N - n)$ edges, and the ratio $\frac{n(N-n)}{n(N-1)} = \frac{N-n}{N-1}$ becomes smallest for $n = N/2$ when N is even, and $n = (N - 1)/2$ for odd N. This then is the value of h in this case.

The fundamental inequality obtained in [4, 61, 82] then says that the constant h controls the eigenvalue λ_2 from above and below, or conversely and better, that this eigenvalue can be used to estimate h, as λ_2 in practice is much easier to compute than h. The essential idea is that the eigenfunction for λ_2 divides V into two parts where it is positive or negative, and that this yields a decomposition that realizes a value for η, which is not too much larger than the minimal value required in the definition of h.

Theorem 4.3.1.

$$1 - \sqrt{1 - h^2} \leq \lambda_2 \leq 2h. \tag{4.3.15}$$

We note that (4.3.15) implies the slightly weaker lower bound

$$\frac{1}{2} h^2 \leq \lambda_2, \tag{4.3.16}$$

which is the bound often quoted in the literature and which is also the type of bound the argument leading to (4.3.10) gives.

In fact, the upper estimate in (4.3.15) holds under rather general conditions, and an appropriate version is also true for the algebraic (non-normalized) graph Laplacian. In contrast, the lower bound crucially needs the normalization.

The *proof* that we shall provide is taken from [24].

We first prove the easier part, the upper bound. We use the variational characterization (4.1.88). Let the vertex set be divided into the two disjoint sets $U, \overline{U}$ of nodes, and let U be the one with the smaller volume. We consider a function u that is $=1$ on all the nodes in U and $= -\alpha$ for some positive α on $\overline{U}$ so that the normalization $\sum_{v \in V} u(v) \deg v = 0$ holds, that is, $\sum_{v \in U} \deg v - \sum_{v \in \overline{U}} \deg v\, \alpha = 0$. Since $\overline{U}$ is the subset with the larger volume $\sum_{v \in \overline{U}} \deg v$, we have $\alpha \leq 1$. Thus, for our choice of u, the quotient in (4.1.88) becomes

$$\lambda_2 = \min \left\{ \frac{\sum_{e=(v,v') \in E} (u(v') - u(v))^2}{\sum_v \deg v\, u(v)^2} : \sum_v \deg v\, u(v) = 0 \right\}$$

$$\leq \frac{(1+\alpha)^2 |E(U,\overline{U})|}{\sum_{v \in U} \deg v + \sum_{v \in \overline{U}} \deg v\, \alpha^2}$$

$$= \frac{(1+\alpha)|E(U,\overline{U})|}{\sum_{v \in U} \deg v}$$

$$\leq 2\frac{|E(U,\overline{U})|}{\sum_{v \in U} \deg v}$$

$$= 2\frac{|E(U,\overline{U})|}{\mathrm{vol}(U)}.$$

Since this holds for all such splittings of our graph Γ, we obtain from (4.3.12) and (4.1.88) the upper bound

$$\lambda_2 \leq 2h. \tag{4.3.17}$$

For the lower bound, we shall use arguments that mainly come from [79] and also contain some generalizations from [61].

Remark 4.3.1. Given $g \colon V \to \mathbb{R}$, let

$$g_+(v) := \begin{cases} g(v), & \text{if } g(v) \geq 0 \\ 0, & \text{if } g(v) < 0 \end{cases}$$

be the *positive part of g*, and let

$$g_-(v) := \begin{cases} |g(v)|, & \text{if } g(v) \le 0 \\ 0, & \text{if } g(v) > 0 \end{cases}$$

be the *negative part of g*. Then, $g(v) = g_+(v) - g_-(v)$, $g(v)^2 = g_+(v)^2 + g_-(v)^2$, and

$$\sum_{v \sim w} \Big(g(v) - g(w)\Big)^2 \ge \sum_{v \sim w} \left[\Big(g_+(v) - g_+(w)\Big)^2 + \Big(g_-(v) - g_-(w)\Big)^2 \right].$$

Therefore, the Rayleigh quotient $\mathcal{R}(g)$ of g is such that

$$
\begin{aligned}
\mathcal{R}(g) &= \frac{\sum_{v \sim w} \Big(g(v) - g(w)\Big)^2}{\sum_{v \in V} \deg v \cdot g(v)^2} \\[2mm]
&\ge \frac{\sum_{v \sim w} \left[\Big(g_+(v) - g_+(w)\Big)^2 + \Big(g_-(v) - g_-(w)\Big)^2 \right]}{\sum_{v \in V} \deg v \cdot (g_+(v)^2 + g_-(v)^2)} \\[2mm]
&\ge \min\{\mathcal{R}(g_+), \mathcal{R}(g_-)\},
\end{aligned}
$$

since $\frac{a+b}{c+d} \ge \min\{\frac{a}{c}, \frac{b}{d}\}$.

Lemma 4.3.1. *Let $g \in L^2(\Gamma)$ with $S(g) := \{v \in V : g(v) > 0\} \ne \emptyset$, let*

$$h(g) := \min_{\emptyset \ne S \subseteq S(g)} \frac{|E(S, \overline{S})|}{\text{vol}(S)}$$

and let g_+ be the positive part of g. Then,

$$1 + \sqrt{1 - h^2(g)} \ge \frac{\sum_{e=(v,v')} (g_+(v) - g_+(v'))^2}{\sum_v \deg v \, g_+(v)^2} \ge 1 - \sqrt{1 - h^2(g)}.$$

The same holds if we consider g_- instead of g_+.

Proof First, we write

$$
\begin{aligned}
W &:= \frac{\sum_{e=(v,v')} (g_+(v) - g_+(v'))^2}{\sum_v \deg v \, g_+(v)^2} \\[2mm]
&= \frac{\sum_{e=(v,v')} (g_+(v) - g_+(v'))^2 \, \sum_{e=(v,v')} (g_+(v) + g_+(v'))^2}{\sum_v \deg v \, g_+(v)^2 \, \sum_{e=(v,v')} (g_+(v) + g_+(v'))^2} \\[2mm]
&=: \frac{I}{II}.
\end{aligned}
$$

With the Cauchy–Schwarz inequality,

$$I^{\frac{1}{2}} \ge \sum_{e=(v,v')} |g_+(v)^2 - g_+(v')^2|$$

$$= \sum_{e=(v,v'):\, g_+(v)\geq g_+(v')} (g_+(v)^2 - g_+(v')^2)$$

$$= 2 \sum_{e=(v,v'):\, g_+(v)\geq g_+(v')} \int_{g_+(v')}^{g_+(v)} t\,dt$$

$$= 2 \int_0^{\infty} \sum_{e=(v,v'):\, g_+(v')\leq t \leq g_+(v)} t\,dt.$$

Now,

$$\sum_{\substack{e=(v,v'):\\ g_+(v')\leq t \leq g_+(v)}} 1 = E(S_t, \overline{S_t}),$$

where $S_t := \{v:\, g_+(v) > t\}$. Thus,

$$I^{\frac{1}{2}} \geq 2h(g) \int_0^{\infty} \mathrm{vol}(S_t)\,t\,dt$$

$$= 2h(g) \int_0^{\infty} \sum_{v:\, g_+(v)>t} \deg v \; t\,dt$$

$$= 2h(g) \sum_{v\in V} \deg v \int_0^{g_+(v)} t\,dt$$

$$= h(g) \sum_v \deg v\, g_+(v)^2$$

and therefore we obtain

$$I \geq h^2(g) \left(\sum_v \deg v\, g_+(v)^2 \right)^2. \tag{4.3.18}$$

Next,

$$II = \sum_v \deg v\, g_+(v)^2 \sum_{e=(v,v')} (g_+(v) + g_+(v'))^2$$

$$= \sum_v \deg v\, g_+(v)^2 \left(\sum_v \deg v\, g_+(v)^2 + 2 \sum_{e=(v,v')} g_+(v)g_+(v') \right)$$

$$= \sum_v \deg v\, g_+(v)^2 \left(2 \sum_v \deg v\, g_+(v)^2 - \sum_{e=(v,v')} (g_+(v) - g_+(v'))^2 \right)$$

$$= (2 - W) \left(\sum_v \deg v\, g_+(v)^2 \right)^2. \tag{4.3.19}$$

From (4.3.18) and (4.3.19), we obtain

$$W \geq \frac{h^2(g)}{(2-W)},$$

and therefore,

$$1 + \sqrt{1 - h^2(g)} \geq W \geq 1 - \sqrt{1 - h^2(g)}.$$

The proof for g_- is analogous. $\square$

We can now conclude the *proof* of Theorem 4.3.1 as follows. Let f be an eigenfunction for λ_2. Then, $\sum_v \deg(v) f(v) = 0$. Thus, f attains both positive and negative values, and the function

$$
\begin{aligned}
p(t) :&= \sum_{v \in V} \deg v \, (f(v) - t)^2 \\
&= \sum_{v \in V} \deg v \left(f(v)^2 + t^2 - 2t f(v) \right) \\
&= \sum_{v \in V} \deg v f(v)^2 + t^2 \sum_{v \in V} \deg v - 2t \sum_{v \in V} \deg v f(v) \\
&= \sum_{v \in V} \deg v f(v)^2 + t^2 \sum_{v \in V} \deg v
\end{aligned}
$$

attains its minimum at $t = 0$.

Now, reorder the vertices so that

$$f(v_1) \leq \ldots \leq f(v_N).$$

Let k be the largest integer such that $\sum_{i=1}^k \deg(v_i) \leq \text{vol}(V)/2$, and let $c := f(v_{k+1})$. Then,

$$\text{vol}(\{v \in V : f(v) < c\}) \leq \sum_{i=1}^k \deg(v_i) \leq \frac{\text{vol}(V)}{2}$$

and, since k is the largest integer such that $\sum_{i=1}^k \deg(v_i) \leq \text{vol}(V)/2$, we have that $\sum_{i=1}^{k+1} \deg(v_i) > \text{vol}(V)/2$, therefore $\sum_{i=k+2}^N \deg(v_i) \leq \text{vol}(V)/2$, implying that

$$\text{vol}(\{v \in V : f(v) > c\}) \leq \sum_{i=k+2}^N \deg(v_i) \leq \frac{\text{vol}(V)}{2}.$$

Therefore,

$$\max\{\mathrm{vol}\,(\{v \in V: f(v) > c\}), \mathrm{vol}\,(\{v \in V: f(v) < c\})\} \le \frac{\mathrm{vol}\,(V)}{2}.$$

Hence, by letting

$$g(v) := f(v) - c \quad \text{for } v \in V,$$

we get that

$$\max\{\mathrm{vol}\,(\{v \in V: g(v) > 0\}), \mathrm{vol}\,(\{v \in V: g(v) < 0\})\} \le \frac{\mathrm{vol}\,(V)}{2}. \quad (4.3.20)$$

Now, since we observed that the function $p(t) = \sum_{v \in V} \deg v\,(f(v) - t)^2$ attains its minimum at $t = 0$, we have that

$$\sum_{v \in V} \deg v f(v)^2 \le \sum_{v \in V} \deg v g(v)^2.$$

Moreover, it follows from the definition of g that

$$\sum_{u \sim v} |f(u) - f(v)|^2 = \sum_{u \sim v} |g(u) - g(v)|^2.$$

This implies that

$$\mathcal{R}(f) = \frac{\sum_{u \sim v} |f(u) - f(v)|^2}{\sum_v \deg(v) f(v)^2} \ge \frac{\sum_{u \sim v} |g(u) - g(v)|^2}{\sum_v \deg(v) g(v)^2} = \mathcal{R}(g).$$

Now, let ϕ be either g_+ or g_-, such that (by Remark 4.3.1)

$$\mathcal{R}(g) \ge \min\{\mathcal{R}(g_+), \mathcal{R}(g_-)\} = \mathcal{R}(\phi).$$

Then, $\mathcal{R}(f) \ge \mathcal{R}(\phi)$. Moreover, let $S(g) := \{v \in V: g(v) > 0\}$ and note that, by (4.3.20), $\mathrm{vol}\, S(g) \le \mathrm{vol}\, \overline{S(g)}$. Therefore, $h \le h(g)$. Together with Lemma 4.3.1, this implies that

$$1 - \sqrt{1 - h^2} \le 1 - \sqrt{1 - h^2(g)} \le \mathcal{R}(\phi) \le \mathcal{R}(f) = \lambda_2.$$

□

Of course, $\lambda_2 = 0$ if and only if $h = 0$, if and only if the graph Γ is not connected, that is, if its vertex set can be decomposed into two nonempty subsets V_1, V_2 such that there are no edges between vertices in different components. More generally, the number of connected components equals the multiplicity of the eigenvalue $\lambda = 0$.

We had seen that for the bistar $S_{q,q}$ Figure 4.9, (Figure 4.14), the smallest positive eigenvalue behaves like $1/q$, and we had determined h as $1/(2q + 1)$ in (4.3.14). Thus, here λ_2 behaves like the upper bound in (4.3.12). For the star graph $K_{1,n}$, we know that the smallest positive eigenvalue is 1, as for every complete bipartite graph, and we had seen that also $h = 1$ in that case.

For the complete graph K_N with even N, we had computed $h = \frac{N}{2(N-1)}$, and since $\lambda_2 = \frac{N}{N-1}$, the upper bound in (4.3.15) is sharp in this case. For odd N, $h = \frac{N+1}{2(N-1)}$, and so, the upper bound here is only asymptotically sharp for $N \to \infty$.

4.4 Dividing a Graph Optimally into Opposite Classes, or the Dual Cheeger Constant and Estimates for the Largest Eigenvalue

In qualitative terms, the Cheeger inequalities (4.3.15) say that λ_2 becomes small when the graph can be easily (that is, by cutting only a few edges) decomposed into two large parts. Thus, λ_2 is small, that is, close to its minimal value of 0, when Γ is similar to a disconnected graph, with equality if and only if Γ is disconnected itself. Similarly, the largest eigenvalue is large, that is, close to its maximal value of 2, when Γ is close to a bipartite graph, with equality precisely if Γ is bipartite itself, by Lemma 4.1.12.

The characteristic property of a bipartite graph is that it can be decomposed into two opposite classes, with connections only between members of different classes. For a general graph, such a perfect division is not possible, but we can ask for a division that comes as close as possible to such a one. Such a division should yield two classes that are almost opposite, in the sense that there are only few, if any, connections inside each of them, plus perhaps a third class that may contain many internal connections. We shall define a constant quantifying this in (4.4.1). We shall interpret that constant as a dual version of the (Pólya–)Cheeger constant (4.3.13), and we shall justify this dual interpretation in this section.

The main purpose of this section, which follows [24], then is a dual version of (4.3.15) for the largest eigenvalue λ_N. More precisely, we shall obtain an estimate for λ_N in terms of a dual version of the Cheeger constant, which we now introduce.

Definition 4.4.1. Let V_1, V_2 and $\overline{V_1 \cup V_2} =: V_3$ be a partition of the vertex set V of the graph Γ into three disjoint sets for which V_1 and V_2 are nonempty. For such a partition V_1, V_2, V_3 of the vertex set V, we define the *dual Cheeger constant* as

$$\overline{h}(\Gamma) := \max_{V_1, V_2} \frac{2|E(V_1, V_2)|}{\text{vol}(V_1) + \text{vol}(V_2)}. \tag{4.4.1}$$

We recall from (4.2.7) that the volume of a vertex set W is defined as $\text{vol}(W) = \sum_{v \in W} \deg v$. $|E(V_1, V_2)|$ as before counts the number of edges between V_1 and V_2. On the other hand, $\text{vol}(V_1)$ also has contributions from edges from V_1 to the remainder set V_3, and so does $\text{vol}(V_2)$.

We should think of $V_1 \cup V_2$ as the (almost) bipartite part of Γ and V_3 as the part of Γ that contains many cycles of odd length, that is, a part of Γ that is different from being bipartite.

Remark 4.4.1. Independently of [24], Trevisan [263] introduced the *bipartiteness ratio*

$$\beta(\Gamma) := \min_{V_1, V_2} \frac{2|E(V_1, V_1)| + 2|E(V_2, V_2)| + |E(V_1 \cup V_2, \overline{V_1 \cup V_2})|)}{\text{vol}(V_1) + \text{vol}(V_2)}, \tag{4.4.2}$$

which by (4.3.11) is equivalent to our $\overline{h}(\Gamma)$, in the sense that

$$\beta(\Gamma) = 1 - \overline{h}(\Gamma), \tag{4.4.3}$$

and with this quantity, he obtained estimates equivalent to (4.4.7).

We first observe that $\overline{h}$ characterizes bipartite graphs.

Lemma 4.4.1. $\overline{h} \leq 1$, and $\overline{h} = 1$ if and only if Γ is bipartite.

Proof With (4.3.11), $\overline{h}$ becomes

$$\overline{h} = \max_{V_1, V_2} \frac{2|E(V_1, V_2)|}{2|E(V_1, V_1)| + 2|E(V_2, V_2)| + 2|E(V_1, V_2)| + |E(V_1, V_3)| + |E(V_2, V_3)|}. \tag{4.4.4}$$

Thus,

$$\overline{h} \leq 1. \tag{4.4.5}$$

If Γ is bipartite, there exists a partition V_1, V_2, V_3 of V with $V_3 = \emptyset$ and without edges within the subsets V_1 and V_2. Thus, $|E(V_1, V_3)| = |E(V_2, V_3)| = |E(V_1, V_1)| = |E(V_2, V_2)| = 0$. By (4.4.4), we have $\overline{h} \geq 1$. Together with (4.4.5), it follows that $\overline{h} = 1$.

Now assume that $\overline{h} = 1$. Equation (4.4.4) implies that there exists a partition V_1, V_2, V_3 of V such that $|E(V_1, V_3)| = |E(V_2, V_3)| = |E(V_1, V_1)| = |E(V_2, V_2)| = 0$. Since Γ is connected, $V_3 = \emptyset$. Thus, Γ is bipartite. $\qquad\square$

Lemma 4.4.2.

$$\frac{1}{2} \leq \overline{h}.$$

Proof If there exists a partition V_1, V_2, and $V_3 = \emptyset$ of the vertex set V with

$$|E(V_1, V_2)| \geq \max_{i=1,2} |E(V_i, V_i)|, \tag{4.4.6}$$

then

$$\bar{h} \geq \max_{V_1, V_2, V_3 = \emptyset} \frac{2|E(V_1, V_2)|}{2|E(V_1, V_2)| + |E(V_1, V_1)| + |E(V_2, V_2)|}$$

$$\geq \max_{V_1, V_2, V_3 = \emptyset} \frac{|E(V_1, V_2)|}{|E(V_1, V_2)| + \max_{i=1,2} |E(V_i, V_i)|}$$

$$\geq \frac{1}{2}.$$

It is easy to find a partition that satisfies (4.4.6). For instance, we can start with an arbitrary partition V_1, V_2. If (4.4.6) is not satisfied, we assume, without loss of generality, that $|E(V_1, V_1)| > |E(V_1, V_2)|$. Then, there has to exist some vertex $v \in V_1$ that has more neighbors in V_1 than in V_2. If we remove the vertex v from V_1 and put it into V_2, then $|E(V_1, V_2)|$ and $|E(V_2, V_2)|$ are increased, while $|E(V_1, V_1)|$ is decreased. Repeating this procedure, if necessary, eventually (4.4.6) has to hold. $\qquad\square$

Remark 4.4.2. Lemma 4.4.2 need not hold for graphs with self-loops.

The dual Cheeger constant $\bar{h}$ controls, or is controlled by, the largest eigenvalue λ_N in the same way that the Cheeger inequalities (4.3.15) estimate h and λ_2 in terms of each other. This is the content of

Theorem 4.4.1.

$$2\bar{h} \leq \lambda_N \leq 1 + \sqrt{1 - (1 - \bar{h})^2}. \tag{4.4.7}$$

For a better comparison with (4.3.15), we can reformulate (4.4.7) as

$$1 - \sqrt{1 - (1 - \bar{h})^2} \leq 2 - \lambda_N \leq 2 - 2\bar{h}. \tag{4.4.8}$$

This implies, in analogy to (4.3.16), the weaker bound

$$\lambda_N \leq 2 - \frac{1}{2}(1 - \bar{h})^2. \tag{4.4.9}$$

This, in particular, implies Lemma 4.1.12 again. Of course, this bound, and the stronger upper bound in (4.4.7), constitutes a quantitative version of that Lemma.

Starting with the *proof* of Theorem 4.4.1, we first prove the easier lower bound. We recall (4.1.50),

$$\lambda_N = \max_{u \neq 0} \frac{(\delta u, \delta u)}{(u, u)}. \tag{4.4.10}$$

Let V_1, V_2, and V_3 be a partition that achieves $\overline{h}$. We construct a function f for which (4.4.10) is not smaller than $2\overline{h}$ as

$$f(v) = \begin{cases} \dfrac{1}{\mathrm{vol}(V_1)} & \text{if } v \in V_1 \\[2mm] \dfrac{-1}{\mathrm{vol}(V_2)} & \text{if } v \in V_2 \\[2mm] 0 & \text{else.} \end{cases}$$

This yields

$$\lambda_N = \sup_{u \neq 0} \frac{(\delta u, \delta u)}{(u.u)} = \sup_{u \neq 0} \frac{\sum_{e=(v,v')}(u(v) - u(v'))^2}{\sum_v \deg v \, u(v)^2}$$

$$\geq \frac{\left(\frac{1}{\mathrm{vol}(V_1)} + \frac{1}{\mathrm{vol}(V_2)}\right)^2 E(V_1, V_2) + \left(\frac{1}{\mathrm{vol}(V_1)}\right)^2 E(V_1, V_3) + \left(\frac{1}{\mathrm{vol}(V_2)}\right)^2 E(V_2, V_3)}{\left(\frac{1}{\mathrm{vol}(V_1)} + \frac{1}{\mathrm{vol}(V_2)}\right)}$$

$$\geq \frac{(\mathrm{vol}(V_1) + \mathrm{vol}(V_2))^2}{2\mathrm{vol}(V_1)\mathrm{vol}(V_2)} \frac{2E(V_1, V_2)}{\mathrm{vol}(V_1) + \mathrm{vol}(V_2)}$$

$$+ \frac{\min(\mathrm{vol}(V_1), \mathrm{vol}(V_2))}{\max(\mathrm{vol}(V_1), \mathrm{vol}(V_2))} \frac{E(V_1 \cup V_2, V_3)}{(\mathrm{vol}(V_1) + \mathrm{vol}(V_2))}$$

$$\geq 2\overline{h} + \frac{\min(\mathrm{vol}(V_1), \mathrm{vol}(V_2))}{\max(\mathrm{vol}(V_1), \mathrm{vol}(V_2))} \frac{E(V_1 \cup V_2, V_3)}{(\mathrm{vol}(V_1) + \mathrm{vol}(V_2))}$$

$$\geq 2\overline{h},$$

using $\frac{(a+b)^2}{2ab} \geq 2$ for $a, b \in \mathbb{R}$.

We now turn to the upper bound $\lambda_N \leq 1 + \sqrt{1 - (1 - \overline{h})^2}$. The graph Laplacian $\Delta = I - P$ and the operator $\Sigma = I + P$ have the same eigenfunctions, and if λ is an eigenvalue of Δ, the corresponding eigenvalue of Σ is $\mu = 2 - \lambda$. In particular,

$$\lambda_N = \mu_1. \tag{4.4.11}$$

Thus, controlling the largest eigenvalue λ_N of Δ from above is equivalent to controlling the smallest eigenvalue μ_1 of Σ from below. The smallest eigenvalue μ_1 of Σ is

$$\mu_1 = \inf_{u \neq 0} \frac{(\Sigma u, u)}{(u, u)}$$

$$= \inf_{u \neq 0} \frac{\frac{1}{2}\sum_{(v,v') \in E}(u(v) + u(v'))^2}{\sum_{v \in V} \deg v \, u(v)^2}$$

In order to prove the lower bound for μ_1, we will use the idea of [78] to construct a bipartite graph Γ' out of Γ for which the quantity $h'(g)$ in Lemma 4.3.1 for Γ' can be controlled by the quantity $1 - \overline{h}$ of Γ.

Let u be an eigenfunction for the eigenvalue μ_1 and define

$$S(u) := \{v \in V : u(v) > 0\}, \quad T(u) := \{v \in V : u(v) < 0\}.$$

Since u is also an eigenfunction for λ_N of Δ, we know that $(u, 1) = 0$ and thus $S(u), T(u) \neq \emptyset$. For constructing the new graph $\Gamma' = (V', E')$ from Γ, we duplicate all vertices in $S(u) \cup T(u)$ and denote the copies by a prime, for example, if $v \in S(u)$, then its copy is $v' \in S'(u)$. The vertex set V' of Γ' is $V' = V \cup S'(u) \cup T'(u)$. Every edge $(v, \hat{v}) \in E(S(u), S(u))$ in Γ is replaced by two edges $(v, \hat{v}')$ and $(\hat{v}, v')$ in Γ' and analogously for the edges in $E(T(u), T(u))$. The edges between $S(u)$ and $T(u)$ remain and are not replaced or duplicated. Thus, $S'(u)$ has no edges to $T(u)$ or $T'(u)$ and analogously for $T'(u)$. In particular, when u has no zeroes, that is, $S(u) \cup T(u) = V$, then Γ' is bipartite, with the two classes $V_1 = S(u) \cup T'(u)$ and $V_2 = T(u) \cup S'(u)$.

We consider the function $g : V' \to \mathbb{R}$,

$$g(v) = \begin{cases} |u(v)| & \text{if} \quad v \in S(u) \cup T(u) \\ 0 & else. \end{cases}$$

Then, since $u(v)$ and $u(\hat{v})$ have opposite signs when $v \in S(u)$ and $\hat{v} \in T(u)$, but the same sign when both of them are either in $S(u)$ or in $T(u)$,

$$\mu_1 = \frac{\sum_{e=(v,\hat{v}) \in E} (u(v) + u(\hat{v}))^2}{\sum_{v \in V} \deg v \, u(v)^2}$$

$$\geq \frac{\sum_{e'=(v,\hat{v}) \in E'} (g(v) - g(\hat{v}))^2}{\sum_{v \in V'} \deg v \, g(v)^2}$$

$$\geq 1 - \sqrt{1 - (h'(g))^2} \qquad (4.4.12)$$

where the last inequality follows from Lemma 4.3.1. For a nonempty subset $W \subseteq S(g) = S(u) \cup T(u)$, we define $S_W = W \cap S(u)$ and $T_W = W \cap T(u)$. Let $\emptyset \neq U \subseteq S(g)$ realize the infimum, that is,

$$h'(g) = \inf_{\emptyset \neq W \subseteq S(g)} \frac{|E'(W, \overline{W})|}{\text{vol}(W)} = \frac{|E'(U, \overline{U})|}{\text{vol}(U)}$$

$$= \frac{2|E(S_W, S_W)| + 2|E(T_W, T_W)| + |E(S_W \cup T_W, \overline{S_W \cup T_W})|}{\text{vol}(S_W) + \text{vol}(T_W)}$$

$$\geq \frac{|E(S_W, S_W)| + |E(T_W, T_W)| + |E(S_W \cup T_W, \overline{S_W \cup T_W})|}{\text{vol}(S_W) + \text{vol}(T_W)}$$

$$= 1 - \frac{2|E(S_W, T_W)|}{\text{vol}(S_W) + \text{vol}(T_W)}$$

$$\geq 1 - \overline{h}. \qquad (4.4.13)$$

(4.4.11), (4.4.12), and (4.4.13) yield

$$2 - \lambda_N = \mu_1 \geq 1 - \sqrt{1 - (1 - \overline{h})^2},$$

and the upper bound in (4.4.7) follows. This completes the proof of Theorem 4.4.1.   □

Corollary 4.4.1. *For complete graphs K_N,*

$$\overline{h}(K_N) \to \frac{1}{2} \quad as\ N \to \infty. \tag{4.4.14}$$

Proof This follows from Lemma 4.1.8 and the lower bound in (4.4.7).   □

In fact, when N is even and we let V_1 and V_2 consist of $N/2$ vertices each, then each vertex has $N/2$ edges to vertices in the other component, while

$$\mathrm{vol}\,(V_1) = \mathrm{vol}\,(V_2) = \frac{N}{2}(N - 1),$$

and so, for this partition

$$\frac{2|E(V_1, V_2)|}{\mathrm{vol}\,(V_1) + \mathrm{vol}\,(V_2)} = \frac{N}{2(N - 1)} = \frac{1}{2}\lambda_N$$

by Lemma 4.1.13, and thus, by Theorem 4.4.1, $2\overline{h} = \lambda_N$, and the lower bound is sharp.

4.5 Neighborhood Graphs

In this section, which is mainly taken from [24], we consider a weighted graph that may also contain self-loops, that is, edges of the form (v, v) for $v \in V$, its vertex set. The weight of the edge (v, v') is denoted by $w_{vv'}$. We recall that the degree of vertex v in such a graph is given by

$$\deg v = \sum_{v'} w_{vv'}. \tag{4.5.1}$$

The neighborhood graph of order l of such a graph, to be defined in a moment, makes two vertices v, v' neighbors when there exists at least one path (with nonvanishing weights) of length l between them in the original graph. The weight of the edge between two such vertices is the sum over the weights of all such paths, and the weight of a path is the product of the weights of the edges contained in it, normalized by the degrees of its vertices. The idea behind this, to be explored in more detail later, is that this weight then equals the probability that a random walker starting at v reaches v' in l steps. And since a random walker may return to its starting position after some steps, these neighborhood graphs in general will possess self-loops.

The main point of this section is to relate the eigenvalues of a graph $\Gamma = (V, E)$ and those of its neighborhood graphs. The spectra are essentially equivalent to each other, and therefore, eigenvalue estimates for a neighborhood graph can be translated into eigenvalue estimates for the original graph, and vice versa. Since the neighborhood graphs encode properties of random walks on Γ, asymptotic ones if $l \to \infty$, we thereby gain a new source of geometric intuition for obtaining eigenvalue estimates.

Definition 4.5.1. The *neighborhood graph* $\Gamma^{[l]} = (V, E^{[l]})$ of order $l \geq 1$ of the weighted graph $\Gamma = (V, E)$ is the graph with the same vertex set V and the weighted edge set $E^{[l]}$ where the weight $w_{vv'}^{[l]}$ of the edge $e^{[l]} = (v, v')$ in $\Gamma^{[l]}$ is given by

$$w_{vv'}^{[l]} = \sum_{v_1, \ldots, v_{l-1}} \frac{1}{\deg v_1} \cdots \frac{1}{\deg v_{l-1}} w_{vv_1} w_{v_1 v_2} \cdots w_{v_{l-1} v'} \qquad (4.5.2)$$

if $l > 1$, and we set $w_{vv'}^{[l]} = w_{vv'}$ if $l = 1$, that is, $\Gamma^{[1]} = \Gamma$.

Thus, v and v' are neighbors in $\Gamma^{[l]}$ if there exists at least one path of length l between v and v' in Γ.

The simplest example is the following:

Example 4.5.1. The graph Γ has two vertices v, v', an edge (vv') of weight 1 and self-loops at v and v' of weight $c \geq 0$. Thus, each vertex has degree $1 + c$, and the adjacency matrix W is

$$W = W^{[1]} = \begin{pmatrix} c & 1 \\ 1 & c \end{pmatrix}.$$

The neighborhood graphs $\Gamma^{[l]}$ of Γ in this example have the same edges, and the weighted adjacency matrices $W^{[l]}$, for $l = 2, 3, 4, 5$, are

$$W^{[2]} = \frac{1}{(1 + c)} \begin{pmatrix} c^2 + 1 & 2c \\ 2c & c^2 + 1 \end{pmatrix},$$

$$W^{[3]} = \frac{1}{(1 + c)^2} \begin{pmatrix} c^3 + 3c & 3c^2 + 1 \\ 3c^2 + 1 & c^3 + 3c \end{pmatrix},$$

$$W^{[4]} = \frac{1}{(1 + c)^3} \begin{pmatrix} (c^2 + 1)^2 + 4c^2 & 4c^3 + 4c \\ 4c^3 + 4c & (c^2 + 1)^2 + 4c^2 \end{pmatrix},$$

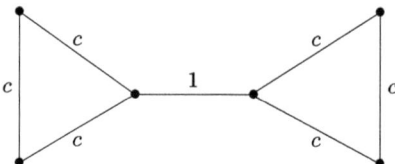

Figure 4.15 A graph for which we want to compute the neighborhood graphs.

$$W^{[5]} = \frac{1}{(1+c)^4} \begin{pmatrix} c(5 + 10c^2 + c^4) & 1 + 10c^2 + 5c^4 \\ 1 + 10c^2 + 5c^4 & c(5 + 10c^2 + c^4) \end{pmatrix},$$

and so on.

Example 4.5.2. For the family of graphs in Figure 4.15,
$W^{[1]}$ is given by

$$\begin{pmatrix} 0 & c & c & 0 & 0 & 0 \\ c & 0 & c & 0 & 0 & 0 \\ c & c & 0 & 1 & 0 & 0 \\ 0 & 0 & 1 & 0 & c & c \\ 0 & 0 & 0 & c & 0 & c \\ 0 & 0 & 0 & c & c & 0 \end{pmatrix},$$

and $W^{[2]}$ is

$$\begin{pmatrix} c/2 + c^2/(1+2c) & c^2/(1+2c) & c/2 \\ c^2/(1+2c) & c/2 + c^2/(1+2c) & c/2 \\ c/2 & c/2 & 1/(1+2c) + c \\ c/(1+2c) & c/(1+2c) & 0 \\ 0 & 0 & c/(1+2c) \\ 0 & 0 & c/(1+2c) \end{pmatrix}$$

$$\begin{pmatrix} c/(1+2c) & 0 & 0 \\ c/(1+2c) & 0 & 0 \\ 0 & c/(1+2c) & c/(1+2c) \\ 1/(1+2c) + c & c/2 & c/2 \\ c/2 & c/2 + c^2/(1+2c) & c^2/(1+2c) \\ c/2 & c^2/(1+2c) & c/2 + c^2/(1+2c) \end{pmatrix}.$$

We first record some simple properties of neighborhood graphs $\Gamma^{[l]}$.

Lemma 4.5.1. (i) *The vertex degrees do not depend on l, that is.*

$$\deg v = \deg v^{[l]} \qquad (4.5.3)$$

for all $v \in V$ and $l \geq 1$. In particular, the scalar products are the same for all l, $(\cdot, \cdot)_\Gamma = (\cdot, \cdot)_{\Gamma^{[l]}}$.

(ii) *If l is even and Γ is connected, then $\Gamma^{[l]}$ consists of exactly two connected components if Γ is bipartite, and if it is not bipartite, then $\Gamma^{[l]}$ is connected.*

(iii) *If l is odd and Γ is connected, then $\Gamma^{[l]}$ is also connected, and $\Gamma^{[l]}$ is bipartite if and only if Γ is bipartite.*

Proof (i).

$$
\begin{aligned}
\deg v^{[l]} &= \sum_{v'} w^{[l]}_{vv'} \\
&= \sum_{v_1,\ldots,v_{l-1}} \frac{1}{\deg v_1} \cdots \frac{1}{\deg v_{l-1}} w_{vv_1} w_{v_1 v_2} \cdots w_{v_{l-2}v_{l-1}} \sum_{v'} w_{v_{l-1} j} \\
&= \sum_{v_1,\ldots,v_{l-2}} \frac{1}{\deg v_1} \cdots \frac{1}{\deg v_{l-2}} w_{vv_1} w_{v_1 v_2} \cdots \sum_{v_{l-1}} w_{v_{l-2}v_{l-1}} \\
&\vdots \\
&= \sum_{v_1} w_{vv_1} = \deg v
\end{aligned}
$$

(ii). When Γ is not bipartite, it contains at least one loop of odd length. When it is connected, therefore, from any vertex v, any other vertex can be reached by some path of even length. Since such a path can be decomposed into paths of length 2, Γ^2 is connected. And since Γ, not being bipartite, contains a path of odd length, so does $\Gamma^{[2]}$, and hence, it is not bipartite either, and we can iterate the argument. When Γ, however, is bipartite, then only other vertices in the same class as v can be reached from v by a path of even length, and since Γ is connected, this is possible for every other vertex in the same class. Therefore, that class yields one component of $\Gamma^{[2]}$, and the other class, the other. Again, the argument iterates.

(iii). When v and v' are connected by an edge in Γ, then one can go from v to v' in any odd number l of steps by going forth and back between v and v' $\frac{l-1}{2}$ times and then again from v to v'. This implies that for odd l, $\Gamma^{[l]}$ is connected if and only if Γ is. And if Γ is bipartite, from a vertex v, one can reach a vertex in the same class only in an even number of steps. Therefore, $\Gamma^{[l]}$ then remains bipartite. □

Remark 4.5.1. We can now understand the observation made in the remark before Definition 4.1.6. For the complete bipartite graph $\Gamma^{[1]} = K_{m,n}$, its neighborhood graph $\Gamma^{[2]}$ is simple the disjoint union of the two complete graphs K_m^0 and K_n^0 with self-loops, and this is the reason why their spectra are so similar. We shall now investigate the spectra of neighborhood graphs systematically.

We shall first express the Laplacian of $\Gamma^{[l]}$ in terms of the Laplacian Δ of Γ.

Lemma 4.5.2. *The graph Laplacian $\Delta^{[l]}$ on $\Gamma^{[l]}$ satisfies*

$$\Delta^{[l]}u = (\mathrm{Id} - (\mathrm{Id} - \Delta)^l)u \tag{4.5.4}$$

for $u \in L^2(\Gamma)$. In particular, the eigenvalues of $\Delta^{[l]}$ are of the form

$$\lambda^{[l]} = 1 - (1 - \lambda)^l, \tag{4.5.5}$$

where λ is an eigenvalue of Δ.

Proof For $u \in L^2(\Gamma)$,

$$(\mathrm{Id} - (\mathrm{Id} - \Delta)^l)u(i) = u(i) - \sum_j \sum_{v_1, \ldots, v_{l-1}} \frac{w_{vv_1}}{\deg v} \cdots \frac{w_{v_{l-1}j}}{\deg v_{l-1}} u(j)$$

$$= \frac{1}{\deg v^{[l]}} \sum_j w_{vv'}^{[l]}(u(i) - u(j))$$

$$= \Delta^{[l]}u(i),$$

by (4.5.2) and since $\deg v = \deg v^{[l]}$ by (4.5.3). ☐

Remark 4.5.2. For $l = 2$, Lemma 4.5.2 says $\lambda^{[2]} = \lambda(2-\lambda)$ for the eigenvalues, and this also implies Corollary 4.1.6.

Corollary 4.5.1. *Let Γ be a graph and $\Gamma^{[l]}$ its neighborhood graph of order l. The eigenvalues of $\Delta^{[l]}$ satisfy*

$$0 = \lambda_1^{[l]} \le \ldots \le \lambda_N^{[l]} \le 1 \qquad \text{for even } l \tag{4.5.6}$$

and

$$0 = \lambda_1^{[l]} \le \ldots \le \lambda_N^{[l]} \le 2 \qquad \text{for odd } l. \tag{4.5.7}$$

Proof All eigenvalues λ_k of Δ satisfy $\lambda_k \in [0,2]$ and thus $1 - (1-\lambda_k)^l \in [0,1]$ if l is even and $1 - (1 - \lambda_k)^l \in [0,2]$ if l is odd. ☐

Remark 4.5.3. By Corollary 4.5.1, for l even, all eigenvalues are ≤ 1 and thus in particular $\lambda_N \le 1$. This is caused by the fact that for even l, $\Gamma^{[l]}$ contains a self-loop at every vertex, since a random walker can return to its starting vertex in two steps, or equivalently, in any graph without isolated vertices, there is a path of length 2 from each vertex to itself. In contrast, we recall that for loopless graphs, we have $1 < \frac{N}{N-1} \le \lambda_N$ by Lemma 4.1.10.

Corollary 4.5.2. *1 is an eigenvalue of $\Delta^{[l]}$ precisely if it is an eigenvalue of Δ, and the multiplicity m_1 of the eigenvalue 1 is the same for all neighborhood graphs, that is, $m_1(\Delta) = m_1(\Delta^{[l]})$ for all $l \geq 1$.*

And if $\lambda \neq 0, 2$ then

$$\lambda^{[l]} = 1 - (1 - \lambda)^l \to 1 \qquad as\ l \to \infty. \tag{4.5.8}$$

Proof Γ and $\Gamma^{[l]}$ have the same vertex set; thus, both Δ and $\Delta^{[l]} = (\mathrm{Id} - (\mathrm{Id} - \Delta)^l)$ have $N = |V|$ eigenvalues. Every eigenfunction u_k for Δ and eigenvalue λ_k is also an eigenfunction for $\Delta^{[l]}$ and the eigenvalue $1 - (1 - \lambda_k)^l$.

We have $1 - (1 - \lambda_k)^l = 1$ if and only if $\lambda_k = 1$.

If $\lambda \neq 0, 2$ is an eigenvalue of Δ, then $|1 - \lambda| < 1$. Hence, $|1 - \lambda|^l \to 0$ as $l \to \infty$, and the result follows from (4.5.4). □

For some eigenvalues, we can make the convergence in (4.5.8) quantitative.

Theorem 4.5.1. *For every Γ with at least three vertices, there is some eigenvalue $\lambda^{[l]}$ of $\Delta^{[l]}$ with*

$$|1 - \lambda^{[l]}| \leq \frac{1}{2^l}. \tag{4.5.9}$$

When l is even, the largest eigenvalue of $\Gamma^{[l]}$ satisfies

$$1 - \frac{1}{2^l} \leq \lambda_N^{[l]} \leq 1, \tag{4.5.10}$$

and both bounds are sharp.

Proof The bound (4.5.9) follows from (4.5.5) and from Theorem 4.1.2, where we have shown that $\min_{\lambda_i} |1 - \lambda_i| \leq 1/2$ for all connected graphs with at least three nodes, and that this bound is realized for the petal and book graphs. By Corollary 4.5.2, when l is even, this concerns the largest eigenvalue. The upper bound in (4.5.10) is already contained in (4.5.6). It is realized whenever Γ possesses the eigenvalue 1, for instance, if it has duplicated nodes. □

Thus, we can control the eigenvalues of $\Delta^{[l]}$ in terms of those of Δ. We shall now derive controls in the opposite direction. That is, we can control eigenvalues of Γ by using geometric properties of its neighborhood graphs $\Gamma^{[l]}$.

Theorem 4.5.2. *If $a^{[l]} \leq \lambda_2^{[l]}$, then*

$$1 - \left(1 - a^{[l]}\right)^{\frac{1}{l}} \leq \lambda_2 \leq \ldots \leq \lambda_N \leq 1 + \left(1 - a^{[l]}\right)^{\frac{1}{l}}\ for\ even\ l, \tag{4.5.11}$$

and

$$1 - \left(1 - a^{[l]}\right)^{\frac{1}{l}} \le \lambda_2 \text{ for odd } l. \tag{4.5.12}$$

Proof Since by Lemma 4.5.2, the eigenvalues of $\Delta^{[l]}$ are $1 - (1 - \lambda_k)^l$ where λ_k are the eigenvalues of Δ, we have

$$1 - (1 - \lambda_k)^l \ge \min_{k \ne 0} 1 - (1 - \lambda_k)^l = \lambda_2^{[l]} \ge a^{[l]}.$$

This implies that for all $k \ne 0$

$$|1 - \lambda_k| \le (1 - a^{[l]})^{\frac{1}{l}}$$

if l is even, and

$$1 - \lambda_k \le (1 - a^{[l]})^{\frac{1}{l}}$$

if l is odd. □

As a concrete example, we use the Cheeger inequality (4.3.12), that is,

$$a^{[l]} = 1 - \sqrt{1 - (h^{[l]})^2} \le \lambda_2^{[l]}.$$

Corollary 4.5.3. *All eigenvalues of $\Delta(\Gamma)$ satisfy*

$$1 - \left(1 - (h^{[l]})^2\right)^{\frac{1}{2l}} \le \lambda_2 \le \ldots \le \lambda_N \le 1 + \left(1 - (h^{[l]})^2\right)^{\frac{1}{2l}} \text{ for even } l, \tag{4.5.13}$$

and

$$1 - \left(1 - (h^{[l]})^2\right)^{\frac{1}{2l}} \le \lambda_2 \text{ for odd } l. \tag{4.5.14}$$

Remark 4.5.4. While Theorem 4.5.1 is concerned with $\min_k |1 - \lambda_k|$, Theorem 4.5.2 yields estimates for $\max_{k \ne 1} |1 - \lambda_k|$.

Remark 4.5.5. For bipartite graphs Γ, we have $\lambda_N = 2$. If l is even, then $h^{[l]} = 0$ since by Lemma 4.5.1, the neighborhood graph $\Gamma^{[l]}$ of a bipartite graph is disconnected. Thus, for bipartite graphs, the upper bound in (4.5.13) is sharp. On the other hand, if l even and the graph is bipartite, then the lower bound in Corollary 4.5.3 yields only the trivial estimate $0 \le \lambda_2$. In contrast, if l is even and Γ is not bipartite or if l is odd, then Corollary 4.5.3 yields a nontrivial lower bound

$$0 < 1 - \left(1 - (h^{[l]})^2\right)^{\frac{1}{2l}} \le \lambda_2$$

for the second smallest eigenvalue. These new lower bounds for λ_2 can actually improve the Cheeger estimate (4.3.12).

In Theorem 4.5.2, we have seen, in particular, that a lower bound for $\lambda_2^{[l]}$ for even l turns into an upper bound for λ_N. Conversely, an upper bound for $\lambda_2^{[l]}$ for even l yields a lower bound for λ_N, as a special case of Theorem 4.5.3.

Theorem 4.5.3. *If $\lambda_2^{[l]} \leq b^{[l]}$, then*

$$\lambda_2 \leq 1 - \left(1 - b^{[l]}\right)^{\frac{1}{l}} \tag{4.5.15}$$

or

$$\lambda_N \geq 1 + \left(1 - b^{[l]}\right)^{\frac{1}{l}} \text{ for even } l, \tag{4.5.16}$$

and

$$\lambda_2 \leq 1 - \left(1 - b^{[l]}\right)^{\frac{1}{l}} \text{ for odd } l. \tag{4.5.17}$$

Proof By Lemma 4.5.1, if l is even, we can assume $b^{[l]} \leq 1$, and hence (4.5.15) and (4.5.16) are well-defined. If $b^{[l]} \geq \lambda_2^{[l]} = \min_{k>1} 1 - (1 - \lambda_k)^l$ (by (4.5.5)), then for at least one eigenvalue $\lambda_i, i \neq 0$, we have

$$(1 - b^{[l]})^{\frac{1}{l}} \leq |1 - \lambda_i|$$

if l is even and

$$(1 - b^{[l]})^{\frac{1}{l}} \leq 1 - \lambda_i$$

if l is odd. $\qquad\square$

Using the Cheeger inequality (4.3.19) for $\Gamma^{[l]}$, we obtain:

Corollary 4.5.4. *Assume that $2h^{[l]} \leq 1$ and l is even; then we have*

$$\lambda_2 \leq 1 - \left(1 - 2h^{[l]}\right)^{\frac{1}{l}} . \tag{4.5.18}$$

or

$$\lambda_N \geq 1 + \left(1 - 2h^{[l]}\right)^{\frac{1}{l}} . \tag{4.5.19}$$

If l is odd, then we have

$$\lambda_2 \leq 1 - \left(1 - 2h^{[l]}\right)^{\frac{1}{l}} .$$

Remark 4.5.6. We now want to discuss under which conditions the estimate (4.5.18) for λ_2 can improve the Cheeger estimate. Comparing Corollary 4.5.3 with (4.3.12), the preceding estimates improve the Cheeger estimate (4.3.12) if

$$h^{[l]} \geq \sqrt{1 - (1 - h^2)^l}, \tag{4.5.20}$$

for some $l \geq 2$. For discussing (4.5.20), let us first look at even l. It cannot hold when Γ is bipartite, since then, by Lemma 4.5.1, $\Gamma^{[l]}$ is disconnected and thus $h^{[l]} = 0$. For graphs that are not bipartite, however, the estimate in (4.5.13) always yields a nontrivial lower bound

$$0 < 1 - \left(1 - (h^{[l]})^2\right)^{\frac{1}{2l}} \leq \lambda_2$$

for the second smallest eigenvalue λ_2. We then need to understand under which conditions $h^{[l]} > h$. We distinguish two cases. If $\lambda_2 \geq 2 - \lambda_N$, then $\lambda_2^{[l]} = 1 - (1 - \lambda_N)^l$, and so it is possible that $\lambda_2^{[l]} < \lambda_2$. If this is the case, we cannot expect (4.5.20). But if $\lambda_2 < 2 - \lambda_N$, then $\lambda_2^{[l]} = 1 - (1 - \lambda_1)^l > \lambda_2$, and so we can expect $h^{[l]} > h$. Thus, if l is even, graphs that are closer to disconnected graphs than to bipartite graphs are good candidates for satisfying (4.5.20).

We now turn to the effect of bounds on the largest eigenvalue $\lambda_N^{[l]}$.

Theorem 4.5.4. *If $c^{[l]} \leq \lambda_N^{[l]}$, then there exists at least one eigenvalue λ_k in the interval*

$$\left[1 - \left(1 - c^{[l]}\right)^{\frac{1}{l}}, 1 + \left(1 - c^{[l]}\right)^{\frac{1}{l}}\right] \tag{4.5.21}$$

if l is even and

$$\lambda_N \geq 1 + \left(1 - c^{[l]}\right)^{\frac{1}{l}}$$

if l is odd.

Proof We recall from Corollary 4.5.1 that if l is even, $c^{[l]} \leq \lambda_N^{[l]} \leq 1$. Recalling (4.5.5) again, we have $c^{[l]} \leq \lambda_N^{[l]} = \max_k(1 - (1 - \lambda_k)^l)$. Thus,

$$\min_k |1 - \lambda_k| \leq \left(1 - c^{[l]}\right)^{\frac{1}{l}}$$

if l is even and

$$\min_k(1 - \lambda_k) = 1 - \max_k \lambda_k \leq \left(1 - c^{[l]}\right)^{\frac{1}{l}}$$

if l is odd. □

In particular, we have from Theorems 4.4.1 and 4.5.4:

Corollary 4.5.5. *Assume that $2\overline{h}^{[l]} \leq 1$ and l is even, then there exists at least one eigenvalue λ_k in the interval*

$$\left[1 - (1 - 2\overline{h}^{[l]})^{\frac{1}{l}}, 1 + (1 - 2\overline{h}^{[l]})^{\frac{1}{l}}\right]. \tag{4.5.22}$$

If l is odd, then

$$\lambda_N \geq 1 + \left(1 - 2\overline{h}^{[l]}\right)^{\frac{1}{l}}.$$

We now turn to the gap phenomenon for eigenvalues, that is, finding some interval that does not contain any eigenvalue.

Theorem 4.5.5. *If $\lambda_N^{[l]} \leq d^{[l]}$, then all eigenvalues λ_k of Δ are contained in the union of intervals*

$$\left[0, 1 - \left(1 - d^{[l]}\right)^{\frac{1}{l}}\right] \bigcup \left[1 + \left(1 - d^{[l]}\right)^{\frac{1}{l}}, 2\right] \qquad (4.5.23)$$

if l is even and

$$\lambda_N \leq 1 - \left(1 - d^{[l]}\right)^{\frac{1}{l}}$$

if l is odd.

The proof is similar to the preceding proofs and hence omitted. Theorem 4.5.5 says that if l is even, then an upper bound $d^{[l]} < 1$ for $\lambda_N^{[l]}$ of $\Delta^{[l]}$ on $\Gamma^{[l]}$ bounds all eigenvalues of Δ on Γ away from 1. In particular, in that case, Γ cannot contain duplicated nodes, as they would cause an eigenvalue 1.

Recalling Lemma 4.1.10 and Theorem 4.1.2, the preceding considerations also imply

Theorem 4.5.6. *Considering the extreme cases for the largest eigenvalue $\lambda_N^{[l]}$, if l is even,*

$$\min_{connected\ graph\ \Gamma} \lambda_N^{[l]}(\Gamma) = 1 - \frac{1}{2^l},$$

and this minimum is realized if and only if Γ is a petal or a book graph.
While if l is odd,

$$\min_{connected\ graph\ \Gamma} \lambda_N^{[l]}(\Gamma) = 1 - \frac{1}{(N-1)^l},$$

and this minimum is realized if and only if Γ is a complete graph.
Moreover, for any l,

$$\max_{connected\ graph\ \Gamma} \lambda_2^{[l]}(\Gamma) = 1 - \frac{1}{(N-1)^l},$$

and this maximum is again realized if and only if Γ is a complete graph.

$\square$

4.6 Signed Graphs

We now turn to graphs with an additional structure, a sign function. This class of graphs, the *signed graphs*, is not only important for several applications but will also naturally occur in our analysis of the Laplacians of simplicial complexes.

In this section, we shall describe the results of [11] (which turn the results of [10, 187] into a general perspective). See also [179, 188] and the references therein, which discuss more general signatures than we do here. We shall only discuss the case where the possible signatures are taken from $\{\pm 1\}$, but for instance in [251, 255] signatures with values in $U(1)$ were also introduced (leading to the notion of a *magnetic Laplacian*), and one can then naturally also allow for signatures with values in other groups like $O(n)$ or $U(n)$ (*connection Laplacian*). For us, however, the signatures ± 1 will become important in Section 8.3.

All graphs in this section are *undirected*.

Definition 4.6.1. A *signed graph* Γ consists of a vertex set V and a set E of undirected edges with a sign function

$$s : E \rightarrow \{+1, -1\} \tag{4.6.1}$$

and a weight function

$$w : E \rightarrow \mathbb{R}_+. \tag{4.6.2}$$

While in some applications, such as neural networks, negative weights would naturally occur, we here consider only positive edge weights, because in the framework developed here, which originally goes back to Harary [124], the signs are carried by the sign function and not by the weights. In social networks [123, 125], a positive sign $s(uv)$ would indicate a positive relation, like friendship, between the individuals represented by the vertices u and v, whereas a negative sign might indicate that they are enemies or opponents.

In the special case where $w(e) = 1$ for every edge, we speak of an unweighted signed graph.

Definition 4.6.2. The *opposite graph* of the signed graph (Γ, s) is the signed graph $(\Gamma, -s)$, that is, the signed graph where all edges have the opposite sign as on (Γ, s).

For $S \subset V$, we denote its complement in V by $\overline{S}$, that is, $\overline{S} = V \setminus S$.

When we speak of a signed graph Γ, we shall usually denote its vertex set by V, its edge set by E, the number of its vertices by N, its sign function by s, and its weight function by w. The edge connecting the vertices u and v is usually written as (uv), and since we are working with undirected graphs in this section, this is the same as (vu). We shall also usually write w_{uv} in place of $w(uv)$. We shall also say that an edge is positive (negative) if its sign is positive (negative).

Some special features of signed graphs are important. A *cycle C_m* of length m in a graph Γ consists of vertices $v_1, \dots, v_m \in V$ and edges $e_1 = (v_1, v_2), \dots, e_{m-1} = (v_{m-1}, v_m), e_m = (v_m, v_1) \in E$.

Definition 4.6.3. The signed cycle C_m is *balanced* if

$$\prod_{i=1}^{m} s(e_i) = 1. \qquad (4.6.3)$$

A signed graph (Γ, s) is *balanced* if every cycle contained in it is balanced. (Γ, s) is *antibalanced* if $(\Gamma, -s)$ is balanced.

We have Harary's characterization of balanced graphs [124].

Lemma 4.6.1 (Harary). *A signed graph Γ is balanced precisely if there is a partition $V = V_1 \cup V_2$, where one of these sets may be empty, such that all edges between points of the same subset have a positive sign and those between points of different subsets have a negative sign.*

Proof The idea of the proof consists in approaching Γ from two ends. For the sufficiency, we take the maximal graph K_N with the same vertex set and delete edges one by one until we are left with the edge set of Γ. For the necessity, we start with a minimally connected graph with the same vertex set, a tree, and add edges one by one until we arrive at the edge set of Γ.

Now the—easy—details. We first observe that a subgraph of a balanced graph is balanced itself, as follows trivially from the definition. For the sufficiency of the condition, we shall therefore show that when this condition holds, we can extend the sign function to the complete graph K_N with the same vertex set V as Γ and make K_N balanced. Thus, for any pair $u, v \in V$, we give the edge (uv) in K_N a positive sign if u and v are in the same subset and a negative sign when they are in different subsets, that is, one is in V_1 and the other in V_2. Since then every cycle in K_N has to contain an even number of edges between V_1 and V_2, K_N with this sign function is balanced, and hence so is our original graph.

Necessity is proved inductively on the number of edges. We may assume that our graph is connected. Therefore, it contains a tree T, that is, a connected graph without cycles, that contains all vertices of V as a subgraph. Take any $v_1 \in V$. Let V_1 be the set of those vertices that are connected in T by a path of positive sign, that is, a path containing an even number of negative edges, and let V_2 be its complement (possibly empty). If we add another edge of Γ to T, it has to be positive when it connects two vertices that are both in V_1 or both in V_2 and negative when it connects vertices in different subsets. We can iterate this and add the remaining edges of Γ one by one to see the necessity of the condition.   □

Definition 4.6.4. Let (Γ, s) be a signed graph. A function $\theta: V \to \{+1, -1\}$ is called a *switching function*. It switches the sign function from s to s^θ by

$$s^\theta(vv') = \theta(v)s(vv')\theta(v'). \tag{4.6.4}$$

(Thus, the sign of an edge is changed precisely if θ has opposite signs on its two vertices.)

The sign functions s and s' are *switching equivalent* if there exists a switching function θ with $s' = s^\theta$. We write $s \approx s^\theta$ and denote the equivalence of s by $[s]$.

Let $S \subset V$. A *switching* of S is the sign function s^S obtained via the switching function

$$\theta^S(v) = \begin{cases} -1 & \text{if } v \in S \\ +1 & \text{if } v \notin S, \end{cases} \tag{4.6.5}$$

that is,

$$s^S(uv) := \begin{cases} -s(uv) & \text{when } u \in S, v \in \overline{S} \\ s(uv) & \text{else.} \end{cases} \tag{4.6.6}$$

(Note that we are discussing undirected graphs here; that is, the edge (uv) is the same as the edge (vu).)

We observe that

$$\theta^S = \prod_{v \in S} \theta^v. \tag{4.6.7}$$

When the vertices of our graph are divided into two groups, and an edge connecting members of different groups has a negative sign, whereas those remaining in the same group are positive, then switching a subset S means that its members switch the group they belong to. Edges connecting them with members of their former group become negative, while those connecting with members of their new group turn positive. (4.6.7) says that we can switch the members of S one by one.

Lemma 4.6.2. *A bipartite graph (Γ, s^+) with the positive sign function s^+, that is, $s^+(e) = 1$ for every edge, which thus is balanced, is also switching equivalent to an antibalanced graph.*

Proof When $V = V_1 \cup V_2$, with no edges inside either subset, we can switch one of them, say V_1, and thereby make every edge negative. The resulting signed bipartite graph then is both balanced and antibalanced. □

In social terminology, we have two groups with two possible opinions, and connections only exist between members of different groups. Switching one of these groups means that all members of that group change their opinion, see [1].

A similar observation will now be used for Zaslavsky's switching lemma [275].

Lemma 4.6.3 (Zaslavsky). *A signed graph* (Γ, s) *is balanced if and only if* s *is switching equivalent to the positive sign function, that is, the one that is* $+1$ *on every edge. It is antibalanced if and only if* s *is switching equivalent to the negative sign function.*

Proof Let (Γ, s) be balanced and take the partition $V = V_1 \cup V_2$ of Lemma 4.6.1. Switching one of these subsets, say V_2, keeps the graph balanced, and s^{V_2} is the positive sign function because it keeps the signs of all edges inside V_2 or V_1 and changes the signs of the edges between these two sets. Conversely, when s is switching equivalent to the positive sign function s^+, we let V_2 be the set of vertices on which the switching function θ is negative. That gives a partition as in Lemma 4.6.1.

The antibalanced case is treated by working with $-s$. □

Definition 4.6.5. Let Γ be a signed graph. Its (normalized) Laplacian is defined by

$$\Delta^s f(v) := f(v) - \frac{1}{\deg v} \sum_{v' \sim v} w_{vv'} s(vv') f(v')$$

$$= \frac{1}{\deg v} \sum_{v' \sim v} w_{vv'} (f(v) - s(vv') f(v')) \tag{4.6.8}$$

with $\deg v = \sum_{v' \sim v} w_{vv'}$, for functions $f : V \to \mathbb{R}$.

The spectrum of a signed graph is the spectrum of Δ^s, with the same conventions as in Section 4.1.

We directly observe

Lemma 4.6.4. λ *is an eigenvalue of the signed graph* (Γ, s) *precisely if* $2 - \lambda$ *is an eigenvalue of the opposite signed graph* $(\Gamma, -s)$. *And the corresponding eigenfunctions are the same.*

In particular, λ is an eigenvalue of the signed graph (Γ, s^+) on which all signs are positive precisely if $2 - \lambda$ is an eigenvalue of the signed graph (Γ, s^-) on which all signs are negative.

Proof If we write the Laplacian of (Γ, s) as

$$\Delta^s = \text{Id} - P, \tag{4.6.9}$$

then

$$\Delta^{-s} = \text{Id} + P, \tag{4.6.10}$$

we see that

$$\Delta^{-s} = 2 - \Delta^s. \tag{4.6.11}$$

This implies the desired relation between the eigenvalues. $\qquad\square$

Let us consider examples; for simplicity, they are unweighted. The signed graph in Figure 4.16 has the eigenvalue $\lambda = 2$ with a constant eigenfunction, $\lambda = 1$ with an eigenfunction that is $+1$ on the left, -1 on the right, 0 on the top vertex, and $\lambda = 0$ with 1 on the two lower vertices and -1 on the top one. This is the same as for the corresponding unsigned graph, the complete bipartite graph $K_{1,2}$, but with the roles of the eigenvalues $\lambda = 2$ and $\lambda = 0$ exchanged. In fact, this is a consequence of Lemma 4.6.4.

If we add an edge and turn the graph of Figure 4.16 into a triangle, as indicated in Figure 4.17, then has the eigenvalue $\lambda = 2$ with a constant eigenfunction and the eigenvalue $\lambda = 1/2$ with multiplicity 2, with an eigenfunction that is $+1$ on one vertex, -1 on another, and 0 on the third. Again, this is a consequence of Lemma 4.6.4 and Lemma 4.1.8.

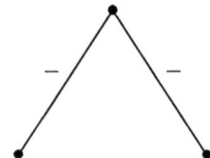

Figure 4.16 Simple signed graph A.

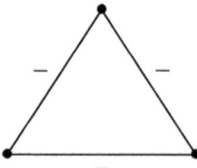

Figure 4.17 A signed graph with a frustrated triangle.

Similarly, the graph of Figure 4.18 now again has $\lambda = 2$ and $\lambda = 1/2$ with multiplicity 3, with an eigenfunction that is ± 1 on the two upper vertices, or on the two lower vertices, and 0 elsewhere, or with 2 at the center and -1 at all other vertices. (Corollary 4.8.3 will provide an explanation for this eigenvalues and its multiplicity.) It has the eigenvalue $\lambda = 3/2$ with an eigenfunction that is $+1$ on the upper, -1 on the lower vertices, and 0 at the center. This follows from Lemma 4.6.4 and the computation of the spectrum of the petal graph in Section 4.1. We move on to the signed graph shown in Figure 4.19.

We again have $\lambda = 2$ with a constant eigenfunctions and $\lambda = 1/2$ with eigenfunctions that are ± 1 either on the two upper or on the two lower vertices. Two further eigenfunctions for that eigenvalue are indicated in Figure 4.20.

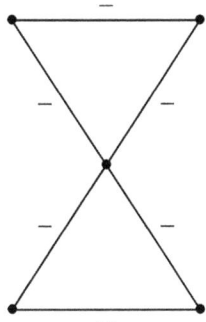

Figure 4.18 A signed graph with two frustrated triangles.

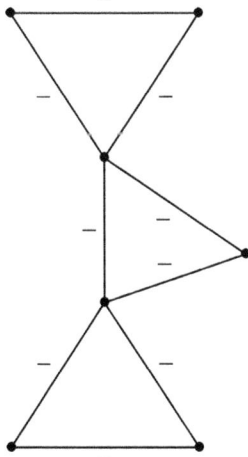

Figure 4.19 A signed graph whose eigenvalues and eigenfunctions will be computed in the main text; see also Figures 4.20 and 4.21.

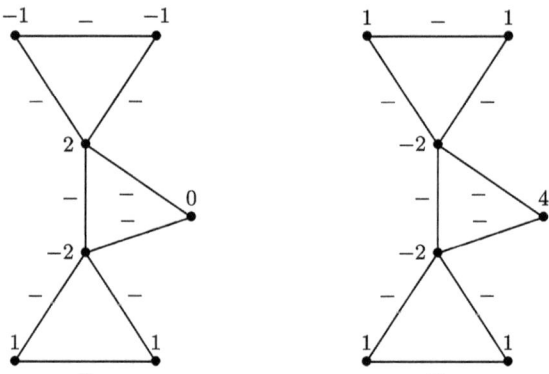

Figure 4.20 Two eigenfunctions on the signed graph of Figure 4.19.

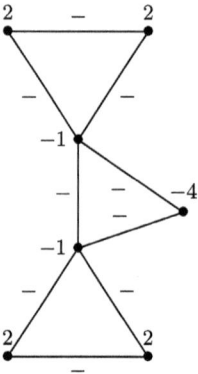

Figure 4.21 A third eigenfunction on the signed graph of Figure 4.19.

Altogether, the eigenvalue $\lambda = 1/2$ has multiplicity 4. We note that, here and in the previous examples, all eigenfunctions for $\lambda = 1/2$ sum to 0 on every loop.

In Figure 4.21, we get the eigenvalue $\lambda = 5/4$ with and in Figure 4.22, the eigenvalue $\lambda = 7/4$ with.

Lemma 4.6.5. Δ^s *is self-adjoint with respect to the product (4.1.32)* $(f,g) = \sum_v \deg(v)\, f(v)g(v)$, *and*

$$(\Delta^s f, g) = \sum_{(vv')} w_{vv'}(f(v) - s(vv')f(v'))(g(v) - s(vv')g(v')) = (f, \Delta^s g).$$

$$(4.6.12)$$

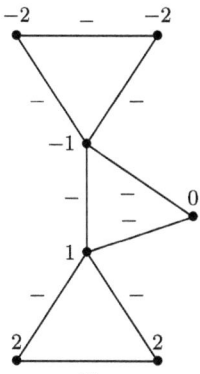

Figure 4.22 A fourth eigenfunction on the signed graph of Figure 4.19.

With the signed adjacency matrix (cf. (4.1.21))

$$A^s := (s(uv)w_{uv})_{u,v \in V} \qquad (4.6.13)$$

and the switching matrix

$$M^\theta := \mathrm{diag}(\theta(v))_{v \in V}, \qquad (4.6.14)$$

we have (cf. (4.1.22), (4.1.36))

$$\Delta^s = \mathrm{Id} - D^{-1} A^s \qquad (4.6.15)$$

and

$$A^{s^\theta} = (M^\theta)^{-1} A^s M^\theta, \qquad (4.6.16)$$

that is,

$$\Delta s^\theta = \mathrm{Id} - (M^\theta)^{-1} D^{-1} A^s M^\theta, \qquad (4.6.17)$$

from which we conclude [276]

Lemma 4.6.6. *The spectrum of a signed graph is switching invariant.*

Thus, for instance, the signed graph of Figure 4.23 has the same spectrum as (Figure 4.18) ($\lambda = 2$ and $= 3/2$ with multiplicity 1 and $= 1/2$ with multiplicity 3), as it is obtained from the latter by switching the central vertex.

From the Lemmata 4.6.2, 4.6.4, and 4.6.6, we obtain another proof of Lemma 4.1.13.

From Theorem 2.6.1 and (4.6.12), we obtain

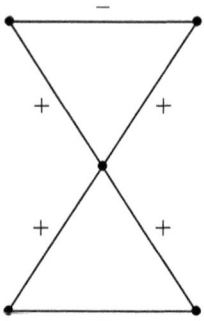

Figure 4.23 The signed graph obtained from that of Figure 4.18 by switching the central node.

Lemma 4.6.7. *The eigenvalues $\lambda_k, k = 1, \ldots, N$ of a signed graph (that is, the eigenvalues of Δ^s) satisfy, when identifying the space of functions $g : V \to \mathbb{R}$ with $\mathbb{R}^N$,*

$$\lambda_k = \min_{g \in \mathbb{R}^N, (g,g_j)=0 \text{ for } j=1,\ldots,k-1} \frac{(\Delta^s g, g)}{(g,g)} = \max_{g \in \mathbb{R}^N, (g,g_\ell)=0 \text{ for } \ell=k+1,\ldots,N} \frac{(\Delta^s g, g)}{(g,g)}$$

$$= \min_{H_k \in \mathcal{H}_k} \max_{g(\neq 0) \in H_k} \frac{(\Delta^s g, g)}{(g,g)} = \max_{H_{m-k+1} \in \mathcal{H}_{N-k+1}} \min_{g(\neq 0) \in H_{N-k+1}} \frac{(\Delta^s g, g)}{(g,g)}$$

$$= \min_{H_k \in \mathcal{H}_k} \max_{g(\neq 0) \in H_k} \frac{\sum_{(vv')} w_{vv'} (g(v) - s(vv')g(v'))^2}{(g,g)}$$

$$= \max_{H_{N-k+1} \in \mathcal{H}_{N-k+1}} \min_{g(\neq 0) \in H_{N-k+1}} \frac{\sum_{(vv')} w_{vv'} (g(v) - s(vv')g(v'))^2}{(g,g)}$$

(4.6.18)

where $\mathcal{H}_k$ is the family of all k-dimensional subspaces of $\mathbb{R}^N$.

Definition 4.6.6. We call the quotient appearing in (4.6.18),

$$R^s(g) := \frac{\sum_{(vv')} w_{vv'} (g(v) - s(vv')g(v'))^2}{\sum_v \deg v \, g(v)^2},$$

(4.6.19)

the *signed Rayleigh quotient* of Γ, and

$$R^{-s}(g) = \frac{\sum_{(vv')} w_{vv'} (g(v) + s(vv')g(v'))^2}{\sum_v \deg v \, g(v)^2}$$

(4.6.20)

its *dual-signed Rayleigh quotient*.

Corollary 4.6.1. *The eigenvalues λ of a signed graph satisfy*

$$0 \leq \lambda \leq 2.$$

(4.6.21)

Proof Because of the characterization of the eigenvalues by Rayleigh quotients, as stated in Lemma 4.6.7, the proof is the same as that of Lemma 4.1.12. □

Lemma 4.6.8. *Let θ be a switching function. Then,*

$$R^{s^\theta}(g) = R^s(\theta g). \tag{4.6.22}$$

Proof We have

$$
\begin{aligned}
R^s(\theta g) &= \frac{\sum_{(vv')} w_{vv'}(\theta(v)g(v) - s(vv')\theta(v')g(v'))^2}{\sum_v \deg v \, g(v)^2} \\
&= \frac{\sum_{(vv')} w_{vv'}(g(v) - \theta(v)s(vv')\theta(v')g(v'))^2}{\sum_v \deg v \, g(v)^2} \\
&= \frac{\sum_{(vv')} w_{vv'}(g(v) - s^\theta(vv')g(v'))^2}{\sum_v \deg v \, g(v)^2} \\
&= R^{s^\theta}(g).
\end{aligned}
$$

□

Lemmata 4.6.7 and 4.6.8 provide another proof of Lemma 4.6.6.

Corollary 4.6.2. *There exists a function g with $R^s(g) = 0$ precisely if s is balanced.*

Proof $R^s(g) = 0$ for a nontrivial g if and only if $g(v) - s(vv')g(v') = 0$ for every pair $v \sim v'$. Without loss of generality, $g(v) = \pm 1$ for all v. We then have the two subsets $V_1 := \{v: g(v) = 1\}$ and $V_2 := \{v: g(v) = -1\}$ satisfying the conditions of Theorem 4.6.1. Conversely, when we have such V_1, V_2, we can put $g(v) = 1 \, (-1)$ on $V_1 \, (V_2)$ to get $R^s(g) = 0$. □

We had observed in Lemma 4.6.4 that

$$\lambda_k(\Delta^{-s}) = 2 - \lambda_{N-k+1}(\Delta^s) \text{ for } k = 1, \ldots, N. \tag{4.6.23}$$

In particular, from Lemma 4.6.7,

Lemma 4.6.9.

$$\lambda_1(\Delta^s) = \min_{g \neq 0} R^s(g) \text{ and } 2 - \lambda_N(\Delta^s) = \lambda_1(\Delta^{-s}) = \min_{g \neq 0} R^{-s}(g). \tag{4.6.24}$$

From Corollary 4.6.2 and Lemma 4.6.9, we have

Corollary 4.6.3. *The following conditions are equivalent for a signed graph (Γ, s).*

1. (Γ, s) *is balanced.*
2. *s is switching equivalent to the positive sign function.*
3. $$\lambda_1 = 0. \tag{4.6.25}$$

Similar to Corollaries 4.6.2 and 4.6.3, we have

Corollary 4.6.4. *There exists a function g with* $R^{-s}(g) = 0$ *precisely if s is antibalanced.*

The following conditions are equivalent for a signed graph (Γ, s).

1. (Γ, s) *is antibalanced.*
2. *s is switching equivalent to the negative sign function.*
3.
$$\lambda_N = 2. \tag{4.6.26}$$

We shall now describe a relation, established in [30], between the spectrum of a 2-covering of a graph and that of a signed graph associated to that covering.

Let Γ be a finite graph with vertex set V, again unweighted, not out of necessity, but for simplicity. Let Γ' be an isomorphic copy of Γ, with vertex v' of Γ' corresponding to vertex v of Γ. For the construction of a 2-covering $\hat{\Gamma}$ of Γ, we can take the union of V and V', and for each pair $u \sim v$ of neighbors in Γ (and $u' \sim v'$ then are neighbors in Γ'), we either connect u and v and hence also u' and v' also in $\hat{\Gamma}$ (parallel connections), or we connect u with v' and v with u' (cross connections). When Γ has N vertices, there are 2^N different possible choices and hence that many different such 2-coverings. In Figure 4.24, we only have parallel connections whereas in Figure 4.25, some of them are replaced by cross connections.

In Figure 4.25, the covering graph is connected.

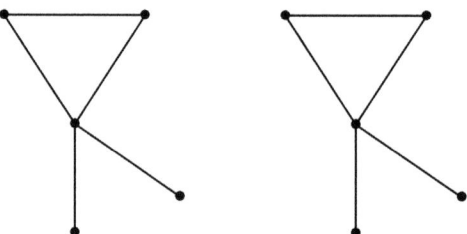

Figure 4.24 Only parallel connections.

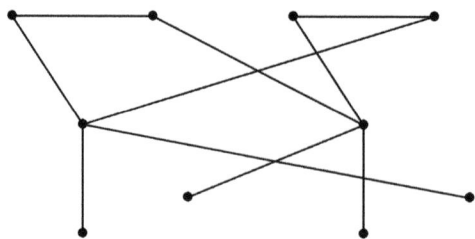

Figure 4.25 Parallel and cross connections.

Theorem 4.6.1. *The spectrum of $\hat{\Gamma}$ is the union of the spectrum of Γ and that of the signed graph (Γ, s) obtained from Γ by assigning $-$ to precisely those edges that are replaced by cross connections.*

In Figure 4.26, this is shown for the example of Figure 4.25.

Proof The operation $\tau \colon V \to V', \tau(v) = v', \tau(v') = v$ for all v, v' induces an automorphism of $\hat{\Gamma}$. As in Section 2.7, we have symmetric and antisymmetric eigenfunctions. The symmetric ones are obtained by taking an eigenfunction f of Γ and putting $f(v') = f(v)$ for all v, v'. This is an eigenfunction of $\hat{\Gamma}$ because if u is connected with v' instead of v, the value of $\Delta f(u)$ does not change.

Likewise, if g is an eigenfunction of (Γ, s), that is,

$$\Delta^s g(v) = g(v) - \frac{1}{\deg v} \sum_{u \sim v} s(uv) g(u) = \lambda g(v),$$

then putting $g(v') = -g(v)$ for all v, v' yields an antisymmetric eigenfunction.

$\square$

For instance, identifying v_i and v_i' for $i = 0, 1, 2$ in Figure 4.27 yields a 2-cover of the triangle, and therefore, also a book graph with an even number of pages is a 2-cover of a petal graph. Hence, by Theorem 4.6.1 and a generalization of the spectral analysis of (Figure 4.23), the spectrum of such a book

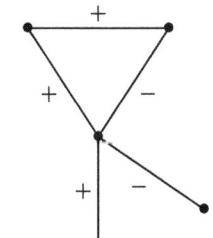

Figure 4.26 An illustration of Theorem 4.6.1.

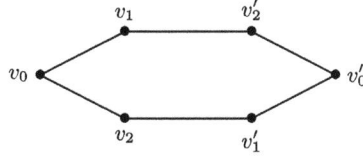

Figure 4.27 A special book graph illustrating the application of Theorem 4.6.1.

graph is obtained from that of the petal graph. It thus consists of the eigenvalues $0, 1/2, 3/2, 2$, as directly determined before Theorem 4.1.2.

Finally, we want to discuss Cheeger constants of signed graphs. Following [11], the constructions and estimates of Section 4.4 and Chapter 8 can be unified by introducing a signed Cheeger constant. For that purpose, let $E^+(V_1, V_2)$ the positively signed and $E^-(V_1, V_2)$ the negatively signed edges between $V_1, V_2 \subset V$. We assume that V_1 and V_2 are not both empty. The *signed Cheeger constant* of the signed graph (Γ, s) then is

$$h^s(\Gamma) :=$$

$$\min_{V_1, V_2} \frac{2|E^+(V_1, V_2)| + 2|E^-(V_1, V_1)| + 2|E^-(V_2, V_2)| + |E(V_1 \cup V_2, \overline{V_1 \cup V_2})|)}{\text{vol}(V_1) + \text{vol}(V_2)}.$$

$$(4.6.27)$$

With the techniques of the proofs of Theorems 4.3.1 and 4.4.1, one can obtain

Theorem 4.6.2. *The smallest eigenvalue of the signed Laplacian Δ^s of the signed graph (Γ, s) satisfies*

$$1 - \sqrt{1 - (h^s)^2} \leq \lambda_1 \leq 2h^s. \qquad (4.6.28)$$

Of course, when the graph is balanced, then by Corollary 4.6.3, $\lambda_1 = 0$, and consequently, also $h^s(\Gamma) = 0$. We can then find vertex sets V_1, V_2 with $V_1 \cup V_2 = V, |E^+(V_1, V_2)| = |E^-(V_1, V_1)| = |E^-(V_2, V_2)| = 0$. For instance, when the sign function is positive on all edges, we can simply take $V_1 = V$. In particular, the signed Cheeger constant on this graph does not reduce to the ordinary Cheeger constant. It rather generalizes the constant h_1 introduced in 4.3.7.

4.7 The General Case: Directed Graphs with Arbitrary Weights

In this section, we shall briefly outline the spectral theory of directed graphs with arbitrary real edge weights. In particular, we also allow the weights to be negative. We do not develop the theory in detail, however, as these general structures will not come up in the rest of the book, except for Chapter 12.

For a directed graph, the spectrum in general is no longer real, and in this book, we only investigate real spectra. As already indicated, there is one exception, the nonbacktracking Laplacian treated in Chapter 12. This Laplacian can be seen as a Laplacian of a directed graph and therefore typically will also possess nonreal eigenvalues. A general spectral theory of Laplacians has been developed by Bauer [23], and we shall reproduce here only the basic setting of that theory, referring to [23] for deeper results and to [25, 26] for applications for the synchronization of dynamical networks. The reader should be alerted that we do not adopt all the conventions of [23]. For instance, there the Laplacian is defined in terms of incoming edges at a node, but here, we use the outgoing edges. The two different versions are of course formally completely analogous.

In this section, $\Gamma = (V, E, w)$ is a weighted directed graph on N vertices. The weight function $w : V \times V \to \mathbb{R}$ is not assumed to be symmetric and can take arbitrary real values. A directed edge from v to v' is denoted by $e = (v, v') \in E$. $w_{vv'}$ is the weight of $e = (v, v')$. $w_{vv'} = 0$ if and only if $e = (v, v') \notin E$. And Γ is undirected precisely if $w_{vv'} = w_{v'v}$ for all v, v'. $W := (w_{vv'})$ is the *weighted adjacency matrix*. A path from v to v' is a sequence of directed edges $(v, v_1), (v_1, v_2), \ldots, (v_k, j)$, that is, if one gets from v to v' along a sequence of directed edges.

The in-degree and the out-degree of vertex v are defined as

$$\deg_{\text{in}} v := \sum_{v'} w_{v'v} \quad \text{and} \quad \deg_{\text{out}} v := \sum_{v'} w_{vv'}. \tag{4.7.1}$$

A graph is called *balanced* if $\deg_{\text{in}} v = \deg_{\text{out}} v$ for all $v \in V$. Obviously, undirected graphs are balanced.

Definition 4.7.1. The vertex v is a *dead end* if $w_{vv'} = 0$ for all $v' \in V$. The vertex v is *ambiguous* if $\deg_{\text{out}} v = \sum_{v'} w_{vv'} = 0$.

Every dead-end vertex is ambiguous, but not vice versa because we are allowing for both positive and negative weights.

Definition 4.7.2. A directed graph Γ is *weakly connected* if for any pair of distinct vertices v and v', there is a path from v to v' or from v' to v in Γ. A directed graph Γ is *strongly connected* if for any pair of distinct vertices v and v', there exist both a path from v to v' and one from v' to v.

Strongly connected graphs are weakly connected, but the converse need not hold. But for undirected graphs, the two notions coincide.

Definition 4.7.3. The *normalized graph Laplacian* for a directed graph Γ operates on a function $f : V \to \mathbb{C}$ via

$$\Delta f(v) := \begin{cases} f(v) - \frac{1}{\deg_{\text{out}} v} \sum_{v'} w_{vv'} f(v') & \text{if } \deg_{\text{out}} v \neq 0. \\ 0 & \text{else.} \end{cases} \tag{4.7.2}$$

The *algebraic graph Laplacian* operates via

$$\Lambda f(v) := (\deg_{\text{out}} v)\, f(v) - \sum_{v'} w_{vv'} f(v') \tag{4.7.3}$$

In the rest of this section, we shall only treat the normalized Laplacian Δ, although Λ enjoys similar results.

If $\deg_{\text{out}} v \neq 0$ for all $v \in V$, then Δ is given by

$$\Delta = \text{Id} - D^{-1}W,$$

where D is the diagonal matrix of vertex out-degrees, and W is the weighted adjacency matrix of the graph Γ. When restricted to undirected graphs with non-negative weights, Definition 4.7.3 reduces to the definitions given earlier in (4.1.2), (4.1.20), (4.1.33), and (4.1.35).

Let $V_R \subseteq V$ be the set of all vertices that are not ambiguous. We can define the reduced Laplacian Δ_R on V_R as

$$\Delta_R = \text{Id}_R - D_R^{-1}W_R,$$

where the subscript R simply denotes the restriction to V_R.

Every graph that is not strongly connected can be decomposed into its strongly connected components. We observe that if Γ_k and Γ_ℓ are two different strongly connected components, then there can exist directed edges between them in only one direction. Namely, if there were both a directed edge from a vertex $v_k \in \Gamma_k$ to a vertex $v_\ell \in \Gamma_\ell$, and one from $w_\ell \in \Gamma_\ell$ to $w_k \in \Gamma_k$, then $\Gamma_k \cup \Gamma_\ell$ would also be strongly connected.

Therefore, using this decomposition, after a permutation of the vertices, the Laplacian Δ can then be written in the Frobenius normal form

$$\Delta = \begin{pmatrix} \Delta_1 & \Delta_{12} & \dots & \Delta_{1r} \\ 0 & \Delta_2 & \dots & \Delta_{2r} \\ \vdots & \vdots & \ddots & \vdots \\ 0 & 0 & \dots & \Delta_r \end{pmatrix}, \tag{4.7.4}$$

where $\Delta_1, \dots \Delta_r$ are square matrices corresponding to the strongly connected components $\Gamma_1, \dots, \Gamma_r$ of Γ. The off-diagonal elements of Δ_k are of the form $\frac{w_{vv'}}{\deg_{\text{out}} v}$ for $v, v' \in \Gamma_k$, and the diagonal elements are either zero (if the out-degree of the corresponding vertex is equal to zero) or one (if it is nonzero). If Γ_k does not contain an ambiguous vertex, then Δ_k is irreducible. Furthermore, the submatrices Δ_{kl}, $1 \leq k < l \leq r$ are determined by the connectivity structure between different strongly connected components. Δ_{kl} contains all elements of the form $\frac{w_{vv'}}{\deg_{\text{out}} v}$ for $v \in \Gamma_k, v' \in \Gamma_l$. We summarize the preceding considerations in

Lemma 4.7.1. *The spectrum of* Δ *consists of the eigenvalues of* Δ_R *and the eigenvalue 0 with multiplicity* $|V \setminus V_R|$, *and it is the union of the spectra of the Laplacians* Δ_{Γ_i} *of the strongly connected components* Γ_i *of* Γ.

Proof The first claim follows from (4.7.2) since $\deg_{\text{out}} = 0$ on $|V \setminus V_R|$. The second claim follows from (4.7.4) since $\Delta_i, i = 1, \ldots, r$, is the Laplacian of the strongly connected component Γ_i. $\qquad\qquad\square$

We shall now present basic spectral properties of the Laplacian Δ, following [23]. While for undirected graphs with non-negative weights, the Laplacian is self-adjoint and non-negative, and hence the eigenvalues of Δ are real and non-negative, the Laplacian considered here is neither self-adjoint nor non-negative, and thus the eigenvalues are in general complex and can have negative real parts.

Theorem 4.7.1.   (i) *The Laplacian* Δ *has always an eigenvalue* $\lambda = 0$ *with eigenvector* $e = (1, \ldots, 1)^{\top}$.
 (ii) *The non-real eigenvalues of* Δ *appear in complex conjugate pairs.*
(iii) *The eigenvalues of* Δ *satisfy*

$$\sum_{i=1}^{N} \lambda_i = \sum_{i=1}^{N} \mathcal{R}(\lambda_i) = |V_R|.$$

 (iv) *The spectrum of* Δ *is invariant under multiplying all weights* $w_{vv'}$ *for some fixed* v *and all* v' *by a nonzero constant* c. *In particular, it is invariant if we multiply all weights by a nonzero constant* c.

Proof   (i) Δ has zero row sums, and so $e = (1, \ldots, 1)^{\top}$ is an eigenvector with eigenvalue 0.
 (ii) Since Δ is a real matrix, the characteristic polynomial is given by

$$\det(\Delta - \lambda I) = a_0 + a_1 \lambda + \ldots + a_{N-1} \lambda^{N-1},$$

with all $a_i \in \mathbb{R}$. Therefore, the roots occur in complex conjugate pairs.
(iii) This follows from (ii) and by considering the trace of Δ.
 (iv) This follows from the definition of Δ and $\deg_{\text{out}}$. $\qquad\qquad\square$

By Lemma 4.7.1, it suffices to consider strongly connected graphs.
We shall now apply Gershgorin's circle theorem [108], formulated here as

Lemma 4.7.2. *Let* $A = (a_{ij})_{i,j=1,\ldots M}$ *be a matrix with entries* $a_{ij} \in \mathbb{C}$ *and*

$$\rho_i = \sum_{j \neq i} |a_{ij}|.$$

Let $D(z,r)$ denote the closed disk in $\mathbb{C}$ with center z and radius r.

Then, every eigenvalue of A lies in one of the Gershgorin discs $D(a_{ii}, \rho_i)$.

Proof Let λ be an eigenvalue of A with eigenvector $x = (x^1, \ldots, x^M)$. Then, for all i,

$$\sum_{j=1}^{M} a_{ij} x^j = \lambda x^i,$$

hence also

$$\sum_{j \neq i} a_{ij} x^j = (\lambda - a_{ii}) x^i.$$

We take a component x^i with the largest absolute value (which is $\neq 0$, as an eigenvector cannot vanish identically). Thus, $|x^j| \leq |x^i|$ for all j, and

$$|\lambda - a_{ii}| = \left| \sum_{j \neq i} \frac{a_{ij} x^j}{x^i} \right| \leq \sum_{j \neq i} \left| \frac{a_{ij} x^j}{x^i} \right| \leq \sum_{j \neq i} |a_{ij}| = R_i.$$

$\square$

Theorem 4.7.2. *Let $V_{R,1}, \ldots, V_{R,r}$ be the strongly connected components of the induced subgraph Γ_R with vertex set V_R. In our formulas, we put $\frac{1}{|\deg_{\text{out}} v|} = 0$ when $\deg_{\text{out}} v = 0$.*

The spectrum of Δ then satisfies

$$\sigma(\Delta) \subseteq D(1, \rho_1) \cup \{0\} \subseteq D(1, \rho_2) \cup \{0\} \subseteq D(1, \rho) \cup \{0\},$$

where $D(1,r)$ is the closed disk in the complex plane with center 1 and radius r, and

$$\rho_1 := \max_{i=1,\ldots,r} \max_{v \in V_{R,i}} \frac{\sum_{v' \in V_{R,i}} |w_{vv'}|}{|\deg_{\text{out}} v|}$$

$$\rho_2 := \max_{v \in V_R} \frac{\sum_{v' \in V_R} |w_{vv'}|}{|\deg_{\text{out}} v|}$$

$$\rho := \max_{v \in V} \rho(v) := \max_{v \in V} \frac{\sum_{v' \in V} |w_{vv'}|}{|\deg_{\text{out}} v|}.$$

Proof $\rho_1 \leq \rho_2 \leq \rho$ and the proof follows from Lemmas 4.7.2 and 4.7.1. $\square$

For undirected graphs with non-negative weights, Theorem 4.7.2 reduces to (4.1.51).

The radius ρ in Theorem 4.7.2 is ≥ 1 if and only if there is at least one vertex that is not ambiguous. And if no vertex is ambiguous, then $\rho = 1$ if and only if for every vertex v the weights $w_{vv'}$ have the same sign for all v'.

From this observation and Theorem 4.7.1 (iv), we conclude that if Γ is a graph without ambiguous vertices, that is, $V_R = V$, and with $\rho = 1$, then the graph Γ^+ where we have replaced all weights by their absolute values $|w_{vv'}|$ has the same spectrum as Γ.

Corollary 4.7.1. *The nonzero eigenvalues of Δ satisfy*

$$1 - \rho \leq \min_{i:\, \lambda_i \neq 0} \Re(\lambda_i) \leq \frac{|V_R|}{N - m_0} \leq \max_{i:\, \lambda_i \neq 0} \Re(\lambda_i) \leq 1 + \rho, \qquad (4.7.5)$$

where m_0 is the multiplicity of the eigenvalue 0. In particular,

$$1 \leq \max_{i:\, \lambda_i \neq 0} \Re(\lambda_i).$$

Proof The estimates follow from Theorem 4.7.1 (iii) and Theorem 4.7.2 and the fact that $N - m_0 \leq |V_R|$. □

4.8 Simplicial Complexes

We return to symmetric operators.

For a simplicial complex, the multiplicities of the eigenvalue 0 of the Laplacians L^q contain topological information. This is the content of Eckmann's Theorem [86], which is a discrete version of the Hodge theorem.

Theorem 4.8.1. *For a simplicial complex Σ,*

$$\ker L^q(\Sigma) \cong H^q(\Sigma).$$

Thus, the multiplicity of the eigenvalue 0 of the operator $L^q(\Sigma)$ is equal to the Betti number b_q, the dimension of $H^q(\Sigma)$.

Proof By (2.5.15),

$$\ker L^q(\Sigma) = \ker \delta^q \cap \ker \delta^{q-1*} \qquad (4.8.1)$$
$$= \ker \delta^q \cap (\operatorname{im} \delta^{q-1})^\perp$$
$$\cong H^q(\Sigma, \mathbb{R}).$$

□

Corollary 4.8.1.

$$C^q = \operatorname{im} \delta^{q-1} \oplus \operatorname{im} (\delta^q)^* \oplus \ker L^q. \qquad (4.8.2)$$

Proof This follows from (2.5.6) and (4.8.1).   □

While cohomology groups are defined as quotients, that is, as equivalence classes of elements of C^q, Theorem 4.8.1 provides us with concrete representatives in C^q of those equivalence classes, the so-called *harmonic cocycles*.

We note that Eckmann's Theorem does not depend on the choice of scalar products on the spaces C^q (although the harmonic cocycles do). That theorem is concerned with the eigenvalue 0 of the Laplacian. We shall now investigate the non zero part of the spectrum.

Since $\delta^q \delta^{q-1} = 0$ and $\delta^{q-1^*} \delta^{q^*} = 0$ (recall (2.5.2)),

$$\operatorname{im} L^q_{down}(\Sigma) \subset \ker L^q_{up}(\Sigma), \tag{4.8.3}$$

$$\operatorname{im} L^q_{up}(\Sigma) \subset \ker L^q_{down}(\Sigma). \tag{4.8.4}$$

Therefore, λ is a nonzero eigenvalue of $L_i(\Sigma)$ if and only if it is a nonzero eigenvalue of either $L^q_{up}(\Sigma)$ or $L^q_{down}(\Sigma)$. Therefore, the nonzero parts of the spectra satisfy

$$\operatorname{spec}_{\neq 0}(L^q(\Sigma)) = \operatorname{spec}_{\neq 0}(L^{up}_i(\Sigma)) \cup \operatorname{spec}_{\neq 0}(L^{down}_i(\Sigma)). \tag{4.8.5}$$

The multiplicity of the eigenvalue 0 may be different, however.

Since $\operatorname{spec}_{\neq 0}(AB) = \operatorname{spec}_{\neq 0}(BA)$, for linear operators A and B on Hilbert spaces, we get the following equality.

$$\operatorname{spec}_{\neq 0}(L^q_{up}(\Sigma)) = \operatorname{spec}_{\neq 0}(L^{q+1}_{down}(\Sigma)). \tag{4.8.6}$$

From (4.8.5) and (4.8.6), we conclude that each of the three families of multisets

$$\{\operatorname{spec}_{\neq 0}(L^q(\Sigma)) \mid 0 \le q \le m\}, \ \{\operatorname{spec}_{\neq 0}(L^q_{up}(\Sigma)) \mid 0 \le q \le m-1\}$$

$$\text{or } \{\operatorname{spec}_{\neq 0}(L^q_{down}(\Sigma)) \mid 1 \le q \le m\}$$

determines the other two. Therefore, it suffices to consider only one of them. In the sequel, we shall often omit the argument Σ from our Laplace operators.

We now apply the constructions of Section 4.6. This is partly based on discussions with Bobo Hua.

The results here are taken from [160]. We recall the formula (2.5.21) for the combinatorial normalized up-Laplacian of a simplicial complex Σ,

$$(L^q_{up}f)([\sigma]) = f([\sigma]) - \frac{1}{\deg \sigma} \sum_{\substack{\sigma' \in \Sigma_q \,:\, \sigma \neq \sigma', \\ \sigma, \sigma' \in \partial \rho}} w_{\sigma \sigma'} s([\sigma], [\sigma']) f([\sigma']),$$

$$\tag{4.8.7}$$

where we have put

$$w_{\sigma\sigma'} := w(\rho) \text{ for } \sigma, \sigma' \in \partial\rho,$$
$$s([\sigma], [\sigma']) := -\operatorname{sgn}([\sigma], \partial[\rho]) \operatorname{sgn}([\sigma'], \partial[\rho]). \tag{4.8.8}$$

Thus, we may express L_{up}^q in terms of the Laplacian $\Delta_{(\Gamma_q, s)}$ for the signed graph (Γ_q, s) with vertex set consisting of the q-simplices of our simplicial complex, and where two different such vertices σ, σ' are connected by an edge, $\sigma \sim \sigma'$, if there exists a $(q + 1)$-simplex ρ in Σ with $\sigma, \sigma' \in \partial\rho$. For instance, when we have a 2-simplex, then the signed graph corresponding to its three boundary edges is given in Figure 4.17, and when we have two adjacent 2-simplices, we get that in Figure 4.18.

Theorem 4.8.2.

$$L_{up}^q = (q + 1)\Delta_{(\Gamma_q, s)} - q \operatorname{Id}. \tag{4.8.9}$$

Proof In order to get that relation, we have to observe that

$$\sum_{\substack{\sigma' \in \Sigma_q : \sigma \neq \sigma', \\ \sigma, \sigma' \in \partial\rho}} w_{\sigma\sigma'} = (q + 1) \deg \sigma, \tag{4.8.10}$$

since according to (2.5.19)

$$\deg \sigma = \sum_{\rho \in \Sigma_{q+1}(\Sigma) : \sigma \in \partial\rho} w(\rho), \tag{4.8.11}$$

and each such ρ has $q + 1$ other facets σ'. Hence,

$$(L_{up}^q f)([\sigma]) = \sum_{\substack{\sigma' \in \Sigma_q : \sigma \neq \sigma', \\ \sigma, \sigma' \in \partial\rho}} \frac{w_{\sigma\sigma'}}{\deg \sigma}(f([\sigma]) - s([\sigma], [\sigma'])f([\sigma'])) - qf([\sigma])$$
$$= (q + 1)\Delta_{(\Gamma_q, s)} f([\sigma]) - qf([\sigma]).$$

$\square$

Remark 4.8.1. The condition (2.5.20) is set up in such a way that it simplifies the up-Laplacian (2.5.17) by making the coefficient of $f([\sigma])$ equal to 1 in (2.5.21). This allows us to derive the relation (4.8.9). Since (2.5.20) does not lead to such a simplification for the down-Laplacian (2.5.18), we do not have a corresponding formula for the down-Laplacian. That would only be obtained if instead of (2.5.20), we had required that the weight of a simplex be equal to the sum of the weights of its facets. Since, however, we already know from (4.8.6) that the nonzero eigenvalues of the down-Laplacians coincide with those of the up-Laplacians, we can concentrate on the latter for investigating the spectra of simplicial complexes.

By Theorem 4.8.2, the eigenvalues μ_j of L_{up}^q and the eigenvalues λ_j of $\Delta_{(\Gamma_q,s)}$ are related by

$$\mu_j = (q+1)\lambda_j - q. \qquad (4.8.12)$$

and we obtain

Corollary 4.8.2. *The eigenvalues of L_{up}^q lie in the interval $[0, q+2]$.*

Proof The eigenvalues are non-negative because L_{up}^q is a non-negative operator, see (2.5.10). The upper bound follows from (4.8.9) and the fact that the spectrum of $\Delta_{(\Gamma_q,s)}$ is contained in the interval $[0,2]$, see (4.6.21). $\qquad \square$

Corollary 4.8.3. *The eigenvalues of $\Delta_{(\Gamma_q,s)}$ are $\geq \frac{q}{q+1}$. Equality holds if and only if there is some nontrivial f with $\partial^q f = 0$. More precisely, the multiplicity of the eigenvalue $\frac{q}{q+1}$ of $\Delta_{(\Gamma_q,s)}$ equals the dimension of the kernel of the coboundary operator δ^q.*

In particular, for $q > 0$, the graph (Γ_q, s) is never balanced.

The geometric intuition for the last statement can be seen in Figure 4.17, which is the signed graph corresponding to the boundary of a 2-simplex. More generally, when Γ_q comes from the boundary of a single $(q+1)$-simplex, then it is the complete graph with $q+1$ vertices and a $-$ sign on every edge.

Proof The inequality follows again from (4.8.9) and the fact that L_{up}^q is a non-negative operator, that is, all its eigenvalues are non-negative. The equality case is handled in Corollary 2.5.2. The last claim follows from Corollary 4.6.4. $\quad \square$

Later, we shall study the eigenvalues of L_{up}^q, and the results are taken from [138, 160].

We now try to characterize those simplicial complexes where the spectrum of L_{up}^q contains the maximal value $q + 2$, following [138]. As before, we use (4.8.9). According to Corollary 4.6.4 or the proof of Corollary 4.6.1, the dual-signed Rayleigh quotient (4.6.20) of $\Delta_{(\Gamma_q,s)}$ must assume the value 0, and for that, we need some function g with (in the notation employed in (4.6.20))

$$g(v) = -s(vv')g(v') \text{ for any pair } v \sim v'. \qquad (4.8.13)$$

In the notation currently employed, see (4.8.7) and (4.8.8), this means that we need

$$\text{sgn}([\sigma], \partial[\rho])f(\sigma) = \text{sgn}([\sigma'], \partial[\rho])f(\sigma') \qquad (4.8.14)$$

whenever σ, σ' are different facets of some ρ. We may then assume that $f(\sigma) = \pm 1$ for all σ, that is, f assigns an orientation to every σ. We observe

Lemma 4.8.1. *The existence of a function f satisfying (4.8.14) means that there exists an orientation on the (q + 1)-simplices of Σ for which any two (q + 1)-simplices intersecting in a common q-face induce the same orientation on the intersecting simplex.*

□

Definition 4.8.1. A pure simplicial complex of dimension p is called a *p-path of length m* if and only if there is an ordering of its p-simplices $\rho_1, \rho_2 \ldots \rho_m$, such that ρ_i and ρ_j are $(p-1)$-down neighbors, that is, share a facet, if and only if $\mid j - l \mid = 1$. When ρ_m coincides with ρ_1, we speak of a *q-circuit* of length $(m - 1)$.

We say that a path or circuit is *orientable* if and only if it is possible to assign an orientation to all its p-simplices in such a way that any two simplices that intersect in an $(p-1)$-face induce a different orientation on that face. We say that such simplices are oriented *coherently*.

Note that some or all the p-simplices of a path or circuit could intersect in some simplices of dimension $\leq p - 2$.

A graph possesses the eigenvalue 2 if and only if it is bipartite, that is, does not contain cycles of odd length; see Lemma 4.1.11. The following is the higher-dimensional generalization of this fact [138].

Theorem 4.8.3. *For a q-connected (that is, any two q-simplices can be connected by a path) simplicial complex Σ, the following statements are equivalent*

1. *The spectrum of L_{up}^q contains the eigenvalue $q + 2$.*
2. *There are no $(q + 1)$-orientable circuits of odd length nor $(q + 1)$-nonorientable circuits of even length in Σ.*

Proof (1) ⇒ (2) proceeds by contradiction: Assume that there exists a $(q+1)$-orientable circuit of odd length, whose q-simplices $\sigma_1, \ldots, \sigma_{2n+1}$ are ordered increasingly, as suggested in Definition 4.8.1. Then, it is possible to orient these simplices in such a way that every two neighboring simplices induce different orientations on their intersecting face. Denote these oriented simplices by $[\sigma_1], \ldots, [\sigma_{2n+1}]$. In order to have the same orientation induced on the intersecting face, we reverse the orientation of every simplex $[\sigma_k]$, for k even. Thus, $[\sigma_l]$ and $-[\sigma_{l+1}]$ induce the same orientation on $[\sigma_l \cap \sigma_{l+1}]$, for every $1 \leq l \leq 2n$. However, $[\sigma_1]$ and $[\sigma_{2n+1}]$ remain coherently oriented, which contradicts Lemma 4.8.1. The analysis for the case of $(q + 1)$-nonorientable circuits is analogous.

(2) $\Rightarrow$ (1): Let σ_1 be an arbitrary $(q + 1)$-face of Σ. Consider its positive orientation $[\sigma_1]$ and call it an *initial* oriented face. Let $[\sigma_{i_1 i_2 \dots i_n}]$ be a $(q + 1)$-face of Σ which shares a q-face with $[\sigma_{i_1 i_2 \dots i_{n-1}}]$, and both faces induce the same orientation on their intersecting face. Now, assume the opposite: The eigenvalue $q + 2$ is not in the spectrum, that is, it is not possible to choose an orientation on the $(q + 1)$-faces of Σ satisfying the conditions of Lemma 4.8.1. This means that after some number of steps in the construction earlier, two faces $[\sigma_{i_1 i_2 \dots i_n}]$, $[\sigma_{i_1 i_2 \dots i_m}]$, which are the same but differently oriented are obtained. Obviously, there exists a circuit containing $[\sigma_{i_1 i_2 \dots i_n}]$ which does not admit an orientation as in Lemma 4.8.1. This is possible only in the case when a circuit is orientable and odd or nonorientable and even. This is a contradiction; hence, $q + 2$ is contained in the spectrum.   □

As a supplement of Theorem 4.8.3, we have:

Proposition 4.8.1. *The spectrum of L_{up}^q contains the eigenvalue $q + 2$ if and only if the signed graph (Γ_q, s) has an antibalanced component. Moreover, the multiplicity of $q + 2$ equals the number of antibalanced components of (Γ_q, s).*

We consider the opposite $(\Gamma_q, -s)$ of the signed graph (Γ_q, s), and we simply use $\Delta_{(\Gamma_q, -s)}$ to denote the Laplacian for the signed graph $(\Gamma_q, -s)$. Then, the eigenvalues of the three Laplacians L_{up}^q, $\Delta_{(\Gamma_q, s)}$ and $\Delta_{(\Gamma_q, -s)}$ satisfy the relation:

Spectrum of L_{up}^q		Spectrum of $\Delta_{(\Gamma_q, s)}$		Spectrum of $\Delta_{(\Gamma_q, -s)}$
0		$\frac{q}{q+1}$		$\frac{q+2}{q+1}$
$\vdots$		$\vdots$		$\vdots$
λ	$\Longleftrightarrow$	$\frac{\lambda+q}{q+1}$	$\Longleftrightarrow$	$\frac{q+2-\lambda}{q+1}$
$\vdots$		$\vdots$		$\vdots$
$q + 2$		2		0

that is, λ is an eigenvalue of L_{up}^q if and only if $\frac{\lambda+q}{q+1}$ is an eigenvalue of $\Delta_{(\Gamma_q, s)}$ if and only if $\frac{q+2-\lambda}{q+1}$ is an eigenvalue of $\Delta_{(\Gamma_q, -s)}$.

In addition, the multiplicity of the eigenvalue 0 of L_{up}^q is larger than or equal to $q + 1$ (when the simplicial complex is pure, the multiplicity of the eigenvalue 0 of L_{up}^q is $q + 1$ if and only if the simplicial complex is a simplex of dimension $q+1$). And, the multiplicity of the eigenvalue $q+2$ of L_{up}^q agrees with the number of balanced components of $\Delta_{(\Gamma_q, -s)}$.

4.9 Oriented Hypergraphs

We consider an (oriented) hypergraph $\Gamma = (V, E, \varphi)$, as defined in Section 2.4.3, and we assume that it has no isolated vertices, that is, all vertices have degree > 0. We let $v_1, \ldots, v_N$ denote its vertices and we let $e_1, \ldots, e_M$ denote its hyperedges. We recall, from Section 2.5.4, that the Laplacian L^0 is the operator

$$L^0 \colon \{f \colon V \to \mathbb{R}\} \to \{f \colon V \to \mathbb{R}\}$$

such that, given $f \colon V \to \mathbb{R}$ and $v \in V$,

$$L^0 f(v) = \frac{\sum_{e_{\text{in}} \colon v \text{ input}} \left(\sum_{v' \text{ input of } e_{\text{in}}} f(v') - \sum_{z' \text{ output of } e_{\text{in}}} f(z') \right)}{\deg v}$$

$$- \frac{\sum_{e_{\text{out}} \colon v \text{ output}} \left(\sum_{\hat{v} \text{ input of } e_{\text{out}}} f(\hat{v}) - \sum_{\hat{z} \text{ output of } e_{\text{out}}} f(\hat{z}) \right)}{\deg v}.$$

Similarly, the Laplacian L^1 is the operator

$$L^1 \colon \{\gamma \colon E \to \mathbb{R}\} \to \{\gamma \colon E \to \mathbb{R}\}$$

such that, given $\gamma \colon E \to \mathbb{R}$ and $e \in E$,

$$L^1 \gamma(e) = \sum_{v_i \text{ input of } e} \frac{\sum_{e_{\text{in}} \colon v_i \text{ input}} \gamma(e_{\text{in}}) - \sum_{e_{\text{out}} \colon v_i \text{ output}} \gamma(e_{\text{out}})}{\deg v_i}$$

$$- \sum_{v^j \text{ output of } e} \frac{\sum_{e'_{\text{in}} \colon v^j \text{ input}} \gamma(e'_{\text{in}}) - \sum_{e'_{\text{out}} \colon v^j \text{ output}} \gamma(e'_{\text{out}})}{\deg v^j}.$$

We want to consider the matrix formulations of L^0 and L^1. We need some preliminary definitions:

Definition 4.9.1. The *degree matrix* of Γ is the $N \times N$ diagonal matrix

$$D := \operatorname{diag}(\deg v_1, \ldots, \deg v_N). \tag{4.9.1}$$

The *incidence matrix* of Γ is the $N \times M$ matrix $\mathcal{I} := (\varphi(v_i, e_j))_{ij}$.

   Given a hyperedge $e \in E$, we say that two vertices v_i and v_j are *co-oriented in e* if

$$\varphi(v_i, e) = \varphi(v_j, e) \neq 0.$$

We say that v_i and v_j are *antioriented in e* if

$$\varphi(v_i, e) = -\varphi(v_j, e) \neq 0.$$

The *adjacency matrix* of Γ is the $N \times N$ matrix $A := (A_{ij})_{ij}$, where $A_{ii} := 0$ for each $i = 1, \ldots, N$ and, for $i \neq j$,

$$A_{ij} := \big|\{\text{hyperedges in which } v_i \text{ and } v_j \text{ are antioriented}\}\big|$$
$$- \big|\{\text{hyperedges in which } v_i \text{ and } v_j \text{ are co-oriented}\}\big|.$$

The above matrices allow us to write

$$L^0 = \text{Id} - D^{-1}A = D^{-1}\mathcal{I}\mathcal{I}^\top$$

and

$$L^1 = \mathcal{I}^\top D^{-1}\mathcal{I},$$

in a matrix form. Moreover, we can also write

$$L^0 f(v_i) = f(v_i) - \frac{1}{\deg v_i} \sum_{j \neq i} A_{ij} f(v_j),$$

for a given function $f \colon V \to \mathbb{R}$ and a vertex v_i.

Remark 4.9.1. We make an important observation. When we consider an oriented hypergraph, if a vertex v belongs to a hyperedge e, then v is either an input or an output for e but not both. This is true also if e is a self-loop, that is, it contains only v. On the other hand, when we consider a self-loop $e = \{v\}$ in Section 4.1, that is, in the case of graphs, the vertex v is treated both as input and as output of e. Therefore, self-loops in this chapter are conceptually different, and, as a consequence, also the Laplacians are different. In particular, while in the context of Section 4.1, the Laplacian matrix L^0 does not only have 1's in the diagonal in the presence of self-loops, the Laplacian L^0 for oriented hypergraphs always has only 1's in the diagonal.

Our aim is to investigate the spectral properties of L^0 (and L^1) and to understand, in particular, which spectral properties from Section 4.1 can be generalized to the hypergraph case and how. We start with the following lemma, which generalizes part of Lemma 4.1.7.

Lemma 4.9.1. *The Laplacians L^0 and L^1 are self-adjoint and non-negative operators. Hence, in particular, their eigenvalues are real and non-negative.*

Proof By the construction in Section 2.5.4, L^0 and L^1 are the two compositions of δ and δ^*, which are adjoint to each other. Therefore, they are both self-adjoint, which implies that their eigenvalues are real.

Moreover, given $f \colon V \to \mathbb{R}$,

$$(L^0 f, f)_V = (\delta^* \delta f, f)_V = (\delta f, \delta f)_E \geq 0.$$

Analogously, for $\gamma \colon E \to \mathbb{R}$,

$$(L^1\gamma, \gamma)_E = (\delta\delta^*\gamma, \gamma)_E = (\delta^*\gamma, \delta^*\gamma)_V \geq 0.$$

Hence, L^0 and L^1 are non-negative operators. In particular, this implies that their eigenvalues are non-negative.

$\square$

Now, as for the graph case, we order the eigenvalues of L^0 as

$$\lambda_1 \leq \ldots \leq \lambda_N$$

and we observe that they always sum to N, since the trace of $L^0 = \mathrm{Id} - D^{-1}A$ is clearly N. Using the min–max principle (Theorem 2.6.1) and the construction of L^0 and L^1 in Section 2.5.4, we can characterize these eigenvalues using the following Rayleigh quotients.

Definition 4.9.2. The *Rayleigh quotient* of $f \colon V \to \mathbb{R}$ with respect to Γ is

$$R(f) := \frac{(\delta f, \delta f)_E}{(f, f)_V}$$

$$= \frac{\sum_{e \in E} \left(\sum_{v_i \text{ input of } e} f(v_i) - \sum_{v^j \text{ output of } e} f(v^j) \right)^2}{\sum_{v \in V} \deg v f(v)^2}.$$

Similarly, the *Rayleigh quotient* of $\gamma \colon E \to \mathbb{R}$ with respect to Γ is

$$R(\gamma) := \frac{(\delta^*\gamma, \delta^*\gamma)_V}{(\gamma, \gamma)_E}$$

$$= \frac{\sum_{v \in V} \frac{1}{\deg v} \cdot \left(\sum_{e_{\mathrm{in}} \colon v \text{ input}} \gamma(e_{\mathrm{in}}) - \sum_{e_{\mathrm{out}} \colon v \text{ output}} \gamma(e_{\mathrm{out}}) \right)^2}{\sum_{e \in E} \gamma(e)^2}.$$

Lemma 4.9.2. *The nonzero eigenvalues of L^0 and L^1 are the same, counted with multiplicity. Moreover, the N eigenvalues of L^0 can be characterized, by the min–max principle, via the Rayleigh quotients $R(f)$ of functions $f \colon V \to \mathbb{R}$. Similarly, the M eigenvalues of L^1 can be characterized, by the min–max principle, via the Rayleigh quotients $R(\gamma)$ of functions $\gamma \colon E \to \mathbb{R}$.*

In particular,

$$\lambda_1 = \min_{f \colon V \to \mathbb{R}, f \neq 0} R(f)$$

and

$$\lambda_N = \max_{f \colon V \to \mathbb{R}} R(f) = \max_{\gamma \colon E \to \mathbb{R}} R(\gamma).$$

Proof It follows from the construction of L^0 and L^1 in Section 2.5.4. $\square$

From here on, following the usual notation, we let $\Delta := L^0$.

Next, we consider some examples of hypergraphs for which we compute the spectrum. As a consequence, we will see that several properties from Section 4.1 that hold for the spectra of graphs do not hold for the general case.

Example 4.9.1. Let $\Gamma = (V, E, \varphi)$ with $V = \{v_1, \ldots, v_N\}$ and $E = \{\{v_1\}, \ldots, \{v_N\}\}$, that is, the vertices are only contained in self-loops. Then, $\Delta = \text{Id}$, implying that 1 is the only eigenvalue in this case.

Example 4.9.2. Assume that $\Gamma = (V, E, \varphi)$ has only one hyperedge, and this contains all vertices as inputs. Then, the Laplacian Δ can be seen as the matrix whose entries are all equal to 1, implying that a nonzero function $f : V \to \mathbb{R}$ is an eigenfunction with eigenvalue λ if and only if

$$f(v_i) = \sum_{j=1}^{N} f(v_j), \quad \forall i = 1, \ldots, N.$$

Hence, N is an eigenvalue with multiplicity 1, and the corresponding eigenfunctions are constant. All other $N-1$ eigenvalues are 0, and the corresponding eigenfunctions satisfy $\sum_{j=1}^{N} f(v_j) = 0$.

The two examples above show, in particular, that the third part of Lemma 4.1.7 does no longer hold in the general case of hypergraphs. In fact, for graphs, 0 is always an eigenvalue, and its multiplicity equals the number of connected components of Γ. In the general case, there could be both cases in which 0 is not an eigenvalue (as in Example 4.9.1) and cases in which the multiplicity of 0 is higher than the number of connected components of Γ (as in Example 4.9.2). Moreover, the third part of Lemma 4.1.7 shows that, for a connected graph, the eigenfunctions of 0 are precisely the constant functions. Interestingly, for the hypergraph in Example 4.9.2, the constant functions are precisely the eigenfunctions of the only nonzero eigenvalue. The underlying reason is that, while the graph Laplacian satisfies a maximum principle, this no longer holds for the hypergraph Laplacian. As we have seen in Section 4.1, the maximum principle says that solutions of $\Delta u = 0$ on a connected graph are constant, and hence, the eigenvalue $\lambda = 0$ has multiplicity 1. On a general hypergraph, due to the possibility of the presence of co-oriented vertices in hyperedges, such a maximum principle can no longer hold for Δ on a hypergraph, and consequently, the multiplicity of $\lambda = 0$ can be different from 1.

Both Examples 4.9.1 and 4.9.2 are particular cases of complete hypergraphs:

Definition 4.9.3. A hypergraph $\Gamma = (V, E, \varphi)$ is *c-complete* if E is the set of all possible $\binom{N}{c}$ subsets of V of cardinality c.

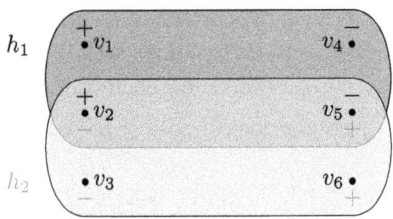

Figure 4.28 A bipartite hypergraph with $V_1 = \{v_1, v_2, v_3\}$ and $V_2 = \{v_4, v_5, v_6\}$.

Clearly, the complete graph is a 2-complete hypergraph, the hypergraph in Example 4.9.1 is 1-complete, and the one in Example 4.9.2 is N-complete.

Now, there are different ways of generalizing the notion of a bipartite graph to the case of hypergraphs. In this chapter, we consider the following definition.

Definition 4.9.4. A hypergraph $\Gamma = (V, E, \varphi)$ is *bipartite* (Figure 4.28) if one can decompose the vertex set as a disjoint union $V = V_1 \sqcup V_2$ such that, for every $e \in E$, either e has all its inputs in V_1 and all its outputs in V_2, or vice versa.

For instance, the underlying hypergraph of any hypergraph is bipartite:

Definition 4.9.5. The *underlying hypergraph* of $\Gamma = (V, E, \varphi)$ is the oriented hypergraph $\Gamma^+ := (V, E, \varphi^+)$ obtained from Γ by letting each vertex be an input for all hyperedges in which it is contained. The *signless Laplacian* of Γ, denoted Δ^+, is the Laplacian of its underlying hypergraph, that is,

$$\Delta^+(\Gamma) := \Delta(\Gamma^+).$$

Remark 4.9.2. If Γ is a simple graph, then

$$\Delta(\Gamma) = \mathrm{Id} - D(\Gamma)^{-1} A(\Gamma)$$

and

$$\Delta^+(\Gamma) = \Delta(\Gamma^+) = \mathrm{Id} + D(\Gamma)^{-1} A(\Gamma) = 2 \cdot \mathrm{Id} - \Delta(\Gamma).$$

Therefore, λ is an eigenvalue for Δ if and only if $2 - \lambda$ is an eigenvalue for Δ^+.

For example, as a consequence of the last remark, we can equivalently reformulate Lemma 4.1.8 by saying that the spectrum of Δ^+ for the complete graph K_N consists of the eigenvalue 2 with multiplicity 1 and the eigenvalue $\frac{N-2}{N-1}$ with multiplicity $N - 1$. This can be generalized as follows.

Lemma 4.9.3. *Let Γ be a c-complete hypergraph, with $c \geq 2$. Then, the spectrum of Δ^+ is given by c, with multiplicity 1, and $\frac{N-c}{N-1}$, with multiplicity $N - 1$.*

Proof Let Γ^+ be the underlying hypergraph of Γ, and let $A^+ := A(\Gamma^+)$.

It is easy to see that the constant functions are eigenfunctions for $\Delta^+(\Gamma) = \Delta(\Gamma^+)$ with eigenvalue c. Now, observe that each vertex has degree $d := \binom{N-1}{c-1}$, while $a := A_{ij}^+ = -\binom{N-2}{c-2}$ is constant for all $i \neq j$. Therefore, $\frac{a}{d} = -\frac{c-1}{N-1}$ and

$$\Delta^+ f(v) = f(v) - \frac{a}{d}\left(\sum_{v' \neq v} f(v')\right) = f(v) + \frac{c-1}{N-1}\left(\sum_{v' \neq v} f(v')\right),$$

for all $v \in V$.

For each $i = 2, \ldots, N$, let $f_i(v_1) := 1$, $f_i(v_i) := -1$ and $f_i := 0$ otherwise. Then,

- $\Delta^+ f_i(v_1) = 1 - \frac{c-1}{N-1} = \frac{N-c}{N-1} \cdot f_i(v_1)$,
- $\Delta^+ f_i(v_i) = -1 + \frac{c-1}{N-1} = \frac{N-c}{N-1} \cdot f_i(v_i)$, and
- $\Delta^+ f_i(v_j) = 0 = \frac{N-c}{N-1} \cdot f_i(v_j)$ for all $j \neq 1, i$.

Therefore, the f_i's are $N-1$ linearly independent eigenfunctions with eigenvalue $\frac{N-c}{N-1}$.   □

We now prove that the spectrum of a bipartite hypergraph coincides with the spectrum of its underlying hypergraph.

Proposition 4.9.1. *If Γ is bipartite, it is isospectral to its underlying hypergraph; therefore, in particular, also to every other bipartite hypergraph that has the same underlying hypergraph as Γ.*

Proof Since Γ is bipartite, up to switching (without loss of generality) the orientations of some hyperedges, we can assume that all the inputs are in V_1 and all the outputs are in V_2, with $V = V_1 \sqcup V_2$. Furthermore, by definition of L^1, which has the same nonzero spectrum as Δ, we can move a vertex from V_1 to V_2 or vice versa by letting it be always an output or always an input without affecting the spectrum. In particular, if we move all vertices to V_1, we obtain the underlying hypergraph of Γ.   □

Now, by Corollary 4.1.4, Lemmas 4.1.12, 4.1.13, and 4.1.10, we have that, if Γ is a connected graph, then

$$\frac{N}{N-1} \leq \lambda_N \leq 2, \tag{4.9.2}$$

and $\lambda_N = \frac{N}{N-1}$ if and only if Γ is complete, while $\lambda_N = 2$ if and only if Γ is bipartite. In Theorem 4.9.1, we generalize this result to the case of hypergraphs. Before that, we give the preliminary Definition 4.9.6.

Definition 4.9.6. A hypergraph $\hat{\Gamma} = (\hat{V}, \hat{E}, \hat{\varphi})$ is a *sub-hypergraph* of $\Gamma = (V, E, \varphi)$, denoted $\hat{\Gamma} \subset \Gamma$, if:

1. $\hat{V} \subseteq V$,
2. $\hat{E} \subseteq \{e \cap \hat{V} : e \in E\}$,
3. $\hat{\varphi}(v,e) = \varphi(v,e)$ for all $v \in \hat{V}$ and $e \in \hat{E}$.

Given a sub-hypergraph $\hat{\Gamma} \subset \Gamma$, we let

$$\eta(\hat{\Gamma}) := \frac{\sum_{v \in \hat{V}} \frac{\deg_{\hat{\Gamma}}(v)^2}{\deg v}}{|\hat{E}|},$$

where $\deg_{\hat{\Gamma}}(v)$ denotes the degree of v in $\hat{\Gamma}$ and $|\hat{E}|$ is the number of hyperedges in $\hat{\Gamma}$.

Theorem 4.9.1. *For every connected hypergraph $\Gamma = (V, E, \varphi)$,*

$$\lambda_N \leq \max_{e \in E} |e|, \tag{4.9.3}$$

with equality if and only if Γ is bipartite and $|e|$ is constant for all $e \in E$, and

$$\lambda_N \geq \max_{\hat{\Gamma} \subset \Gamma \ bipartite} \eta(\hat{\Gamma}). \tag{4.9.4}$$

Proof We first prove (4.9.3). Let $f : V \to \mathbb{R}$ be an eigenfunction for λ_N. Then,

$$\lambda_N = \frac{\sum_{e \in E} \left(\sum_{v_i \ input \ of \ e} f(v_i) - \sum_{v^j \ output \ of \ e} f(v^j) \right)^2}{\sum_{v \in V} \deg v f(v)^2}$$

$$\leq \frac{\sum_{e \in E} \left(\sum_{v \in e} |f(v)| \right)^2}{\sum_{i \in V} \deg v f(v)^2},$$

with equality if and only if f has its nonzero values on a bipartite sub-hypergraph. Now, for each $e \in E$,

$$\left(\sum_{v \in e} |f(v)| \right)^2 = \sum_{v \in e} f(v)^2 + \sum_{\{v,v'\} : v \neq v' \in e} 2 \cdot |f(v)| \cdot |f(v')|$$

$$\leq \sum_{v \in e} f(v)^2 + \sum_{\{v,v'\} : v \neq v' \in e} \left(f(v)^2 + f(v')^2 \right)$$

$$= \sum_{v \in e} f(v)^2 + \sum_{v \in e} (|e| - 1) f(v)^2$$

$$= |e| \cdot \sum_{v \in e} f(v)^2,$$

with equality if and only if $|f|$ is constant on all $v \in e$. Therefore,

$$\frac{\sum_{e \in E}\left(\sum_{v \in h}|f(v)|\right)^2}{\sum_{i \in V}\deg v f(v)^2} \leq \frac{\sum_{e \in E}\sum_{v \in e}|e| \cdot f(v)^2}{\sum_{i \in V}\deg v f(v)^2}$$

$$= \frac{\sum_{v \in V}\sum_{e \ni v}|e| \cdot f(v)^2}{\sum_{i \in V}\deg v f(v)^2}$$

$$\leq \left(\max_{e \subset E}|e|\right) \cdot \frac{\sum_{v \in V}\deg v f(v)^2}{\sum_{i \in V}\deg v f(v)^2}$$

$$= \max_{e \in E}|e|,$$

where the first inequality is an equality if and only if $|f|$ is constant (since we are assuming that Γ is connected), and the last inequality is an equality if and only if $|e|$ is constant for all e. Putting everything together, we have that

$$\lambda_N \leq \max_{e \in E}|e|,$$

with equality if and only if $|e|$ is constant for all $|e|$, while $|f|$ is constant, and it is defined on a bipartite sub-hypergraph (i.e., $|f|$ is constant and Γ is bipartite). This proves the first claim.

It is left to prove (4.9.4). Given a bipartite sub-hypergraph $\hat{\Gamma} \subset \Gamma$, let $\gamma' : E \to \mathbb{R}$ be 1 on $\hat{E}$ and 0 otherwise. Then, up to changing (without loss of generality), the orientations of the hyperedges,

$$\lambda_N = \max_{\gamma : E \to \mathbb{R}}\frac{\sum_{v \in V}\frac{1}{\deg v} \cdot \left(\sum_{e_{in}: v \text{ input}}\gamma(e_{in}) - \sum_{e_{out}: v \text{ output}}\gamma(e_{out})\right)^2}{\sum_{e \in E}\gamma(h)^2}$$

$$\geq \frac{\sum_{v \in V}\frac{1}{\deg v} \cdot \left(\sum_{e_{in}: v \text{ input}}\gamma'(e_{in}) - \sum_{e_{out}: v \text{ output}}\gamma'(e_{out})\right)^2}{\sum_{e \in E}\gamma'(e)^2}$$

$$\geq \frac{\sum_{v \in \hat{V}}\frac{1}{\deg v} \cdot \left(\sum_{e_{in}: v \text{ input}}\gamma'(e_{in}) - \sum_{e_{out}: v \text{ output}}\gamma'(e_{out})\right)^2}{\sum_{e \in E}\gamma'(e)^2}$$

$$= \frac{\sum_{v \in \hat{V}}\frac{\deg_{\hat{\Gamma}}(v)^2}{\deg v}}{|\hat{E}|}.$$

Since the above inequality is true for all $\hat{\Gamma}$, this proves (4.9.4).   □

Remark 4.9.3. Theorem 4.9.1 generalizes the inequalities in (4.9.2). In fact, while this is clear for the upper bound, to see it for the lower bound, consider a graph Γ. Fix a vertex v and let $\hat{\Gamma}$ be the bipartite sub-graph of Γ given by the edges that have v as endpoint. Then, (4.9.4) implies

$$\lambda_N \geq \eta(\hat{\Gamma}) = 1 + \sum_{v' \sim v} \frac{1}{\deg v' \cdot \deg v}$$

$$\geq 1 + \sum_{v' \sim v} \frac{1}{(N-1) \cdot \deg v} = 1 + \frac{1}{N-1} = \frac{N}{N-1}.$$

An immediate consequence of Theorem 4.9.1 is the following

Corollary 4.9.1. *If Γ is c-uniform and bipartite, then c is an eigenvalue for the Laplacian Δ, and its multiplicity equals the number of connected components of Γ.*

If Γ is c-uniform, then c is an eigenvalue for the signless Laplacian Δ^+, and its multiplicity equals the number of connected components of Γ.

Notably, Corollary 4.9.1 can be seen as a generalization of the fact that the multiplicity of 0 for Δ, in the case of simple graphs, equals the number of connected components of Γ. In fact, if Γ is a simple graph, then by Remark 4.9.2, λ is an eigenvalue for Δ if and only if $2 - \lambda$ is an eigenvalue for Δ^+. Hence, the fact that the multiplicity of 0 for Δ, in this case, equals the number of connected components of Γ, can be equivalently reformulated in terms of the multiplicity of 2 for Δ^+.

We now generalize the notion of duplicate vertices to the case of hypergraphs.

Definition 4.9.7. Two vertices v_i and v_j are *duplicates* of each other if the corresponding rows/columns of the adjacency matrix are the same, that is,

$$A_{il} = A_{jl} \quad \text{for each } l = 1, \ldots, N.$$

In particular, $A_{ij} = A_{jj} = 0$.

Lemma 4.9.4. *Let v_i and v_j be duplicates of each other. Let $f : V \to \mathbb{R}$ satisfy $f(v_i) = -f(v_j) \neq 0$ and $f = 0$ otherwise. Then, 1 is an eigenvalue for Δ and f is a corresponding eigenfunction.*

Proof By definition of f, we have that:

- $\Delta f(v_i) = f(v_i)$,
- $\Delta f(v_j) = f(v_j)$, and
- For each $l \neq i, j$,

$$\Delta f(v_l) = -\frac{1}{\deg v_l}(A_{li}f(v_i) + A_{lj}f(v_j))$$

$$= -\frac{1}{\deg v_l}(A_{li}f(v_i) - A_{li}f(v_i))$$

$$= 0 = f(v_l).$$

Therefore, $\Delta f = f$, that is, 1 is an eigenvalue for Δ and f is a corresponding eigenfunction.

□

Corollary 4.9.2. *If there are n duplicate vertices, 1 is an eigenvalue with multiplicity at least $n - 1$.*

Proof Assume, without loss of generality, that $v_1, \ldots, v_n$ are duplicates of each other. For each $i = 1, \ldots, n-1$, let $f_i : V \to \mathbb{R}$ such that $f_i(v_i) = 1$, $f_i(v_{i+1}) = -1$, and $f_i = 0$ otherwise. Then, by Lemma 4.9.4, the f_i's are eigenfunctions corresponding to the eigenvalue 1. Also, $\dim(\text{span}(f_1, \ldots, f_{n-1})) = n - 1$; therefore, the multiplicity of 1 is at least $n - 1$.   □

Similarly, we introduce twin vertices:

Definition 4.9.8. Two vertices v_i and v_j are *twins* of each other if they belong exactly to the same hyperedges, with the same orientations. In particular, $A_{ij} = -\deg v_i = -\deg v_j$ and $A_{il} = A_{jl}$ for all $l \neq i, j$.

Note that, while duplicate vertices exist for graphs, twin vertices cannot exist for graphs. Also, two vertices v_i and v_j cannot be both duplicates and twins of each other. In fact, if they are duplicates, then $A_{ij} = 0$, while, if they are twins, then $A_{ij} < 0$.

We now compute the spectrum of the signless Laplacian for the hyperflowers, which generalize star graphs.

Definition 4.9.9. Given $t \geq 1$, a hypergraph $\Gamma = (V, E, \varphi)$ is a *hyperflower with t twins* if (Figure 4.29):

- The vertex set can be decomposed as $V = C \sqcup P$, where C is the *core*, containing c vertices, and P is given by tM *peripheral vertices* $v_{11}, \ldots, v_{t1}, \ldots, v_{1M}, \ldots, v_{tM}$;
- The hyperedges are

$$e_j = C \sqcup \bigcup_{i=1}^{t} v_{ij}, \quad \text{for } j = 1, \ldots, M.$$

Thus, the hyperflower contains a core of c vertices and M leaves with t vertices each. Each hyperedge contains the core and one of the leaves.

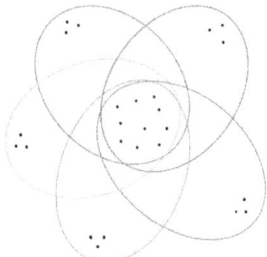

Figure 4.29 A hyperflower with 3 twins, 5 hyperedges, and 10 central vertices.

Proposition 4.9.2. *Given $t \geq 1$, let Γ be a hyperflower with t twins. Then, the spectrum of Δ^+ is given by:*

- *0, with multiplicity $N - M$;*
- *t, with multiplicity $M - 1$;*
- *$N - tM + t$, with multiplicity 1.*

Proof We want to compute the spectrum of $\Delta(\Gamma^+)$. Since all hyperedges have cardinality $N - tM + t$, by Theorem 4.9.1,

$$\lambda_N = N - tM + t,$$

and the constant functions are the corresponding eigenfunctions.

Furthermore, t is an eigenvalue with multiplicity at least $M - 1$. To see this, consider the functions $f \colon V \to \mathbb{R}$ that are $= +1$ on one leaf, $= -1$ on another leaf, and 0 elsewhere. This construction gives $M - 1$ linearly independent functions for the eigenvalue t.

We have therefore listed M eigenvalues whose sum is N. Since $\Delta(\Gamma^+)$ has, in total, N eigenvalues whose sum is N, this implies that 0 has multiplicity $N - M$. □

5

Eigenvalues and Eigenfunctions of the Laplace Operator via Floer Theory

By considering Rayleigh quotients as Morse functions, we obtain eigenvalues as critical values and eigenfunctions as critical points. This allows us to apply Floer's theory of gradient flows to derive homological information from the relations between different eigenfunctions.

5.1 Rayleigh Quotients, Eigenfunctions, and Eigenvalues

Earlier, we had obtained the eigenvalues of the Laplace operator via Rayleigh quotients, see Section 2.6.1. The general theorem was the Min-Max Theorem 2.6.1. In this section, we shall present an alternative method. That method also works in the general case and can therefore replace Theorem 2.6.1, but for simplicity, we carry it out here only for the (normalized) graph Laplacian, leaving the general case to the reader.

Thus, we consider the space H of functions $f \colon V \to \mathbb{R}$ on the vertex set V of a weighted graph equipped with the scalar product (4.1.32)

$$(f,g) = \sum_v \deg v \, f(v)g(v), \text{ with } \deg v = \sum_{v' \sim v} w_{vv'}. \tag{5.1.1}$$

We let

$$H_1 := \{ f \in H \colon (f,f) = 1 \} \tag{5.1.2}$$

be the unit sphere in H.

As before, our operator is the weighted normalized combinatorial Laplace operator (4.1.35) given by

$$L(f)(v) = f(v) - \frac{1}{\deg v} \sum_{v' \sim v} w_{vv'} f(v') = \frac{1}{\deg v} \sum_{v' \sim v} (f(v) - f(v')). \tag{5.1.3}$$

We also write (see (2.5.16))

$$(\delta f, \delta g) = \sum_{v,v'} w_{vv'}(f(v) - f(v'))(g(v) - g(v')) = (L(f), g) = (g, L(f)),$$
(5.1.4)

and put

$$D(f) := (\delta f, \delta f).$$
(5.1.5)

Our project then consists of studying D as a function on the unit sphere H_1 and investigating its critical points. Equivalently (recalling the more general constructions of Section 2.6.2), we can introduce a Lagrange multiplier λ and study the critical points of

$$D(f, \lambda) := D(f) - \lambda((f, f) - 1)$$
(5.1.6)

on the space H. Criticality with respect to λ simply means that $(f, f) = 1$, that is, $f \in H_1$, and f is a critical point of this functional (5.1.6) if for all $\phi \in H$

$$0 = dD(f, \lambda)(\phi) := \frac{d}{dt} D(f + t\phi, \lambda)_{|t=0} = 2(\delta f, \delta \phi) - 2\lambda(f, \phi)$$

$$= 2(Lf, \phi) - \lambda(f, \phi)),$$
(5.1.7)

that is,

$$Lf = \lambda f,$$
(5.1.8)

that is, f is an eigenfunction of L for the eigenvalue λ. Here, in the last step, we have used the formula (2.5.10). Conversely, if f is an eigenfunction, it satisfies (5.1.7) for its eigenvalue λ. We have thus shown

Theorem 5.1.1. *The eigenfunctions f of L with the normalization $(f, f) = 1$ are precisely the critical points of the functional D, given by $D(f) = (\delta f, \delta f)$, see (5.1.6), on the sphere H_1. The corresponding eigenvalue arises as the Lagrange multiplier for the constraint $(f, f) = 1$.*

□

Also, by putting $\phi = f$ in (5.1.7), we obtain

$$(\delta f, \delta f) = \lambda(f, f)$$
(5.1.9)

for an eigenfunction f with eigenvalue λ. Thus, the critical values of the functional D on H_1 are precisely the eigenvalues of L. In particular, when we order the eigenfunctions by increasing values of D, this is the same as ordering them according to their eigenvalues. The lowest eigenvalue is 0, realized by a constant function as eigenfunction.

We next consider the second variation of $D(f, \lambda)$ at such a critical point f. This is given by

$$d^2 D(f, \lambda)(\phi, \psi) = 2(\delta\phi, \delta\psi) - 2\lambda(\phi, \psi). \qquad (5.1.10)$$

(Since λ enters linearly into (5.1.6), we do not need to bother about second derivatives in the λ-direction.) Since we only consider directions tangent to H_1, we can restrict to those functions ρ that satisfy

$$(f, \rho) = 0. \qquad (5.1.11)$$

Polarizing, we have

$$d^2 D(f, \lambda)(g, g) <, =, > 0 \text{ if and only if } D(g) <, =, > \lambda(g, g). \qquad (5.1.12)$$

Theorem 5.1.1 can also be turned into a constructive scheme for the eigenfunctions and eigenvalues of L. By starting with the minimum of $D(f)$ on H_1, achieved by $f_1 \equiv const$, iterating and in each iteration step restricting to those f that are orthogonal to all those from preceding steps, and ending up with the maximum, we see that we can find N critical points (where N is the number of vertices of our graph and hence the dimension of H) $f_k, k = 1, \ldots, N$ of our functional, that is, N mutually orthogonal eigenfunctions

$$L f_k = \lambda_k f_k, \qquad (5.1.13)$$

with $\lambda_{k-1} \leq \lambda_k$ for $k = 2, \ldots, N$. But then it suffices in (5.1.12) to consider only $g = f_j$ for some eigenfunctions, as those span H, and we get

$$d^2 D(f_k, \lambda_k)(f_j, f_j) <, =, > 0 \text{ if and only if } \lambda_j <, =, > \lambda_k. \qquad (5.1.14)$$

The number of negative directions of $d^2 D$ at a critical point f_k, which is also called the index of the function D, equals the dimension of the space of orthogonal eigenfunctions with eigenvalues $\lambda_j < \lambda_k$, and the nullity, which is the dimension of the space of zero directions, equals the dimension -1 of the eigenspace of λ_k. Also, since the dimension of H_1 is $N - 1$, we can then compute the number of positive directions.

In general terms, a twice differentiable function S on a Riemannian manifold is called a *Morse function* if all its critical points are nondegenerate, that is, their nullity vanishes. In our situation, this is the case when all eigenvalues have multiplicity 1, that is, when different critical points f_j have different eigenvalues λ_j.

In fact, if f is an eigenfunction with $(f, f) = 1$, then so is $-f$, and more generally $D(-f) = D(f)$. Thus, saying that f_k and f_j are different eigenfunctions really means that $f_j \neq \pm f_k$. Therefore, identifying antipodal points on the $(N-1)$-dimensional sphere H_1, we may consider $D(f,\lambda)$ as a functional on $(N-1)$-dimensional projective space.

5.2 Gradient Flow

In any case, there is more interesting geometry. For a functional S on H_1, we can define the gradient of S via

$$(\text{grad } S(f),g) = dS(f)(g). \tag{5.2.1}$$

Then,

$$(\text{grad } D(f,\lambda),g) = 2(\delta f,\delta g) - 2\lambda(f,g) = 2(Lf - \lambda f,g). \tag{5.2.2}$$

In particular,

$$\text{grad } D(f,\lambda) = 0 \text{ if and only if } Lf = \lambda f, \tag{5.2.3}$$

which, of course, is simply the relation (5.1.7) and (5.1.8).

The gradient of a function gives the direction of its steepest ascent, and the negative of the gradient then is the direction of steepest descent. One therefore looks at the (negative) gradient flow

$$\dot{x}(t) = -\text{grad } S(x(t)) \tag{5.2.4}$$
$$x(0) = x$$

for functions $x\colon \mathbb{R} \to H_1$ with some initial value $x \in H_1$. (As such, this is an initial value problem for an ODE, and one may naturally consider it for $t \geq 0$. Subsequently, however, it will be useful to also extend it for negative values of t.)

When we consider such a flow for our functional D, we can of course no longer keep the value λ fixed. We rather need to project the gradient grad $_H D = 2L$ onto H_1 to get

$$\text{grad }_{H_1} D(f) = \text{grad }_H D(f) - (\text{grad }_H D(f),f)f$$
$$= 2(L(f) - (L(f),f)f) = 2(L(f) - D(f)f), \tag{5.2.5}$$

using (5.1.4).

In any case, since L is a linear operator, the gradient flow (5.2.4) for D then proceeds along a circle in H_1, the intersection of the sphere with a two-dimensional linear subspace of H.

We now assume that D is a Morse function on H_1, that is, as explained, that all eigenvalues of L are nondegenerate. The index of a critical point then is the number of negative directions of the Hessian d^2D, and these are the directions in which the value of D decreases or as they are also called, the unstable directions. Conversely, the stable directions are the others, in which the value of D increases. The sum of the dimensions of those spaces then is $N - 1$ because we are assuming that there are no null directions. If we then have two different critical points v_1, v_2, of indices $i_1 < i_2$, then the stable space of v_1 and the unstable space of v_2 intersect in an $(i_2 - i_1)$-dimensional subspace. In particular, when the index difference is one, $i_2 = i_1 + 1$, the intersection is one-dimensional, that is, a line. When we consider the gradient flow $x(t)$ from (5.2.4) going through a point in this line, then $x(-\infty) := \lim_{t \to =-\infty} x(t) = v_2$ and $x(\infty) = v_1$. Thus, the gradient flow connects these two critical points of index difference 1. In fact, in our situation, there is a single such flow line from v_2 to v_1 and another in the opposite direction, to $-v_1$.

In conclusion, we have a simple geometric picture of the critical points, that is, the eigenfunctions of L, and the flow lines between them. From each of the two maxima $\pm f_N$, we have two flow lines to $\pm f_{N-1}$, from those points two flow lines to f_{N-2}, until we reach the two minima $\pm f_1$.

By Floer's Theory [97], the flow lines between critical points of index difference 1 can be used to define a boundary operator and to reconstruct the homology of the underlying space (see [150, 244] for systematic expositions). This requires the condition that all critical points are nondegenerate, which, as explained above, simply means that all eigenvalues have multiplicity 1. We need one technical preparation. For a vector space W of dimension n, an orientation is given by an ordered choice of a basis $(e_1, e_2, \dots, e_n)$. Interchanging two of those basis vectors yields the opposite orientation. We also call the original orientation positive, that obtained by such a permutation negative. More generally then, an odd permutation of the basis vectors changes the orientation, whereas an even one preserves it. Also, a linear change of basis with positive determinant preserves the orientation, whereas one with negative determinant changes it. Let thus W carry an orientation, and let $V \subset W$ be a linear subspace, with a complement V', that is $V \oplus V' = W$ and $V \cap V' = \{0\}$. When we choose an arbitrary orientation on V, realized by a basis $(e_1, \dots, e_k)$, say, this then induces an orientation on V', by declaring a basis $(e_{k+1}, \dots, e_n)$ of V' positively oriented if the combined basis $(e_1, \dots, e_n)$ of V yields the orientation of V.

Since the sphere H_1 is an orientable manifold, we can choose a consistent orientation on all the tangent spaces $T_g H_1$ for $g \in H_1$. In our setting, such a tangent space is simply given by $\{v \in H : (v, g) = 0\}$, the space orthogonal to the position vector g. When we move from g_0 to g_1 in H_1 along a continuous path $g(t), t \in [0, 1]$, we can also continuously change a basis of $T_{g_0} H_1$ into one of $T_{g_1} H_1$ along that path, and we can arrange the orientations in such a way that when we transform an oriented basis of the first space into the second one, it yields the latter's orientation. And this does not depend on the chosen path. See [150] for details.

Equipped with these concepts, we choose an orientation on the unstable space of every critical point f of D. As explained, this unstable space is simply the linear subspace of the tangent space $T_f H_1$ on which the second variation is negative. As explained, this then also induces an orientation on the complementary space, the stable space of that critical point.

Let us now consider two critical points f, g of index difference 1, that is, the index $i(f)$ of f satisfies $i(f) = i(g) + 1$. This means that the dimension of the unstable space of f equals the dimension of that of g plus 1. Since the unstable space is spanned by the eigenfunctions with lower eigenvalues, g belongs to the unstable space of f, whereas f belongs to the stable space of g, while the remaining stable and unstable directions of f and g coincide. Therefore, the intersection of the unstable space of f with the stable space of g is one-dimensional. This is a general fact for critical points of nondegenerate Morse functions (under a slight additional technical condition), even in situations that are no longer linear as that considered here, and this is the basis of Floer theory. The fact alluded to is that there are finitely many flow lines for the gradient flow (5.2.4) between critical points of index difference 1. But let us continue in our particular setting, although the constructions generalize to the general case. The sum of the stable space of f and the unstable space of g thus has codimension 1, and its complement is the intersection mentioned earlier. Thus, the sum of these two spaces, which carry orientations by our construction, also induces an orientation on that intersection. We then count this intersection (in fact, the gradient flow line from f to g) positive when that orientation is the same as that of the direction from f to g, negative else. Thus, every such intersection $i(f, g)$ has a sign $\sigma(i(f, g)) = \pm 1$. We can now define the Floer boundary operator

$$\delta^F(f) = \sum_{g\,:\,i(f)=i(g)+1} \sigma(i(f, g))g. \qquad (5.2.6)$$

When we let C_j be the free abelian group spanned by the critical points of index j, that is, of objects of the form

$$\sum_{i(f)=j} n_f f, \text{ with } n_f \in \mathbb{Z}, \qquad (5.2.7)$$

we can linearly extend δ^F as an operator

$$\delta_j^F : C_j \to C_{j-1}. \qquad (5.2.8)$$

One then checks the fundamental relation

$$\delta^F \circ \delta^F = 0, \tag{5.2.9}$$

which arises in our case from the simple fact that when $i(f) = i(g) + 1$, there is a flow line from f to g, as well as one from f to $-g$, and when we then look at flow lines from $\pm g$ to $\pm h$, where $i(g) = i(h) + 1$, the combinations of the signed flow lines from f via g or $-g$ to h or $-h$ cancel in pairs. In fact, the flow line from f to h via g comes with the orientation opposite to that from f to h via $-g$.

As before, we can then define homology groups

$$H_j := \ker \delta_j^F / \mathrm{im}\, \delta_{j+1}^F. \tag{5.2.10}$$

As a test, we can recover the homology of H_1 which is an $(N-1)$-dimensional sphere. For the index $N-1$, we have f_N and $-f_N$, the eigenfunctions for the highest eigenvalue. Depending on our choices of orientation of the unstable manifolds, we have either $\delta_{N-1}^F(f_N + (-f_N)) = 0$ or $\delta_{N-1}^F(f_N - (-f_N)) = 0$ (note that the first $\pm$ signs here are formal ones as in (5.2.7), whereas the $-$sign in front of f_N refers to the vector space H). Thus, since the image of δ_N^F is trivial, as $\pm f_N$ are the critical points of highest index, H_{N-1} is one-dimensional. Next, for $0 < j < N-1$, $H_j = 0$, since the kernel $f_j \pm (-f_j)$ of δ_j^F agrees with the image of δ_{j+1}^F. Finally, H_0 is one-dimensional again, as both f_0 and $-f_0$ are in the kernel of δ_0^F, as that operator is trivial.

PART II

Eigenvalues and Eigenfunctions on
Simplicial Complexes and Hypergraphs

6

Lovász Extensions

In this chapter, we introduce the theory of Lovász extensions as a powerful tool for studying eigenvalue problems and, in particular, Cheeger-type inequalities.

As a fundamental tool in discrete mathematics, the Lovász extension has been deeply connected to submodular analysis [59, 189] and has been applied in many areas such as combinatorial optimization, game theory, matroid theory, stochastic processes, electrical networks, computer vision, and machine learning [101]. There are many generalizations, such as the disjoint-pair Lovász extension and the Lovász extension on distributive lattices [101, 221]. Recent developments include quasi-Lovász extension on some algebraic structures and fuzzy mathematics, applications of Lovász extensions to graph cut problems and computer science, as well as Lovász-softmax loss in deep learning.

In this chapter, we mainly focus on the spectral theory regarding a pair of Lovász extensions, which offers a new perspective on the eigenvalue problems for discrete structures. The contents are largely taken from [159] and [160], which also provide convenient tools for Chapter 8 on Cheeger inequalities. However, here we also provide new examples and different formulations.

For convenience, in this chapter, we shall use slightly different notations than in the rest of the book. For instance, V will be used to denote a general set, which is not necessarily the vertex set of a graph, and its elements will be denoted by $1, \ldots, N$. Also, the notation for oriented hypergraphs will be different.

6.1 Extensions and Eigenvalue Problems

Before embarking on that, however, let us describe the strategy in abstract terms. We had obtained eigenvalues and eigenfunctions as critical values and critical points of Rayleigh quotients, for instance, in Section 2.6.1,

$$\frac{(Au, u)}{(u, u)}, \tag{6.1.1}$$

where u is in some Hilbert space with a scalar product $(\cdot,\cdot)$. In general, when we seek critical values λ and points x_0 of some quotient

$$\frac{F(x)}{G(x)}, \tag{6.1.2}$$

then, when F and G are differentiable, we obtain the equation

$$\nabla F(x_0) - \lambda \nabla G(x_0) = 0, \tag{6.1.3}$$

where

$$\lambda = \frac{F(x_0)}{G(x_0)} \tag{6.1.4}$$

is the critical value. In the case of (6.1.1), this leads to the eigenvalue equation

$$Au_0 = \lambda u_0. \tag{6.1.5}$$

Now, in the Cheeger-type inequalities, we had compared these eigenvalues, that is, the critical values of the Rayleigh quotient (6.1.1), with the extremal values of other quotients $f(S)/g(S)$ that were defined combinatorially, for $S \in \mathcal{P}(V)$, where V is the set of vertices of some geometric structure. The idea that we shall pursue in this chapter is to extend f and g also to functions on some Euclidean space and translate the combinatorial optimization problem into one on that Euclidean space that we can approach with the tools of calculus. If V has N elements $v_1, \ldots, v_N$, we can identify $S = \{v_{i_1}, \ldots, v_{i_k}\} \in \mathcal{P}(V)$ with the vector in $\mathbb{R}^N$ whose j-th component is 1 if $j \in \{i_1, \ldots, i_k\}$ and 0 else, that is, a vertex of the unit cube in $\mathbb{R}^N$. Thus, the functions f and g are naturally defined on those vertices of the unit cube, and we can then extend them to all of $\mathbb{R}^N$. Of course, many extensions are possible, but we should take one that, on one hand, translates qualitative properties of the discrete functions f, g into suitable qualitative properties of their extensions and, on the other hand, is amenable to continuous optimization tools. That will be achieved by the Lovász extension. It will extend f and g to certain piecewise linear functions. These functions in general will not be smooth, but only Lipschitz, and so, optimization will require some extension of the usual differentiable calculus; see, for instance, [88, 239]. The required concepts have already been introduced at the end of Section 2.6.2. We equip $\mathbb{R}^N$ with its Euclidean product $\langle \cdot, \cdot \rangle$. Recalling Definition 2.6.4, a function $F \colon \mathbb{R}^N \to \mathbb{R} \cup \{\pm\infty\}$ is *subdifferentiable* at $\mathbf{x}$ if there exists some affine linear

$$L(\mathbf{y}) = \langle \mathbf{y} - \mathbf{x}, \mathbf{x}^* \rangle + F(\mathbf{x}) \tag{6.1.6}$$

$$\text{with } L(\mathbf{z}) \le F(\mathbf{z}) \text{ for all } \mathbf{z}, \tag{6.1.7}$$

and the set of all $\mathbf{x}^*$ with this property is called the *subgradient*, which will be denoted by $\nabla F(\mathbf{x})$. Of course, when F is not subdifferentiable at $\mathbf{x}$, this set is empty. Thus, the subgradient of F at $\mathbf{x}$, if it exists, consists of the slopes of all affine linear functions that are not larger than F and agree with F at $\mathbf{x}$. As such, this is a global condition, as the inequality is required for all $\mathbf{z} \in \mathbb{R}^N$, but one can also conceive a local version. For our purposes, however, the global condition will suffice. In this setting, the eigenvalue equation (6.1.3) will become

$$0 \in \nabla F(\mathbf{x}_0) - \lambda \nabla G(\mathbf{x}_0). \tag{6.1.8}$$

Later, we shall need the following easy result whose proof we omit.

Lemma 6.1.1. *For a piecewise linear function $F : \mathbb{R}^N \to \mathbb{R}$ with finite pieces, if F is linear on a convex subset Ω, then $\nabla F(\mathbf{x}) \subset \nabla F(\mathbf{y})$ for any relative interior point $\mathbf{x}$ of Ω and any relative boundary point $\mathbf{y}$ of $\overline{\Omega}$.*

6.2 Definitions and Basic Properties of Lovász Extensions

In this section, we present the various definitions and important properties of Lovász extensions. We shall start by looking at the original Lovász extension. For simplicity, we shall work throughout this chapter with a finite and nonempty set $V = \{1, \ldots, N\}$ and its power set $\mathcal{P}(V)$. We identify every $A \in \mathcal{P}(V)$ with its indicator vector $\mathbf{1}_A \in \{0, 1\}^N \subset \mathbb{R}^N$. These indicator vectors can be identified with the vertices of the unit cube $[0, 1]^N$ in $\mathbb{R}^N$. A function $f : \mathcal{P}(V) \to \mathbb{R}$ with $f(\emptyset) = 0$ then can be considered as a function on those indicator vectors, that is, on the vertices of the unit cube, and we can then extend its domain to all of $\mathbb{R}^N$. Of course, some prescription is necessary for having an extension that, on one hand, preserves qualitative properties of f and, on the other hand, possesses a simple and convenient mathematical structure. In particular, we want to transfer optimization problems from the discrete set $\mathcal{P}(V)$ to $\mathbb{R}^N$ with its Euclidean structure where we can apply the tools of calculus.

This is achieved by the *Lovász extension*, which extends the domain of f to the whole Euclidean space[1] $\mathbb{R}^V$ in a particular manner.

We shall assume

$$f(\emptyset) = 0, \quad g(\emptyset) = 0 \tag{6.2.1}$$

---

[1] Some other versions in the literature only extend the domain to the cube $[0, 1]^V$ or the non-negative orthant $\mathbb{R}^V_{\geq 0}$. In fact, many works on Boolean lattices identify $\mathcal{P}(V)$ with the discrete cube $\{0, 1\}^N$.

for all functions f, g occurring in the sequel.

Definition 6.2.1. For $\mathbf{x} = (x_1, \ldots, x_N) \in \mathbb{R}^N$, let $\sigma \colon V \cup \{0\} \to V \cup \{0\}$ be a bijection such that $x_{\sigma(1)} \leq x_{\sigma(2)} \leq \ldots \leq x_{\sigma(N)}$ and $\sigma(0) = 0$, where $x_0 := 0$. The *Lovász extension* of f is defined by

$$f^L(\mathbf{x}) = \sum_{i=0}^{N-1} (x_{\sigma(i+1)} - x_{\sigma(i)}) f(V^{\sigma(i)}(\mathbf{x})), \qquad (6.2.2)$$

where $V^0(\mathbf{x}) = V$ and $V^{\sigma(i)}(\mathbf{x}) := \{j \in V \colon x_j > x_{\sigma(i)}\}$, for $i = 1, \ldots, N-1$.

There are several equivalent expressions. We can write (6.2.2) in an integral form as

$$f^L(\mathbf{x}) = \int_{\min\limits_{1 \leq i \leq N} x_i}^{\max\limits_{1 \leq i \leq N} x_i} f(V^t(\mathbf{x})) dt + f(V) \min_{1 \leq i \leq N} x_i \qquad (6.2.3)$$

$$= \int_{-\infty}^0 (f(V^t(\mathbf{x})) - f(V)) dt + \int_0^{+\infty} f(V^t(\mathbf{x})) dt, \qquad (6.2.4)$$

where $V^t(\mathbf{x}) = \{i \in V \colon x_i > t\}$. If we apply the Möbius transformation, this becomes

$$f^L(\mathbf{x}) = \sum_{A \subset V} \sum_{B \subset A} (-1)^{\#A - \#B} f(B) \bigwedge_{i \in A} x_i, \qquad (6.2.5)$$

where $\bigwedge\limits_{i \in A} x_i$ is the minimum over $\{x_i \colon i \in A\}$.

Note that by these expressions, $f^L(\mathbf{1}_A) = f(A)$ for any $A \subset V$. This is why we call f^L an "extension", and moreover, f^L is actually a piecewise linear extension of f.

We now discuss some simple **examples**. We take $N = 2$ and, as indicated above, we identify $\emptyset$ with $(0,0)$, $\{1\}$ with $(1,0)$, $\{2\}$ with $(0,1)$, and $\{1,2\}$ with $(1,1)$ in $\mathbb{R}^2$.

1. We consider the function defined by $f_1(0,0) = 0$, $f_1(1,0) = 0$, $f_1(0,1) = 1$, and $f_1(1,1) = 0$. For $\mathbf{x} = (x_1, x_2)$, we have

$$f_1^L(\mathbf{x}) = \begin{cases} 0 & \text{if } x_1 \geq x_2, \\ x_2 - x_1 & \text{if } x_2 \geq x_1. \end{cases}$$

We observe that f_1^L is convex.

2. Let $f_2(0,0) = 0$, $f_2(1,0) = 0$, $f_2(0,1) = 0$, and $f_2(1,1) = 1$. For $\mathbf{x} = (x_1, x_2)$, we have

$$f_2^L(\mathbf{x}) = \begin{cases} x_2 & \text{if } x_1 \geq x_2, \\ x_1 & \text{if } x_2 \geq x_1. \end{cases}$$

We observe that f_2^L is not convex, even though the function f_2 also possesses some convex extension. The reason is, as we shall see in a moment, that f_2, in contrast to f_1, is not submodular.

Definition 6.2.2. A function $f : \mathcal{P}(V) \to \mathbb{R}$ is *modular* if

$$f(A) + f(B) = f(A \cup B) + f(A \cap B),$$

and it is *submodular* if

$$f(A) + f(B) \geq f(A \cup B) + f(A \cap B),$$

for all $A, B \in \mathcal{P}(V)$.

Some further **examples** that are important for us:

3. Suppose that f is additive in the sense that $f(i_1,\ldots,i_N) = \sum_j f(0,\ldots,i_j,0,\ldots,0)$, where $i_j \in \{0,1\}$ (and, as always, $f(0,\ldots,0) = 0$), and we have identified $\mathcal{P}(V)$ with $\{0,1\}^V$. Then,

$$f^L(\mathbf{x}) = \sum_j f(0,\ldots,i_j,0,\ldots,0)x_j.$$

To prove this, by symmetry it suffices to consider the case

$$0 < x_1 < x_2 < \ldots < x_N.$$

Then, the contribution of x_j comes from

$$(x_{j+1} - x_j)f(0,\ldots,0,1_{j+1},\ldots 1_N) + (x_j - x_{j-1})f(0,\ldots,0,1_j,\ldots 1_N),$$

where 1_k indicates the entry 1 in position k. By additivity, the contribution of x_j therefore is $f(0,\ldots,0,1_j,0,\ldots,0)x_j$. In particular, when f is defined on $\mathcal{P}(V)$ for the vertex set V of a graph, and $f(A) = \operatorname{vol} A$, then

$$f^L(\mathbf{x}) = \sum_i \deg i\, x_i, \quad \text{for } f(A) = \operatorname{vol} A. \tag{6.2.6}$$

4. Another example that is relevant in our context is $f(A) = |E(A, V \setminus A)|$, the number of edges between a subset A of the vertex set V of a graph and its complement $V \setminus A$. We note that this f is symmetric in the sense that $f(A) = f(V \setminus A)$. In particular, $f(V) = f(\emptyset) = 0$. When $x_i < x_j$, then when $i \sim j$, a contribution from that edge appears in $V^t(\mathbf{x})$ when $t > x_i$ and disappears when $t > x_j$. Therefore, from (6.2.4), we obtain

$$f^L(\mathbf{x}) = \sum_{i \sim j} |x_j - x_i|, \quad \text{for } f(A) = |E(A, V \setminus A)|. \tag{6.2.7}$$

5. Given a constant function $f \equiv c$, we use (6.2.3) to derive

$$f^L(\mathbf{x}) = c \min_{1 \le i \le N} x_i + \int_{\min_{1 \le i \le N} x_i}^{\max_{1 \le i \le N} x_i} c \, dt = c \max_{1 \le i \le N} x_i.$$

6. As a reformulation of (6.2.2), we have

$$f^L(\mathbf{x}) = (x_{\overline{\sigma}(1)} - x_{\overline{\sigma}(2)}) f(V^{(2)}(\mathbf{x})) + \cdots$$
$$+ (x_{\overline{\sigma}(N-1)} - x_{\overline{\sigma}(N)}) f(V^{(N)}(\mathbf{x})) + x_{\overline{\sigma}(N)} f(V),$$

where $\overline{\sigma} : V \to V$ is a permutation such that $x_{\overline{\sigma}(1)} \ge \ldots \ge x_{\overline{\sigma}(N)}$, and

$$V^{(k)}(\mathbf{x}) := \left\{ j \in V : x_j > x_{\overline{\sigma}(k)} \right\},$$

for $k = 2, \ldots, N$. Now, given $k \in \{1, \ldots, N\}$, let $f : \mathcal{P}(V) \to \mathbb{R}$ be defined as

$$f(A) := \begin{cases} 1, & \text{if } |A| \ge k, \\ 0, & \text{otherwise.} \end{cases}$$

Then, by letting $x_{\overline{\sigma}(k)}$ indicate the k-th largest component of $\mathbf{x}$, we have

$$f^L(\mathbf{x}) = (x_{\overline{\sigma}(k)} - x_{\overline{\sigma}(k+1)}) f(V^{(k+1)}(\mathbf{x})) + \cdots$$
$$+ (x_{\overline{\sigma}(N-1)} - x_{\overline{\sigma}(N)}) f(V^{(N)}(\mathbf{x})) + x_{\overline{\sigma}(N)}$$
$$= (x_{\overline{\sigma}(k)} - x_{\overline{\sigma}(k+1)}) + \cdots + (x_{\overline{\sigma}(N-1)} - x_{\overline{\sigma}(N)}) + x_{\overline{\sigma}(N)}$$
$$= x_{\overline{\sigma}(k)}.$$

7. Finally, we leave it for the reader to check (or to consult [50, 132]) that

$$f^L(\mathbf{x}) = \min_{t \in \mathbb{R}} \|\mathbf{x} - t\mathbf{1}\|_{L^1}, \quad \text{for } f(A) = \min(\text{vol } A, \text{vol } (V \setminus A)). \quad (6.2.8)$$

We now list some fundamental properties of the Lovász extension [189].

Proposition 6.2.1. (a) f^L is positively one-homogeneous, PL (piecewise linear), and Lipschitz continuous.

(b) For all $t \in \mathbb{R}$ and for all $\mathbf{x} \in \mathbb{R}^N$,

$$f^L(\mathbf{x} + t\mathbf{1}_V) = f^L(\mathbf{x}) + t f(V).$$

Thus, f^L is linear along the direction $\mathbf{1}_V$.

(c) f^L is convex if and only if $f : \mathcal{P}(V) \to \mathbb{R}$ is submodular.

(d) A continuous function $F : \mathbb{R}^N \to \mathbb{R}$ is the Lovász extension of some $f : \mathcal{P}(V) \to \mathbb{R}$ if and only if $F(\mathbf{x} + \mathbf{y}) = F(\mathbf{x}) + F(\mathbf{y})$ whenever $\mathbf{x}$ and $\mathbf{y}$ are comonotonic, meaning that

$$(x_i - x_j)(y_i - y_j) \ge 0,$$

for all $i, j \in V$.

(e) A one-homogeneous function $F: \mathbb{R}^N \to \mathbb{R}$ is the Lovász extension of some submodular function if and only if

$$F(\mathbf{x} + t\mathbf{1}_V) = F(\mathbf{x}) + tF(\mathbf{1}_V),$$

for all $t \in \mathbb{R}$ and for all $\mathbf{x} \in \mathbb{R}^N$, and

$$F(\mathbf{x}) + F(\mathbf{y}) \geq F(\mathbf{x} \vee \mathbf{y}) + F(\mathbf{x} \wedge \mathbf{y}),$$

where the ith components of $\mathbf{x} \vee \mathbf{y}$ and $\mathbf{x} \wedge \mathbf{y}$ are $(\mathbf{x} \vee \mathbf{y})_i = \max\{x_i, y_i\}$ and $(\mathbf{x} \wedge \mathbf{y})_i = \min\{x_i, y_i\}$, respectively.

A useful variant of the Lovász extension is the *disjoint-pair Lovász extension*, which was introduced by Qi [231] and has been systematically investigated by Fujishige [101] and Murota [221] in the context of discrete convex analysis (or submodular analysis). Specifically, with $\mathcal{P}(V \times V)$ being the power set of $V \times V$, for a function $f: \mathcal{P}(V \times V) \to \mathbb{R}$, its disjoint-pair Lovász extension is defined by the integral formulation

$$f^L(\mathbf{x}) = \int_0^{\|\mathbf{x}\|_\infty} f(V_+^t(\mathbf{x}), V_-^t(\mathbf{x}))dt, \tag{6.2.9}$$

where $V_\pm^t(\mathbf{x}) = \{i \in V: \pm x_i > t\}$, for all $t \geq 0$. If we naturally assume

$$f(\emptyset, \emptyset) = 0, \tag{6.2.10}$$

we can write (6.2.9) as

$$f^L(\mathbf{x}) = \int_0^\infty f(V_+^t(\mathbf{x}), V_-^t(\mathbf{x}))dt. \tag{6.2.11}$$

Equivalently,

$$f^L(\mathbf{x}) = \sum_{i=0}^{N-1}(|x_{\sigma(i+1)}| - |x_{\sigma(i)}|)f(V_{\sigma(i)}^+(\mathbf{x}), V_{\sigma(i)}^-(\mathbf{x})), \tag{6.2.12}$$

where $\sigma: V \cup \{0\} \to V \cup \{0\}$ is a bijection such that $|x_{\sigma(1)}| \leq |x_{\sigma(2)}| \leq \dots \leq |x_{\sigma(N)}|$ and $\sigma(0) = 0$, where $x_0 := 0$, and

$$V_{\sigma(i)}^\pm(\mathbf{x}) := \{j \in V: \pm x_j > |x_{\sigma(i)}|\}, \quad i = 0, 1, \dots, N-1.$$

Note that to define f^L, we only use the information on disjoint pairs $(A, B) \subset V \times V$, not a general subset $S \subset V \times V$. Thus, we shall focus on

$$\mathcal{P}_2(V) := \{(A, B): A, B \subset V, A \cap B = \emptyset\},$$

a subfamily of $\mathcal{P}(V \times V)$. We regard $\mathcal{P}_2(V)$ as $\{-1, 0, 1\}^N$ by identifying the disjoint pair (A, B) with the ternary (indicator) vector $\mathbf{1}_A - \mathbf{1}_B$.

Analogously to (d) in Proposition 6.2.1, we have

Proposition 6.2.2. *A continuous function F is the disjoint-pair Lovász extension of some function $f: \mathcal{P}_2(V) \to \mathbb{R}$, if and only if $F(\mathbf{x}) + F(\mathbf{y}) = F(\mathbf{x} + \mathbf{y})$ whenever $\mathbf{x}$ and $\mathbf{y}$ are absolutely comonotonic, that is, $x_i y_i \geq 0$ for all i, and*

$$(|x_i| - |x_j|)(|y_i| - |y_j|) \geq 0$$

for all i, j.

Proof From (6.2.12), we have that F is the disjoint-pair Lovász extension of some function $f: \mathcal{P}_2(V) \to \mathbb{R}$ if and only if

$$\lambda F(\mathbf{x}) + (1 - \lambda)F(\mathbf{y}) = F(\lambda \mathbf{x} + (1 - \lambda)\mathbf{y}),$$

for all absolutely comonotonic vectors $\mathbf{x}$ and $\mathbf{y}$ and for all $\lambda \in [0, 1]$. Therefore, we only need to prove sufficiency.

Given $\mathbf{x} \in \mathbb{R}^V$, since $s\mathbf{x}$ and $t\mathbf{x}$ are absolutely comonotonic for $s, t \geq 0$, we have that

$$F(s\mathbf{x}) + F(t\mathbf{x}) = F((s + t)\mathbf{x}),$$

which yields a Cauchy equation on the half-line. Thus, the continuity assumption implies the linearity of F on the ray $\mathbb{R}^+\mathbf{x}$, which implies the property $F(t\mathbf{x}) = tF(\mathbf{x})$, for all $t \geq 0$, and hence

$$\lambda F(\mathbf{x}) + (1 - \lambda)F(\mathbf{y}) = F(\lambda \mathbf{x} + (1 - \lambda)\mathbf{y}),$$

for any absolutely comonotonic vectors $\mathbf{x}$ and $\mathbf{y}$, and for all $\lambda \in [0, 1]$. □

6.3 Eigenvalue Problem for a Pair of Lovász Extensions

As explained in Section 6.1, we now reformulate our eigenvalue problem.

Definition 6.3.1. $\lambda \in \mathbb{R}$ is called an *eigenvalue*, and $\mathbf{x} \in \mathbb{R}^N \setminus \{0\}$ a corresponding *eigenvector*, of the function pair (f^L, g^L), if

$$\mathbf{0} \in \nabla f^L(\mathbf{x}) - \lambda \nabla g^L(\mathbf{x}).$$

We also call $(\lambda, \mathbf{x})$ an *eigenpair*.

Proposition 6.3.1. *Given $f, g: \mathcal{P}(V) \to \mathbb{R}$, for any eigenvalue λ of (f^L, g^L) there exists $A \in \mathcal{P}(V) \setminus \{\emptyset\}$ such that $\lambda = f(A)/g(A)$ and $\mathbf{1}_A$ is a corresponding eigenvector. In addition, given any two distinct real numbers a and b, every eigenvalue has an eigenvector in $\{a, b\}^N$. Moreover, we have that:*

- *If $g(V) \neq 0$, then $\lambda = f(V)/g(V)$ is the only possible eigenvalue of (f^L, g^L);*

- If $g(V) = 0$ and (f^L, g^L) has at least one eigenvalue, then $f(V) = 0$, and, in this case, (f^L, g^L) may have many distinct eigenvalues.

Proof Let $(\lambda, \mathbf{x})$ be an eigenpair of (f^L, g^L). Since f^L is one-homogeneous and linear along the direction $\mathbf{1}$, without loss of generality we can assume that $\mathbf{x}$ lies in the relative interior of the simplex $\triangle$ with vertices $\mathbf{1}_{A_1}, \ldots, \mathbf{1}_{A_k}$, where $A_1, \ldots, A_k$ are upper-level sets of $\mathbf{x}$. By applying Lemma 6.1.1 to the piecewise linear functions f^L and g^L, we infer that

$$\mathbf{0} \in \nabla f^L(\mathbf{x}) - \lambda \nabla g^L(\mathbf{x}) \subset \nabla f^L(\mathbf{1}_{A_i}) - \lambda \nabla g^L(\mathbf{1}_{A_i}),$$

for any $i = 1, \ldots, k$, meaning that each $\mathbf{1}_{A_i}$ is an eigenvector of (f^L, g^L).

By definition of Lovász extension, we can also check that, for all $\mathbf{v} \in \nabla f^L(\mathbf{x})$,

$$\sum_{i \in V} v_i = f(V).$$

Therefore, for all $\mathbf{x}$,

$$\nabla f^L(\mathbf{x}) \cdot \mathbf{1} = f(V).$$

The same identity holds for g, that is, $\nabla g^L(\mathbf{x}) \cdot \mathbf{1} = g(V)$. This implies that, for any eigenpair $(\lambda, \mathbf{x})$,

$$f(V) - \lambda g(V) = (\nabla f^L(\mathbf{x}) - \lambda \nabla g^L(\mathbf{x})) \cdot \mathbf{1} = 0.$$

Therefore $\lambda = f(V)/g(V)$ if $g(V) \neq 0$, and $f(V) = 0$ if $g(V) = 0$.

For the case when $f(V) = g(V) = 0$, we may consider the following example. Let $V = \{1, 2\}$ and consider two functions $f, g : \mathcal{P}(V) \to \mathbb{R}$ defined by $f(\{1\}) = g(\{2\}) = 2$, $f(\{2\}) = g(\{1\}) = 1$, and $f(\{1, 2\}) = f(\emptyset) = g(\{1, 2\}) = g(\emptyset) = 0$. Then, one can check that

$$f^L(\mathbf{x}) = \begin{cases} x_2 - x_1, & \text{if } x_1 \leq x_2, \\ 2(x_1 - x_2), & \text{if } x_1 \geq x_2, \end{cases} \quad \text{and } g^L(\mathbf{x}) = \begin{cases} 2(x_2 - x_1), & \text{if } x_1 \leq x_2, \\ x_1 - x_2, & \text{if } x_1 \geq x_2. \end{cases}$$

Moreover, it can be verified that $(2, \mathbf{1}_{\{1\}} - \mathbf{1}_{\{2\}})$ and $(1/2, \mathbf{1}_{\{2\}} - \mathbf{1}_{\{1\}})$ are two eigenpairs of (f^L, g^L). Hence, in this example, the function pair (f^L, g^L) has at least two distinct eigenvalues: $1/2$ and 2.   □

The case where $g(V) = 0$ will be the interesting one for us. Recall the quantity

$$\eta(S) = \frac{|E(S, V \setminus S)|}{\min\{\text{vol}(S), \text{vol}(V \setminus S)\}} = \frac{f(S)}{g(S)}$$

from (4.3.12), for a given $S \subset V$. In this case, the pair (f^L, g^L) will have several distinct eigenvalues. Since the quantity $\eta(S)$ considers edges between pairs of sets, namely S and $V\backslash S$, the eigenvectors of (f^L, g^L) directly correspond to subsets of V, and therefore, the first nontrivial eigenvector of (f^L, g^L) relates to the subset S realizing the minimum of $\eta(\cdot)$. In particular, we need to work with disjoint pair extensions, and this is addressed in Theorem 6.3.1.

Lemma 6.3.1. *Let C and Ω be two convex cones in $\mathbb{R}^N$ that satisfy*

$$\Omega \cap ((-C) \cup C) = \{\mathbf{0}\}.$$

Let also

$$C^* := \{\mathbf{x} \in \mathbb{R}^n : (\mathbf{x}, \mathbf{y}) \le 0 \,\forall \mathbf{y} \in C\}$$

be the polar cone of C. Then, $\Omega^* \cap C^* \ne \{\mathbf{0}\}$ and $\Omega^* \cap (-C^*) \ne \{\mathbf{0}\}$.

Proof We have that $\Omega^* \cap C^* = (\Omega \cup C)^* \supset (\Omega + C)^*$, where $\Omega + C$ is the Minkowski summation of C and Ω. If $\Omega + C = \mathbb{R}^N$, then for any $-\mathbf{c} \in (-C)\backslash\{\mathbf{0}\}$, there exist $\mathbf{a} \in \Omega \backslash \{\mathbf{0}\}$ and $\mathbf{c}' \in C \backslash \{\mathbf{0}\}$ such that $\mathbf{a} + \mathbf{c}' = -\mathbf{c}$. This implies that $\mathbf{a} = -\mathbf{c}' - \mathbf{c} \in -C \backslash \{\mathbf{0}\}$, which contradicts the condition that $\Omega \cap (-C) = \{\mathbf{0}\}$. Therefore, the convex cone $\Omega + C$ is not the whole space $\mathbb{R}^N$, which implies that $(\Omega + C)^* \ne \{\mathbf{0}\}$. Hence, $\Omega^* \cap C^* \ne \{\mathbf{0}\}$, and similarly, $\Omega^* \cap (-C^*) \ne \{\mathbf{0}\}$. $\quad\square$

Theorem 6.3.1. *Given $f, g : \mathcal{P}_2(V) \to \mathbb{R}$, every eigenvalue of (f^L, g^L) has an eigenvector of the form $\mathbf{1}_A - \mathbf{1}_B$. Moreover,*

1. *If, for any $(A, B) \in \mathcal{P}_2(V) \backslash \{(\emptyset, \emptyset)\}$ we have that*

   $$2f(A, B) = f(A, V\backslash A) + f(V\backslash B, B) \text{ and } 2g(A, B) = g(A, V\backslash A) + g(V\backslash B, B),$$

   then every eigenvalue of (f^L, g^L) has an eigenvector of the form $\mathbf{1}_A - \mathbf{1}_{V\backslash A}$.
2. *If g is a constant function, then for any $A \subset V$, $\mathbf{1}_A - \mathbf{1}_{V\backslash A}$ is an eigenvector.*

Proof The general result on the disjoint-pair Lovász extension is similar to Proposition 6.3.1, and thus, we omit the proof. We prove the other statements in a sequence of steps.

1. We first prove that, in this case, for any $A_1 \subset \ldots \subset A_k$ with $(A_1, A_k) \ne (\emptyset, V)$, f^L is linear on $\text{conv}\{\mathbf{1}_{A_i} - \mathbf{1}_{V\backslash A_i} : i = 1, \ldots, k\}$.
   Proposition 6.2.2 implies that f^L is linear on $\text{conv}\{\mathbf{1}_{A_i} - \mathbf{1}_{B_i} : i = 1, \ldots, k\}$ whenever $A_1 \subset \ldots \subset A_k$, $B_1 \subset \ldots \subset B_k$, and $A_k \cap B_k = \emptyset$. Thus, for any $\sigma, \tau : \{1, \ldots, k\} \to \{1, \ldots, k\}$ with

   $$\sigma(1) \le \ldots \le \sigma(k) \le \tau(k) \le \ldots \le \tau(1),$$

   f^L is linear on $\text{conv}\{\mathbf{1}_{A_{\sigma(i)}} - \mathbf{1}_{V\backslash A_{\tau(i)}} : i = 1, \ldots, k\}$. Observe that

   $$\text{conv}\{\mathbf{1}_{A_i} - \mathbf{1}_{V\backslash A_i} : i = 1, \ldots, k\}$$

$$= \bigcup_{\sigma,\tau} \text{conv}\{\mathbf{1}_{A_{\sigma(i)}} - \mathbf{1}_{V \setminus A_{\tau(i)}} : i = 1, \ldots, k\}$$

$$= \bigcup_{l=1}^{k} \bigcup_{\sigma(1)=1, \sigma(k)=\tau(k)=l, \tau(1)=k} \text{conv}\{\mathbf{1}_{A_{\sigma(i)}} - \mathbf{1}_{V \setminus A_{\tau(i)}} : i = 1, \ldots, k\},$$

where in the second line, we can further assume that each

$$\text{conv}\{\mathbf{1}_{A_{\sigma(i)}} - \mathbf{1}_{V \setminus A_{\tau(i)}} : i = 1, \ldots, k\}$$

is a $(k-1)$-dimensional simplex, and there are exactly $\sum_{l=1}^{k} \binom{k-1}{l-1} = 2^{k-1}$ simplexes of dimension $(k-1)$. Since, furthermore,

$$2f^L(\mathbf{1}_{A_{\sigma(i)}} - \mathbf{1}_{V \setminus A_{\tau(i)}}) = 2f(A_{\sigma(i)}, V \setminus A_{\tau(i)})$$

$$= f(A_{\sigma(i)}, V \setminus A_{\sigma(i)}) + f(A_{\tau(i)}, V \setminus A_{\tau(i)})$$

$$= f^L\left(\mathbf{1}_{A_{\sigma(i)}} - \mathbf{1}_{V \setminus A_{\sigma(i)}}\right) + f^L\left(\mathbf{1}_{A_{\tau(i)}} - \mathbf{1}_{V \setminus A_{\tau(i)}}\right)$$

and

$$\mathbf{1}_{A_{\sigma(i)}} - \mathbf{1}_{V \setminus A_{\sigma(i)}} + \mathbf{1}_{A_{\tau(i)}} - \mathbf{1}_{V \setminus A_{\tau(i)}} = 2(\mathbf{1}_{A_{\sigma(i)}} - \mathbf{1}_{V \setminus A_{\tau(i)}}),$$

f^L must be linear on the segment

$$\text{conv}\{\mathbf{1}_{A_{\sigma(i)}} - \mathbf{1}_{V \setminus A_{\sigma(i)}}, \mathbf{1}_{A_{\tau(i)}} - \mathbf{1}_{V \setminus A_{\tau(i)}}\}.$$

Therefore, one can check that f^L is linear on the simplex

$$\text{conv}\{\mathbf{1}_{A_i} - \mathbf{1}_{V \setminus A_i} : i = 1, \ldots, k\}.$$

Now, given an eigenvalue λ, let $\mathbf{1}_A - \mathbf{1}_B$ be a corresponding eigenvector, for some $(A, B) \in \mathcal{P}_2(V) \setminus \{(\emptyset, \emptyset)\}$. By the above argument, it can be verified that both $\mathbf{1}_A - \mathbf{1}_{V \setminus A}$ and $\mathbf{1}_{V \setminus B} - \mathbf{1}_B$ are eigenvectors for λ.

2. We shall prove that, for any $f : \mathcal{P}_2(V) \to [0, +\infty)$ and for any $A \subset V$, the vector $\mathbf{1}_A - \mathbf{1}_{V \setminus A}$ is an eigenvector of $(f^L, \|\cdot\|_\infty)$. In particular, by definition of the eigenvalue problem on $(f^L, \|\cdot\|_\infty)$, we only need to prove that

$$\nabla f^L(\mathbf{x}) \cap \left((-\mathbb{R}^N_{\text{sign}(\mathbf{x})}) \cup \mathbb{R}^N_{\text{sign}(\mathbf{x})}\right) \neq \emptyset$$

for any $\mathbf{x} \in \{-1, 1\}^N$, where

$$\mathbb{R}^N_{\text{sign}(\mathbf{x})} := \{\mathbf{y} \in \mathbb{R}^N : y_i x_i \geq 0 \ \forall i\} = \text{cone}(\nabla \|\mathbf{x}\|_\infty).$$

We first assume that f is positive-definite, that is, $f(A, B) > 0$ whenever $(A, B) \neq (\emptyset, \emptyset)$, and we shall apply Lemma 6.3.1 to this case.

Assume by contradiction that, for some $\mathbf{x} \in \{-1, 1\}^N$,

$$\nabla f^L(\mathbf{x}) \cap ((-\mathbb{R}^N_{\text{sign}(\mathbf{x})}) \cup \mathbb{R}^N_{\text{sign}(\mathbf{x})}) = \emptyset.$$

Then,

$$\text{cone}(\nabla f^L(\mathbf{x})) \cap ((-\mathbb{R}^N_{\text{sign}(\mathbf{x})}) \cup \mathbb{R}^N_{\text{sign}(\mathbf{x})}) = \{\mathbf{0}\}$$

and, by Lemma 6.3.1,

$$\text{cone}^*(\nabla f^L(\mathbf{x})) \cap \mathbb{R}^N_{\text{sign}(\mathbf{x})} = \text{cone}^*(\nabla f^L(\mathbf{x})) \cap (-\mathbb{R}^N_{\text{sign}(\mathbf{x})})^* \neq \{\mathbf{0}\}.$$

Now, fix a set $S \subset \mathbb{R}^N$. The *normal cone* of S at $\mathbf{x}$ is

$$\mathsf{N}_{\mathbf{x}}(S) := \{\mathbf{y} \in \mathbb{R}^N : \langle \mathbf{y}, \mathbf{x}' - \mathbf{x} \rangle \leq 0, \forall \mathbf{x}' \in S\}.$$

The *tangent cone* $\mathsf{T}_{\mathbf{x}}(S)$ of S at $\mathbf{x}$ is the closure of the set

$$\{\mathbf{z} \in \mathbb{R}^N : \mathbf{x} + \varepsilon \mathbf{z} \in S, \text{ for some } \varepsilon > 0\}.$$

Since f is positive-definite, if f is further assumed to be submodular, from the fact that f^L is convex and positive-definite, we have

$$\text{cone}(\nabla f^L(\mathbf{x})) = \mathsf{N}_{\mathbf{x}}(\{\mathbf{y} : f^L(\mathbf{y}) \leq f^L(\mathbf{x})\}).$$

It follows that

$$\text{cone}^*(\nabla f^L(\mathbf{x})) \subset \mathsf{T}_{\mathbf{x}}(\{\mathbf{y} : f^L(\mathbf{y}) \leq f^L(\mathbf{x})\}),$$

implying that

$$\mathsf{T}_{\mathbf{x}}(\{\mathbf{y} : f^L(\mathbf{y}) \leq f^L(\mathbf{x})\}) \cap \mathbb{R}^N_{\text{sign}(\mathbf{x})} \neq \{\mathbf{0}\}.$$

The above relation still holds if f is just positive-definite.

Now, since $\mathbf{x} \in \{-1, 1\}^N$, we can write $\mathbf{x} = \mathbf{1}_{A_N} - \mathbf{1}_{B_N}$, with $A_N \sqcup B_N = V$. Also, for any given permutation $\sigma : \{1, \ldots, N\} \to \{1, \ldots, N\} = V = A_N \sqcup B_N$, we define $A_i := A_N \setminus \{\sigma(i+1), \ldots, \sigma(N)\}$ and $B_i := B_N \setminus \{\sigma(i+1), \cdots, \sigma(N)\}$, for $i = 1, \ldots, N-1$.

Now, for $i = 1, \ldots, N$, let

$$\mathbf{x}^i := \frac{\mathbf{1}_{A_i} - \mathbf{1}_{B_i}}{f(A_i, B_i)}.$$

Since $f(A_i, B_i) > 0$, we have $f^L(\mathbf{x}^i) = 1$. Therefore,

$$\mathsf{T}_{\mathbf{x}^N}(\{\mathbf{y} : f^L(\mathbf{y}) \leq 1\}) = \mathsf{T}_{\mathbf{x}}(\{\mathbf{y} : f^L(\mathbf{y}) \leq f^L(\mathbf{x})\}).$$

Hence, without loss of generality, we may assume that $\mathbf{x} = \mathbf{x}^N$.

Now, the definition of f^L implies that

$$\text{conv}(\mathbf{0}, \mathbf{x}^1, \ldots, \mathbf{x}^N) \subset \{\mathbf{y} : f^L(\mathbf{y}) \leq 1\}.$$

We write $\triangle_\sigma := \mathrm{conv}(\mathbf{0}, \mathbf{x}^1, \ldots, \mathbf{x}^N)$, since the construction of $\mathbf{x}^1, \ldots, \mathbf{x}^N$ depends on the permutation σ. For any

$$\mathbf{y} := \sum_{i=1}^{N} t_i \mathbf{x}^i \in \mathrm{conv}(\mathbf{0}, \mathbf{x}^1, \ldots, \mathbf{x}^N) \setminus \{\mathbf{x}\},$$

we have that

$$(\mathbf{y} - \mathbf{x})_{\sigma(N)} x_{\sigma(N)} = -(1 - t_N) x_{\sigma(N)}^2 < 0,$$

and thus, $\mathbf{y} - \mathbf{x} \notin \mathbb{R}_{\mathrm{sign}(\mathbf{x})}^N$. Hence,

$$\mathsf{T}_{\mathbf{x}}(\triangle_\sigma) \cap \mathbb{R}_{\mathrm{sign}(\mathbf{x})}^N = \{\mathbf{0}\}.$$

Since

$$\mathsf{T}_{\mathbf{x}}(\{\mathbf{y} : f^L(\mathbf{y}) \le 1\}) = \bigcup_\sigma \mathsf{T}_{\mathbf{x}}(\triangle_\sigma),$$

it follows that

$$\mathsf{T}_{\mathbf{x}}(\{\mathbf{y} : f^L(\mathbf{y}) \le 1\}) \cap \mathbb{R}_{\mathrm{sign}(\mathbf{x})}^N = \{\mathbf{0}\}.$$

This is a contradiction. Therefore, we have proved that, for any positive-definite function f on $\mathcal{P}_2(V)$, every vector $\mathbf{x} \in \{-1, 1\}^N$ is an eigenvector of the pair $(f^L, \|\cdot\|_\infty)$, that is, there exists $\mathbf{u} \in \nabla\|\mathbf{x}\|_\infty$ and $\lambda \in \mathbb{R}$ such that $\lambda\mathbf{u} \in \nabla f^L(\mathbf{x})$. Applying Euler's identity on one-homogeneous convex functions, we have that $\langle \mathbf{u}, \mathbf{x} \rangle = \|\mathbf{x}\|_\infty$ and $\langle \lambda\mathbf{u}, \mathbf{x} \rangle = f^L(\mathbf{x})$, which implies that

$$\lambda = \frac{f^L(\mathbf{x})}{\|\mathbf{x}\|_\infty} > 0.$$

We now consider the general case where $f \ge 0$. Fix a sequence $\{f_n\}_{n \ge 1}$ of positive-definite functions on $\mathcal{P}_2(V)$ such that $f_n \to f$, as n tends to $+\infty$. It can then be verified that, for any $\mathbf{v}_n \in \nabla f_n^L(\mathbf{x})$, all limit points of $\{\mathbf{v}_n\}_{n \ge 1}$ belong to $\nabla f^L(\mathbf{x})$. We know that there exist $\mathbf{u}_n \in \nabla\|\mathbf{x}\|_\infty$ and

$$\lambda_n = \frac{f_n^L(\mathbf{x})}{\|\mathbf{x}\|_\infty} > 0$$

such that $\lambda_n \mathbf{u}_n \in \nabla f_n^L(\mathbf{x})$. This implies that, for any limit point $\mathbf{u}$ of $\{\mathbf{u}_n\}_{n \ge 1}$, $\mathbf{u} \in \nabla\|\mathbf{x}\|_\infty$ and $\lambda\mathbf{u} \in \nabla f^L(\mathbf{x})$, where $\lambda := \lim_{n \to +\infty} \lambda_n$. Therefore, $(\lambda, \mathbf{x})$ is an eigenpair of $(f^L, \|\cdot\|_\infty)$.

$\square$

We now introduce the (f,g)–*Cheeger constant* corresponding to a given pair (f,g) of functions $f \colon \mathcal{P}(V) \to \mathbb{R}$ and $g \colon \mathcal{P}(V) \to \mathbb{R}_+$ as

$$\mathrm{Ch}(f,g) := \min_{A \in \mathcal{P}(V) \setminus \{\emptyset, V\}} \frac{f(A)}{\min\{g(A), g(V \setminus A)\}}. \tag{6.3.1}$$

Theorem 6.3.2. *If $f \colon \mathcal{P}(V) \to \mathbb{R}$ is symmetric, that is, $f(A) = f(V \setminus A)$ for all A, while $g \colon \mathcal{P}(V) \to \mathbb{R}_+$ is submodular and nondecreasing, then*

$$\mathrm{Ch}(f,g) = \min_{(A,B) \in \mathcal{P}_2(V) \setminus \{(\emptyset, \emptyset)\}} \max \left\{ \frac{f(A)}{g(A)}, \frac{f(B)}{g(B)} \right\}. \tag{6.3.2}$$

Proof From the symmetry of f, it follows that

$$\frac{f(A)}{\min\{g(A), g(V \setminus A)\}} = \max \left\{ \frac{f(A)}{g(A)}, \frac{f(V \setminus A)}{g(V \setminus A)} \right\},$$

and this yields

$$\mathrm{Ch}(f,g) \geq \min_{(A,B) \in \mathcal{P}_2(V) \setminus \{(\emptyset, \emptyset)\}} \max \left\{ \frac{f(A)}{g(A)}, \frac{f(B)}{g(B)} \right\}.$$

To see that (6.3.2) is equivalent to (6.3.1), we need to show that, for any $A, B \neq \emptyset$ with $A \cap B = \emptyset$,

$$\max \left\{ \frac{f(A)}{g(A)}, \frac{f(B)}{g(B)} \right\} \geq \min \left\{ \frac{f(A)}{\min\{g(A), g(V \setminus A)\}}, \frac{f(B)}{\min\{g(B), g(V \setminus B)\}} \right\}.$$

To see this, assume the contrary, by contradiction. Then, $g(A) > g(V \setminus A)$ and $g(B) > g(V \setminus B)$, implying that $g(A) + g(B) > g(V \setminus A) + g(V \setminus B)$. Since $A \subset V \setminus B$ and g is nondecreasing, this implies that $g(A) \leq g(V \setminus B)$. Similarly, $g(B) \leq g(V \setminus A)$, which leads to a contradiction. This proves the claim. □

For a simple graph G, by letting $f(A) = |\partial A|$ and $g(A) = \mathrm{vol}(A)$, the Cheeger constant $h(G)$ equals $\mathrm{Ch}(f,g)$.

Theorem 6.3.3. *Let $f_s, g_s \colon \mathcal{P}_2(V) \to \mathbb{R}$ be defined by $f_s(A,B) = f(A) + f(B)$ and $g_s(A,B) = g(A) + g(B)$. Then,*

$$\mathrm{Ch}(f,g) = \lambda_2(f_s^L, g_s^L),$$

that is, the (f,g)–Cheeger constant equals the second eigenvalue of (f_s^L, g_s^L).

Proof Let $g_m(A) := \min\{g(A), g(V \setminus A)\}$. Since g is nondecreasing, $g(V^t(\mathbf{x}))$ must be nonincreasing with respect to $t \in \mathbb{R}$, that is, $g(V^{t_1}(\mathbf{x})) \geq g(V^{t_2}(\mathbf{x}))$ if $t_1 \leq t_2$. Hence, the function $t \mapsto g(V^t(\mathbf{x})) - g(V \setminus V^t(\mathbf{x}))$ is nonincreasing and left-continuous on $t \in \mathbb{R}$. Since

$$\lim_{t \to -\infty} (g(V^t(\mathbf{x})) - g(V \setminus V^t(\mathbf{x}))) = g(V) - g(\emptyset) \geq 0$$

and

$$\lim_{t \to +\infty} (g(V^t(\mathbf{x})) - g(V \setminus V^t(\mathbf{x}))) = g(\emptyset) - g(V) \le 0,$$

there exists $t_0 \in \mathbb{R}$ such that

$$g(V^t(\mathbf{x})) \ge g(V \setminus V^t(\mathbf{x})), \quad \forall t < t_0,$$

and

$$g(V^t(\mathbf{x})) \le g(V^{t_0}(\mathbf{x})) \le g(V \setminus V^t(\mathbf{x})), \quad \forall t \ge t_0.$$

Therefore, the Lovász extension g_m^L of g_m satisfies

$$
\begin{aligned}
g_m^L(\mathbf{x}) &= \int_{\min \mathbf{x}}^{\max \mathbf{x}} g_m(V^t(\mathbf{x}))dt + \min \mathbf{x} \, g_m(V) \\
&= \int_{\min \mathbf{x}}^{t_0} g(V \setminus V^t(\mathbf{x}))dt + \int_{t_0}^{\max \mathbf{x}} g(V^t(\mathbf{x}))dt \quad (\text{as } g(\emptyset) = 0) \\
&= \int_{\min(\mathbf{x}-t_0\mathbf{1})}^{0} g(V \setminus V^t(\mathbf{x}-t_0\mathbf{1}))dt + \int_0^{\max(\mathbf{x}-t_0\mathbf{1})} g(V^t(\mathbf{x}-t_0\mathbf{1}))dt \\
&= \int_{-\|\mathbf{x}-t_0\mathbf{1}\|_\infty}^{0} g(V \setminus V^t(\mathbf{x}-t_0\mathbf{1}))dt + \int_0^{\|\mathbf{x}-t_0\mathbf{1}\|_\infty} g(V^t(\mathbf{x}-t_0\mathbf{1}))dt \\
&= \int_0^{\|\mathbf{x}-t_0\mathbf{1}\|_\infty} g(V^t(\mathbf{x}-t_0\mathbf{1})) + g(V \setminus V^{-t}(\mathbf{x}-t_0\mathbf{1}))dt \\
&= g_s^L(\mathbf{x}-t_0\mathbf{1}) \\
&= \min_{t \in \mathbb{R}} g_s^L(\mathbf{x}-t\mathbf{1}).
\end{aligned}
$$

Since f is symmetric, this implies that $f_s^L(\mathbf{x}) = f^L(\mathbf{x}) = f_m^L(\mathbf{x})$.

Moreover, since g is positive, submodular, and nondecreasing, it follows that g_s is bisubmodular. Therefore, $g_s(A, B)$ is submodular in A for any fixed B, and it is submodular in B for any fixed A. Thus, as the Lovász extension of a submodular function, g_s^L is a convex function. Therefore, we have

$$\min_{\mathbf{x} \perp \mathbf{1}} \frac{f_s^L(\mathbf{x})}{\min_{t \in \mathbb{R}} g_s^L(\mathbf{x}-t\mathbf{1})} = \min_{\mathbf{x} \in \mathbb{R}_+^N : \min \mathbf{x}=0} \frac{f_m^L(\mathbf{x})}{g_m^L(\mathbf{x})} = \min_{A \ne \emptyset, V} \frac{f_m(A)}{g_m(A)} = \mathrm{Ch}(f,g),$$

which is the second eigenvalue of the pair (f_s^L, g_s^L). $\qquad\qquad \square$

If we remove the symmetry assumption of f in Theorem 6.3.2, we can then express the (f,g)–Cheeger constant as

$$\mathrm{Ch}(f,g) = \min_{A \in \mathcal{P}(V) \setminus \{\emptyset, V\}} \frac{\min\{f(A), f(V \setminus A)\}}{\min\{g(A), g(V \setminus A)\}}.$$

From here on, we assume for simplicity that $g : \mathcal{P}(V) \to \mathbb{R}_+$ is modular.

Theorem 6.3.4. *Let $g^L(|\cdot|)$ be the function defined by $\mathbf{x} \mapsto g^L(|\mathbf{x}|)$, for all $\mathbf{x} \in \mathbb{R}^N$, where $|\mathbf{x}| := (|x_1|,\ldots,|x_N|)$. If $f(V) = 0$, then*

$$\mathrm{Ch}(f,g) = \lambda_2(f^L, g^L(|\cdot|)),$$

that is, the (f,g)–Cheeger constant equals the second eigenvalue of $(f^L, g^L(|\cdot|))$.

Proof Since $f \ge 0$, f^L is non-negative on $\mathbb{R}^N$. By Proposition 6.3.1 and Theorem 6.3.1, we may take an eigenvector of the form $\mathbf{1}_A - \mathbf{1}_B$ corresponding to the second eigenvalue. Let now $\hat{f}(A,B) := f(A \cup B)$ and

$$\hat{g}(A,B) := \min\{g(A \cup B), g(V \setminus (A \cup B))\}.$$

Then, $\hat{f}^L(\mathbf{x}) = f^L(|\mathbf{x}|)$ and $\hat{g}^L(\mathbf{x}) = \min_{t \in \mathbb{R}} g^L(||\mathbf{x}| - t|)$, where $||\mathbf{x}| - t| := (||x_1| - t|,\ldots,||x_N| - t|)$ for $\mathbf{x} := (x_1,\ldots,x_N)$. Therefore,

$$\mathrm{Ch}(f,g) = \min_{A \in \mathcal{P}(V) \setminus \{\emptyset, V\}} \frac{f(A)}{\min\{g(A), g(V \setminus A)\}}$$

$$= \min_{(A,B) \in \mathcal{P}_2(V), A \cup B \neq V} \frac{\hat{f}(A,B)}{\hat{g}(A,B)}$$

$$= \inf_{\mathbf{x} \in \mathbb{R}^N \,:\, |\mathbf{x}| \text{ nonconstant}} \frac{\hat{f}^L(\mathbf{x})}{\hat{g}^L(\mathbf{x})}$$

$$= \inf_{\mathbf{x} \in \mathbb{R}^N \,:\, |\mathbf{x}| \text{ nonconstant}} \frac{f^L(|\mathbf{x}|)}{\min_{t \in \mathbb{R}} g^L(||\mathbf{x}| - t|)}$$

$$= \inf_{\mathbf{x} \in \mathbb{R}^N_+ \,:\, \mathbf{x} \text{ nonconstant}} \frac{f^L(\mathbf{x})}{\min_{t \in \mathbb{R}} g^L(|\mathbf{x} - t|)}$$

$$= \inf_{\mathbf{x} \in \mathbb{R}^N \,:\, \mathbf{x} \text{ nonconstant}} \frac{f^L(\mathbf{x})}{\min_{t \in \mathbb{R}} g^L(|\mathbf{x} - t|)}$$

is the second eigenvalue of $(f^L, g^L(|\cdot|))$, where we use the fact that f^L is translation invariant along $\mathbf{1}$. □

Remark 6.3.1. In Section 6.2, we have seen that, for a simple graph G, by letting $f(A) = |\partial A|$ and $g(A) = \mathrm{vol}\,(A)$, we have

$$f^L(\mathbf{x}) = f_s^L(\mathbf{x}) = \sum_{ij \in E} |x_i - x_j|,$$

$$g^L(\mathbf{x}) = \sum_{i \in V} \deg i \, x_i$$

and

$$g_s^L(\mathbf{x}) = g^L(|\mathbf{x}|) = \sum_{i \in V} \deg i \, |x_i|.$$

Both Theorems 6.3.3 and 6.3.4 say, in this case, that $h(G)$ equals the second eigenvalue of the function pair $(\sum_{ij \in E} |x_i - x_j|, \sum_{i \in V} \deg i \, |x_i|)$. As we shall see in Section 7.2, this is the second eigenvalue of the graph 1-Laplacian.

As a slight generalization of an oriented hypergraph, a *chemical hypergraph* further allows for *catalysts*, that is, vertices that are both inputs and outputs of a hyperedge. We shall now work on a chemical hypergraph satisfying $e_{in} \neq \emptyset \neq e_{out}$ and $\#(e_{in} \cup e_{out}) \geq 2$ for all $e \in E$, where e_{in} denotes the set of all the inputs of e, and e_{out} is the set of the outputs of e.

We introduce the *Cheeger constant* on a chemical hypergraph as

$$h := \min_{A \in \mathcal{P}(V) \setminus \{\emptyset, V\}} \frac{\#(\partial A)}{\min\{\mu(A), \mu(V \setminus A)\}}, \qquad (6.3.3)$$

where $\mu(A) := \sum_{i \in A} \mu(i)$ with $\mu(i) > 0$, and the boundary set is

$$\partial A := \{e \in E : e_{in} \cap A \neq \emptyset \neq e_{out} \setminus A \text{ or } e_{out} \subset A \subset V \setminus e_{in}\}. \qquad (6.3.4)$$

We note that the above boundary set is not necessarily symmetric, that is, $\partial A \neq \partial(V \setminus A)$ in general.

Let $G = (V, E)$ be a simple graph, and let $\mathcal{N}(i)$ denote the set of neighbors of i.

- If $e_{in}^i = \mathcal{N}(i)$, $e_{out}^i = \{i\}$ and $\mu(A) = \#A$, then the boundary set (6.3.4) has the same cardinality as the *external vertex boundary*

$$\partial_{ext} A := \{j \in V \setminus A \mid \{j, i\} \in E \text{ for some } i \in A\}$$

from Section 4.3. Moreover, the corresponding Cheeger constant (6.3.3) equals

$$\inf_{\langle \mathbf{x}, \mathbf{1} \rangle = 0, \mathbf{x} \neq 0} \frac{\sum_{i \in V} \left(\max\limits_{j \in \{i\} \cup \mathcal{N}(i)} x_j - x_i \right)}{\min\limits_{t \in \mathbb{R}} \|\mathbf{x} - t\mathbf{1}\|_1}.$$

- If $e_{in}^i = \{i\}$ and $e_{out}^i = \mathcal{N}(i)$, then the boundary set (6.3.4) has the same cardinality as the *internal vertex-boundary*

$$\partial_{int} A := \{i \in A \mid \{i, j\} \in E \text{ for some } j \in V \setminus A\}.$$

- If $e_{in}^i = e_{out}^i = \mathcal{N}(i)$ for all $i \in V$, then the boundary set (6.3.4) has the same cardinality as the *vertex-boundary*

$$\partial_{ver} A := \partial_{ext} A \cup \partial_{int} A = V(E(A, V \setminus A)) = V(\partial A).$$

6.4 Lovász Extension and Cheeger Inequalities

In Theorem 6.3.4, we have seen that the (f,g)–Cheeger constant

$$\text{Ch}(f,g) = \min_{A \in \mathcal{P}(V) \setminus \{\emptyset, V\}} \frac{\min\{f(A), f(V \setminus A)\}}{\min\{g(A), g(V \setminus A)\}}$$

equals the second eigenvalue $\lambda_2(f^L, g^L(|\cdot|))$ of a pair of Lovász extensions. We now want to consider pth powers and derive inequalities between the Cheeger constant and the second eigenvalue. For $p = 1$, these will reduce to the equalities derived above. For $p > 1$, we need to utilize Hölder-type inequalities.

The cases of interest for us are covered by functions f and g of the form

$$f(A) := \sum_{e \in E} w_e f_e(A) \qquad \text{and} \qquad g(A) := \sum_{e \in E} g_e(A),$$

where $\{w_e\}_{e \in E}$ is a family of positive real numbers, while

$$\{f_e : \mathcal{P}(V) \to \mathbb{R}_{\geq 0}\}_{e \in E} \quad \text{and} \quad \{g_e : \mathcal{P}(V) \to \mathbb{R}_{\geq 0}\}_{e \in E}$$

are two families of functions satisfying

$$f_e(\emptyset) = f_e(V) = 0 = g_e(\emptyset) \quad \text{and} \quad g_e(A) = \sum_{i \in A} g_e(\{i\}), \quad \forall A \subset V.$$

We also let

$$F_p(\mathbf{x}) := \sum_{e \in E} (f_e^L(\mathbf{x}))^p \quad \text{and} \quad G_p(\mathbf{x}) := \sum_{e \in E} g_e^L(|\mathbf{x}|^p),$$

where $|\mathbf{x}|^p := (|x_1|^p, \ldots, |x_N|^p)$. By Theorem 6.3.1, the second eigenvalue of the function pair (F_p, G_p) is

$$\lambda_2(F_p, G_p) = \inf_{\mathbf{x} \text{ nonconstant}} \frac{\sum_{e \in E} (f_e^L(\mathbf{x}))^p}{\min_{c \in \mathbb{R}} \sum_{e \in E} g_e^L(|\mathbf{x} - c\mathbf{1}|^p)}. \tag{6.4.1}$$

If we assume that $g(\{i\}) := \sum_{e \in E} g_e(\{i\}) > 0$ for any $i \in V$, the following Cheeger inequalities hold.

Theorem 6.4.1. *In the above setting, we have*

$$\left(\frac{2}{c}\right)^{p-1} \frac{h^p}{p^p} \leq \lambda \leq 2^{p-1} C \cdot h, \tag{6.4.2}$$

where $h := \text{Ch}(f,g)$, $\lambda := \lambda_2(F_p, G_p)$, $C := \max_{e, A} f_e(A)^{p-1}$,

$$c := \max_{i \in V} \frac{\sum_{e \in E_i} w_e}{g(i)}, \quad \text{and} \quad E_i := \{e \in E : \exists S \subset V : f_e(S \setminus \{i\}) \neq f_e(S)\}.$$

Proof Let $A \in \mathcal{P}(V) \setminus \{\emptyset, V\}$ be such that

$$h = \frac{f(A)}{\min\{g(A), g(V \setminus A)\}}.$$

By taking the nonconstant vector $\mathbf{x} = \mathbf{1}_A$ in (6.4.1), we get

$$\lambda \leq \frac{\sum_{e \in E} w_e (f_e^L(\mathbf{1}_A))^p}{\min_{c \in \mathbb{R}} \sum_{e \in E} g_e^L(|\mathbf{1}_A - c\mathbf{1}|^p)}.$$

Since g_e is modular and $g = \sum_{e \in E} g_e$, we have

$$\sum_{e \in E} g_e^L(|\mathbf{1}_A - c\mathbf{1}|^p) = \sum_{i \in V} g(i)|(\mathbf{1}_A)_i - c|^p,$$

and

$$\min_{c \in \mathbb{R}} \sum_{i \in V} g(i)|(\mathbf{1}_A)_i - c|^p = \frac{g(A)g(V \setminus A)}{(g(A)^{\frac{1}{p-1}} + g(V \setminus A)^{\frac{1}{p-1}})^{p-1}}.$$

Therefore,

$$\begin{aligned}
\lambda &\leq \frac{\sum_{e \in E} w_e (f_e^L(\mathbf{1}_A))^p}{g(A)g(V \setminus A)/(g(A)^{\frac{1}{p-1}} + g(V \setminus A)^{\frac{1}{p-1}})^{p-1}} \\
&= \sum_{e \in E} w_e (f_e(A))^p \left(\sqrt[p-1]{\frac{1}{g(A)}} + \sqrt[p-1]{\frac{1}{g(V \setminus A)}} \right)^{p-1} \\
&\leq \max_{e,A} f_e(A)^{p-1} \sum_{e \in E} w_e f_e(A) 2^{p-1} \frac{1}{\min\{g(A), g(V \setminus A)\}} \\
&= \frac{2^{p-1} C f(A)}{\min\{g(A), g(V \setminus A)\}} \\
&= 2^{p-1} C \cdot h.
\end{aligned}$$

We shall now prove the lower bound for λ in (6.4.2). For simplicity, we identify a vector $\mathbf{x} \in \mathbb{R}^V$ with the function $\mathbf{x} \colon V \to \mathbb{R}$. Given $j \in V$, let $\deg(j) := \sum_{e \in E_j} w_e$. Then, for all j, we have that $\deg(j) \leq cg(j)$.

Also, we rewrite the *Lovász extension* (6.2.2) as

$$f^L(\mathbf{x}) = \sum_{i=0}^{N-1} (x_{(i+1)} - x_{(i)}) f(V_i(\mathbf{x})), \tag{6.4.3}$$

where $x_{(0)} := 0$ and $V_0(\mathbf{x}) := V$, and equivalently, as

$$f^L(\mathbf{x}) = \sum_{i=0}^{k-1} (x_{[i+1]} - x_{[i]}) f(V_{[i]}(\mathbf{x})),$$

where

$$k := 1 + \sum_{i=0}^{N-1} \text{sign}(|f(V_i(\mathbf{x})) - f(V_{i+1}(\mathbf{x}))|),$$

while $\{[1], \ldots, [k]\} \subset V$ satisfy:

- $x_{[0]} = 0$,
- $x_{[1]} < \ldots < x_{[k]}$,
- $V_{[0]}(\mathbf{x}) := V$,
- $V_{[i]}(\mathbf{x}) := \{j \in V : x_j > x_{[i]}\}$, for $i \geq 1$,
- $f(V_{[i]}(\mathbf{x})) \neq f(V_{[i+1]}(\mathbf{x}))$, for all $i = 0, \ldots, k - 1$.

We call the set $\{[1], \ldots, [k]\}$ a *simple index set* for f at $\mathbf{x}$. For a fixed vector $\mathbf{x}$, we also let $\text{si}(f) := \{i : f(V_{[i]}(\mathbf{x})) \neq 0\}$. One can then check that

$$f^L(\mathbf{x}) = \sum_{i \in \text{si}(f)} (x_{[i+1]} - x_{[i]}) f(V_{[i]}(\mathbf{x})). \tag{6.4.4}$$

Given $p \geq 1$ and $\mathbf{x} \in \mathbb{R}_{\geq 0}^N$, by (6.4.4) and since f_e is non-negative, we have

$$\sum_{e \in E} w_e f_e^L(\mathbf{x}^p) \tag{6.4.5}$$

$$= \sum_{e \in E} w_e \sum_{i \in \text{si}(f_e)} (x_{[i+1]_e}^p - x_{[i]_e}^p) f_e(V_{[i]_e}(\mathbf{x}^p))$$

$$\leq \sum_{e \in E} w_e \sum_{i \in \text{si}(f_e)} p(x_{[i+1]_e} - x_{[i]_e}) \left(\frac{x_{[i+1]_e}^p + x_{[i]_e}^p}{2} \right)^{\frac{1}{p'}} f_e(V_{[i]_e}(\mathbf{x})) \tag{6.4.6}$$

$$\leq \frac{p}{2^{\frac{1}{p'}}} \left(\sum_{e \in E} w_e \sum_{i \in \text{si}(f_e)} f_e(V_{[i]_e}(\mathbf{x}))^p (x_{[i+1]_e} - x_{[i]_e})^p \right)^{\frac{1}{p}}$$

$$\left(\sum_{e \in E} w_e \sum_{i \in \text{si}(f_e)} (x_{[i+1]_e}^p + x_{[i]_e}^p) \right)^{\frac{1}{p'}} \tag{6.4.7}$$

$$\leq \frac{p}{2^{\frac{1}{p'}}} \left(\sum_{e \in E} w_e \left(\sum_{i \in \text{si}(f_e)} f_e(V_{[i]_e}(\mathbf{x}))(x_{[i+1]_e} - x_{[i]_e}) \right)^p \right)^{\frac{1}{p}} \left(\sum_{j \in V} \widetilde{\deg}(j) x_j^p \right)^{\frac{1}{p'}}$$

$$\leq \left(\frac{c}{2} \right)^{\frac{1}{p'}} p \left(\sum_{e \in E} w_e (f_e^L(\mathbf{x}))^p \right)^{\frac{1}{p}} \left(\sum_{j \in V} g(j) x_j^p \right)^{\frac{1}{p'}}, \tag{6.4.8}$$

where p' is the Hölder conjugate of p, that is, $\frac{1}{p} + \frac{1}{p'} = 1$; $\{[1]_e, \ldots, [i]_e, \ldots\}$ is a simple index set for f_e at $\mathbf{x}$;

$$\widetilde{\deg}(j) := \sum_{e \in \widetilde{E}_j} w_e, \quad \text{and} \quad \widetilde{E}_j := \{e \in E : j=[i]_e \text{ or } j=[i+1]_e \text{ for some } i \in \mathrm{si}(f_e)\}.$$

The first above inequality (6.4.6) follows from the inequality in Lemma 3 in [5]. The second inequality (6.4.7) uses Hölder's inequality, and the last inequality (6.4.8) follows from the fact that

$$\sum_{j \in V : x_j \geq t} \widetilde{\deg}(j) \leq \sum_{j \in V : x_j \geq t} \deg(j) \leq c \sum_{j \in V : x_j \geq t} g(j), \quad \text{for any } t \in \mathbb{R}.$$

$$(6.4.9)$$

In fact, for any $e \in \widetilde{E}_j$, we may assume that $j = [i]_e$, which implies that

$$f_e(V_{[i-1]_e}(\mathbf{x})) \neq f_e(V_{[i]_e}(\mathbf{x})).$$

Clearly, if $j' \neq j$ and $x_{[i]_e} \leq x_{j'} < x_{[i+1]_e}$, then $e \notin \widetilde{E}_{j'}$, since $j' \neq [i']_e$ for any i'. That is, e is counted exactly once in $\widetilde{E}_j$, for each j such that $x_j \in [x_{[i]_e}, x_{[i+1]_e})$. We now want to show that there exists j' such that

$$e \in E_{j'} \quad \text{and} \quad x_{[i]_e} \leq x_{j'} < x_{[i+1]_e}.$$

To see this, suppose the contrary. Then, for each j' such that $x_{[i]_e} \leq x_{j'} < x_{[i+1]_e}$, we have that $e \notin E_{j'}$. Consider

$$\{j_1, \ldots, j_l\} = \{j : x_{[i]_e} \leq x_j < x_{[i+1]_e}\} = V_{[i-1]_e} \setminus V_{[i]_e}.$$

Since $e \notin E_{j_1} \cup \ldots \cup E_{j_l}$, it follows that

$$f(V_{[i-1]_e}(\mathbf{x})) = f(V_{[i-1]_e}(\mathbf{x}) \setminus \{j_1\}) = \ldots = f(V_{[i-1]_e}(\mathbf{x}) \setminus \{j_1, \ldots, j_l\})$$
$$= f(V_{[i]_e}(\mathbf{x})),$$

which is a contradiction. As a consequence, e is counted at least once in E_j, for all j such that $x_j \in [x_{[i]_e}, x_{[i+1]_e})$. This proves the inequality in (6.4.9). Therefore,

$$\sum_{j \in V} \widetilde{\deg}(j) x_j^p = \int_0^\infty \sum_{j \in V : x_j^p \geq t} \widetilde{\deg}(j) dt \leq c \int_0^\infty \sum_{j \in V : x_j^p \geq t} g(j) dt = c \sum_{j \in V} g(j) x_j^p,$$

which implies (6.4.8).

We infer that, for $\mathbf{x} \in \mathbb{R}_{\geq 0}^N \setminus \{0\}$,

$$\frac{\sum_{e \in E} w_e f_e^L(\mathbf{x}^p)}{\|\mathbf{x}\|_{p,g}^p} := \frac{\sum_{e \in E} w_e f_e^L(\mathbf{x}^p)}{\sum_{j \in V} g(j) x_j^p} \leq p \left(\frac{c}{2}\right)^{\frac{1}{p'}} \left(\frac{\sum_{e \in E} w_e (f_e^L(\mathbf{x}))^p}{\sum_{j \in V} g(j) x_j^p}\right)^{\frac{1}{p}}.$$

Similarly, for $\mathbf{x} \in \mathbb{R}^N_{\leq 0} \setminus \{\mathbf{0}\}$, if we let $|\mathbf{x}| = (|x_1|, \ldots, |x_N|)$ and we take $\tilde{f}_e(S) = f_e(V \setminus S)$, for all $S \subset V$, we have that

$$\frac{\sum\limits_{e \in E} w_e \tilde{f}^L_e(|\mathbf{x}|^p)}{\|\mathbf{x}\|^p_{p,g}} = \frac{\sum\limits_{e \in E} w_e f^L_e(-|\mathbf{x}|^p)}{\| - |\mathbf{x}|\|^p_{p,g}}$$

$$= \frac{\sum\limits_{e \in E} w_e f^L_e(-|\mathbf{x}|^p)}{\sum_{j \in V} g(j)|x_j|^p}$$

$$\leq p \left(\frac{c}{2}\right)^{\frac{1}{p'}} \left(\frac{\sum\limits_{e \in E} w_e (f^L_e(\mathbf{x}))^p}{\sum_{j \in V} g(j)|x_j|^p}\right)^{\frac{1}{p}}.$$

Now, for $\mathbf{x} \in \mathbb{R}^N \setminus (\mathbb{R}^N_{\geq 0} \cup \mathbb{R}^N_{\leq 0})$, write $\mathbf{x} = \mathbf{x}_+ + \mathbf{x}_-$, with $(x_+)_i = \max\{x_i, 0\}$ and $(x_-)_i = \min\{x_i, 0\}$ for all $i \in V$. We derive that

$$\frac{\sum\limits_{e \in E} w_e (f^L_e(\mathbf{x}))^p}{\sum_{j \in V} g(j)|x_j|^p} = \frac{\sum\limits_{e \in E} w_e |f^L_e(\mathbf{x}_+) + f^L_e(\mathbf{x}_-)|^p}{\sum_{j \in V} g(j)|x_{+,j}|^p + \sum_{j \in V} g(j)|x_{-,j}|^p}$$

$$\geq \min \left\{ \frac{\sum\limits_{e \in E} w_e (f^L_e(\mathbf{x}_+))^p}{\|\mathbf{x}_+\|^p_{p,g}}, \frac{\sum\limits_{e \in E} w_e (f^L_e(\mathbf{x}_-))^p}{\|\mathbf{x}_-\|^p_{p,g}} \right\}$$

$$\geq \frac{1}{p^p} \left(\frac{2}{c}\right)^{\frac{p}{p'}} \min \left\{ \frac{\sum\limits_{e \in E} w_e f^L_e(\mathbf{x}^p_+)}{\|\mathbf{x}^p_+\|_{1,g}}, \frac{\sum\limits_{e \in E} w_e \tilde{f}^L_e(|\mathbf{x}_-|^p)}{\|\mathbf{x}^p_-\|_{1,g}} \right\}^p$$

$$\tag{6.4.10}$$

$$\geq \frac{(2/c)^{p-1}}{p^p} \min \left\{ \frac{f(A_+)}{g(A_+)}, \frac{\tilde{f}(A_-)}{g(A_-)} \right\}^p$$

$$\geq \frac{(2/c)^{p-1}}{p^p} \min \left\{ \frac{\hat{f}(A_+)}{g(A_+)}, \frac{\hat{f}(A_-)}{g(A_-)} \right\}^p,$$

for some nonempty subset $A_\pm \subset \operatorname{supp}(\mathbf{x}_\pm)$, where

$$\hat{f}(S) := \min\{f(S), f(V \setminus S)\} = \min\{f(S), \tilde{f}(S)\}.$$

For any nonconstant vector $\hat{\mathbf{x}}$, since g is a counting measure function, there exists $c \in \{\hat{x}_1, \ldots, \hat{x}_N\} \subset \mathbb{R}$ such that the vector $\mathbf{x} = \hat{\mathbf{x}} - c\mathbf{1}$ satisfies

$$g(\operatorname{supp}(\mathbf{x}_+)) \leq \frac{1}{2} g(V) \quad \text{and} \quad g(\operatorname{supp}(\mathbf{x}_-)) \leq \frac{1}{2} g(V).$$

Therefore, since $f_e(V) = 0$, we obtain that

$$f^L_e(\mathbf{x}) = f^L_e(\hat{\mathbf{x}}) - c f_e(V) = f^L_e(\hat{\mathbf{x}}).$$

Therefore,

$$\frac{\sum_{e \in E} w_e (f_e^L(\hat{\mathbf{x}}))^p}{\min_{c \in \mathbb{R}} \sum_{e \in E} g_e^L(|\hat{\mathbf{x}} - c\mathbf{1}|^p)} \geq \frac{\sum_{e \in E} w_e (f_e^L(\mathbf{x}))^p}{\sum_{e \in E} g_e^L(|\mathbf{x}|^p)} \geq \frac{(2/c)^{p-1}}{p^p} \cdot \frac{\hat{f}(A)^p}{g(A)^p},$$

for some nonempty subset $A \subset \operatorname{supp}(\mathbf{x}_+) \subset \operatorname{supp}(\mathbf{x})$ or $A \subset \operatorname{supp}(\mathbf{x}_-) \subset \operatorname{supp}(\mathbf{x})$. As a consequence, $g(A) \leq \frac{1}{2}g(V)$, or equivalently, $g(A) \leq g(V \setminus A)$. One can then check that

$$\min_{A \neq \emptyset, V} \frac{\hat{f}(A)}{\min\{g(A), g(V \setminus A)\}} = \min_{A \neq \emptyset, V} \frac{\min\{f(A), f(V \setminus A)\}}{\min\{g(A), g(V \setminus A)\}}$$

$$= \min_{A \neq \emptyset, V} \frac{f(A)}{\min\{g(A), g(V \setminus A)\}}$$

$$= h.$$

Therefore,

$$\lambda \geq \frac{(2/c)^{p-1}}{p^p} h^p.$$

$\square$

In order to relate Theorem 6.4.1 to our earlier estimates (4.3.15) and (4.3.16), we need particular choices of the functions f_e and g_e. Given a simple graph $G = (V, E)$, consider the functions $f_e, g_e : \mathcal{P}(V) \to \mathbb{R}$ defined by

$$f_e(A) := \begin{cases} 1, & \text{if } e \text{ has one end point in } A \text{ and the other in } V \setminus A, \\ 0, & \text{otherwise}, \end{cases} \tag{6.4.11}$$

and $g_e(A) := \#(e \cap A)$. By the original Lovász extension in Section 6.2, we have $f_e^L(\mathbf{x}) = |x_i - x_j|$ and $g_e^L(\mathbf{x}) = x_i + x_j$, where $\{i, j\} = e \in E$. Therefore,

$$\sum_{e \in E} (f_e^L(\mathbf{x}))^p = \sum_{\{i,j\} \in E} |x_i - x_j|^p$$

and

$$\sum_{e \in E} g_e^L(|\mathbf{x}|^p) = \sum_{\{i,j\} \in E} (|x_i|^p + |x_j|^p) = \sum_{i \in V} \deg(i)|x_i|^p.$$

Hence, in this case, the eigenvalue problem of the function pair (F_p, G_p) coincides with the eigenvalue problem of the normalized p-Laplacian on a graph (see Section 2.6.2 and Chapter 7).

Now, as in Section 6.3, we consider a chemical hypergraph satisfying $e_{in} \neq \emptyset \neq e_{out}$ and $\#(e_{in} \cup e_{out}) \geq 2$, for all $e \in E$. Let $f_e : \mathcal{P}(V) \to \mathbb{R}$ be defined by

$$f_e(A) := \begin{cases} 1, & \text{if } e_{in} \cap A \neq \emptyset \neq e_{out} \setminus A \text{ or } e_{out} \subset A \subset V \setminus e_{in}, \\ 0, & \text{otherwise.} \end{cases} \quad (6.4.12)$$

Then, the Lovász extension of f_e is

$$f_e^L(\mathbf{x}) = \left| \max_{i \in e_{in}} x_i - \min_{j \in e_{out}} x_j \right|.$$

Moreover,

$$\lambda_2(\Delta_p) := \inf_{\mathbf{x} \text{ non-constant}} \frac{\sum_{e \in E} \left| \max_{i \in e_{in}} x_i - \min_{j \in e_{out}} x_j \right|^p}{\min_{t \in \mathbb{R}} \sum_{i \in V} \deg(i) |x_i - t|^p}$$

is the second eigenvalue of the pair

$$\left(\sum_{e \in E} \left| \max_{i \in e_{in}} x_i - \min_{j \in e_{out}} x_j \right|^p, \sum_{i \in V} \deg(i) |x_i|^p \right).$$

Let now $\mu(i) := \deg(i)$ in (6.3.3), for all $i \in V$, and let h denote the Cheeger constant in (6.3.3). Then, we have

Theorem 6.4.2 (Cheeger inequality for chemical hypergraphs).

$$2^{p-1} \frac{h^p}{p^p} \leq \lambda_2(\Delta_p) \leq 2^{p-1} h. \quad (6.4.13)$$

In particular, for $p = 1$, we infer that $h = \lambda_2(\Delta_1)$, and this generalizes the identity relating the second eigenvalue of the graph 1-Laplacian to the Cheeger constant on a graph. If we consider, for example, the complete graph of order 4, then the Cheeger constant is $h = 2/3$, while the second eigenvalue of the p-Laplacian is $\lambda_2(\Delta_p) = 2^p/3$ for any $p \in [1, 2]$. Clearly,

$$2^{p-1} \frac{(2/3)^p}{p^p} \leq \frac{2^p}{3} = 2^{p-1} \frac{2}{3}.$$

Hence, in this case, the upper bound $2^{p-1} h$ for $\lambda_2(\Delta_p)$ is tight for any $p \in [1, 2]$. Also, if we let $p \to 1$, then the lower bound $2^{p-1} h^p / p^p$ for $\lambda_2(\Delta_p)$ tends to $2/3$, which coincides with $\lim_{p \to 1} \lambda_2(\Delta_p) = \lambda_2(\Delta_1)$.

7

Discrete p-Laplacians

This chapter provides an introduction to graph p-Laplacians and their analogues on hypergraphs. It describes them in variational terms and derives many spectral properties of these nonlinear operators.

As a discrete version of the p-Laplacian treated in Section 3.3, the graph p-Laplacian has been successfully used in various applications, including data and image processing problems and spectral clustering. Many recent works indicate that the graph p-Laplacians, in particular the case $p = 1$, may enhance the performance of classical algorithms that are based on the standard graph Laplacian [39, 130]. This has contributed to progress on both the theoretical and the numerical aspects of p-Laplacians on graphs and networks. A remarkable development is that the second eigenvalue has a mountain-pass characterization, and thus it is a min–max eigenvalue, and more importantly, the second eigenvalue satisfies a Cheeger inequality that turns into an equality for $p = 1$.

This chapter develops the spectral properties and estimates for the spectra of various discrete versions of the p Laplacian, which not only extends the discussions in Section 3.3 to discrete settings but also provides some important applications of the Lovász extension in Chapter 6. In addition, the contents in this chapter establish the foundations for Cheeger-type inequalities in Chapter 8 and also nodal domain theory in Chapter 9. Some general introduction has been given in Section 2.6.2, but we shall repeat the material necessary for the developments in the present chapter.

7.1 Graph p-Laplacians ($p > 1$)

For simplicity and because this fits best into our general framework (recall, for instance, Chapter 4), we mainly discuss the normalized p-Laplacian introduced by Amghibech [5]. Let $C(V)$ denote the collection of real functions on the vertex set $V = \{v_1, \ldots, v_N\}$ of a simple graph. Clearly, we can identify $C(V)$ with $\mathbb{R}^N$ via a standard linear isomorphism. For $1 < p < \infty$, the *normalized p-Laplacian* $\Delta_p : C(V) \to C(V)$ is defined by

$$\Delta_p f(v) = \frac{1}{\deg v} \sum_{v' \in V : v' \sim v} |f(v) - f(v')|^{p-2}(f(v) - f(v')),$$

for $v \in V$ and $f \in C(V)$. The eigenvalue problem for Δ_p consists of finding an eigenfunction $f \neq 0$ and its corresponding eigenvalue $\lambda \in \mathbb{R}$ such that

$$\Delta_p f(v) = \lambda |f(v)|^{p-2} f(v), \quad \forall v \in V. \tag{7.1.1}$$

Just like the continuous p-Laplacian $-\mathrm{div}(|\nabla u|^{p-2}\nabla u)$ in (3.3.2), we can express the graph p-Laplacian in terms of the incidence matrix $\mathcal{I}$ and the degree matrix D as

$$\Delta_p f = D^{-1}\mathcal{I}(|\mathcal{I}^\top f|^{p-2}\mathcal{I}^\top f). \tag{7.1.2}$$

Moreover, the eigenproblem for Δ_p can be rewritten as $\Delta_p f = \lambda |f|^{p-2} f$.

Remark 7.1.1. The operator Δ_p is *not* the subgradient of the functional

$$f \mapsto \sum_{\{v,v'\} \in E} |f(v) - f(v')|^p.$$

The latter is the numerator of the relevant nonlinear Rayleigh quotient (7.1.3). That would rather be the operator

$$\widehat{\Delta}_p f(v) = \deg v \, \Delta_p f(v),$$

for $v \in V$ and $f \in C(V)$. With this operator, the eigenvalue problem (7.1.1) becomes

$$\widehat{\Delta}_p f(v) = \lambda \deg v |f(v)|^{p-2} f(v),$$

and this then leads to a variational problem for the nonlinear Rayleigh quotient below. However, in order to be consistent with the normalization that we have used for $p = 2$ in this book, we always use Δ_p and its eigenproblem (7.1.1).

Having clarified that issue, we can now see how the eigenvalues and eigenfunctions of Δ_p correspond to the critical values and critical points of the Rayleigh quotient R_p defined by

$$R_p(f) := \frac{\sum_{\{v,v'\} \in E} |f(v) - f(v')|^p}{\sum_{v \in V} \deg v \cdot |f(v)|^p}. \tag{7.1.3}$$

We shall use the general convention

$$\phi_p(t) := |t|^{p-2}t, \tag{7.1.4}$$

so that (7.1.1) becomes

$$\Delta_p f(v) = \lambda \phi_p(f(v)), \ \forall v \in V \tag{7.1.5}$$

which can be reformulated as

$$\sum_{v' \in V : v' \sim v} \phi_p(f(v) - f(v')) = \lambda \deg v \, \phi_p(f(v)), \ \forall v \in V. \tag{7.1.6}$$

Since the eigenvalue equation for $p \neq 2$ is no longer linear, sums of eigenfunctions corresponding to the same eigenvalue no longer need to be eigenfunctions, and thus, the eigenspaces in general are not linear.

In order to develop some intuition, we compute the eigenvalues of the graph p-Laplacian for general p for some examples. As we shall see, this turns out to be more difficult than in the case $p = 2$, even for simple graphs like a path or a star graph.

For the path graph P_3 with vertex set $\{v_1, v_2, v_3\}$ and edge set $\{\{v_1, v_2\}, \{v_2, v_3\}\}$, the p-Laplacian eigenvalue problem requires to solve

$$\begin{cases} \phi_p(f(v_1) - f(v_2)) = \lambda \phi_p(f(v_1)), \\ \phi_p(f(v_2) - f(v_1)) + \phi_p(f(v_2) - f(v_3)) = 2\lambda \phi_p(f(v_2)), \\ \phi_p(f(v_3) - f(v_2)) = \lambda \phi_p(f(v_3)). \end{cases}$$

By the preceding equations, we obtain

$$\begin{cases} f(v_1) - f(v_2) = \lambda^{\frac{1}{p-1}} f(v_1), \\ f(v_3) - f(v_2) = \lambda^{\frac{1}{p-1}} f(v_3), \\ \lambda(\phi_p(f(v_1)) + 2\phi_p(f(v_2)) + \phi_p(f(v_3))) = 0. \end{cases}$$

If $\lambda = 0$, then $f(v_1) = f(v_2) = f(v_3)$. For $\lambda > 0$, we obtain

$$\phi_p(f(v_1)) + 2\phi_p(f(v_2)) + \phi_p(f(v_3)) = 0.$$

If $\lambda = 1$, then $f(v_2) = 0$ and $f(v_1) = -f(v_3)$. Finally, if $\lambda > 0$ and $\lambda \neq 1$, then

$$f(v_1) = f(v_3) = (1 - \lambda^{\frac{1}{p-1}})^{-1} f(v_2) \neq 0$$

and thus

$$0 = (\phi_p(f(v_1)) + 2\phi_p(f(v_2)) + \phi_p(f(v_3)))\phi_p^{-1}(f(v_2)) = 2(1 - \lambda^{\frac{1}{p-1}})^{1-p} + 2,$$

which implies $\lambda = 2^{p-1}$. Therefore, the whole spectrum is $\{0, 2^{p-1}, 1\}$.

We now look at the path graph P_4 with vertex set $\{v_1, v_2, v_3, v_4\}$ and edge set $\{\{v_1, v_2\}, \{v_2, v_3\}, \{v_3, v_4\}\}$. In this case, the p-Laplacian eigenvalue problem requires to solve the equations

$$\begin{cases} \phi_p(f(v_1) - f(v_2)) = \lambda\phi_p(f(v_1)), \\ \phi_p(f(v_2) - f(v_1)) + \phi_p(f(v_2) - f(v_3)) = 2\lambda\phi_p(f(v_2)), \\ \phi_p(f(v_3) - f(v_2)) + \phi_p(f(v_3) - f(v_4)) = 2\lambda\phi_p(f(v_3)), \\ \phi_p(f(v_4) - f(v_3)) = \lambda\phi_p(f(v_4)). \end{cases}$$

Since $\lambda = 0$ has, as always, a constant eigenfunction, we now look at the case $\lambda > 0$. Similarly to the case of P_3, we derive that

$$\begin{cases} f(v_1) - f(v_2) = \lambda^{\frac{1}{p-1}} f(v_1), \\ f(v_4) - f(v_3) = \lambda^{\frac{1}{p-1}} f(v_4), \\ \phi_p(f(v_1)) + 2\phi_p(f(v_2)) + 2\phi_p(f(v_3)) + \phi_p(f(v_4)) = 0. \end{cases}$$

Moreover, since it is easy to check that 1 is not an eigenvalue, we obtain

$$\begin{cases} f(v_1) = (1 - \lambda^{\frac{1}{p-1}})^{-1} f(v_2), \\ f(v_4) = (1 - \lambda^{\frac{1}{p-1}})^{-1} f(v_3), \\ \left(2 + \phi_p((1 - \lambda^{\frac{1}{p-1}})^{-1})\right)(\phi_p(f(v_2)) + \phi_p(f(v_3))) = 0. \end{cases}$$

One can then verify that

$$\lambda = (1 + 2^{\frac{1}{1-p}})^{p-1} \text{ and } f = (-2^{\frac{1}{p-1}}, 1, 1, -2^{\frac{1}{p-1}})$$

form an eigenpair. Now, if $\lambda \neq (1 + 2^{\frac{1}{1-p}})^{p-1}$, then $f(v_2) = -f(v_3)$ and

$$\begin{aligned} 2^{p-1}\phi_p(f(v_2)) &= \phi_p(f(v_2) - f(v_3)) \\ &= 2\lambda\phi_p(f(v_2)) + \lambda\phi_p(f(v_1)) \\ &= \lambda\left(2 + \phi_p((1 - \lambda^{\frac{1}{p-1}})^{-1})\right)\phi_p(f(v_2)), \end{aligned}$$

which implies

$$\lambda\left(2 + \phi_p((1 - \lambda^{\frac{1}{p-1}})^{-1})\right) = 2^{p-1}.$$

This can be reformulated as

$$\begin{cases} \lambda(2 + (1 - \lambda^{\frac{1}{p-1}})^{1-p}) = 2^{p-1} & , \text{ for } \lambda \in (0, 1), \\ \lambda(2 - (\lambda^{\frac{1}{p-1}} - 1)^{1-p}) = 2^{p-1} & , \text{ for } \lambda \in [1, +\infty). \end{cases} \tag{7.1.7}$$

One can then check that this has exactly two roots in $(0, +\infty)$: one is $\lambda = 2^{p-1}$ and the other one lies in the interval $(0, 1)$. In summary, the whole spectrum is

$$\{0, \lambda_2, (1 + 2^{\frac{1}{1-p}})^{p-1}, 2^{p-1}\},$$

where λ_2 is the unique root of (7.1.7) in the interval $(0, 1)$.

The next example is the star graph S_3 with vertex set $\{v_1, v_2, v_3, v_4\}$ and edge set $\{\{v_1, v_2\}, \{v_1, v_3\}, \{v_1, v_4\}\}$. Its p-Laplacian eigenvalue problem requires to solve the equations

$$\begin{cases} \phi_p(f(v_2) - f(v_1)) = \lambda \phi_p(f(v_2)), \\ \phi_p(f(v_3) - f(v_1)) = \lambda \phi_p(f(v_3)), \\ \phi_p(f(v_4) - f(v_1)) = \lambda \phi_p(f(v_4)), \\ \phi_p(f(v_1) - f(v_2)) + \phi_p(f(v_1) - f(v_3)) + \phi_p(f(v_1) - f(v_4)) = 3\lambda \phi_p(f(v_1)). \end{cases}$$

Again, we only need to consider the case $\lambda > 0$; therefore, the above eigenequation can be simplified as

$$\begin{cases} f(v_2) - f(v_1) = \lambda^{\frac{1}{p-1}} f(v_2), \\ f(v_3) - f(v_1) = \lambda^{\frac{1}{p-1}} f(v_3), \\ f(v_4) - f(v_1) = \lambda^{\frac{1}{p-1}} f(v_4), \\ 3\phi_p(f(v_1)) + \phi_p(f(v_2)) + \phi_p(f(v_3)) + \phi_p(f(v_4)) = 0. \end{cases}$$

For $\lambda = 1$, the corresponding eigenfunctions are the nonzero functions satisfying $f(v_1) = 0$ and

$$\phi_p(f(v_2)) + \phi_p(f(v_3)) + \phi_p(f(v_4)) = 0.$$

Thus, the multiplicity of $\lambda = 1$ is 2. If $\lambda \neq 1$ and $\lambda > 0$, then

$$f(v_2) = f(v_3) = f(v_4) = (1 - \lambda^{\frac{1}{p-1}})^{-1} f(v_1),$$

therefore

$$(3 + 3\phi_p((1 - \lambda^{\frac{1}{p-1}})^{-1}))\phi_p(f(v_1)) = 0,$$

which implies $(1 - \lambda^{\frac{1}{p-1}})^{-1} = -1$ and thus $\lambda = 2^{p-1}$. In summary, we conclude that the whole spectrum of the star graph S_3 is $\{0, 1, 2^{p-1}\}$.

From the preceding examples, we can already see that the spectrum of the graph p-Laplacian, for general $p \neq 2$, is quite difficult to determine.

We shall now present some first properties of the spectrum of graph p-Laplacians. In contrast to the difficulties with the examples, the proofs are very similar to the case $p = 2$. Before that, we need to clarify the concept of the multiplicity of an eigenvalue of the p-Laplacian. For an eigenvalue λ of Δ_p, its multiplicity is defined to be the genus of the set of eigenfunctions corresponding to λ, which is formally written as

$$\text{genus}\{f \colon (\lambda, f) \text{ is an eigenpair of } \Delta_p, \|f\|_2 = 1\}.$$

In the linear case $p = 2$, this agrees with the usual algebraic multiplicity of λ.

Proposition 7.1.1. *The spectrum of Δ_p satisfies the following properties:*

- *The eigenvalues lie in the interval $[0, 2^{p-1}]$.*
- *The multiplicity of the eigenvalue 0 equals the number of connected components of the graph.*
- *The largest eigenvalue is 2^{p-1} if and only if the graph has a bipartite connected component. The multiplicity of 2^{p-1} equals the number of bipartite connected components.*

The proof of the above proposition is easy, and thus, we omit it.

Proposition 7.1.2. *Let (λ, f) be an eigenpair of Δ_p. Then, for any $A \subset V$,*

$$\sum_{v \in A,\, v' \notin A,\, v' \sim v} \phi_p(f(v) - f(v')) = \lambda \sum_{v \in A} \deg v \cdot \phi_p(f(v)).$$

In particular, if f is an eigenfunction corresponding to a positive eigenvalue, then

$$\sum_{v \in V} \deg v \cdot \phi_p(f(v)) = 0.$$

Moreover, if $A = \{v \in V : f(v) > 0\} \neq \emptyset$, then

$$\lambda \geq \frac{\sum_{v \in A} \deg_{V \setminus A}(v) \cdot (f(v))^{p-1}}{\sum_{v \in A} \deg v \cdot (f(v))^{p-1}},$$

where $\deg_{V \setminus A}(v)$ denotes the number of edges connecting v with the vertices in $V \setminus A$.

Proof Since f and λ satisfy the eigenequation (7.1.6), we have

$$\lambda \sum_{v \in A} \deg v \cdot \phi_p(f(v)) = \sum_{v \in A,\, v' \in V,\, v' \sim v} \phi_p(f(v) - f(v'))$$

$$= \sum_{v \in A,\, v' \notin A,\, v' \sim v} \phi_p(f(v) - f(v'))$$

where we used the fact that

$$\phi_p(f(v) - f(v')) + \phi_p(f(v') - f(v)) = 0$$

when $v \in A$, $v' \in A$, and $v' \sim v$. If $\lambda > 0$ and $A = V$, we simply obtain

$$\sum_{v \in A} \deg v \cdot \phi_p(f(v)) = \sum_{v \in V} \deg v \cdot \phi_p(f(v)) = 0.$$

If

$$A = \{v \in V : f(v) > 0\} \neq \emptyset,$$

then $f(v) - f(v') \geq f(v) > 0$ for any $v \in A$ and $v' \notin A$ such that $v' \sim v$. Therefore,

$$\sum_{v \in A, \, v' \notin A, \, v' \sim v} \phi_p(f(v) - f(v')) \geq \sum_{v \in A, \, v' \notin A, \, v' \sim v} \phi_p(f(v))$$

$$= \sum_{v \in A} \deg_{V \setminus A}(v) \cdot (f(v))^{p-1}.$$

□

Proposition 7.1.3. *For a connected graph, the second eigenvalue λ_2 of Δ_p is positive, and it satisfies*

$$\lambda_2 = \min_{f \text{ nonconstant}} \frac{\sum_{\{v,v'\} \in E} |f(v) - f(v')|^p}{\min_{t \in \mathbb{R}} \sum_{v \in V} \deg v \cdot |f(v) - t|^p} \tag{7.1.8}$$

$$= \inf_{\substack{\text{curve } \gamma: \, [-1,1] \to \mathbb{R}^N \setminus \{0\}, \, f \in \gamma([-1,1]) \\ \gamma(\pm 1) = \pm 1}} \sup R_p(f). \tag{7.1.9}$$

The *proof* is the same as for the case $p = 2$, which was treated in Section 4.1.

For more elaborated results, we now recall the variational characterization of the eigenvalues of p-Laplacians in terms of the Krasnoselskii $\mathbb{Z}_2$ genus, which we introduced in Definition 2.6.2. The Krasnoselskii genus of a centrally symmetric set S in $\mathbb{R}^N \setminus \{0\}$ was defined as the smallest k for which there exists an odd continuous map from S into the sphere $\mathbb{S}^{k-1}$. To start with a trivial example, the genus of a sphere $\mathbb{S}^{k-1}$ is k. If we consider the centrally symmetric subset

$$S = \{(x_1, x_2, x_3) \in \mathbb{R}^3 : |x_1|^3 + |x_2|^3 + |x_3|^3 = 1, \, x_1 + x_2 + x_3 = 0\},$$

then, S is homeomorphic to $\mathbb{S}^1$ and genus(S) = genus$(\mathbb{S}^1)$ = 2. Another interesting centrally symmetric set is

$$S(p) := \{\vec{x} \in \mathbb{R}^4 : \|\vec{x}\|_2 = 1, \, x_1 = 0, \, \phi_p(x_2) + \phi_p(x_3) + \phi_p(x_4) = 0\},$$

for $p > 1$. The set $S(p)$ is also homeomorphic to $\mathbb{S}^1$ and genus$(S(p))$ = 2. If we consider again the above example where we have computed the p-Laplacian spectrum of the star graph S_3, we conclude that the genus of the eigenspace corresponding to the eigenvalue 1 of the p-Laplacian on S_3 is 2.

For $k \geq 1$, let now

$$\text{Gen}_k := \{S \subset \mathbb{R}^n : S \text{ centrally symmetric with genus}(S) \geq k\}.$$

As mentioned in Section 2.6.2, for $k = 1, \ldots, N$, the constants

$$\lambda_k(\Delta_p) := \inf_{S \in \text{Gen}_k} \sup_{f \in S \setminus \{0\}} R_p(f) \tag{7.1.10}$$

are eigenvalues of Δ_p. They satisfy

$$\lambda_1 \leq \ldots \leq \lambda_N$$

and, if $\lambda = \lambda_{k+1} = \ldots = \lambda_{k+l}$ for $0 \leq k < k + l \leq N$, then

$$\text{genus}(\{\text{eigenfunctions corresponding to } \lambda\}) \geq l.$$

In addition, $\{\lambda_k(\Delta_p)\}_{k=1}^N$ defines a sequence of N critical values of the Rayleigh quotient R_p.

We now define the independence number of a graph as follows.

Definition 7.1.1. A subset $U \subseteq V$ of the vertex set of a graph $\Gamma = (V, E)$ is *independent* if $\#(U \cap e) \leq 1$ for all $e \in E$. The *independence number* of Γ is

$$\alpha := \max\{\#U : U \subseteq V \text{ is independent}\}.$$

Thus, a subset of the vertex set is independent if it contains at most one endpoint for each edge. Among all connected graphs with N vertices, the star graph $S_{N-1} = K_{1,N-1}$ has the largest independence number, $\alpha = N - 1$. More generally, a complete bipartite graph $K_{m,n}$ has independence number $\alpha = \max(m, n)$. The complete graph K_N has the smallest independence number, $\alpha = 1$.

We now prove two results on the independence number of a graph, in relation with the eigenvalues of its p-Laplacian.

Proposition 7.1.4. *Let α be the independence number of a graph. Then,*

$$\alpha \leq \min\{\#\{i : \lambda_i(\Delta_p) \leq 1\}, \#\{i : \lambda_i(\Delta_p) \geq 1\}\}, \tag{7.1.11}$$

and the inequality is sharp.

The *proof* of this nonlinear inertia bound is a special case of the proof of Theorem 7.4.1; hence, we do not present it here.

Proposition 7.1.5. *Let Γ be a graph and let α be its independence number. If Γ has $n \geq 2$ duplicate vertices, then for any $p > 1$, 1 is an eigenvalue of Δ_p with multiplicity at least*

$$\max\{n - 1, 2\alpha - N\}.$$

Proof If $2\alpha - N > 0$, then $\alpha \geq N - \alpha + 1$ and therefore

$$\lambda_{N-\alpha+1} \leq \ldots \leq \lambda_\alpha.$$

Combining this with the inequalities $1 \leq \lambda_{N-\alpha+1}$ and $\lambda_\alpha \leq 1$, which are shown in (7.1.11), we infer that

$$\lambda_{N-\alpha+1} = \ldots = \lambda_\alpha = 1,$$

and this implies that the multiplicity of the eigenvalue 1 is at least $\alpha - (N - \alpha + 1) + 1 = 2\alpha - N$. Now, let $v_1, \ldots, v_n$ be duplicate vertices and let

$$X = \left\{ f \in C(V) \colon |f(v_1)|^{p-2} f(v_1) + \ldots + |f(v_n)|^{p-2} f(v_n) = 0 \right.$$
$$\left. \mathrm{supp}(f) \subset \{v_1, \ldots, v_n\} \right\}.$$

Then, it is not difficult to check that f is an eigenfunction corresponding to the eigenvalue 1 and that $\mathrm{genus}(X) = n - 1$. This implies that the multiplicity of the eigenvalue 1 is at least $n - 1$. $\qquad\square$

If Γ has no duplicate vertices and its independence number is smaller than or equal to $N/2$, then Γ may have a large spectral gap around 1. For instance, as we have seen in Theorem 4.1.2 for $p = 2$, petal graphs (see Figure 4.6) and book graphs (see Figure 4.7) do not have duplicate vertices, they have independence number $\alpha = [N/2]$, and they have a large spectral gap around 1.

Remark 7.1.2. For $p \neq 2$ and for some graphs, there may exist eigenvalues of the p-Laplacian that are not min–max (variational) eigenvalues. That is, the variational scheme does not necessarily produce all the eigenvalues. We refer to [278] for a refined analysis of such nonvariational eigenvalues of graph p-Laplacians. In this reference it is also shown that, for $k = 1, \ldots, N$,

- The function

$$p \mapsto p(2\lambda_k(\Delta_p))^{\frac{1}{p}}$$

  is increasing on $[1, +\infty)$;
- The function

$$p \mapsto 2^{-p} \lambda_k(\Delta_p)$$

  is decreasing on $[1, +\infty)$.

7.2 The Graph 1-Laplacian

The 1-Laplacian on graphs is an analogue of the 1-Laplacian in partial differential equations discussed in Section 3.3. The latter has been successfully applied for modelling the interface of two different media and the segmentation in image processing. The spectral theory for the graph 1-Laplacian was proposed by Hein and Bühler [130] for 1-spectral clustering and was later studied by Chang [52] from a variational point of view. In some sense, the graph 1-Laplacian is much closer to the essence of a graph than the linear 2-Laplacian. For example, the Cheeger constant, which has only some upper and lower bounds in linear spectral theory, equals the second eigenvalue of the graph 1-Laplacian [52, 130].

Let us thus present the definition of the graph 1-Laplacian that we shall utilize. This definition is in line with the more abstract setting developed in Section 2.6.2, but here we can identify the relevant set as the interval $[-1, 1]$. The reason is essentially that $\frac{f}{|f|}$ becomes undetermined for $f = 0$, and we therefore admit the entire range of values from that interval. This issue did not arise for the p-Laplacian for $p > 1$, because in that case, the quantity $|f|^{p-2}f$ can be safely set to 0 when $f = 0$.

The 1-Laplacian Δ_1 is the set-valued map defined by

$$\Delta_1 f(v) = \left\{ \frac{1}{\deg v} \sum_{v' \in V : v' \sim v} z_{vv'} \, \middle| \, z_{vv'} \in \mathrm{Sgn}(f(v) - f(v')), \, z_{vv'} = -z_{v'v} \right\},$$

in which

$$\mathrm{Sgn}(t) := \begin{cases} \{1\} & \text{if } t > 0, \\ [-1, 1] & \text{if } t = 0, \\ \{-1\} & \text{if } t < 0. \end{cases}$$

The eigenvalue problem for Δ_1 is to find $\lambda \in \mathbb{R}$ and $f \neq 0$ such that

$$\Delta_1 f(v) \bigcap \lambda \, \mathrm{Sgn}(f(v)) \neq \emptyset, \quad \forall v \in V,$$

which can be interpreted as the differential inclusion $0 \in \Delta_1 f - \lambda \, \mathrm{Sgn}(f)$, where '$-$' is understood in the sense of Minkowski addition. As in (7.1.3), we have the Rayleigh quotient

$$R_1(f) := \frac{\sum_{\{v,v'\} \in E} |f(v) - f(v')|}{\sum_{v \in V} \deg v \cdot |f(v)|}. \tag{7.2.1}$$

The Lusternik–Schnirelman theory (see for instance [152, 277]) leads to the sequence of min–max eigenvalues of Δ_1:

$$\lambda_k(\Delta_1) := \inf_{\mathrm{genus}(S) \geq k} \sup_{f \in S} \frac{\sum_{\{v,v'\} \in E} |f(v) - f(v')|}{\sum_{v \in V} \deg v |f(v)|}, \quad k = 1, 2, \ldots \tag{7.2.2}$$

where $\mathrm{genus}(S)$ is the Krasnoselskii genus of a centrally symmetric compact subset S in $C(V) \cong \mathbb{R}^V$ (see Definition 2.6.2).[1] But we now have an additional powerful tool at our disposition, the Lovász extension. In fact, the graph 1-Laplacian is a typical example of a Lovász extension. Recalling the discussion in Chapter 6, the denominator of the Rayleigh quotient $R_1(\cdot)$ is the disjoint-pair

---

[1] We should point out, however, that in addition to the variational eigenvalues λ_k as defined here, there may exist other eigenvalues, sitting between those λ_k, that do not arise by the scheme in (7.2.2). An example will be given below.

Lovász extension of the volume function $(A, B) \mapsto \text{vol}\,(A) + \text{vol}\,(B)$, while the numerator of the Rayleigh quotient is the disjoint-pair Lovász extension of the boundary measure function $(A, B) \mapsto |\partial A| + |\partial B|$. Theorems 6.3.1, 6.3.3, and 6.3.4 then cover many important properties of Δ_1. For example,

$$\lambda_k(\Delta_1) = \inf_{S \subset 1^\perp, \text{genus}(S) \geq k-1} \sup_{f \in S} \frac{\sum_{\{v,v'\} \in E} |f(v) - f(v')|}{\min_{t \in \mathbb{R}} \sum_{v \in V} \deg v |f(v) - t|}, \quad k = 2, 3, \ldots$$

Just like the continuous 1-Laplacian $-\text{div}\,(\nabla u / |\nabla u|)$ in (3.3.7), we can express the graph 1-Laplacian Δ_1 as a normalized gradient-like operator, that is, a nonlinear operator involving the incidence matrix:

$$\Delta_1 f = D^{-1} \mathcal{I} \cdot \frac{\mathcal{I}^\top f}{|\mathcal{I}^\top f|}$$

is analogous to (7.1.2). To clarify the meaning of $\frac{\mathcal{I}^\top f}{|\mathcal{I}^\top f|}$ when $\mathcal{I}^\top f = 0$, we use the set-valued function $t \mapsto \text{Sgn}(t)$ to replace the function $t \mapsto \frac{t}{|t|}$. Finally, we define

$$\Delta_1 f = D^{-1} \mathcal{I} \cdot \text{Sgn}(\mathcal{I}^\top f),$$

which coincides with the definition in coordinate form above. In fact, the graph 1-Laplacian agrees with the subgradient of the functional

$$f \mapsto \sum_{\{v,v'\} \in E} |f(v) - f(v')|,$$

up to the inverse of the degree matrix, D^{-1}. Another remarkable relation between the cases of $p > 1$ and $p = 1$ is that $\Delta_1 f$ is actually the collection of all the limit points of $\Delta_p \widehat{f}$ as p tends to 1^+ and $\widehat{f}$ converges to f. The situation for the eigenvalues is more subtle, however.

In fact, while already the properties of the eigenvalues and eigenfunctions of Δ_p for $p > 1$ were different from those of Δ_2 when $p \neq 2$, when we turn to Δ_1, almost everything ceases to hold, as we shall see in this section and also in Section 9.2. The smallest eigenvalue is still 0, with a constant eigenfunction when Γ is connected. This will be established in Theorem 7.2.1, among with other basic properties, for instance that the largest eigenvalue is at most 1. However, the number of eigenvalues need no longer be equal to the number N of vertices, but can be substantially larger, and the sum of the eigenvalues will no longer be N. Eigenfunctions f no longer satisfy orthogonality relations, and in particular, we no longer have the relation $\sum_v \deg v f(v) = 0$ for nonconstant eigenfunctions. In fact, eigenfunctions need not even change sign. Importantly, and very usefully, when f is an eigenfunctions, so is $\frac{f}{|f|}$, as follows for instance from the characterization (7.2.2) by a Rayleigh quotient. That means that eigenfunctions can be characteristic functions of sets, and we can then in turn try to identify those sets whose characteristic functions are eigenfunctions. In Section 9.2, we shall consider these sets as nodal domains, realizing Cheeger cuts, that is, optimal decompositions of the underlying graph. Thus, not only will λ_2 of Δ_1 equal the Cheeger constant, see Theorem 7.2.2 below, but even better, the corresponding eigenfunction will yield an optimal decomposition. So, let us start with the observation that eigenfunctions can be assumed to be characteristic functions of vertex sets.

Proposition 7.2.1. *For any eigenfunction f corresponding to an eigenvalue λ of Δ_1,*

$$(\lambda, \mathbf{1}_{\{f>0\}}) \quad (\text{respectively, } (\lambda, \mathbf{1}_{\{f<0\}}))$$

is also an eigenpair if $\{f>0\} \neq \emptyset$ (respectively, $\{f<0\} \neq \emptyset$). In particular, every eigenvalue of Δ_1 has an eigenfunction of the form $\mathbf{1}_A$ for some $A \subset V$.

Proof This is a consequence of the proof of Proposition 6.3.1 and Theorem 6.3.1.   □

This result is of course different from what holds in the case $p > 1$, and it is an indication that the 1-Laplacian, in spite of the additional technical difficulties, possesses some very useful properties. In particular, since eigenfunctions can be chosen as indicator functions of vertex sets, we can then relate the properties of such sets to the eigenvalues.

Corollary 7.2.1. *The eigenvalue corresponding to the eigenfunction $\mathbf{1}_A$ of Proposition 7.2.1 is*

$$\lambda = \frac{|\partial A|}{\operatorname{vol} A}. \tag{7.2.3}$$

Proof This follows from the variational characterization (7.2.2) in terms of the Rayleigh quotient (7.2.1).   □

Theorem 7.2.1. *The spectrum of Δ_1 satisfies the following properties.*

- *The eigenvalues lie in the interval $[0, 1]$.*
- *The multiplicity of the eigenvalue 0 equals the number of connected components of the graph.*
- *The largest eigenvalue is 1, and it is realized by eigenvectors of the form $\mathbf{1}_v$, where v is a vertex.*

The last property, in particular, will be very useful for evaluating the Cheeger constant, which is one of our main topics.

Proof We split the proof into four parts.

- Observe that, for any $z_{vv'} \in \text{Sgn}(f(v) - f(v'))$ with $z_{vv'} = -z_{v'v}$, we have

$$\sum_{v \in V} f(v) \sum_{v' \in V : v' \sim v} z_{vv'} = \sum_{v \in V} \sum_{v' \in V : v' \sim v} z_{vv'} f(v)$$

$$= \sum_{\{v,v'\} \in E} (z_{vv'} f(v) + z_{v'v} f(v'))$$

$$= \sum_{\{v,v'\} \in E} z_{vv'} (f(v) - f(v'))$$

$$= \sum_{\{v,v'\} \in E} |f(v) - f(v')|,$$

and

$$\sum_{v \in V} f(v) \deg v \, \text{Sgn}(f(v)) = \sum_{v \in V} \deg v \, f(v) \text{Sgn}(f(v)) = \sum_{v \in V} \deg v \, |f(v)|.$$

Thus,

$$\langle \Delta_1 f, Df \rangle = \langle D\Delta_1 f, f \rangle = \sum_{\{v,v'\} \in E} |f(v) - f(v')|$$

and

$$\langle \text{Sgn}(f), Df \rangle = \langle D\text{Sgn}(f), f \rangle = \sum_{v \in V} \deg v \, |f(v)|,$$

where $\langle \cdot, \cdot \rangle$ denotes the inner product. Given any eigenpair (λ, f) of Δ_1, by taking the inner product of both sides of the eigenequation $0 \in \Delta_1 f - \lambda D\text{Sgn}(f)$ with Df, we then obtain

$$\sum_{\{v,v'\} \in E} |f(v) - f(v')| = \lambda \sum_{v \in V} \deg v \, |f(v)|,$$

which implies $\lambda \geq 0$. Also, since

$$\sum_{\{v,v'\} \in E} |f(v) - f(v')| \leq \sum_{\{v,v'\} \in E} (|f(v)| + |f(v')|) = \sum_{v \in V} \deg v \, |f(v)|,$$

(7.2.4)

we must have $\lambda \leq 1$. Hence, the spectrum of Δ_1 is contained in the interval $[0,1]$.

- The eigenfunctions corresponding to the eigenvalue 0 of Δ_1 are determined by the equation

$$\sum_{\{v,v'\} \in E} |f(v) - f(v')| = 0.$$

Its solutions form the linear subspace $\mathrm{span}(\mathbf{1}_{V_1}, \ldots, \mathbf{1}_{V_k})$, where $V_1, \ldots, V_k$ are the connected components of the vertex set. Consequently, the multiplicity of the eigenvalue 0 equals k, the number of connected components of the graph.

- The vector $\mathbf{1}_v$ is an eigenvector corresponding to the eigenvalue 1 of Δ_1 because it realizes the supremum in (7.2.2). Since, by the first statement, no eigenvalues are larger than 1, we conclude that the maximum eigenvalue is 1.

$\square$

Now, given a graph $\Gamma = (V, E)$, a nonempty proper subset $A \subset V$ is called a *Cheeger set* of Γ if $\mathrm{vol}\, A \le \mathrm{vol}\, A^c$ and

$$ h = \frac{|\partial A|}{\min\{\mathrm{vol}\, A, \mathrm{vol}\, A^c\}}. $$

In this case, (A, A^c) is called a *Cheeger cut* of Γ.

Theorem 7.2.2. *For any graph, $h = \lambda_2(\Delta_1)$. Moreover,*

- *If A is a Cheeger set, then, $\mathbf{1}_A$ is an eigenfunction corresponding to $\lambda_2(\Delta_1)$.*
- *If f is an eigenfunction corresponding to $\lambda_2(\Delta_1)$, then either $\{f > 0\}$ or $\{f < 0\}$ is a Cheeger set.*

Proof The fact that the second eigenvalue coincides with the Cheeger constant is a direct consequence of Theorem 6.3.1.

For the results on eigenfunctions, by Proposition 7.2.1, we only need to prove that A is a Cheeger set if and only if $\mathbf{1}_A$ is an eigenfunction corresponding to $\lambda_2(\Delta_1)$. In fact, by the third statement of Theorem 6.3.1 (or Theorems 6.3.3 and 6.3.4),

$$ \lambda_2(\Delta_1) = \min_{f \text{ nonconstant}} \frac{\sum_{\{v,v'\} \in E} |f(v) - f(v')|}{\min_{t \in \mathbb{R}} \sum_{v \in V} |f(v) - t|} = \min_{f \text{ nonconstant}} \max_{t \in \mathbb{R}} R_1(f - t) $$

$$ = \min_{\emptyset \neq A \subsetneq V} \frac{|\partial A|}{\min\{\mathrm{vol}\, A, \mathrm{vol}\, A^c\}} = h \tag{7.2.5} $$

and thus, any nonconstant function f satisfying $R_1(f) = \max_{t \in \mathbb{R}} R_1(f - t) = \lambda_2(\Delta_1)$ is an eigenfunction corresponding to $\lambda_2(\Delta_1)$. Moreover, note that for any nonempty proper subset $A \subset V$,

$$ \max_{t \in \mathbb{R}} R_1(\mathbf{1}_A - t) = \frac{|\partial A|}{\min\{\mathrm{vol}\, A, \mathrm{vol}\, A^c\}}. \tag{7.2.6} $$

The equalities (7.2.5) and (7.2.6) imply that A is a Cheeger set if and only if $\mathbf{1}_A$ is an eigenfunction corresponding to $\lambda_2(\Delta_1)$. $\square$

The above theorem is closely related to some more general results in Chapter 9, for instance, Theorem 9.2.1 and its consequences.

Using the above results, one can compute the entire spectrum of the 1-Laplacian for graphs, although the process can be very complicated. Interestingly, the variational eigenvalues of the 1-Laplacian have deep relations with the multiway Cheeger constants,

$$h_k(\Gamma) := \min_{\text{disjoint } S_1,\dots,S_k} \max_{1 \le i \le k} \frac{|\partial S_i|}{\text{vol}(S_i)}, \qquad (7.2.7)$$

for $k = 1,\dots,N$. By Theorem 2 in [54], the 1-Laplacian spectrum of the path graph P_N is

$$\left\{ 0, 1, \frac{1}{3}, \dots, \frac{1}{2[\frac{N}{2}]-1}, \frac{1}{2}\cdot\frac{1}{4}, \dots, \frac{1}{2[\frac{N-2}{4}]} \right\}.$$

Notably, the min–max eigenvalues of the 1-Laplacian for P_N coincide with the multiway Cheeger constants [76]. For example, when $N = 6$, we have

$$h_1(P_6) = \lambda_1(P_6) = 0,$$

$$h_2(P_6) = \lambda_2(P_6) = \frac{1}{5},$$

$$h_3(P_6) = \lambda_3(P_6) = \frac{1}{2},$$

$$h_4(P_6) = h_5(P_6) = h_6(P_6) = \lambda_4(P_6) = \lambda_5(P_6) = \lambda_6(P_6) = 1.$$

Here, we simply use $\lambda_k(\Gamma)$ or λ_k to denote the k-th min–max eigenvalue of the 1-Laplacian on Γ. Since the set of eigenvalues is

$$\left\{ 0, \frac{1}{5}, \frac{1}{3}, \frac{1}{2}, 1 \right\},$$

this implies that $1/3$ is a nonvariational eigenvalue.

Theorem 7.2.3. *Let Γ be a connected graph, and let f be an eigenfunction corresponding to λ_2. Then,*

1. *Either $\{f \ge 0\}$ or $\{f < 0\}$ induces a connected subgraph of Γ;*
2. *Either $\{f \le 0\}$ or $\{f > 0\}$ induces a connected subgraph of Γ;*
3. *Either $\{f \le 0\}$ or $\{f \ge 0\}$ induces a connected subgraph of Γ.*

Proof We divide the proof into three parts.

1. We first prove that either $\{f \ge 0\}$ or $\{f < 0\}$ induces a connected subgraph of Γ. To do so, assume that $\{f \ge 0\}$ is not connected. Then, $\{f < 0\} \neq \emptyset$, and the set $\{f \ge 0\}$ can be expressed as the disjoint union of two nonadjacent, nonempty subsets V_1 and V_2. In particular, the sets V_1, V_2, $\{f < 0\} \cup V_1$ and

$\{f < 0\} \cup V_2$ are all nonempty. Since f is an eigenfunction corresponding to λ_2, by Proposition 7.2.1, we have that $\mathbf{1}_{\{f<0\}}$ is also an eigenfunction corresponding to λ_2. Also, $\{f < 0\}$ is a Cheeger set with

$$\frac{|\partial\{f < 0\}|}{\text{vol}(\{f < 0\})} = \frac{|\partial\{f < 0\}|}{\min\{\text{vol}(\{f < 0\}), \text{vol}(\{f \geq 0\})\}} = h.$$

Since $V_1 \cup V_2$ is the complement of $\{f < 0\}$ and the sets V_1 and V_2 are nonadjacent, we also have that

$$|\partial(V_1)| + |\partial(V_2)| = |\partial\{f < 0\}|.$$

Furthermore,

$$\min\{\text{vol}(\{f < 0\} \cup V_2), \text{vol}(V_1)\} + \min\{\text{vol}(\{f < 0\} \cup V_1), \text{vol}(V_2)\}$$
$$\geq \text{vol}(\{f < 0\}).$$

Accordingly,

$$h = \frac{|\partial\{f < 0\}|}{\text{vol}(\{f < 0\})}$$
$$\geq \frac{|\partial(V_1)| + |\partial(V_2)|}{\min\{\text{vol}(\{f < 0\} \cup V_2), \text{vol}(V_1)\} + \min\{\text{vol}(\{f < 0\} \cup V_1), \text{vol}(V_2)\}}$$
$$\geq \min\left\{\frac{|\partial(V_1)|}{\min\{\text{vol}(V_1), \text{vol}(\{f < 0\} \cup V_2)\}}, \frac{|\partial(V_2)|}{\min\{\text{vol}(V_2), \text{vol}(\{f < 0\} \cup V_1)\}}\right\}$$
$$\geq h.$$

Therefore, we must have equality everywhere in the above inequalities, and this implies that

$$\text{vol}(\{f < 0\}) = \text{vol}(V_1 \cup V_2). \tag{7.2.8}$$

Now, our goal is to prove that $\{f < 0\}$ induces a connected subgraph of Γ. Assume, by contradiction, that is not true. Then, $\{f < 0\}$ can be written as the disjoint union of two nonadjacent, nonempty subsets V_3 and V_4. By (7.2.8), we then have $\text{vol}(V_3 \cup V_4) = \text{vol}(V_1 \cup V_2)$. Also, by the same reasoning as above,

$$\partial(V_1 \cup V_3) + \partial(V_1 \cup V_4) = \partial(V_1 \cup V_2)$$

and

$$\min\{\text{vol}(V_1 \cup V_3), \text{vol}(V_2 \cup V_4)\} + \min\{\text{vol}(V_1 \cup V_4), \text{vol}(V_2 \cup V_3)\}$$
$$> \text{vol}(V_1 \cup V_2).$$

Therefore,

$$h = \frac{|\partial(V_1 \cup V_2)|}{\text{vol}\,(V_1 \cup V_2)}$$

$$> \frac{|\partial(V_1 \cup V_3)| + |\partial(V_1 \cup V_4)|}{\min\{\text{vol}\,(V_1 \cup V_3), \text{vol}\,(V_2 \cup V_4)\} + \min\{\text{vol}\,(V_1 \cup V_4), \text{vol}\,(V_2 \cup V_3)\}}$$

$$\geq \min\left\{ \frac{|\partial(V_1 \cup V_3)|}{\min\{\text{vol}\,(V_1 \cup V_3), \text{vol}\,(V_2 \cup V_4)\}}, \frac{|\partial(V_1 \cup V_4)|}{\min\{\text{vol}\,(V_1 \cup V_4), \text{vol}\,(V_2 \cup V_3)\}} \right\}$$

$$> 0,$$

and this contradicts the definition of the Cheeger constant. This proves the first claim.

2. The proof is analogous to that of the first claim.
3. Finally, we want to show that either $\{f \geq 0\}$ or $\{f \leq 0\}$ induces a connected subgraph on Γ. Suppose the contrary. Then, by the same reasoning as above, we obtain

$$\text{vol}\,(\{f \geq 0\}) = \text{vol}\,(\{f < 0\}) \text{ and vol}\,(\{f \leq 0\}) = \text{vol}\,(\{f > 0\})$$

implying that

$$\{f = 0\} = \emptyset, \{f > 0\} = \{f \geq 0\} \text{ and vol}\,(\{f < 0\}) = \text{vol}\,(\{f > 0\}),$$

but, this contradicts the first statement of the theorem. □

As a consequence of the above theorem, for any Cheeger cut (A, A^c) of a connected graph, either A or A^c induces a connected subgraph[2]. As we shall see, a more interesting consequence is a nodal domain property presented in Proposition 9.2.2 in Chapter 9.

A graph Γ is *perfect* [60] if for every induced subgraph Γ' of Γ, the chromatic number of Γ' equals the size of the largest clique of Γ'.

Proposition 7.2.2. *Fix a graph Γ and let α be its independence number. Then,*

1. *The multiplicity of the eigenvalue 1 for Δ_1 is at least α and at most 2α.*
2. *If Γ is a triangle-free perfect graph, then the multiplicity of the eigenvalue 1 for Δ_1 is α.*

Proof 1. Let t denote the multiplicity of the eigenvalue 1 for Δ_1. To prove that $t \geq \alpha$, we need to show that $\lambda_k = 1$ for all $k \geq N - \alpha + 1$.

---

[2] This property is folklore, and such formulation presented here was first proposed by Zhang in [131] with a proof in [53]. The proof of Theorem 7.2.3 in this book provides new insights into this folklore property.

Note that, if $N - k + 1 \leq \alpha$, then there exist at least $N - k + 1$ pairwise nonadjacent vertices in Γ. We denote them by $v_1, \ldots, v_{N-k+1}$. Let $\mathbf{1}_v \in C(V)$ be the indicator function on v, that is, $\mathbf{1}_v(v) = 1$ and $\mathbf{1}_v(v') = 0$ for all $v' \neq v$. Let also

$$X_{N-k+1} := \mathrm{span}(\mathbf{1}_{v_1}, \ldots, \mathbf{1}_{v_{N-k+1}}).$$

Note that, for each f in X_{N-k+1}, there exists $(l_1, \ldots, l_{N-k+1}) \neq \mathbf{0}$ such that

$$f = l_1 \mathbf{1}_{v_1} + \ldots + l_{n-k+1} \mathbf{1}_{v_{N-k+1}}.$$

Also, if $f(v)f(v') \neq 0$, then $v \nsim v'$. Thus, each nodal domain of f must be a singleton set, which implies that $R_1(f) = 1$. Therefore, the intersection property of the genus derived in Lemma 2.6.3 implies that

$$A \cap X_{N-k+1} = A \cap \mathrm{span}(\mathbf{1}_{v_1}, \ldots, \mathbf{1}_{v_{N-k+1}}) \neq \emptyset$$

for any A with $\mathrm{genus}(A) \geq k$. Hence,

$$\sup_{f \in A} R_1(f) \geq \inf_{f \in X_{N+1-k}} R_1(f) = 1,$$

implying that $\lambda_k = 1$. This shows that $t \geq \alpha$.

To prove that $t \leq 2\alpha$, it suffices to show that $\lambda_k < 1$ for all $k \leq N - 2\alpha$. Given $k \leq N - 2\alpha$, let $g_i \in C(V)$ be defined by $g_i(v_v) := v^{i-1}$ for $v \in \{1, \ldots, N\}$ and $i \in \{1, 2, \ldots, k\}$. Let also $X_k := \mathrm{span}(g_1, \ldots, g_k)$. Then, X_k is a closed symmetric set with $\mathrm{genus}(X_k) = k$. Now, for any $f \not\equiv 0$ in X_k, there exists $(l_1, l_2, \ldots, l_k) \neq \mathbf{0}$ such that $f = l_1 g_1 + \ldots + l_k g_k$. By the properties of the Vandermonde matrix, not every element of the form $l_i g_i$, for $i = 1, \ldots, k$, can vanish on k vertices simultaneously. This implies that the number of zeros of f is at most $(k - 1)$. Therefore, the support of f has cardinality at least $N + 1 - k$. By the pigeonhole principle, it follows that at least $\lceil \frac{N+1-k}{2} \rceil$ vertices with a nonzero f-value have the same sign, where $\lceil \frac{N+1-k}{2} \rceil$ is the smallest integer larger than or equal to $\frac{N+1-k}{2}$. Since $\alpha < \lceil \frac{N+1-k}{2} \rceil$, there exist two adjacent vertices whose values for f have the same sign, which implies that $R_1(f) < 1$. Note that R_1 is continuous on the compact set $X_k \cap \mathbb{S}^{N-1}$, and we have proved that $R_1(f) < 1$ for any $f \in X_k$. Since every continuous function on a compact set attains its maximum, this implies that

$$\lambda_k \leq \sup_{f \in X_k} R_1(f) = \max_{f \in X_k \cap \mathbb{S}^{N-1}} R_1(f) < 1.$$

2. By [278], any triangle-free graph is such that

$$\alpha \leq t \leq \kappa,$$

where κ denotes the clique covering number of the graph, that is, the smallest possible number of cliques that can cover Γ. Moreover, if Γ is a perfect graph, then $\alpha = \kappa$. By putting everything together, we obtain that $\alpha = t = \kappa$, if Γ is both triangle-free and perfect.

$\square$

We refer to [278] for more results on the multiplicity of the eigenvalue 1 for Δ_1.

Proposition 7.2.3. *For a simple graph* $\Gamma = (V, E)$ *on N nodes, the 1-Laplacian* Δ_1 *has at most* vol $(V)^2/4 + 1$ *distinct eigenvalues. Hence, in particular,* Δ_1 *has at most* $(N^2(N-1)^2)/4 + 1$ *distinct eigenvalues, and counting multiplicity, the number of the eigenvalues of* Δ_1 *does not exceed* $O(N^5)$.

Proof Let λ and $\tilde{\lambda}$ be two distinct eigenvalues. By Proposition 7.2.1 and Corollary 7.2.1, there exist $A, B \subset V$ with vol (A), vol $(B) \leq$ vol $(V)/2$, such that

$$\lambda = \frac{|\partial A|}{\text{vol}\,(A)} \quad \text{and} \quad \tilde{\lambda} = \frac{|\partial B|}{\text{vol}\,(B)}.$$

We have that

$$\begin{aligned}
|\lambda - \tilde{\lambda}| &= \left| \frac{|\partial A|}{\text{vol}\,(A)} - \frac{|\partial B|}{\text{vol}\,(B)} \right| \\
&= \frac{||\partial A|\text{vol}\,(B) - |\partial B|\text{vol}\,(A)|}{\text{vol}\,(A)\text{vol}\,(B)} \\
&\geq \frac{1}{\text{vol}\,(A)\text{vol}\,(B)} \\
&\geq \frac{4}{\text{vol}\,(V)^2},
\end{aligned}$$

which implies that the distance between two distinct eigenvalues is at least $4/\text{vol}\,(V)^2$. Since the eigenvalues lie in $[0, 1]$, this implies that there are at most vol $(V)^2/4 + 1$ distinct eigenvalues of Δ_1. This proves the first claim. The second claim follows from the fact that vol $(V) \leq N(N-1)$ and that the multiplicity of any eigenvalue does not exceed N. $\square$

The bound $O(N^5)$ from Proposition 7.2.3 may seem excessive, but the graph 1-Laplacian can indeed have many eigenvalues. For example, let $\Gamma = (V, E)$ be the graph displayed in Figure 7.1 with vertex set $V = \{v_1 \ldots, v_6\}$: Then, as shown in [278], the spectrum of Δ_1 is

$$\left\{ 0, \frac{2}{5}, \frac{5}{9}, \frac{3}{5}, \frac{2}{3}, \frac{5}{7}, \frac{3}{4}, \frac{7}{9}, 1 \right\}.$$

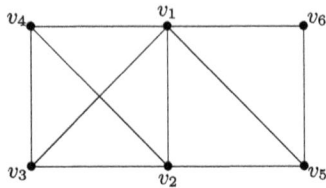

Figure 7.1 A graph with six vertices and nine distinct 1-Laplacian eigenvalues.

Hence, the graph has six nodes, and its 1-Laplacian has nine distinct eigenvalues.

7.3 The Graph ∞-Laplacian

In this section, we consider the ∞-Laplacian on graphs, which has been studied, for instance, in [35, 40, 77, 157, 186]. As we shall see, the spectrum of this operator is related to the maximal radius that allows us to inscribe a fixed number of disjoint balls in the graph.

As the 1-Laplacian, also the ∞-Laplacian Δ_∞ on a graph $\Gamma = (V, E)$ is the subgradient of a functional. (Recall (6.1.6), (6.1.7) for the definition of the subgradient.) Here, such functional is $f \mapsto \sum_{\{v,v'\} \in E} |f(v) - f(v')|$, that is,

$$\Delta_\infty f := \nabla_f \max_{\{v,v'\} \in E} |f(v) - f(v')|. \tag{7.3.1}$$

In this case, we are interested in the nonlinear eigenvalue problem

$$0 \in \Delta_\infty f - \lambda \nabla \|f\|_\infty, \tag{7.3.2}$$

and we say that (λ, f) is an eigenpair of the ∞-Laplacian Δ_∞ on Γ if (7.3.2) holds. Thus, (λ, f) is an eigenpair of Δ_∞ if there exist $z_{vv'}$ and z_v with

$$\sum_{v' \in V : v' \sim v} z_{vv'} = \lambda z_v, \quad \forall v \in V,$$

$$\begin{cases} z_{vv'}(f(v) - f(v')) \geq 0, z_{v'v} = -z_{vv'} \\ \sum_{vv' \in E} |z_{vv'}| = 1 \\ z_{vv'} = 0 \text{ if } |f(v) - f(v')| < \max_{\hat{v}\bar{v} \in E} |f(\hat{v}) - f(\bar{v})| \end{cases} \quad \text{and} \quad \begin{cases} z_v f(v) \geq 0 \\ \sum_{v \in V} |z_v| = 1 \\ z_v = 0 \text{ if } |f(v)| \\ \quad < \|f\|_\infty. \end{cases}$$

$$\tag{7.3.3}$$

The following characterization of the first nonzero eigenvalue of Δ_∞ has been given in [186].

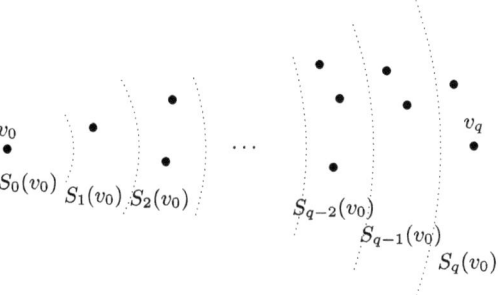

Figure 7.2 A sequence of concentric discrete spheres, used in the proof of Proposition 7.3.1.

Proposition 7.3.1. *For any connected graph* Γ,

$$\lambda_2(\Delta_\infty) = \frac{2}{\text{diam}(\Gamma)}.$$

Proof Similarly to the proofs of (6.4.1) and Theorem 6.3.1 or by applying Theorem 2.1 in [160] to the functionals $f \mapsto \sum_{\{v,v'\}\in E} |f(v) - f(v')|$ and $f \mapsto \|f\|_\infty$ directly, the second eigenvalue $\lambda_2(\Delta_\infty)$ can be written as

$$\min_{f \text{ nonconstant}} \frac{\max_{\{v,v'\}\in E} |f(v) - f(v')|}{\min_{t\in\mathbb{R}} \max_{v\in V} |f(v) - t|} = \min_{\substack{\max f=1 \\ \min f=-1}} \max_{\{v,v'\}\in E} |f(v) - f(v')|.$$

$$(7.3.4)$$

Now, let q be the diameter of Γ, let v_0 and v_q be a pair of vertices that have distance q, and let $v_0, v_1, \ldots, v_q$ form a shortest path between v_0 and v_q. Let also $f: V \to \mathbb{R}$ be defined by

$$f(v) := 1 - \frac{2}{q} \cdot \text{dist}(v_0, v), \quad \forall v \in V, \qquad (7.3.5)$$

where $\text{dist}(v_0, v)$ denotes the distance between v_0 and v. Then, the maximum value of f is 1, and the minimum value of f is -1. For $k \geq 0$, we also let

$$S_k(v_0) := \{v \in V : \text{dist}(v_0, v) = k\} \qquad (7.3.6)$$

be the spheres that are centered at v_0, and we observe that any edge $\{v, v'\} \in E$ intersects at most two adjacent spheres $S_k(v_0)$ and $S_{k+1}(v_0)$, for some $k \geq 0$. This is illustrated in Figure 7.2.

Since f is constant on each sphere and $|f(v) - f(v')| = 2/q$ whenever $v \in S_k(v_0)$ and $v' \in S_{k+1}(v_0)$ for some $k \geq 0$, we can infer that, for any $\{v, v'\} \in E$,

$$|f(v) - f(v')| \leq \frac{2}{q},$$

which implies that

$$\lambda_2(\Delta_\infty) \leq \frac{2}{q}.$$

Now, fix f such that the maximum value of f is 1, the minimum value of f is -1, and $\|f\|_\infty = 1$. Let also $v_0, v_1, \ldots, v_l$ be a shortest path satisfying $f(v_0) = 1$ and $f(v_l) = -1$. Then, $l \leq q$ and

$$
\begin{aligned}
\max_{\{v,v'\}\in E} |f(v) - f(v')| &\geq \max_{i=0,\ldots,l-1} |f(v_i) - f(v_{i+1})| \\
&\geq \frac{1}{l} \sum_{i=0}^{l-1} |f(v_i) - f(v_{i+1})| \\
&\geq \frac{1}{l} |f(v_0) - f(v_l)| \\
&= \frac{2}{l} \\
&\geq \frac{2}{q},
\end{aligned}
$$

which implies that $\lambda_2(\Delta_\infty) \geq 2/q$. □

7.4 *p*-Laplacians on Oriented Hypergraphs

We now extend the previous concepts and considerations to oriented hypergraphs. This section covers the entire range $1 \leq p \leq \infty$, that is, also includes the limiting cases $p = 1$ and $p = \infty$ that had been giving special treatments above. While some aspects are straightforward, also some new phenomena and difficulties arise. The first issue is the definition of the *p*-Laplacian, and in fact, several notions of *p*-Laplacians on hypergraphs have been proposed and investigated in the literature. For instance, the hypergraph *p*-Laplacians which are introduced in [133] are related to the Lovász extension. The corresponding functionals for the *p*-Laplacians in [133] are of the form

$$f \mapsto \sum_{e\in E} \max_{v,v'\in e} |f(v) - f(v')|^p,$$

and they are nonsmooth in general, even for $p = 2$.

In this section, we consider the *p*-Laplacian on oriented hypergraphs proposed in [157], which is a natural extension of the hypergraph Laplacian defined in Section 2.5.4, the case of $p = 2$.

Let $\Gamma = (V, E, \varphi)$ be an oriented hypergraph. Recall that, from Definition 2.4.7, if $\varphi(v, e) = 1$ (respectively $\varphi(v, e) = -1$), then v is said to be an *output* (respectively *input*) for e. Also, if $v \neq v'$ and

$$\varphi(v,e) = \varphi(v',e) \neq 0 \quad (\text{respectively } \varphi(v,e) = -\varphi(v',e) \neq 0),$$

the vertices v and v' are *co-oriented* (respectively *antioriented*) in e. We also let

$$o_e(v,v') := \begin{cases} -1, & \text{if } v \text{ and } v' \text{ are co-oriented in } e \\ 1, & \text{if } v \text{ and } v' \text{ are antioriented in } e \\ 0, & \text{otherwise.} \end{cases}$$

For $p \in \mathbb{R}_{>1}$, the *p-Laplacian* $\Delta_p : C(V) \to C(V)$ is defined by

$$\Delta_p f(v) := \frac{1}{\deg v} \sum_{e \ni v} |\delta f(e)|^{p-2} \left(\sum_{v' \in e:\, o_e(v,v')=-1} f(v') - \sum_{v' \in e:\, o_e(v,v')=1} f(v') \right),$$

where, recalling (2.5.37),

$$\delta f(e) = \sum_{v:\, \varphi(v,e)=1} f(v) - \sum_{v:\, \varphi(v,e)=-1} f(v).$$

We define its *eigenvalue problem* as

$$\Delta_p f = \lambda |f|^{p-2} f. \tag{7.4.1}$$

A nonzero function f and a real number λ satisfying (7.4.1) are called an *eigenfunction* and the corresponding *eigenvalue* for Δ_p.

The 1-*Laplacian* is the set-valued operator which, for $f \in C(V)$, is defined by

$$\Delta_1 f :=$$

$$\left\{ \sum_{v \in V} \frac{1}{\deg v} \sum_{e \ni v} z_{ve} \mathbf{1}_v \left| \begin{array}{l} z_{ve} \in \text{Sgn}\left(\sum_{v' \in e:\, o_e(v,v')=-1} f(v') - \sum_{v' \in e:\, o_e(v,v')=1} f(v') \right) \\ z_{ve} = -o_e(v,v') z_{v'e} \end{array} \right. \right\}$$

where $\mathbf{1}_1, \ldots, \mathbf{1}_N$ form the orthonormal basis of $\mathbb{R}^N$ and

$$\text{Sgn}(t) := \begin{cases} \{1\} & \text{if } t > 0, \\ [-1,1] & \text{if } t = 0, \\ \{-1\} & \text{if } t < 0. \end{cases}$$

The *eigenvalue problem* of Δ_1 is to find the eigenpairs (λ, f) for which there exist

$$z_{ve} \in \text{Sgn}\left(\sum_{v' \in e:\, o_e(v,v')=-1} f(v') - \sum_{v' \in e:\, o_e(v,v')=1} f(v') \right)$$

with $z_{ve} = -o_e(v,v') z_{v'e}$ for $v, v' \in e$, and $z_v \in \text{Sgn}(f(v))$ satisfying

$$\sum_{e \ni v} z_{ve} = \lambda \deg v \cdot z_v, \quad \forall v \in V.$$

A shorter coordinate form of the eigenvalue problem for the 1-Laplacian is

$$\exists z_v \in Sgn(f(v)) \text{ and } z_e \in Sgn\left(\sum_{v':\,\varphi(v',e)=1} f(v') - \sum_{v':\,\varphi(v',e)=-1} f(v')\right)$$

$$\text{such that} \sum_{e:\,\varphi(v,e)=1} z_e - \sum_{e:\,\varphi(v,e)=-1} z_e = \lambda \deg v \cdot z_v, \quad \forall v \in V. \qquad (7.4.2)$$

Moreover, we can rewrite (7.4.2) as

$$\exists z_e \in Sgn\left(\sum_{v':\,\varphi(v',e)=1} f(v') - \sum_{v':\,\varphi(v',e)=-1} f(v')\right) \text{ such that}$$

$$\sum_{e:\,\varphi(v,e)=1} z_e - \sum_{e:\,\varphi(v,e)=-1} z_e \in \lambda \deg v \cdot Sgn(f(v)), \quad \forall v \in V. \qquad (7.4.3)$$

Similar to the graph ∞-Laplacian presented in Section 7.3, we also define the ∞-Laplacian Δ_∞ on an oriented hypergraph as the subgradient of the functional $f \mapsto \|\delta f\|_\infty$. The kth min–max eigenvalue $\lambda_k(\Delta_\infty)$ is similarly defined as

$$\inf_{S \in \mathrm{Gen}_k} \sup_{f \in S \setminus \{0\}} \frac{\|\delta f\|_\infty}{\|f\|_\infty}.$$

Then, we have

Proposition 7.4.1. *For any $k \geq 1$,*

$$\lim_{p \to +\infty} (\lambda_k(\Delta_p))^{\frac{1}{p}} = \lambda_k(\Delta_\infty).$$

Proof We have that

$$\frac{|E|^{\frac{1}{p}} \max_{e \in E} |\delta f(e)|}{\max_{v \in V} |f(v)|} \geq \frac{\left(\sum_{e \in E} |\delta f(e)|^p\right)^{\frac{1}{p}}}{\left(\sum_{v \in V} \deg v |f(v)|^p\right)^{\frac{1}{p}}} \geq \frac{\max_{e \in E} |\delta f(e)|}{\mathrm{vol}\,(V)^{\frac{1}{p}} \max_{v \in V} |f(v)|},$$

which implies

$$|E|^{\frac{1}{p}} \frac{\|\delta f\|_\infty}{\|f\|_\infty} \geq (RQ_p(f))^{\frac{1}{p}} \geq \mathrm{vol}\,(V)^{-\frac{1}{p}} \frac{\|\delta f\|_\infty}{\|f\|_\infty}.$$

By definition of min–max eigenvalues, we then obtain

$$|E|^{\frac{1}{p}} \lambda_k(\Delta_\infty) \geq (\lambda_k(\Delta_p))^{\frac{1}{p}} \geq \mathrm{vol}\,(V)^{-\frac{1}{p}} \lambda_k(\Delta_\infty).$$

Therefore, $(\lambda_k(\Delta_p))^{\frac{1}{p}}$ tends to $\lambda_k(\Delta_\infty)$, as $p \to +\infty$. $\qquad \square$

The *independence number* of a hypergraph Γ is defined in the same manner as that for a graph in Definition 7.1.1. Thus, a subset $U \subseteq V$ is *independent* if $\#(U \cap e) \leq 1$ for all $e \in E$, and the independence number is

$$\alpha := \max\{\#U \colon U \subseteq V \text{ independent}\}.$$

Theorem 7.4.1. *Let $p \in [1, \infty]$. Then,*

$$\alpha \leq \min\{\#\{i \colon \lambda_i(\Delta_p) \leq 1\}, \#\{i \colon \lambda_i(\Delta_p) \geq 1\}\}, \tag{7.4.4}$$

and the inequality is sharp.

Proof We first consider the case where $p < \infty$. Let $U \subseteq V$ be an independent set such that $\#U = \alpha$. Then, for any $f \in C(V)$ with $\mathrm{supp}(f) \subset U$,

$$\sum_{e \in E} |\delta f(e)|^p = \sum_{e \in E} \sum_{v \in e} |f(v)|^p = \sum_{v \in V} \sum_{e \in E, e \ni v} |f(v)|^p = \sum_{v \in V} \deg v \cdot |f(v)|^p,$$

which implies that $\mathcal{R}_p(f) = 1$. Now, the linear subspace

$$X := \{f \in C(V) \colon \mathrm{supp}(f) \subset U\}$$

has dimension α, and for any $f \in X$, $\mathcal{R}_p(f) = 1$. Thus,

$$\lambda_1(\Delta_p) \leq \ldots \leq \lambda_\alpha(\Delta_p) = \inf_{X' \in \mathrm{Gen}_\alpha} \sup_{f \in X'} \mathcal{R}_p(f) \leq \sup_{f \in X} \mathcal{R}_p(f) = 1.$$

Similarly, by the intersection property of the $\mathbb{Z}_2$-genus (see Lemma 2.6.3), for any $X' \in \mathrm{Gen}_{N-\alpha+1}$, we have that $X' \cap X \neq \emptyset$. Hence,

$$\lambda_N(\Delta_p) \geq \ldots \geq \lambda_{N-\alpha+1}(\Delta_p) = \inf_{X' \in \mathrm{Gen}_{N-\alpha+1}} \sup_{f \in X'} \mathcal{R}_p(f) \geq \inf_{f \in X} \mathcal{R}_p(f) = 1.$$

Hence, there are at least α eigenvalues which are smaller (respectively, larger) than 1. This proves the first claim. To see that (7.4.4) is sharp, assume that Γ is such that $V = \{v_1, \ldots, v_N\}$ and $E = \{\{v_1\}, \ldots, \{v_N\}\}$. Then, clearly, $\alpha = N$ and all eigenvalues are equal to 1. This proves the claim for $p < \infty$.

In the case when $p = \infty$, assume, without loss of generality, that

$$\lambda_1(\Delta_\infty) \leq \ldots \leq \lambda_k(\Delta_\infty) < 1 = \lambda_{k+1}(\Delta_\infty) = \ldots$$
$$= \lambda_{k+r}(\Delta_\infty) < \lambda_{k+r+1}(\Delta_\infty) \leq \ldots$$

By Proposition 7.4.1, $\lambda_i(\Delta_p)^{\frac{1}{p}} \to \lambda_i(\Delta_\infty)$ as p tends to $+\infty$, which implies that

$$\lambda_1(\Delta_p) \leq \ldots \leq \lambda_k(\Delta_p) < 1 < \lambda_{k+r+1}(\Delta_p) \leq \ldots \leq \lambda_N(\Delta_p)$$

for a sufficiently large p. Therefore, by applying (7.4.4) to a sufficiently large $p < \infty$, it follows that

$$\alpha \leq \min\{\#\{i \colon \lambda_i(\Delta_p) \leq 1\}, \#\{i \colon \lambda_i(\Delta_p) \geq 1\}\}$$

$$\leq \min\{k + r, N - k\}$$
$$= \min\{\#\{i \colon \lambda_i(\Delta_\infty) \leq 1\}, \#\{i \colon \lambda_i(\Delta_\infty) \geq 1\}\}.$$

$\square$

A *clique* of an oriented hypergraph Γ is a set of vertices $U \subseteq V$ such that, for any $\{v, v'\} \subset U$, there exists a hyperedge $e \supset \{v, v'\}$. The *clique covering number* of Γ is

$$\kappa := \min\{k \colon \exists k \text{ cliques } C_1, \ldots, C_k \text{ s.t. } V = \cup_{i=1}^k C_i\}.$$

Theorem 7.4.2 is a sandwich theorem for hypergraphs with only inputs which involves the independence number, the clique covering number, and the multiplicity of the eigenvalue 1 for Δ_1.

Theorem 7.4.2. *Let Γ be an oriented hypergraph with only inputs, and let η denote the multiplicity of the eigenvalue 1 for the 1-Laplacian. Then,*

$$\alpha \leq \eta \leq \kappa.$$

Proof By Theorem 7.4.1 and since 1 is the largest eigenvalue of Δ_1,

$$\alpha \leq \#\{i \colon \lambda_i(\Delta_1) \geq 1\} = \#\{i \colon \lambda_i(\Delta_1) = 1\} \leq \eta.$$

This proves the first inequality. Now, given a clique U, let $f \in C(V)$ with $\mathrm{supp}(f) \subset U$. Since every hyperedge has only inputs,

$$R_1(f) = 1 \iff f(v)f(v') \geq 0 \ \forall v, v' \in U \iff \text{either } f \geq 0 \text{ or } f \leq 0.$$

Now, one can check that the map

$$\{f \in C(V) \setminus \{0\} \text{ with } \mathrm{supp}(f) \subset U \colon f \geq 0 \text{ or } f \leq 0\}\} \to \{-1, 1\}$$

defined by

$$f \mapsto \begin{cases} 1, & \text{if } f \geq 0 \\ -1, & \text{if } f \leq 0 \end{cases}$$

is an odd continuous map. Therefore, the $\mathbb{Z}_2$-genus of the set

$$\{f \in C(V) \setminus \{0\} \colon \mathrm{supp}(f) \subset U, R_1(f) = 1\}$$

is 1. Now, let $U_1, \ldots, U_\kappa$ be pairwise disjoint cliques with $\cup_{i=1}^\kappa U_i = V$, and let f be such that $R_1(f) = 1$. Then, clearly $f = \sum_{j=1}^\kappa f|_{U_j}$. Also, from

$$TV_1(f) \leq \sum_{j=1}^\kappa TV_1(f|_{U_j}) \quad \text{and} \quad \sum_{v \in V} \deg v \cdot |f(v)| = \sum_{j=1}^\kappa \sum_{v \in V} \deg v \cdot |f|_{U_j}(v)|,$$

it follows that

$$1 = R_1(f) \le \frac{\sum_{j=1}^{K} TV_1(f|_{U_j})}{\sum_{j=1}^{K} \sum_{v \in V} \deg v \cdot |f|_{U_j}(v)|} \le \max_{j=1,\dots,K} R_1(f|_{U_j}) \le 1.$$

Hence, $R_1(f|_{U_j}) = 1$, for all $j = 1,\dots,K$. Together with the fact that R_1 is zero-homogeneous, this implies that $R_1(N \cdot f|_{U_i}) = R_1(f|_{U_i}) = 1$, for $i = 1,\dots,K$. Thus, by the fact that

$$f = \sum_{i=1}^{K} f|_{U_i} = \frac{1}{N} \sum_{i=1}^{K} N \cdot f|_{U_i}$$

and

$$N \cdot f|_{U_i} \in \{f \ne 0 : \operatorname{supp}(f) \subset U_i, R_1(f) = 1\},$$

we have

$$\{f \ne 0 : R_1(f) = 1\} \subset \underset{i=1}{\overset{K}{\bowtie}} \{f \ne 0 : \operatorname{supp}(f) \subset U_i, R_1(f) = 1\},$$

where $\bowtie$ is the simplicial join operator defined by

$$\underset{i=1}{\overset{K}{\bowtie}} S_i := \Big\{ t_1 s_1 + \dots + t_K s_K : s_i \in S_i, t_i \ge 0, \sum_{i=1}^{K} t_i = 1 \Big\},$$

for subsets $S_1,\dots,S_K$ of a linear space.

The $\mathbb{Z}_2$-genus is monotonic, as established in Lemma 2.6.3, and subadditive with respect to the join operator, which follows from the fact that $S^{i_1-1} \times \cdots \times S^{i_k-1}$ naturally injects into $S^{i_1+\dots+i_k-1}$. Therefore, we have

$$\eta = \operatorname{genus}\{f \ne 0 : R_1(f) = 1\}$$

$$\le \operatorname{genus}\Big(\underset{i=1}{\overset{K}{\bowtie}} \{f \ne 0 : \operatorname{supp}(f) \subset U_i, R_1(f) = 1\} \Big)$$

$$\le \sum_{i=1}^{K} \operatorname{genus}\{f \ne 0 : \operatorname{supp}(f) \subset U_i, R_1(f) = 1\}$$

$$= K.$$

$\square$

8

Cheeger Inequalities

This chapter systematically treats Cheeger type inequalities. Our theory combines higher-order Cheeger inequalities, Cheeger inequalities for signed graphs, and Cheeger inequalities for the 1-Laplacian. We introduce a new definition of a Cheeger constant for simplicial complexes. This leads us to a solution of the higher-order Cheeger problem on simplicial complexes.

Cheeger inequalities on discrete structures play an important role in expander theory, spectral graph theory, and spectral clustering methods. In the first two sections of this chapter, we present (multiway) Cheeger inequalities for simple graphs and signed graphs. In the third section, we then derive generalized inequalities for simplicial complexes, following mainly [160, 161]. Finally, in Section 8.4, we generalize the results in [214], and we prove new Cheeger-type inequalities for p-Laplacians on hypergraphs, which appear here for the first time in this general form.

Some Cheeger inequalities already occurred in Chapter 4. We first need to recall some of these results, in order to put them into the systematic perspective developed in the present chapter and also to build upon them.

8.1 Cheeger Inequalities on Graphs

We start by recalling the Cheeger constant (4.3.13) of a simple graph $\Gamma = (V, E)$,

$$h := \min_{\emptyset \neq S \subsetneq V} \frac{|E(S, S^c)|}{\min\{\text{vol}(S), \text{vol}(S^c)\}},$$

as well as the Cheeger inequalities from Theorem 4.3.1:

Theorem 8.1.1 (Cheeger inequalities on a graph)**.** *Let $\Gamma = (V, E)$ be a simple graph. Let h be its Cheeger constant, and let $\lambda_2(\Delta_2)$ be the second eigenvalue of the normalized Laplacian. Then,*

$$\frac{h^2}{2} \le \lambda_2(\Delta_2) \le 2h.$$

In fact, this is also a special case of Theorem 6.4.2, which shows Cheeger inequalities for p-Laplacians on chemical hypergraphs.

We point out that if (A, A^c) is a Cheeger cut of a connected graph Γ, then either A or A^c induces a connected subgraph. Moreover, there exists a Cheeger cut of Γ such that both A and A^c are connected. For the latter statement, a version for compact Riemannian manifolds was first proved by Yau [274], and it is collected in Buser's book [43, Lemma 8.3.6]. The idea of the original proof in the geometric setting is to use mathematical induction on the number of connected components of the boundary of A, but this idea does not work in the case of graphs. A combinatorial proof for the graph Cheeger cut was given by Mohar [211]. We refer to Theorem 7.2.3 for our new proof, which works in the graph case.

Miclo [206] introduced the higher-order Cheeger constant (7.2.7) and asked whether Cheeger-type inequalities can be proved for this constant. Lee, Oveis Gharan, and Trevisan [180] answered Miclo's question as follows.

Theorem 8.1.2. *There exists a universal constant C such that, for every graph, and for every natural number k,*

$$\frac{h_k^2}{Ck^4} \le \lambda_k(\Delta_2) \le 2h_k.$$

8.2 Cheeger Inequalities on Signed Graphs

Let now (Γ, s) be a signed graph, with $\Gamma = (V, E)$. Given two disjoint sets $V_1, V_2 \subset V$, let

$$E^+(V_1, V_2) := \{\{u, v\} \in E : u \in V_1, v \in V_2, s(uv) = 1\},$$

and let

$$E^-(V_1) := \{\{u, v\} \in E : u, v \in V_1, s(uv) = -1\}.$$

The *signed bipartiteness ratio* is defined as

$$\beta^s(V_1, V_2) := \frac{2\Big(|E^-(V_1)| + |E^-(V_2)| + |E^+(V_1, V_2)|\Big) + |\partial(V_1 \sqcup V_2)|}{\text{vol}(V_1 \sqcup V_2)}.$$

We recall that the *signed Cheeger constant* of (Γ, s) (see also (4.6.27)) is

$$h^s := \min_{(V_1, V_2) \neq (\emptyset, \emptyset)} \beta^s(V_1, V_2),$$

where the minimum is taken over all possible sub-bipartitions of V. Moreover, by Theorem 4.6.2,

Theorem 8.2.1. *Given a signed graph (Γ, s), we have*

$$\frac{\lambda_1(\Delta^s)}{2} \leq h^s \leq \sqrt{2\lambda_1(\Delta^s)}.$$

More generally, the *k-way signed Cheeger constant* has been introduced in [11] as

$$h_k^s := \min_{\{(V_{2i-1}, V_{2i})\}_{i=1}^k} \max_{1 \leq i \leq k} \beta^s(V_{2i-1}, V_{2i}),$$

where the minimum is taken over the set of all possible k pairs of disjoint sub-bipartitions $(V_1, V_2), (V_3, V_4), \ldots, (V_{2k-1}, V_{2k})$. In particular, we note that h_k^s is switching invariant.

With these concepts, following [11], we can generalize and put into perspective the multiway Cheeger inequalities for simple graphs by Lee, Oveis Gharan, and Trevisan [180]. Before stating the main result, we need the following preliminary lemmata.

Lemma 8.2.1. *Let (Γ, s) be a signed graph. Let $k \in \{1, \ldots, N\}$ and let $f_1, \ldots, f_k$ be an orthonormal system of eigenfunctions corresponding to $\lambda_1(\Delta^s), \ldots, \lambda_k(\Delta^s)$. Given a vertex v, let*

$$F(v) := (f_1(v), \ldots, f_k(v)) \in \mathbb{R}^k,$$

and let $V_F := \{v \in V : F(v) \neq (0, \ldots, 0)\}$. Let $\| \cdot \|$ denote the Euclidean norm in $\mathbb{R}^k$, and consider the pseudometric

$$d_F(v, v') := \min \left\{ \left\| \frac{F(v)}{\|F(v)\|} - \frac{F(v')}{\|F(v')\|} \right\|, \left\| \frac{F(v)}{\|F(v)\|} + \frac{F(v')}{\|F(v')\|} \right\| \right\}.$$

Then, there exist k nonempty, mutually disjoint subsets $S_1, \ldots, S_k \subset V_F$, and a universal constant $C > 0$ (which is the same for all signed graphs), such that $d_F(S_i, S_j) \geq k^{-\frac{5}{2}}/C$, for all $i \neq j$, and

$$\sum_{v \in S_i} \deg v \cdot \|F(v)\|^2 \geq \frac{1}{2k} \sum_{v \in V} \deg v \cdot \|F(v)\|^2,$$

for all $i \in \{1, \ldots, k\}$.

For the *proof*, which is based on random partition theory, we refer to [11, 187].

Lemma 8.2.2. *Consider the setting of Lemma 8.2.1. For any subset $S \subset V$, and for any $\varepsilon > 0$, we define the cut-off function $\eta_S : V \to \mathbb{R}$ by*

$$\eta_S(v) := \begin{cases} 0, & \text{if } F(v) = 0, \\ \max\left\{0, 1 - \frac{1}{\varepsilon} d_F(v, S)\right\}, & \text{otherwise.} \end{cases}$$

Then, for any $0 < \varepsilon < 2$, and for any $\{v, v'\} \in E$, we have that

$$\|\eta_S(v)F(v) - s(vv')\eta_S(v')F(v')\| \leq \left(1 + \frac{2}{\varepsilon}\right) \|F(v) - s(vv')F(v')\|.$$

For the *proof* we refer to [11].

Theorem 8.2.2. *There exists a universal constant C such that, for any signed graph (Γ, s) and for any $k \in \{1, \ldots, N\}$,*

$$\frac{\lambda_k(\Delta^s)}{2} \leq h_k^s \leq Ck^3 \sqrt{\lambda_k(\Delta^s)}.$$

Proof We first prove the lower bound. Let $\{(V_{2i-1}, V_{2i})\}_{i=1}^k$ be a k-sub-bipartition of V that realizes h_k^s, that is,

$$\max_{1 \leq i \leq k} \beta^s(V_{2i-1}, V_{2i}) = h_k^s,$$

and let $\mathbf{1}_{V_i}$ be the indicator function of V_i. For any

$$g \in \mathrm{span}(\mathbf{1}_{V_1} - \mathbf{1}_{V_2}, \ldots, \mathbf{1}_{V_{2k-1}} - \mathbf{1}_{V_{2k}}),$$

we can write

$$g(v) = \sum_{i=1}^k t_i(\mathbf{1}_{V_{2i-1}}(v) - \mathbf{1}_{V_{2i}}(v)),$$

for some $t_1, \ldots, t_k \in \mathbb{R}$. We then have

$$\begin{aligned}
&(g(v) - s(vv')g(v'))^2 \\
&\leq 2\left|g(v)^2\mathrm{sign}(g(v)) - s(vv')g(v')^2\mathrm{sign}(g(v'))\right| \\
&= 2\left|\sum_{i=1}^m t_i^2\mathrm{sign}(t_i)\left(\mathbf{1}_{V_{2i-1}}(v) - \mathbf{1}_{V_{2i}}(v) - s(vv')(\mathbf{1}_{V_{2i-1}}(v') - \mathbf{1}_{V_{2i}}(v'))\right)\right| \\
&\leq 2\sum_{i=1}^m t_i^2\left|\mathbf{1}_{V_{2i-1}}(v) - \mathbf{1}_{V_{2i}}(v) - s(vv')(\mathbf{1}_{V_{2i-1}}(v') - \mathbf{1}_{V_{2i}}(v'))\right|,
\end{aligned}$$

where the first inequality follows from the fact that

$$(a \pm b)^2 \leq 2|a^2\mathrm{sign}(a) \pm b^2\mathrm{sign}(b)|.$$

Therefore,

$$R(g) \leq 2 \cdot \frac{\sum_{i=1}^{m} t_i^2 \sum_{\{v,v'\}\in E} \left| \mathbf{1}_{V_{2i-1}}(v) - \mathbf{1}_{V_{2i}}(v) - s(vv')(\mathbf{1}_{V_{2i-1}}(v') - \mathbf{1}_{V_{2i}}(v')) \right|}{\sum_{i=1}^{m} t_i^2 \sum_{v\in V} \left| \mathbf{1}_{V_{2i-1}}(v) - \mathbf{1}_{V_{2i}}(v) \right|}$$

$$\leq 2 \max_{i=1,\ldots,k} \frac{\sum_{\{v,v'\}\in E} \left| \mathbf{1}_{V_{2i-1}}(v) - \mathbf{1}_{V_{2i}}(v) - s(vv')(\mathbf{1}_{V_{2i-1}}(v') - \mathbf{1}_{V_{2i}}(v')) \right|}{\sum_{v\in V} \left| \mathbf{1}_{V_{2i-1}}(v) - \mathbf{1}_{V_{2i}}(v) \right|}$$

$$= 2 \max_{1\leq i \leq k} \beta^s(V_{2i-1}, V_{2i})$$

$$= 2 h_k^s.$$

By the min–max principle, we obtain that $\lambda_k^s \leq 2 h_k^s$. This proves the lower bound,

$$\frac{\lambda_k(\Delta^s)}{2} \leq h_k^s.$$

To prove the upper bound, consider the setting of Lemma 8.2.1. We have that

$$\lambda_k(\Delta^s) = \max_{i=1,\ldots,k} R(f_i)$$

$$\geq \frac{\sum_{i=1}^{k} \sum_{v\sim v'} (f_i(v) - s(vv') f_i(v'))^2}{\sum_{i=1}^{k} \sum_{v\in V} \deg(v) f_i(v)^2}$$

$$= \frac{\sum_{v\sim v'} \|F(v) - s(vv') F(v')\|^2}{\sum_{v\in V} \deg(v) \|F(v)\|^2}$$

$$:= R(F).$$

Let now $\varepsilon := k^{-\frac{5}{2}}/C$. Then, by Lemma 8.2.2,

$$\sum_{\{v,v'\}\in E} \|\eta_{S_i}(v) F(v) - s(vv') \eta_{S_i}(v') F(v')\|^2$$

$$\leq \left(1 + \frac{2}{\varepsilon}\right)^2 \sum_{\{v,v'\}\in E} \|F(v) - s(vv') F(v')\|^2.$$

Moreover, by Lemma 8.2.1,

$$\sum_{v\in V} \deg v \cdot \|\eta_{S_i}(v) F(v)\|^2 \geq \sum_{v\in S_i} \deg v \cdot \|F(v)\|^2 \geq \frac{1}{2k} \sum_{v\in V} \deg v \cdot \|F(v)\|^2.$$

Therefore, for $i = 1,\ldots,k$,

$$R(\eta_{S_i} F) \leq 2k(1 + C k^{\frac{5}{2}})^2 R(F) \leq C' k^6 R(F) \leq C' k^6 \lambda_k(\Delta^s).$$

We now write

$$\eta_{S_i} F = (g_i^1,\ldots,g_i^k) : V \to \mathbb{R}^k,$$

and we observe that we can always find $j_i \in \{1, \ldots, k\}$ such that $R(g_i^{j_i}) \leq R(\eta_{S_i} F)$, for all $i = 1, \ldots, k$. Now, similarly to (6.4.10) in the proof of Theorem 6.4.1, one can show that there exist disjoint subsets $\Omega_i^+, \Omega_i^- \subset \mathrm{supp}(g_i^{j_i})$ such that

$$\beta^s(\Omega_i^+, \Omega_i^-) \leq \sqrt{2R(g_i^{j_i})}.$$

As a consequence,

$$
\begin{aligned}
h_k^s &\leq \max_{i=1,\ldots,k} \beta^s(\Omega_i^+, \Omega_i^-) \\
&\leq \max_{i=1,\ldots,k} \sqrt{2R(g_i^{j_i})} \\
&\leq \max_{i=1,\ldots,k} \sqrt{2R(\eta_{S_i} F)} \\
&\leq C k^3 \sqrt{\lambda_k(\Delta^s)},
\end{aligned}
$$

for some constant $C > 0$. □

8.3 Cheeger Inequalities on Simplicial Complexes

This section addresses the fundamental problem of generalizing the classical Cheeger inequalities that are known for graphs to simplicial complexes. Estimating the first nontrivial eigenvalue of the Hodge Laplacian on a simplicial complex is a major long-standing open problem that has been formulated in the field of expander graph theory. Also, the analogue in Riemannian geometry, which is to find Cheeger inequalities for differential k-forms, is still far from being understood and solved.

The first issue in any attempt of extending Cheeger-type inequalities to higher-order structures, like simplicial complexes or differential k-forms, is to identify the right quantities to be estimated. Here, we shall focus on this problem and build upon Sections 4.8 and 8.2. The key is to convert a Cheeger problem for higher-dimensional simplices into one for signed graphs. For this purpose, we shall describe a relation between simplicial complexes and signed graphs that will allow us to interpret the Laplace operator Δ_d^{up} as the Laplacian of a signed graph.

We first recall some constructions from Section 2.5.1. We consider a simplicial complex Σ, and we denote the collection of its d-dimensional simplices by Σ_d. When σ and σ' are two orientations on the same underlying d-simplex, we write $\mathrm{sgn}(\sigma, \sigma') = \pm 1$ when those orientations agree/differ. And $\partial \sigma$ is the *cellular boundary* of a simplex; that is, the collection of its facets. When the simplices ρ carry weights $w(\rho)$, the *degree* of a d-simplex σ of Σ is

$$\deg \sigma := \sum_{\rho \in \Sigma_{d+1}(\Sigma) : \sigma \in \partial \rho} w(\rho). \tag{8.3.1}$$

Moreover, since we want to work with the normalized up-Laplacian, we assume that the weight function satisfies

$$w(\sigma) = \deg \sigma, \tag{8.3.2}$$

for every $\sigma \in \Sigma_d$, and that the weights of the facets of Σ are equal to 1. Under these conditions, the normalized combinatorial up-Laplacian, according to (2.5.21), is given by

$$(\Delta_{up}^d f)([\sigma]) = f([\sigma])$$

$$+ \frac{1}{\deg \sigma} \sum_{\sigma' \in \Sigma_d : \sigma' \neq \sigma, \sigma, \sigma' \in \partial \rho} w(\rho) \operatorname{sgn}([\sigma], \partial[\rho]) \operatorname{sgn}([\sigma'], \partial[\rho]) f([\sigma']). \tag{8.3.3}$$

Recalling the constructions of Section 4.8, we can express Δ_d^{up} in terms of the Laplacian of an associated signed graph, rewriting it as

$$(\Delta_d^{up} f)([\sigma]) = f([\sigma]) - \frac{1}{\deg \sigma} \sum_{\substack{\sigma' \in \Sigma_d : \sigma \neq \sigma', \\ \sigma, \sigma' \in \partial \rho}} w(\rho) s([\sigma], [\sigma']) f([\sigma']), \tag{8.3.4}$$

with

$$s([\sigma], [\sigma']) := -\operatorname{sgn}([\sigma], \partial[\rho]) \operatorname{sgn}([\sigma'], \partial[\rho]). \tag{8.3.5}$$

This naturally suggests a signed graph (Γ_d, s). Its vertex set consists of the d-simplices of our simplicial complex, and there is an edge $\sigma \sim \sigma'$ between vertices $\sigma \neq \sigma'$ if there exists a $(d+1)$-simplex ρ in Σ with $\sigma, \sigma' \in \partial \rho$. By Theorem 4.8.2 and by evaluating the weights $w(\rho)$ in (8.3.4), we have that the Laplacian Δ_d^{up} in (8.3.4) is related to the Laplacian $\Delta_{(\Gamma_d, s)}$ of the signed graph (Γ_d, s) via

$$\Delta_d^{up} = (d+1)\Delta_{(\Gamma_d, s)} - d \operatorname{Id}. \tag{8.3.6}$$

Remark 8.3.1. A similar construction is used in [98] to define the signed adjacency matrix of a triangulation on a surface.

By (8.3.6), in particular, the eigenvalues μ_j of Δ_d^{up} and the eigenvalues λ_j of $\Delta_{(\Gamma_d, s)}$ are related by

$$\mu_j = (d+1)\lambda_j - d. \tag{8.3.7}$$

Hence, since the eigenvalues of $\Delta_{(\Gamma_d, s)}$ lie in the interval $[0, 2]$, those of Δ_d^{up} lie in the interval $[0, d + 2]$. In fact, since $\mu_j \geq 0$ in (8.3.7), the eigenvalues of $\Delta_{(\Gamma_d, s)}$ are $\geq d/(d + 1)$. Equality holds if and only if there is some nontrivial f with $\delta^d f = 0$. More precisely, the multiplicity of the eigenvalue $d/(d + 1)$ of $\Delta_{(\Gamma_d, s)}$ equals the dimension of the kernel of the coboundary operator δ^d. In particular, for $d > 0$, the graph (Γ_d, s) is never balanced.

Now, given disjoint $A, A' \subset \Sigma_d$, let

$$|E^+(A, A')| := \#\{\{\tau, \tau'\} : \tau \in A, \tau' \in A', \mathrm{sgn}([\tau], [\tau']) = 1\},$$

and let

$$|E^-(A)| := \#\{\{\tau, \tau'\} : \tau, \tau' \in A, \mathrm{sgn}([\tau], [\tau']) = -1\}.$$

Let also

$$\beta(A, A') := \frac{2\left(|E^-(A)| + |E^-(A')| + |E^+(A, A')|\right) + |\partial(A \sqcup A')|}{\mathrm{vol}\,(A \sqcup A')},$$

where $|\partial A|$ is the number of the edges of (Γ_d, s) that cross A and $\Sigma_d \setminus A$,

$$\mathrm{vol}\,(A) := \sum_{\tau \in A} \deg \tau,$$

and $\deg \tau := \#\{\sigma \in \Sigma_{d+1} : \tau \subset \sigma\}$. We define the kth *Cheeger constant* of Σ_d as

$$h_k(\Sigma_d) := \min_{\substack{A_1, \ldots, A_{2k} \\ \text{disjoint, in } \Sigma_d}} \max_{1 \leq i \leq k} \beta(A_{2i-1}, A_{2i}).$$

In order to have $h_k(\Sigma_d) = 0$, we need to find $2k$ disjoint sets $A_1, \ldots, A_{2k}$ such that $|E^-(A_i)| = 0$ and $|E^+(A_{2i-1}, A_{2i})| = 0$ for all of them. Thus, all edges inside any A_i have to be positive, whereas all edges between A_{2i-1} and A_{2i} have to be negative, and also $A_{2i-1} \cup A_{2i}$ has to be disconnected from the rest of the graph so that there are no boundary terms. By Lemma 4.6.3, this means that the graph $G_{up}^-(\Sigma_d) := (\Gamma_d, -s)$ with the opposite signs has at least k balanced connected components.

Remark 8.3.2. For $d = 0$, the constant $h_k(\Sigma_0)$ reduces to the k-way Cheeger constant (7.2.7) of a graph [180].

Theorem 8.3.1. *For every simplicial complex Σ and for every $d \geq 0$,*

$$\frac{h_1(\Sigma_d)^2}{2(d + 1)} \leq d + 2 - \lambda_n(\Delta_d^{up}) \leq 2h_1(\Sigma_d), \qquad (8.3.8)$$

where $n = \#\Sigma_d$. Moreover, there exists a universal constant C such that, for any simplicial complex Σ and for any $k \geq 1$,

$$\frac{h_k(\Sigma_d)^2}{Ck^6(d+1)} \leq d + 2 - \lambda_{n+1-k}(\Delta_d^{up}) \leq 2h_k(\Sigma_d). \qquad (8.3.9)$$

Proof We first observe that

$$d + 2 - \frac{\sum\limits_{\sigma \in \Sigma_{d+1}} \left(\sum_{\tau \in \Sigma_d, \tau \subset \sigma} \operatorname{sgn}([\tau], \partial[\sigma])f(\tau)\right)^2}{\sum_{\tau \in \Sigma_d} \deg_\tau f(\tau)^2}$$

$$= (d+1)\frac{\sum_{[\tau] \sim [\tau']} (f(\tau) - \operatorname{sgn}(\tau, \tau')f(\tau')))^2}{\sum_{\tau \in \Sigma_d} \widetilde{\deg}_\tau f(\tau)^2},$$

where $[\tau] \sim [\tau']$ represents an edge in $G_{up}^-(\Sigma_d)$ and $\widetilde{\deg}_\tau := (d+1)\deg_\tau$ is the degree of τ in $G_{up}^-(\Sigma_d)$. It follows that

$$d + 2 - \lambda_{n-i+1}(\Delta_d^{up}) = (d+1)\lambda_i(\Delta(G_{up}^-(\Sigma_d))),$$

for $i = 1, \ldots, n$. Moreover, since $h_k(\Sigma_d)/(d+1)$ is the kth Cheeger constant of the signed graph $G_{up}^-(\Sigma_d)$, by Theorem 8.2.1, we have

$$\frac{\lambda_1(\Delta(G_{up}^-(\Sigma_d)))}{2} \leq \frac{h_1(\Sigma_d)}{d+1} \leq \sqrt{2\lambda_1(\Delta(G_{up}^-(\Sigma_d)))}.$$

And by Theorem 8.2.2, there exists a universal constant C such that, for any signed graph and for any $k \geq 1$,

$$\frac{\lambda_k(\Delta(G_{up}^-(\Sigma_d)))}{2} \leq \frac{h_k(\Sigma_d)}{d+1} \leq Ck^3\sqrt{\lambda_k(\Delta(G_{up}^-(\Sigma_d)))}.$$

Altogether, we obtain

$$\frac{d + 2 - \lambda_n(\Delta_d^{up})}{2} \leq h_1(\Sigma_d) \leq \sqrt{2(d+1)(d+2-\lambda_n(\Delta_d^{up}))},$$

and

$$\frac{d + 2 - \lambda_{n+1-k}(\Delta_d^{up})}{2} \leq h_k(\Sigma_d) \leq Ck^3\sqrt{(d+1)(d+2-\lambda_{n+1-k}(\Delta_d^{up}))}.$$

This completes the proof of (8.3.8) and (8.3.9). $\qquad\square$

As a consequence of Theorem 8.3.1, we have that $\lambda_n(\Delta_d^{up}) = d + 2$ if and only if $h_1(\Sigma_d) = 0$ and if and only if $G_{up}^-(\Sigma_d)$ has a balanced component.

Moreover, Theorem 8.3.1 provides the estimates for the spectral gap from $d + 2$. The more important estimate, however, is the one for the spectral gap from 0, namely, the Cheeger-type estimates for the first nontrivial eigenvalue of the Hodge Laplacian. This was proposed as an open problem in the original paper by Cheeger [56], in the setting of differential k-forms on a closed Riemannian manifold. Its discrete version, a Cheeger-type inequality on simplicial complexes, has also been a long-standing open problem in the area of high-dimensional expanders [84, 121, 194, 225, 254].

While some partial answers have been proposed and developed in the literature, it seems that none of them gives a complete solution to the problem. In fact, there are many different definitions of Cheeger constants on simplicial complexes. For example, an upper Cheeger-like bound for the Cheeger constant has been suggested by Parzanchevski, Rosenthal, and Tessler [121, 225]. In particular, in the field of higher-dimensional expanders, people use the so-called $\mathbb{Z}_2$-*expander* to indicate the Cheeger constants on a simplicial complex.

Steenbergen, Klivans, and Mukherjee [254] introduced the Cheeger constant

$$h^d(\Sigma) := \min_{\phi \in C^d(\Sigma, \mathbb{Z}_2) \setminus \mathrm{im}\,\delta} \frac{\|\delta\phi\|}{\min_{\psi \in \mathrm{im}\,\delta} \|\phi + \psi\|},$$

for all $d \geq 0$. Here, $\|\cdot\|$ is the Hamming norm on $C^d(\Sigma, \mathbb{Z}_2)$, that is, the l^1-norm on $\mathbb{Z}_2^n$, where $n = \#\Sigma_d$. This constant satisfies

$$h^d(\Sigma) = 0 \iff \tilde{H}^d(\Sigma, \mathbb{Z}_2) \neq 0.$$

According to the examples and theorems in [84, 121, 122, 184, 205, 225, 254], all Cheeger constants defined using cohomology (or homology) with $\mathbb{Z}_2$-coefficients cannot satisfy both Cheeger inequalities as in the graph setting. This follows from the fact that, on the one hand, we have the well-known relation

$$\lambda(\Delta_d^{up}) = 0 \iff \lambda(L_d^{up}) = 0 \iff \tilde{H}^d(\Sigma, \mathbb{R}) \neq 0,$$

for $d \geq 0$, while on the other hand, for $d \geq 1$, $\tilde{H}^d(\Sigma, \mathbb{R}) \neq 0$ is not equivalent to $\tilde{H}^d(\Sigma, \mathbb{Z}_2) \neq 0$.

We shall now present a Cheeger constant that appears in [160, 161] and solves the problem formulated above of generalizing both Cheeger inequalities. In particular, we shall propose four equivalent definitions, denoted by (D1)–(D4). In Theorem 8.3.2, we shall then prove that they are all equivalent.

(D1) A *(generalized) multiset* is a pair $\mu = (S, m)$, where S is a set (given by distinct elements), and $m: S \to \mathbb{Z}$ is an integer-valued function, which gives the *multiplicity* of the elements in the multiset. For simplicity, in this section, we shall simply call μ a *multiset*. We also let

$$|\mu| := \sum_{s \in S} |m(s)|$$

denote the *size* of μ.

In our context, we consider multisets on $S = \Sigma_d$, and we use the notation $\mu \subset_M \Sigma_d$ to indicate that $\mu = (\Sigma_d, m)$ and the multiplicities are in $\{-M, \ldots, 0, \ldots, M\}$. We also define the *coboundary*[1] $\partial^*_{d+1}\mu$ of μ as the multiset of all $(d+1)$-simplices that have a member of μ in its boundary, together with the corresponding multiplicities. Thus, an element $\sigma \in \Sigma_{d+1}$ has multiplicity

$$\sum_{\tau \in \Sigma_d} m(\tau)\mathrm{sgn}([\tau], \partial[\sigma]),$$

where $m(\tau)$ is the multiplicity of τ in μ. And the *support* of $\partial^*_{d+1}\mu$ then consists of all such simplices with nonzero multiplicity. We also define the *volume* of μ as

$$\mathrm{vol}\,(\mu) := \sum_{\tau \in \Sigma_d} \deg \tau |m(\tau)|. \tag{8.3.10}$$

Moreover, for $d \geq 0$, we let

$$h(\Sigma_d) := \min_{\substack{\mu \subset_M \Sigma_d: \\ \mu \neq \partial^*_d(\nu), \\ \forall \nu \subset_M \Sigma_{d-1}}} \frac{|\partial^*_{d+1}\mu|}{\min_{\substack{\mu' \neq 0: \\ \partial^*_{d+1}\mu' = \partial^*_{d+1}\mu}} \mathrm{vol}\,(\mu')}. \tag{8.3.11}$$

We have that $h(\Sigma_d)$ is constant when M is sufficiently large. For such sufficiently large M, we call $h(\Sigma_d)$ the *Cheeger constant* of Σ_d.

Remark 8.3.3. In this definition, we need to look at quotients of the form $|\partial^*_{d+1}\mu|/\mathrm{vol}\,(\mu')$, where μ has to be homologically nontrivial, that is, it cannot be of the form $\partial^*_d(\nu)$ and where $\partial^*_{d+1}\mu' = \partial^*_{d+1}\mu$. The new aspect here, compared to the graph case, is that we need to consider multisets of simplices. In fact, when we were looking at graphs, all edges had weight or degree 1, but now $(d+1)$-dimensional simplices may have degrees $\neq 1$.

(D2) We now describe the Cheeger constant in terms of the $\mathbb{Z}$-expander. In particular, we let

$$h(\Sigma_d) := \min_{\phi \in C^d(K, \mathbb{Z}) \setminus \mathrm{im}\,\delta} \frac{\|\delta\phi\|_1}{\min_{\psi \in \mathrm{im}\,\delta} \|\phi + \psi\|_{1, \deg}},$$

---

[1] This is different from the notion of coboundary used above when we defined the coboundary as the dual of the boundary operator in homology theory.

where

$$\|\phi\|_{1,\deg} := \sum_{\tau \in \Sigma_d} \deg \tau |\phi(\tau)|.$$

Remark 8.3.4. Compared to the definition of $h^d(\Sigma)$, here we use $\mathbb{Z}$-instead of $\mathbb{Z}_2$-coefficients and we use the (weighted) l^1-norm instead of the Hamming norm.

(D3) We define the *up 1-Laplacian eigenproblem* on Σ_d as the eigenvalue problem of the function pair $(\|B_{d+1}^\top \cdot \|_1, \| \cdot \|_{1,\deg})$, which consists in finding λ and $\mathbf{x}$ such that

$$\mathbf{0} \in \partial \|B_{d+1}^\top \mathbf{x}\|_1 - \lambda \partial \|\mathbf{x}\|_{1,\deg}. \tag{8.3.12}$$

Moreover, let

$$I_d := \dim \operatorname{im}(B_d^\top) + 1 = \operatorname{rank}(B_d) + 1,$$

and let $\lambda_{I_d}(\Delta_{d,1}^{up})$ be the smallest nontrivial eigenvalue of the up 1-Laplacian. Given a norm $\| \cdot \|$ on a real linear space with an inner product $\langle \cdot, \cdot \rangle$, we say that $\mathbf{x}$ is $\| \cdot \|$-*orthogonal* to $\mathbf{y}$ if there exists $\mathbf{u} \in \nabla \|\mathbf{x}\|$ satisfying $\langle \mathbf{u}, \mathbf{y} \rangle = 0$. We say $\mathbf{x}$ is $\| \cdot \|$-*orthogonal* to a nonempty set Y if $\mathbf{x}$ is $\| \cdot \|$-orthogonal to all $\mathbf{y} \in Y$.

We define

$$h(\Sigma_d) := \lambda_{I_d}(\Delta_{d,1}^{up}) = \min_{\mathbf{x} \perp^1 \operatorname{im}(B_d^\top)} \frac{\|B_{d+1}^\top \mathbf{x}\|_1}{\|\mathbf{x}\|_{1,\deg}},$$

where $\mathbf{x} \perp^1 \operatorname{im}(B_d^\top)$ indicates that $\mathbf{x}$ is $\| \cdot \|_{1,\deg}$-orthogonal to $\operatorname{im}(B_d^\top)$, that is, $\mathbf{u} \in \operatorname{im}(B_d^\top)^\perp$, for some $\mathbf{u} \in \nabla \|\mathbf{x}\|_{1,\deg}$.

Remark 8.3.5. In this definition, we draw upon the insights developed in Sections 6.4 and 7.2, which suggest to consider L^1- instead of L^2-quantities in order to get equalities instead of inequalities when relating Cheeger-type constants and eigenvalues.

Remark 8.3.6. If $\| \cdot \| = \| \cdot \|_2$ is the standard l^2-norm, then the $\| \cdot \|_2$-orthogonality coincides with the classical orthogonality with respect to the standard inner product.

(D4) The norm $\| \cdot \|_{1,\deg}$ on $C^d(\Sigma)$ induces a quotient norm on $C^d(\Sigma)/\operatorname{im}(\delta_{d-1})$, which we denote by $\| \cdot \|$, for simplicity. More precisely, given an equivalence class $[\mathbf{x}] \in C^d(\Sigma)/\operatorname{im}(\delta_{d-1})$, we let

$$\|[\mathbf{x}]\| := \inf_{\mathbf{x}' \in [\mathbf{x}]} \|\mathbf{x}'\|_{1,\deg}.$$

We use it to define

$$h(\Sigma_d) := \min_{0 \neq [\mathbf{x}] \in C^d(\Sigma)/\operatorname{im}(\delta_{d-1})} \frac{\|\delta_d \mathbf{x}\|_1}{\|[\mathbf{x}]\|}$$

$$= \min_{0 \neq [\mathbf{x}] \in C^d(\Sigma,\mathbb{Z})/\operatorname{im}(\delta_{d-1})} \frac{\|\delta_d \mathbf{x}\|_1}{\|[\mathbf{x}]\|}.$$

In the case when $\tilde{H}^d(\Sigma,\mathbb{R}) = 0$, we then have

$$h(\Sigma_d) = \min_{\mathbf{y} \in \operatorname{im}(\delta_d)} \frac{\|\mathbf{y}\|_1}{\|\mathbf{y}\|_{\mathrm{fil}}}$$

$$= \frac{1}{\max\limits_{\mathbf{y} \in \operatorname{im}(\delta_d)} \|\mathbf{y}\|_{\mathrm{fil}}/\|\mathbf{y}\|_1}$$

$$= \frac{1}{\|\delta_d^{-1}\|_{\mathrm{fil}}},$$

where

$$\|\mathbf{y}\|_{\mathrm{fil}} := \inf_{\mathbf{x} \in \delta_d^{-1}(\mathbf{y})} \|\mathbf{x}\|_{1,\deg}$$

is the filling norm of $\mathbf{y}$, and

$$\|\delta_d^{-1}\|_{\mathrm{fil}} := \max_{\mathbf{y} \in \operatorname{im}(\delta_d)} \frac{\|\mathbf{y}\|_{\mathrm{fil}}}{\|\mathbf{y}\|_1}$$

is called the *filling profile* by Gromov (see Section 2.3 in [117]).

Remark 8.3.7. This definition has a perspective that is in some sense opposite to (D1). Here, instead of first considering a set S and then looking at its boundary or coboundary, we first consider a boundary Y, and we then try to find the smallest set for which Y is a boundary.

Theorem 8.3.2. *The four definitions in (D1)–(D4) are all equivalent.*

Proof The idea of the proof is to show that (D1), (D2), and (D4) are all equivalent to (D3). In order to do so, we need to apply some technical results from [160].

Since $\operatorname{im}(B_d^{\top}) \subset \ker(B_{d+1}^{\top})$, by Theorem 2.1 in [160], we have that

$$\lambda_{I_d}(\Delta_{d,1}^{up}) = \inf_{\mathbf{x} \in \mathbb{R}^n \setminus \operatorname{im}(B_d^{\top})} \frac{\|B_{d+1}^{\top} \mathbf{x}\|_1}{\inf\limits_{\mathbf{z} \in \operatorname{im}(B_d^{\top})} \|\mathbf{x} + \mathbf{z}\|_{1,\deg}} \tag{8.3.13}$$

$$= \inf_{[\mathbf{x}] \in \mathbb{R}^n / \operatorname{im}(B_d^{\top})} \frac{\|B_{d+1}^{\top} \mathbf{x}\|_1}{\|[\mathbf{x}]\|} \tag{8.3.14}$$

$$= \inf_{\substack{\mathbf{x} \in \mathbb{R}^n: \\ \nabla \|\mathbf{x}\|_{1,\deg} \cap \operatorname{im}(B_d^\top)^\perp \neq \emptyset}} \frac{\|B_{d+1}^\top \mathbf{x}\|_1}{\|\mathbf{x}\|_{1,\deg}}, \tag{8.3.15}$$

where $n := \#\Sigma_d$,

$$[\mathbf{x}] := \left\{ \mathbf{y} \in \mathbb{R}^n : \mathbf{y} - \mathbf{x} \in \operatorname{im}(B_d^\top) \right\},$$

and

$$\|[\mathbf{x}]\| := \inf_{\mathbf{x}' \in [\mathbf{x}]} \|\mathbf{x} + \mathbf{z}\|_{1,\deg}. \tag{8.3.16}$$

In fact, the definition of the norm $\|\cdot\|$ on the quotient space $\mathbb{R}^n / \operatorname{im}(B_d^\top)$ implies that

$$\|[\mathbf{x}]\| = \inf_{\mathbf{z} \in \operatorname{im}(B_d^\top)} \|\mathbf{x} + \mathbf{z}\|_{1,\deg}.$$

Moreover, Proposition 2.3 in [160] yields

$$\|[\mathbf{x}]\| = \|\mathbf{x}\|_{1,\deg} \iff \nabla \|\mathbf{x}\|_{1,\deg} \bigcap \operatorname{im}(B_d^\top)^\perp \neq \emptyset.$$

Hence, the minimization problem

$$\inf_{\mathbf{x}' \in \mathbb{R}^n : \mathbf{x}' - \mathbf{x} \in \operatorname{im}(B_d^\top)} \|\mathbf{x}'\|_{1,\deg}$$

reaches its minimum for some points in the set

$$\left\{ \mathbf{x} \in \mathbb{R}^n : \nabla \|\mathbf{x}\|_{1,\deg} \bigcap \operatorname{im}(B_d^\top)^\perp \neq \emptyset \right\}.$$

This shows that (8.3.13), (8.3.14), and (8.3.15) coincide.

Now, let $\mathbf{x} \perp^1 \operatorname{im}(B_d^\top)$ denote that $\mathbf{u} \perp \operatorname{im}(B_d^\top)$ for some $\mathbf{u} \in \nabla \|\mathbf{x}\|_{1,\deg}$. Then,

$$\left\{ \mathbf{x} \in \mathbb{R}^n : \nabla \|\mathbf{x}\|_{1,\deg} \bigcap \operatorname{im}(B_d^\top)^\perp \neq \emptyset \right\}$$

in (8.3.15) can be reduced to

$$\left\{ \mathbf{x} \in \mathbb{R}^n : \mathbf{x} \perp^1 \operatorname{im}(B_d^\top) \right\},$$

as shown in (D3).

Similar to the proof of Proposition 6.3.1, we can apply Lemma 6.1.1 to derive that every eigenvalue of the function pair $(\|B_{d+1}^\top \cdot \|_1, \| \cdot \|_{1,\deg})$ has an eigenvector in the set of the extreme points associated with $(\|B_{d+1}^\top \cdot \|_1, \| \cdot \|_{1,\deg})$, since both $\|B_{d+1}^\top \cdot \|_1$ and $\| \cdot \|_{1,\deg}$ are piecewise linear.

Now, let k be the smallest integer such that the unit l^1-sphere,

$$\left\{ \mathbf{x} \in \mathbb{R}^n : \|\mathbf{x}\|_{1,\deg} = 1 \right\},$$

can be represented as the union $P_1 \cup \ldots \cup P_k$ of k convex polytopes of dimension $(n-1)$, on which both $\|B_{d+1}^\top \cdot \|_1$ and $\| \cdot \|_{1,\deg}$ are linear. Moreover, for $i = 1, \ldots, k$, let $\mathrm{Ext}(\|B_{d+1}^\top \cdot \|_1, \| \cdot \|_{1,\deg})$ be the vertex set of P_i. Then, $\mathrm{Ext}(\|B_{d+1}^\top \cdot \|_1, \| \cdot \|_{1,\deg})$ is a finite set, and its elements are called the *extreme points* determined by the function pair $(\|B_{d+1}^\top \cdot \|_1, \| \cdot \|_{1,\deg})$.

Since all entries of the matrix $B_{d+1}^\top$ and the degrees are rational numbers, by the theory of systems of linear equations, we have that

$$\mathrm{Ext}(\|B_{d+1}^\top \cdot \|_1, \| \cdot \|_{1,\deg}) \subset \mathbb{Q}^n.$$

Now, let M be a natural number that is greater than the least common multiple of all the denominators of the components of all points in $\mathrm{Ext}(\|B_{d+1}^\top \cdot \|_1, \| \cdot \|_{1,\deg})$. Then,

$$\mathrm{Ext}(\|B_{d+1}^\top \cdot \|_1, \| \cdot \|_{1,\deg}) \subset \{t\mathbf{x} : t \geq 0 \text{ and } \mathbf{x} \in \{-M, \ldots, -1, 0, 1, \ldots, M\}^n\},$$

and thus every eigenvalue has an eigenvector in

$$\{t\mathbf{x} : t \geq 0 \text{ and } \mathbf{x} \in \{-M, \ldots, -1, 0, 1, \ldots, M\}^n\}.$$

Since both $\|B_{d+1}^\top \cdot \|_1$ and $\| \cdot \|_{1,\deg}$ are positively one-homogeneous, we further derive that every eigenvalue has an eigenvector in the set

$$\{-M, \ldots, -1, 0, 1, \ldots, M\}^n.$$

Moreover, the minimizations in (8.3.13) and (8.3.15) can reach their minima at some points in $\{-M, \ldots, -1, 0, 1, \ldots, M\}^n$, while the minimization problem in (8.3.14) achieves its minima at some equivalence class $[\mathbf{x}]$, for some

$$\mathbf{x} \in \{-M, \ldots, -1, 0, 1, \ldots, M\}^n.$$

This implies that we can use $\{-M, \ldots, -1, 0, 1, \ldots, M\}^n$ instead of $\mathbb{R}^n$ in the constraints of the minimization problems in (8.3.13), (8.3.14), and (8.3.15). Since, furthermore,

$$\{-M, \ldots, -1, 0, 1, \ldots, M\}^n \subset \mathbb{Z}^n \subset \mathbb{R}^n,$$

it follows that we can also replace $\mathbb{R}^n$ by $\mathbb{Z}^n$, in the constraints of these three minimization problems.

By using $\mathbb{Z}^n$ instead of $\mathbb{R}^n$ in (8.3.13) and by equivalently replacing $\mathbb{Z}^n$ by $C^d(\Sigma, \mathbb{Z})$ and $\mathrm{im}(B_d^\top)$ by $\mathrm{im}\,\delta$, we obtain that (D2) is a reformulation of (8.3.13). Similarly, (D4) is a reformulation of (8.3.14). Moreover, if $\tilde{H}^d(\Sigma, \mathbb{R}) = 0$, then $\mathrm{im}(\delta_{d-1}) = \ker(\delta_d)$, which implies that

$$C^d(\Sigma)/\mathrm{im}(\delta_{d-1}) = C^d(\Sigma)/\ker(\delta_d) \cong \mathrm{im}(\delta_d).$$

Therefore, by considering the equivalence classes, we have that $\delta_d^{-1}(\mathbf{y}) = [\mathbf{x}]$ when $\mathbf{y} = \delta_d \mathbf{x}$. Note that, by (8.3.16), the filling norm

$$\|\mathbf{y}\|_{\mathrm{fil}} := \inf_{\mathbf{x} \in \delta_d^{-1}(\mathbf{y})} \|\mathbf{x}\|_{1,\deg}$$

coincides with $\|[\mathbf{x}]\|$, and

$$\|\mathbf{y}\|_1 = \|\delta_d \mathbf{x}\|_1 = \|B_{d+1}^\top \mathbf{x}\|_1.$$

Therefore, (D4) and (8.3.14) coincide.

By using $\{-M, \dots, -1, 0, 1, \dots, M\}^n$ instead of $\mathbb{R}^n$ in (8.3.13), we can similarly identify every multiset $\mu \subset_M \Sigma_d$ with a unique

$$\mathbf{x} \in \{-M, \dots, -1, 0, 1, \dots, M\}^n,$$

by identifying x_τ with $m(\tau)$ for any $\tau \in \Sigma_d$, where $m(\tau)$ is the multiplicity of τ in μ. For such μ and $\mathbf{x}$, we then have vol $(\mu) = \|\mathbf{x}\|_{1,\deg}$, and

$$|\partial_{d+1}^* \mu| = \|B_{d+1}^\top \mathbf{x}\|_1.$$

If $\tilde{H}^d(\Sigma, \mathbb{R}) \neq 0$, then $\mathrm{im}(B_d^\top)$ is a proper subset of $\ker(B_{d+1}^\top)$, and thus (8.3.13) is zero. In this case, there exists $\mu' \neq \emptyset$ such that

$$\partial_{d+1}^* \mu' = \partial_{d+1}^* \mu = \emptyset,$$

which means that (8.3.11) equals zero. If $\tilde{H}^d(\Sigma, \mathbb{R}) = 0$, then $\mathrm{im}(B_d^\top) = \ker(B_{d+1}^\top)$, and thus for such μ and $\mathbf{x}$ with $\mathbf{x} \notin \ker(B_{d+1}^\top)$, we have

$$\inf_{\mathbf{z} \in \mathrm{im}(B_d^\top)} \|\mathbf{x} + \mathbf{z}\|_{1,\deg} = \inf_{\substack{\mathbf{x}' \in \mathbb{R}^n : \\ \mathbf{x}' - \mathbf{x} \in \ker(B_{d+1}^\top)}} \|\mathbf{x}'\|_{1,\deg}$$

$$= \min_{\substack{S' \neq \emptyset : \\ \partial_{d+1}^* S' = \partial_{d+1}^* S}} \mathrm{vol}\,(S').$$

Therefore, (8.3.11) and (8.3.13) represent the same quantity, which has been denoted by $h(\Sigma_d)$. This proves the equivalence between (D1), (D2), (D3), and (D4). $\square$

Hence, the four equivalent definitions in (D1)–(D4) represent the same Cheeger constant $h(\Sigma_d)$ from different viewpoints. In particular,

1. (D1) uses the language of multisets and provides a combinatorial formulation of $h(\Sigma_d)$.
2. (D2) presents $h(\Sigma_d)$ as a $\mathbb{Z}$-expander and makes it clear that

$$h(\Sigma_d) = 0 \iff \tilde{H}^d(\Sigma, \mathbb{R}) \neq 0,$$

for all $d \geq 0$. In particular, as we have discussed, when the Cheeger constant is defined as a $\mathbb{Z}_2$-expander, then the Cheeger inequalities cannot hold on general simplicial complexes. However, as a $\mathbb{Z}$-expander, it is possible to get Cheeger inequalities.

3. (D3) shows that $h(\Sigma_d)$ coincides with the smallest nontrivial 1-Laplacian eigenvalue. This was already known in both the graph and the domain settings, which are treated in Sections 7.2 and 3.3, respectively.

4. Finally, (D4) reveals the nonobvious fact that $h(\Sigma_d)$ has a deep relation with Gromov's filling profile.

In addition, for sufficiently large numbers $M \in \mathbb{Z}_+$, if $\tilde{H}^d(\Sigma, \mathbb{R}) = 0$, then

$$h(\Sigma_d) = \min_{\substack{S \subset_M \Sigma_d \\ \partial_{d+1}^* S \neq \emptyset}} \frac{|\partial_{d+1}^* S|}{\min_{S': \, \partial_{d+1}^* S' = \partial_{d+1}^* S} \mathrm{vol}\,(S')} > 0.$$

For the case of $d = 0$, we can take $M = 1$, and then $h(\Sigma_0)$ reduces to the usual Cheeger constant on graphs.

For example, consider the usual tetrahedron Σ, which is viewed as a three-dimensional simplicial complex on $V = \{1, 2, 3, 4\}$. It is not difficult to check $h(\Sigma_0) = \frac{2}{3}$, $h(\Sigma_1) = 1$, and $h(\Sigma_2) = 1$.

The following result shows that the constant $h(\Sigma_d)$ satisfies Cheeger-type inequalities and therefore provides a solution to the problem that was formulated at the beginning of this section.

Theorem 8.3.3. *If* $\deg \tau > 0$ *for all* $\tau \in \Sigma_d$, *then*

$$\frac{h^2(\Sigma_d)}{\#\Sigma_{d+1}} \leq \lambda_{I_d}(\Delta_d^{up}) \leq \mathrm{vol}\,(\Sigma_d) h(\Sigma_d),$$

and, for all $p \geq 1$,

$$\frac{h^p(\Sigma_d)}{|\#\Sigma_{d+1}|^{p-1}} \leq \lambda_{I_d}(\Delta_{d,p}^{up}) \leq \mathrm{vol}\,(\Sigma_d)^{p-1} h(\Sigma_d).$$

Proof For simplicity, we let $h = h(\Sigma_d)$ and $\lambda = \lambda_{I_d}(\Delta_d^{up})$. To prove the first statement, it suffices to prove that

$$\frac{\min\limits_{\tau \in \Sigma_d} \deg \tau}{\#\Sigma_{d+1}} h^2 \leq \lambda \leq \mathrm{vol}\,(\Sigma_d) h^2, \tag{8.3.17}$$

by using the fact that, for all $\tau \in \Sigma_d$, we have $h \leq 1 \leq \deg \tau$.

Note that λ and h are the I_dth min–max eigenvalues of the dth normalized up Laplacian and the dth up 1-Laplacian, respectively. One can show that

$$\min_{\tau} \deg \tau \leq \frac{\|\mathbf{x}\|_{1,\deg}^2}{\|\mathbf{x}\|_{2,\deg}^2} \leq \sum_{\tau \in \Sigma_d} \deg \tau \quad \text{and} \quad 1 \leq \frac{\|B_{d+1}^\top \mathbf{x}\|_1^2}{\|B_{d+1}^\top \mathbf{x}\|_2^2} \leq \#\Sigma_{d+1} \tag{8.3.18}$$

by applying Cauchy's inequality

$$\|\mathbf{x}\|_{2,\deg}^2 = \sum_{\tau \in \Sigma_d} \deg \tau \cdot x_\tau^2 \geq \frac{\left(\sum_{\tau \in \Sigma_d} \deg \tau \, |x_\tau|\right)^2}{\sum_{\tau \in \Sigma_d} \deg \tau} = \frac{\|\mathbf{x}\|_{1,\deg}^2}{\sum_{\tau \in \Sigma_d} \deg \tau},$$

as well as

$$\#\Sigma_{d+1} \|B_{d+1}^\top \mathbf{x}\|_2^2 \geq \|B_{d+1}^\top \mathbf{x}\|_1^2 \geq \|B_{d+1}^\top \mathbf{x}\|_2^2,$$

and the inequality

$$\left(\sum_{\tau \in \Sigma_d} \deg \tau \, |x_\tau|\right)^2 \geq \sum_{\tau \in \Sigma_d} (\deg \tau)^2 \, x_\tau^2 \geq \min_\tau \deg \tau \sum_{\tau \in \Sigma_d} \deg \tau \cdot x_\tau^2.$$

By (8.3.18), we then obtain

$$\frac{1}{\sum\limits_{\tau \in \Sigma_d} \deg \tau} \frac{\|B_{d+1}^\top \mathbf{x}\|_2^2}{\|\mathbf{x}\|_{2,\deg}^2} \leq \frac{\|B_{d+1}^\top \mathbf{x}\|_1^2}{\|\mathbf{x}\|_{1,\deg}^2} \leq \frac{\#\Sigma_{d+1}}{\min\limits_\tau \deg \tau} \frac{\|B_{d+1}^\top \mathbf{x}\|_2^2}{\|\mathbf{x}\|_{2,\deg}^2}. \tag{8.3.19}$$

Since the kth min–max eigenvalue of the dth up 1-Laplacian is

$$\lambda_k(\Delta_{d,1}^{up}) = \inf_{S \in \mathrm{Gen}_k} \sup_{\mathbf{x} \in S \setminus \{\mathbf{0}\}} \frac{\|B_{d+1}^\top \mathbf{x}\|_1}{\|\mathbf{x}\|_{1,\deg}},$$

while the kth smallest eigenvalue of dth normalized up Laplacian is

$$\lambda_k(\Delta_d^{up}) = \inf_{S \in \mathrm{Gen}_k} \sup_{\mathbf{x} \in S \setminus \{\mathbf{0}\}} \frac{\|B_{d+1}^\top \mathbf{x}\|_2^2}{\|\mathbf{x}\|_{2,\deg}^2},$$

we can apply (8.3.19) to derive

$$\frac{1}{\sum\limits_{\tau \in \Sigma_d} \deg \tau} \lambda_k(\Delta_d^{up}) \leq \left(\lambda_k(\Delta_{d,1}^{up})\right)^2 \leq \frac{\#\Sigma_{d+1}}{\min\limits_\tau \deg \tau} \lambda_k(\Delta_d^{up}). \tag{8.3.20}$$

This is a general inequality relating $\lambda_k(\Delta_{d,1}^{up})$ and $\lambda_k(\Delta_d^{up})$. By taking $k = I_d$ in (8.3.20), and by observing that $h = h(\Sigma_d) = \lambda_{I_d}(\Delta_{d,1}^{up})$, we finally obtain

$$\frac{1}{\mathrm{vol}\,(\Sigma_d)} \lambda = \frac{1}{\sum\limits_{\tau \in \Sigma_d} \deg \tau} \lambda$$

$$\leq h^2$$

$$\leq \frac{\#\Sigma_{d+1}}{\min\limits_\tau \deg \tau} \lambda,$$

which completes the proof of (8.3.17). The case of $\Delta_{d,p}^{up}$ is similar, and thus we omit the proof of the second statement. □

For $d \geq 1$, we also define the *down Cheeger constant* as

$$h_{down}(\Sigma_d) := \min_{\mathbf{x} \perp^1 \text{im}(B_{d+1})} \frac{\|B_d\mathbf{x}\|_1}{\|\mathbf{x}\|_{1,\deg}} = \lambda_{I_{d+1}}(\Delta_{d,1}^{down}),$$

where

$$I_{d+1} := \dim \text{im}(B_{d+1}) + 1 = \text{rank}(B_{d+1}) + 1.$$

Remark 8.3.8. Let Σ be a d-dimensional *combinatorial manifold*, that is, a topological manifold that has a triangulation as a simplicial complex. As a manifold, we assume that Σ is connected and has no boundary. We then have that B_{d+1} is a $|\Sigma_d| \times 1$ matrix of rank 1, therefore

$$I_{d+1} = \text{rank}(B_{d+1}) + 1 = 1 + 1 = 2.$$

Therefore, in particular, $\lambda_{I_{d+1}} = \lambda_2$. Let now $\sigma \overset{down}{\sim} \sigma'$ indicate that σ and σ' are *down adjacent*, that is, they share a common facet. Then, the down adjacency relation induces a graph on Σ_d, and we can infer the following Cheeger inequality:

$$\frac{h_{down}^2(\Sigma_d)}{2} \leq \lambda_2(\Delta_d^{down}) \leq 2h_{down}(\Sigma_d).$$

Definition 8.3.1. Let M be a d-dimensional, orientable, compact, and closed Riemannian manifold, and let $c > 1$. A triangulation T of M is *c-uniform* if, for any two d-simplexes $\triangle$ and $\triangle'$ in the triangulation T,

$$\frac{1}{c} \leq \frac{\text{diam}(\triangle)}{\text{diam}(\triangle')} \leq c \qquad (8.3.21)$$

and

$$\frac{1}{c} \leq \frac{\text{diam}(\triangle)}{\text{vol}(\triangle)^{\frac{1}{d}}} \leq c. \qquad (8.3.22)$$

A set $\mathcal{T}$ of triangulations of M is *uniform* if there exist $N > 1$ and $c > 1$ such that, for each triangulation in $\mathcal{T}$, either its number of vertices is smaller than N, or it is c-uniform. In this case, these triangulations are simply said to be *uniform*, and the constants N and c are called the *uniform parameters* of these triangulations.

Thus, (8.3.21) requires that all the diameters of the simplices of the triangulation are compatible with each other, while (8.3.22) asks that they do not become too thin. To illustrate the concept of uniform triangulations, we consider the following example. Let $S \subset \mathbb{R}^3$ be the unit sphere with center at $\mathbf{0}$, and let T_1 be a regular tetrahedron inscribed in the sphere. Using central projection from $\mathbf{0}$ onto S, the one-dimensional skeleton of T_1 can be projected onto the sphere to obtain a geodesic triangulation of the sphere, which we denote by S_1.

By induction, having constructed T_n, we obtain T_{n+1} by connecting the mid-points of the edges of each triangle, and by centrally projecting the resulting triangulation onto the sphere, we obtain a triangulation S_{n+1}. The geometric simplicial complexes $\{T_n\}_{n=1}^\infty$ are then $2/\sqrt[4]{3}$-uniform, and the resulting trian-gulations $\{S_n\}_{n=1}^\infty$ are c-uniform, for some $c \leq 4$. Of course, this is a family of particularly nice triangulations, as uniform as they could possibly be. Other families of triangulations that one may encounter may not be so nice, but the concept of uniformity excludes degenerations within families that would make our estimates invalid.

Theorem 8.3.4. *Let M be an orientable, compact, and closed Riemannian manifold of dimension $(d + 1)$. There exists a constant C such that, for all sim-plicial complexes Σ that are combinatorially equivalent to some given uniform triangulations of M,*

$$\frac{h^2(\Sigma_d)}{C} \leq \lambda_{I_d}(\Delta_d^{up}) \leq C \cdot h(\Sigma_d). \tag{8.3.23}$$

Moreover, $h(\Sigma_d) > 0$ if and only if $H_1(\Sigma) = 0$ (or equivalently, $H_1(M) = 0$).

Proof By Theorem 8.3.3, we have that

$$\lambda_{I_d}(\Delta_d^{up}) = 0 \iff h(\Sigma_d) = 0.$$

Therefore, it suffices to assume that $h(\Sigma_d) > 0$, that is,

$$\tilde{H}^d(M) = \tilde{H}^d(\Sigma) = 0.$$

Since M and Σ are of dimension $(d + 1)$, Poincaré duality implies that

$$\tilde{H}_1(M) = \tilde{H}^d(M) = 0.$$

Moreover, since there are only finitely many simplicial complexes with less than a given number of vertices, the existence of the constant $C > 0$ in (8.3.23) for these simplicial complexes follows immediately from Theorem 8.3.3.

We may assume, without loss of generality, that M is simply connected. Since M is compact, the triangulation is c-uniform for some $c > 1$, and Σ_d has n elements for some positive integer n. Also, for any $\varepsilon > 0$, there exists $N > 0$ such that any c-uniform triangulation with at least N facets satisfies

$$\frac{1}{2c}\varepsilon' < \text{diam}(\triangle) < \varepsilon', \tag{8.3.24}$$

for all $\triangle$, and for some $\varepsilon' \in (0, \varepsilon)$. This follows since for sufficiently large N, there must be one simplex of small diameter, and then by (8.3.21), all diam-eters must be compatible with that one. We shall also consider the uniform triangulation as a uniform ε-net.

Now, we let $\varepsilon := \max_{\triangle \in T} \operatorname{diam}(\triangle)$ denote the maximum diameter of the facets in the triangulation T. In other words, we fix a small $\varepsilon > 0$ and a sufficiently large $N > 0$ in (8.3.24); we consider any simplicial complex Σ, which is combinatorially equivalent to a c-uniform triangulation with at least N faces, and we let $\varepsilon := \max_{\sigma \in \Sigma_{d+1}} \operatorname{diam}(|\sigma|)$.

We split the rest of the proof into multiple steps.

1. The down Cheeger constant satisfies

$$\frac{d+2}{4} h^2_{down}(\Sigma_{d+1}) \le \lambda_{I_d}(\Delta_d^{up}) \le (d+2) h_{down}(\Sigma_{d+1}).$$

This is derived by the Cheeger inequalities from Remark 8.3.8,

$$\frac{h^2_{down}(\Sigma_{d+1})}{2} \le \lambda_2(\Delta_{d+1}^{down}) \le 2 h_{down}(\Sigma_{d+1}),$$

and by the equality

$$\lambda_{I_d}(\Delta_d^{up}) = \frac{1}{2}\lambda_{I_d}(L_d^{up}) = \frac{1}{2}\lambda_2(L_{d+1}^{down}) = \frac{d+2}{2}\lambda_2(\Delta_{d+1}^{down}). \quad (8.3.25)$$

Here, $\lambda_{I_d}(L_d^{up})$ and $\lambda_2(L_{d+1}^{down})$ denote the smallest nonzero eigenvalues of the unnormalized up-Laplacian L_d^{up} and the unnormalized down-Laplacian L_{d+1}^{down}, respectively. By (4.8.6), we infer the second equality in (8.3.25). Moreover, the first and the last equality in (8.3.25) are based on the fact that the degrees appearing in the expression of the dth normalized up-Laplacian Δ_d^{up} are all equal to 2, while the degrees in the expression of the $(d+1)$-th normalized down-Laplacian Δ_{d+1}^{down} are all equal to $d+2$. In fact, since M is a compact manifold without boundary, each d-face is contained in exactly two $(d+1)$-faces, and every $(d+1)$-face contains exactly $(d+2)$ different d-faces.

2. The Cheeger constant $h(\Sigma_d)$ and the down Cheeger constant $h_{down}(\Sigma_{d+1})$ satisfy $h(\Sigma_d) \sim h_{down}(\Sigma_{d+1})$, that is, there exists a uniform constant $C > 1$ such that

$$\frac{1}{C} h_{down}(\Sigma_{d+1}) \le h(\Sigma_d) \le C h_{down}(\Sigma_{d+1}).$$

The proof of this step is further divided into the following two claims:

(a)

$$\frac{1}{\varepsilon} h_{down}(\Sigma_{d+1}) \sim h(M).$$

Proof Let G be the graph on $n := \#\Sigma_{d+1}$ vertices, which is located in the barycenters of all $(d+1)$-simplexes and is such that two vertices form an edge if and only if the corresponding two $(d+1)$-simplexes are down adjacent. We call G the *underlying graph of the triangulation*, and we observe that $h_{down}(\Sigma_{d+1})$ coincides with the Cheeger constant of G. Now, an approximation approach developed in [264, 265] implies that the Cheeger constant of a uniform triangulation approximates the Cheeger constant of the manifold when we equip the edges of the underlying graph of the triangulation with appropriate weights (related to ε). In fact, since G is the underlying graph of the uniform triangulation, we may assume that G is embedded in the manifold M, and that the distribution of the vertices of G is uniform. Then, according to such approximation results, by putting appropriate weights[2] on the edges of G, the Cheeger constant of G together with those edge weights approximates $h(M)$. (More precisely, the difference between $h(M)$ and the Cheeger constant of the weighted graph G is bounded by $h(M)/2$, whenever ε is sufficiently small.) We can then adopt such an approximation approach to derive that

$$\frac{1}{\varepsilon}h_{down}(\Sigma_{d+1}) \sim h(M).$$

(b) If $H_1(M) = 0$, then

$$\frac{1}{\varepsilon}h(\Sigma_d) \sim h(M).$$

Proof By Poincaré duality,

$$H_1(M) = 0 \iff H^d(M) = 0 \iff \ker(\delta_d) = \operatorname{im}(\delta_{d-1}),$$

since M is a compact closed manifold of dimension $(d + 1)$. Thus, by $\operatorname{im}(B_d^\top) = \ker(B_{d+1}^\top)$, $\delta_d := B_{d+1}^\top$, and by applying (8.3.13), we have that

$$h(\Sigma_d) = \min_{x \notin \ker(\delta_d)} \frac{\sum\limits_{\sigma \in \Sigma_{d+1}}\left|\sum\limits_{\tau \in \Sigma_d} \operatorname{sgn}(\tau, \partial\sigma)x_\tau\right|}{\min\limits_{z \in \ker(\delta_d)}\sum\limits_{\tau \in \Sigma_d} 2|x_\tau + z_\tau|} > 0.$$

Now, note that $h(\Sigma_d)$ equals the smallest nonzero eigenvalue of $\Delta_{d,1}^{up}$; therefore, the smallest nonzero eigenvalue of the nonlinear eigenproblem in (8.3.12). By the Duality Theorem on nonlinear eigenvalue problems (Lemma 2.5 in [160], Theorem 1 in [267]), the nonzero spectrum of the eigenproblem (8.3.12) coincides with the nonzero spectrum

---

[2] The weight of an edge $\{u, v\}$ is determined by the distance of u and v in M, which is about $O(\varepsilon)$.

of its *dual* eigenproblem:

$$\mathbf{0} \in \partial \|B_{d+1}\mathbf{y}\|^*_{1,\deg} - \lambda \partial \|\mathbf{y}\|^*_1, \tag{8.3.26}$$

where $\|\cdot\|^*_{1,\deg}$ indicates the dual norm of $\|\cdot\|_{1,\deg}$. Since every d-face is a facet of exactly two $(d+1)$-faces, we have that $\deg \equiv 2$, and (8.3.26) is equivalent to

$$\mathbf{0} \in \partial \frac{1}{2}\|B_{d+1}\mathbf{y}\|_\infty - \lambda \partial \|\mathbf{y}\|_\infty, \tag{8.3.27}$$

which can also be written as

$$\mathbf{0} \in \partial \frac{1}{2} \max_{\sigma \overset{\text{down}}{\sim} \sigma'} |y_\sigma - y_{\sigma'}| - \lambda \partial \max_{\sigma \in \Sigma_{d+1}} |y_\sigma|. \tag{8.3.28}$$

where we recall that $\sigma \overset{\text{down}}{\sim} \sigma'$ indicates that σ and σ' are down adjacent, that is, they share a common facet. By Theorem 2.1 in [160], the smallest nonzero eigenvalue of (8.3.28) can then be formulated as

$$\min_{\mathbf{y} \text{ nonconstant}} \frac{\displaystyle\max_{\sigma \overset{\text{down}}{\sim} \sigma'} \frac{1}{2}|y_\sigma - y_{\sigma'}|}{\displaystyle\min_{t \in \mathbb{R}} \max_{\sigma \in \Sigma_{d+1}} |y_\sigma + t|}.$$

As a consequence, $h(\Sigma_d)$ is the smallest nonzero eigenvalue of both the 1-Laplacian eigenvalue problem (8.3.12) and its dual eigenproblem (8.3.28). Therefore,

$$h(\Sigma_d) = \min_{\mathbf{y} \text{ nonconstant}} \frac{\displaystyle\max_{\sigma \overset{\text{down}}{\sim} \sigma'} \frac{1}{2}|y_\sigma - y_{\sigma'}|}{\displaystyle\min_{t \in \mathbb{R}} \max_{\sigma \in \Sigma_{d+1}} |y_\sigma + t|}.$$

Now, we have that

$$\min_{t \in \mathbb{R}} \max_{\sigma \in \Sigma_{d+1}} |y_\sigma + t| = \max_{\sigma \in \Sigma_{d+1}} |y_\sigma + \tilde{t}|,$$

where

$$\tilde{t} := -\frac{\displaystyle\max_{\sigma \in \Sigma_{d+1}} y_\sigma + \min_{\sigma \in \Sigma_{d+1}} y_\sigma}{2}.$$

By taking $\widetilde{y_\sigma} = y_\sigma + \tilde{t}$, one can check that $\min_\sigma \widetilde{y_\sigma} + \max_\sigma \widetilde{y_\sigma} = 0$. Hence,

$$h(\Sigma_d) = \min_{\min_\sigma \widetilde{y_\sigma} + \max_\sigma \widetilde{y_\sigma} = 0} \frac{\displaystyle\max_{\sigma \overset{\text{down}}{\sim} \sigma'} |\widetilde{y_\sigma} - \widetilde{y_{\sigma'}}|}{2 \max_\sigma |\widetilde{y_\sigma}|} = \frac{1}{\operatorname{diam}(G)},$$

where diam(G) indicates the diameter of G. Hence, we can rewrite $h(\Sigma_d)$ as $1/2$ times the smallest nontrivial eigenvalue of the ∞-Laplacian, which is equal to $1/\text{diam}(G)$. This argument is similar to Lemma 1.5 in [163].

Finally, since the triangulation is C-uniform, we have that

$$\frac{1}{\varepsilon} h(\Sigma_d) = \frac{1}{\varepsilon \cdot \text{diam}(G)} \sim \frac{1}{\text{diam}(M)}.$$

Hence,

$$\frac{1}{\varepsilon} h(\Sigma_d) \sim h(M).$$

The proof then follows by combining all the steps above. □

Remark 8.3.9. We conclude this section with some observations regarding the last theorem.

- The constant C in Theorem 8.3.4 depends on the parameters of the given triangulations of the ambient manifold. In future work, it would be good to find a new constant that only depends on the dimension d.
- Under the same condition as in Theorem 8.3.4, with the same proof, one can show that

$$\frac{\lambda_{k_d}(\Delta_{d,1}^{up})^2}{C} \le \lambda_{k_d}(\Delta_d^{up}) \le C \lambda_{k_d}(\Delta_{d,1}^{up}),$$

where $k_d := \dim \ker(\delta_d) + 1$. This inequality coincides with the Cheeger inequality in Theorem 8.3.4 if and only if $H_1(M) = 0$.
- With a modification of the proof of Theorem 8.3.4, we can show that $1/\text{diam}(G) \sim \lambda_2(G)$ whenever G can be uniformly embedded into such a manifold, where $\lambda_2(G)$ is the second smallest eigenvalue of the normalized Laplacian of G.
- Inspired by the approximation theory for Laplacians on triangulations of manifolds that has been proposed by Dodziuk in [81] and Dodziuk and Patodi in [83], we hope that it is possible to develop an approximation theory for Cheeger constants on triangulations of manifolds.

8.4 Cheeger Inequalities on Hypergraphs

In this section, we shall derive Cheeger inequalities for hypergraph p-Laplacians, with a theory that is very different from that for simplicial complexes that we developed in the previous section.

Fix a hypergraph $\Gamma = (V, E)$ on N vertices. Let $\Theta := \max_{e \in E} |e|$ and, given $v \in V$, let $d_v := \sum_{e \ni v} |e|/\Theta$. The Laplacian $\tilde{\Delta}_2$ that we consider in this section is the operator $\tilde{\Delta}_2 \colon C(V) \to C(V)$ such that, for given $f \colon V \to \mathbb{R}$ and $v \in V$,

$$\tilde{\Delta}_2 f(v) := \frac{1}{d_v} \sum_{e \ni v} \sum_{v' \in e} f(v'). \tag{8.4.1}$$

In the case of uniform hypergraphs, $\tilde{\Delta}_2$ coincides with the Laplacian for uniform oriented hypergraphs with only inputs from Chapters 2 and 4. In the general case, the two operators might be different. We let $\lambda_{N-1}(\tilde{\Delta}_2)$ denote the second largest eigenvalue of $\tilde{\Delta}_2$. Given a vertex measure $\mu_v > 0$ for each $v \in V$, we let $\mu(S) := \sum_{v \in S} \mu_v$ for each $S \subseteq V$, and we let

$$h := \min_{S \neq \emptyset, V} \frac{\sum_{e \in E} |e \cap S| \cdot |e \setminus S|}{\min\{\mu(S), \mu(V \setminus S)\}}.$$

Note that the quantity $\sum_{e \in E} |e \cap S| \cdot |e \setminus S|$ was introduced in [279] as a *hypergraph cut measure*, and it is also called the *clique expansion* [133].

In particular, if the hypergraph is Θ-uniform, for some integer $\Theta \geq 2$, and μ_v is the usual vertex degree, then h coincides with the Cheeger constant for uniform hypergraphs that were introduced in [214]. The main result in [214] states that, in this case,

$$\frac{h^2}{2(\Theta - 1)} \leq \Theta - \lambda_{N-1}(\tilde{\Delta}_2) \leq 2h. \tag{8.4.2}$$

And even though this might seem counterintuitive at first, the above inequalities generalize the classical Cheeger inequalities for simple graphs. In fact, the classical Cheeger inequalities involve the second eigenvalue $\lambda_2(\Delta)$ of the graph Laplacian Δ. But, by Definition 4.9.5 and Remark 4.9.2, we know that $\lambda_2(\Delta) = 2 - \lambda_{N-1}(\Delta^+)$, where Δ^+ is the signless Laplacian of the graph Γ, or, equivalently, the Laplacian of the oriented graph Γ^+, which is obtained from Γ by letting each vertex be an input for all edges in which it is contained. Hence, in particular, the classical Cheeger inequalities can be equivalently rewritten in terms of $\lambda_{N-1}(\Delta^+)$, which coincides with our $\lambda_{N-1}(\tilde{\Delta}_2)$.

We now generalize the inequalities in (8.4.2) as follows.

Theorem 8.4.1. *Let* $\Gamma = (V, E)$ *be a hypergraph. Then,*

$$ch^2 \leq \Theta - \lambda_{N-1}(\tilde{\Delta}_2) \leq Ch, \tag{8.4.3}$$

where

$$c := \min_{v \in V} \frac{\sum_{e \ni v} (|e| - 1)}{2 \sum_{e \ni v} |e|/\Theta} \cdot \min_{v \in V} \frac{\mu_v^2}{\left(\sum_{e \ni v} (|e| - 1)\right)^2},$$

and

$$C := \max_{v \in V} \frac{\sum\limits_{e \ni v} (|e| - 1)}{\sum\limits_{e \ni v} |e|/\Theta} \cdot \max_{v \in V} \frac{2\mu_v}{\sum\limits_{e \ni v} (|e| - 1)}.$$

Proof Let G be the weighted graph that has vertex set V, edge weight

$$w_{vv'} := \#\{e \in E : e \supset \{v, v'\}\},$$

and vertex weight

$$w_v = \sum_{v' \in V} w_{vv'} = \sum_{e \ni v}(|e| - 1).$$

Let $\hat{\lambda}_2$ denote the second eigenvalue of the normalized Laplacian on G and let $\hat{h}$ denote the standard Cheeger constant on G. Then,

$$\Theta - \lambda_{N-1}(\tilde{\Delta}_2) = \Theta - \max_{\dim(X) \geq 2} \min_{f \in X} \frac{\sum_{e \in E}(\sum_{v \in e} f(v))^2}{\sum_{v \in V} d_v f(v)^2}$$

$$= \min_{\dim(X) \geq 2} \max_{f \in X} \frac{\Theta \sum_{v \in V} d_v f(v)^2 - \sum_{e \in E}(\sum_{v \in e} f(v))^2}{\sum_{v \in V} d_v f(v)^2}$$

$$= \min_{\dim(X) \geq 2} \max_{f \in X} \frac{\sum_{v \in V} \sum_{e \ni v} |e| f(v)^2 - \sum_{e \in E}(\sum_{v \in e} f(v))^2}{\sum_{v \in V} d_v f(v)^2}$$

$$= \min_{\dim(X) \geq 2} \max_{f \in X} \frac{\sum_{e \in E} \sum_{v \in e} |e| f(v)^2 - \sum_{e \in E}(\sum_{v \in e} f(v))^2}{\sum_{v \in V} d_v f(v)^2}$$

$$= \min_{\dim(X) \geq 2} \max_{f \in X} \frac{\sum_{e \in E}(|e| \sum_{v \in e} f(v)^2 - (\sum_{v \in e} f(v))^2)}{\sum_{v \in V} d_v f(v)^2}$$

$$= \min_{\dim(X) \geq 2} \max_{f \in X} \frac{\sum_{e \in E} \sum_{\{v,v'\} \subset e}(f(v) - f(v'))^2}{\sum_{v \in V} d_v f(v)^2}$$

$$- \min_{\dim(X) \geq 2} \max_{f \in X} \frac{\sum_{\{v,v'\} \subset V} w_{vv'}(f(v) - f(v'))^2}{\sum_{v \in V} d_v f(v)^2}$$

$$= \min_{\dim(X) \geq 2} \max_{f \in X} \frac{\sum_{\{v,v'\} \subset V} w_{vv'}(f(v) - f(v'))^2}{\sum_{v \in V} w_v f(v)^2} \cdot \frac{\sum_{v \in V} w_v f(v)^2}{\sum_{v \in V} d_v f(v)^2}$$

$$\in [m\hat{\lambda}_2, M\hat{\lambda}_2] \subset [m\hat{h}^2/2, M2\hat{h}] \subset [m\tilde{c}^2 h^2/2, 2MC\tilde{h}],$$

where

$$m := \min_v \frac{w_v}{d_v}, \quad M := \max_v \frac{w_v}{d_v}, \quad \tilde{c} := \min_v \frac{\mu_v}{w_v}, \quad \text{and} \quad \tilde{C} := \max_v \frac{\mu_v}{w_v}.$$

In the preceding formulae, we use the inequality

$$m \leq \frac{\sum_{v \in V} w_v x_v^2}{\sum_{v \in V} d_v x_v^2} \leq M,$$

the Cheeger inequality $\hat{h}^2/2 \le \hat{\lambda}_2 \le 2\hat{h}$ for the graph G, and the relation $\tilde{c}h \le \hat{h} \le \tilde{C}h$, which follows by

$$
\hat{h} = \min_{S \neq \emptyset, V} \frac{\sum\limits_{v \in S, v' \notin S} w_{vv'}}{\min\left\{\sum\limits_{v \in S} w_v, \sum\limits_{v \in V \setminus S} w_v\right\}}
$$

$$
= \min_{S \neq \emptyset, V} \frac{\sum\limits_{e \in E} |e \cap S| \cdot |e \setminus S|}{\min\left\{\sum\limits_{v \in S} \mu_v, \sum\limits_{v \in V \setminus S} \mu_v\right\}} \cdot \frac{\min\left\{\sum\limits_{v \in S} \mu_v, \sum\limits_{v \in V \setminus S} \mu_v\right\}}{\min\left\{\sum\limits_{v \in S} w_v, \sum\limits_{v \in V \setminus S} w_v\right\}}.
$$

The claim follows by taking $c := m\tilde{c}^2/2$ and $C := 2M\tilde{C}$. □

Now, we also define a multiway version of the hypergraph Cheeger constant, as follows. Given $S \subseteq V$, let

$$
\psi(S) := \frac{\sum_{e \in E} |e \cap S| \cdot |e \setminus S|}{\mu(S)}.
$$

We define the *lth Cheeger constant* as

$$
h_l := \min_{\substack{S_1, \ldots, S_l \\ \text{pairwise disjoint}}} \max_{i=1,\ldots,l} \psi(S_i).
$$

By the higher-order Cheeger inequalities for graphs from Lee–Oveis Gharan–Trevisan [180], and by using a similar technique as in the preceding proof, we obtain:

Theorem 8.4.2. *There exists a universal constant C such that, for any $l = 1, \ldots, N$,*

$$
c_l h_l^2 \le \Theta - \lambda_{N-l+1} \le C_l h_l,
$$

where

$$
C_l := \max_v \frac{\sum\limits_{e \ni v} (|e| - 1)}{\sum\limits_{e \ni v} |e|/\Theta} \cdot \max_v \frac{\mu_v}{2 \sum\limits_{e \ni v} (|e| - 1)}, \quad and
$$

$$
c_l := \min_v \frac{\sum\limits_{e \ni v} (|e| - 1)}{C^2 l^4 \sum\limits_{e \ni v} |e|/\Theta} \cdot \min_v \frac{\mu_v^2}{\left(\sum\limits_{e \ni v} (|e| - 1)\right)^2}.
$$

In order to prove Theorem 8.4.3, we need the following preliminary lemma.

Lemma 8.4.1. *Let* $\Gamma = (V, E)$ *be a* Θ-*uniform hypergraph. Fix* $p \in (1, +\infty)$ *and fix* $e \in E$. *Then, there exist* $m_{p,\Theta} > 0$ *(if* $p \geq 2$*) and* $M_{p,\Theta} > 0$ *(if* $1 < p \leq 2$*) such that, for any* $\mathbf{x} \in \mathbb{R}^n$,

$$m_{p,\Theta} \cdot \sum_{\{v,v'\} \subset e} |x_v - x_{v'}|^p \leq |e|^{p-1} \sum_{v \in e} |x_v|^p - \left| \sum_{v \in e} x_v \right|^p,$$

$$M_{p,\Theta} \cdot \sum_{\{v,v'\} \subset e} |x_v - x_{v'}|^p \geq |e|^{p-1} \sum_{v \in e} |x_v|^p - \left| \sum_{v \in e} x_v \right|^p.$$

Proof Since Γ is Θ-uniform, we have $|e| = \Theta$ for any $e \in E$. And for a fixed $e \in E$, we can assume without loss of generality that $e = \{1, \cdots, \Theta\}$. Then

$$\sum_{\{v,v'\} \subset e} |x_v - x_{v'}|^p = \frac{1}{2} \sum_{i,j=1}^{\Theta} |x_i - x_j|^p.$$

It suffices to prove that

$$m'_{p,\Theta} := \inf_{\mathbf{x} \text{ nonconstant}} \frac{\Theta^{p-1} \sum_{i=1}^{\Theta} |x_i|^p - \left| \sum_{i=1}^{\Theta} x_i \right|^p}{\sum_{i,j=1}^{\Theta} |x_i - x_j|^p} > 0,$$

for $p \geq 2$, and

$$M'_{p,\Theta} := \sup_{\mathbf{x} \text{ nonconstant}} \frac{\Theta^{p-1} \sum_{i=1}^{\Theta} |x_i|^p - \left| \sum_{i=1}^{\Theta} x_i \right|^p}{\sum_{i,j=1}^{\Theta} |x_i - x_j|^p} < +\infty,$$

for $p \in (1, 2]$. In fact, we can then simply take $m_{p,\Theta} := 2m'_{p,\Theta}$ and $M_{p,\Theta} := 2M'_{p,\Theta}$ to prove Lemma 8.4.1.

Clearly,

$$\sum_{i,j-1}^{\Theta} |x_i - x_j|^p > 0 \iff \mathbf{x} \text{ is nonconstant.}$$

By Hölder's inequality,

$$\Theta^{p-1} \sum_{i=1}^{\Theta} |x_i|^p \geq \left| \sum_{i=1}^{\Theta} x_i \right|^p,$$

and equality holds if and only if $\mathbf{x}$ is constant. Therefore,

$$\sum_{i,j=1}^{\Theta} |x_i - x_j|^p > 0 \iff \Theta^{p-1} \sum_{i=1}^{\Theta} |x_i|^p - \left| \sum_{i=1}^{\Theta} x_i \right|^p > 0.$$

Now, given $\delta > 0$, for any vector $\mathbf{x}$ satisfying

$$\max_i x_i - \min_i x_i = \delta > 0,$$

we have that

$$\delta^p \leq \sum_{i,j=1}^{\Theta} |x_i - x_j|^p \leq \Theta^2 \delta^p.$$

Given $\mathbf{x} \neq \mathbf{0}$ such that $\mathbf{x} \perp \mathbf{1}$, $t \in \mathbb{R}$, and $p \geq 1$, let now

$$g(\mathbf{x},t,p) := \Theta^{p-1} \sum_{i=1}^{\Theta} |x_i + t|^p - \left| \sum_{i=1}^{\Theta} x_i + \Theta t \right|^p.$$

Since $\mathbf{x}$ is nonconstant and $p > 1$, by Hölder's inequality, we have that $g(\mathbf{x},t,p) > 0$. Moreover,

$$\partial_t g(\mathbf{x},t,p) = p\Theta^{p-1} \sum_{i=1}^{\Theta} |x_i + t|^{p-1} \text{sign}(x_i + t) - p\Theta |\Theta t|^{p-1} \text{sign}(\Theta t),$$

where we used the assumption $\mathbf{x} \perp \mathbf{1}$, that is, $\sum_{i=1}^{\Theta} x_i = 0$. Fix $p \geq 2$. If $t > \delta$, then $x_i + t > 0$ for any $i = 1,\ldots,\Theta$, and by Hölder's inequality, we have that $\partial_t g(\mathbf{x},t,p) \geq 0$. Similarly, if $t < -\delta$, then $\partial_t g(\mathbf{x},t,p) \leq 0$. Therefore, $\mathbb{R} \ni t \mapsto g(\mathbf{x},t,p)$ reaches its minimum at some point in $[-\delta,\delta]$, implying that

$$\min_{t \in \mathbb{R}} g(\mathbf{x},t,p) = \min_{-\delta \leq t \leq \delta} g(\mathbf{x},t,p)$$

is a continuous function on the compact set

$$\left\{ \mathbf{x} \in \mathbb{R}^{\Theta} : \sum_{i=1}^{\Theta} x_i = 0, \max_i x_i - \min_i x_i = \delta \right\}.$$

Hence,

$$\min_{\substack{\mathbf{x} \perp \mathbf{1}, \\ \max_i x_i - \min_i x_i = \delta}} \min_{t \in \mathbb{R}} g(\mathbf{x},t,p) > 0.$$

Thus,

$$\inf_{\substack{\mathbf{x} \text{ nonconstant}}} \frac{\Theta^{p-1} \sum_{i=1}^{\Theta} |x_i|^p - |\sum_{i=1}^{\Theta} x_i|^p}{\sum_{i,j=1}^{\Theta} |x_i - x_j|^p} = \inf_{\substack{\mathbf{x} \perp \mathbf{1}, \\ \max_i x_i - \min_i x_i = \delta}} \min_{t \in \mathbb{R}} \frac{g(\mathbf{x},t,p)}{\sum_{i,j=1}^{\Theta} |x_i - x_j|^p}$$

$$\geq \frac{1}{\Theta^2 \delta^p} \min_{\substack{\mathbf{x} \perp \mathbf{1}, \\ \max_i x_i - \min_i x_i = \delta}} \min_{t \in \mathbb{R}} g(\mathbf{x},t,p)$$

$$> 0.$$

Now,

$$g(\mathbf{x},t,2) = \sum_{\{i,j\} \subset \{1,\ldots,\Theta\}} (x_i - x_j)^2 \quad \text{and} \quad g(\mathbf{x},t,1) \geq 0.$$

Furthermore,

$$\partial_p g(\mathbf{x},t,p) = \frac{1}{\Theta} \sum_{i=1}^{\Theta} |\Theta x_i + \Theta t|^p \ln |\Theta x_i + \Theta t| - \left| \sum_{i=1}^{\Theta} x_i + \Theta t \right|^p \ln \left| \sum_{i=1}^{\Theta} x_i + \Theta t \right|.$$

Since $s \mapsto s^p \ln s$ is a convex and increasing function on $s \in (1,+\infty)$, by Jensen's inequality for convex functions, we have that $\partial_p g(\mathbf{x},t,p) > 0$ whenever $|t| > \delta + 1/\Theta$ and $p > 1$. Therefore,

$$g(\mathbf{x},t,1) < g(\mathbf{x},t,p) \le \sum_{\{i,j\} \subset \{1,\dots,\Theta\}} (x_i - x_j)^2,$$

whenever $|t| > \delta + 1/\Theta$ and $1 < p \le 2$. Consequently,

$$\sup_{\substack{\mathbf{x} \text{ nonconstant}}} \frac{\Theta^{p-1} \sum_{i=1}^{\Theta} |x_i|^p - |\sum_{i=1}^{\Theta} x_i|^p}{\sum_{i,j=1}^{\Theta} |x_i - x_j|^p} = \sup_{\substack{\mathbf{x} \perp \mathbf{1}, \\ \max_i x_i - \min_i x_i = \delta}} \max_{t \in \mathbb{R}} \frac{g(\mathbf{x},t,p)}{\sum_{i,j=1}^{\Theta} |x_i - x_j|^p}$$

$$\le \frac{1}{\delta^p} \max_{\substack{\mathbf{x} \perp \mathbf{1}, \\ \max_i x_i - \min_i x_i = \delta}} \max_{t \in \mathbb{R}} g(\mathbf{x},t,p)$$

$$< +\infty.$$

$\square$

We need Lemma 8.4.1 to show an inequality involving the Cheeger constant for uniform hypergraphs that was introduced in [214] and that we presented in the beginning of this section, as well as the hypergraph p-Laplacian $\tilde{\Delta}_p$ that was introduced in [157] and discussed in Section 7.4. In particular, we consider

$$\tilde{\lambda}_{N-1}(\tilde{\Delta}_p) = \sup_{\text{gen}(A) \ge 2} \inf_{\mathbf{x} \in A} \frac{\sum_{e \in E} |\sum_{v \in e} x_v|^p}{\sum_{v \in V} \deg(v) |x_v|^p},$$

that is, the second largest max–min eigenvalue of $\tilde{\Delta}_p$.

Theorem 8.4.3. *Let $\Gamma = (V,E)$ be a Θ-uniform hypergraph. Given $p \in (1,2]$, we have*

$$\Theta^{p-1} - \tilde{\lambda}_{N-1}(\tilde{\Delta}_p) \le C_{p,\Theta} h,$$

and given $p \in [2,+\infty)$, we have

$$c_{p,\Theta} h^p \le \Theta^{p-1} - \tilde{\lambda}_{N-1}(\tilde{\Delta}_p),$$

where the universal positive constants $c_{p,\Theta}$ and $C_{p,\Theta}$ only depend on p and Θ. In particular, in the case of $p = 2$, one can let $C_{2,\Theta} = 2$ and $c_{2,\Theta} = 2/2(\Theta - 1)$.

Proof Let G be the weighted graph that has vertex set V, edge weight

$$w_{vv'} := \#\{e \in E : e \supset \{v,v'\}\},$$

and vertex weight

$$w_v := \sum_{v' \in V} w_{vv'} = \sum_{e \ni v} (|e| - 1) = (\Theta - 1) \deg(v).$$

Let $\hat{\lambda}_2(\Delta_p)$ denote the second eigenvalue of the p-Laplacian of G and let $\hat{h}$ denote the Cheeger constant of G. Since $|e| = \Theta$ for all $e \in E$, we have

$$
\begin{aligned}
\Theta^{p-1} - \widetilde{\lambda}_{N-1}(\tilde{\Delta}_p) &= \Theta^{p-1} - \sup_{\text{gen}(A) \geq 2} \inf_{x \in A} \frac{\sum_{e \in E} | \sum_{i \in e} x_v|^p}{\sum_{v \in V} \deg(v)|x_v|^p} \\
&= \inf_{\text{gen}(A) \geq 2} \sup_{x \in A} \frac{\Theta^{p-1} \sum_{v \in V} \deg(v)|x_v|^p - \sum_{e \in E} | \sum_{v \in e} x_v|^p}{\sum_{v \in V} \deg(v)|x_v|^p} \\
&= \inf_{\text{gen}(A) \geq 2} \sup_{x \in A} \frac{\Theta^{p-1} \sum_{e \in E} \sum_{v \in e} |x_v|^p - \sum_{e \in E} | \sum_{v \in e} x_v|^p}{\sum_{v \in V} \deg(v)|x_v|^p} \\
&= \inf_{\text{gen}(A) \geq 2} \sup_{x \in A} \frac{\sum_{e \in E} \left(|e|^{p-1} \sum_{v \in e} |x_v|^p - | \sum_{v \in e} x_v|^p \right)}{\sum_{v \in V} \deg(v)|x_v|^p}.
\end{aligned}
$$

By Lemma 8.4.1,

$$
\begin{aligned}
\Theta^{p-1} - \widetilde{\lambda}_{N-1}(\tilde{\Delta}_p) &\leq \inf_{\text{gen}(A) \geq 2} \sup_{x \in A} \frac{\sum_{e \in E} M_{p,\Theta} \sum_{\{v,v'\} \subset e} |x_v - x_{v'}|^p}{\sum_{v \in V} \deg(v)|x_v|^p} \\
&= M_{p,\Theta}(\Theta - 1) \inf_{\text{gen}(A) \geq 2} \sup_{x \in A} \frac{\sum_{\{v,v'\} \subset V} w_{vv'} |x_v - x_{v'}|^p}{\sum_{v \in V} w_v |x_v|^p} \\
&= M_{p,\Theta}(\Theta - 1) \hat{\lambda}_2(\Delta_p) \\
&\leq M_{p,\Theta}(\Theta - 1) 2^{p-1} \hat{h} \\
&= M_{p,\Theta} 2^{p-1} h.
\end{aligned}
$$

Similarly,

$$
\begin{aligned}
\Theta^{p-1} - \widetilde{\lambda}_{N-1}(\tilde{\Delta}_p) &\geq m_{p,\Theta}(\Theta - 1) \hat{\lambda}_2(\Delta_p) \\
&\geq m_{p,\Theta}(\Theta - 1) 2^{p-1} \frac{\hat{h}^p}{p^p} \\
&= m_{p,\Theta} \cdot \frac{h^p}{p^p} \left(\frac{2}{\Theta - 1} \right)^{p-1}.
\end{aligned}
$$

Therefore, we can choose

$$c_{p,\Theta} = \frac{m_{p,\Theta} 2^{p-1}}{p^p (\Theta - 1)^{p-1}} \quad \text{and} \quad C_{p,\Theta} = 2^{p-1} M_{p,\Theta}.$$

For the case of $p = 2$ (see also Theorem 8.4.1), since

$$|e| \sum_{v \in e} x_v^2 - \left(\sum_{v \in e} x_v \right)^2 = \sum_{\{v,v'\} \subset e} (x_v - x_{v'})^2,$$

it follows that we can choose $m_{2,\Theta} = M_{2,\Theta} = 1$, $C_{2,\Theta} = 2$, and $c_{2,\Theta} = 1/2$ $(\Theta - 1)$. $\qquad\square$

Remark 8.4.1. For $p = 2$, Theorem 8.4.3 enhances the main result in [214]. Moreover, for $p > 2$,

$$\sup_{x \text{ nonconstant}} \frac{\Theta^{p-1} \sum_{i=1}^{\Theta} |x_i|^p - |\sum_{i=1}^{\Theta} x_i|^p}{\sum_{i,j=1}^{\Theta} |x_i - x_j|^p} = +\infty,$$

and for $1 < p < 2$,

$$\inf_{x \text{ nonconstant}} \frac{\Theta^{p-1} \sum_{i=1}^{\Theta} |x_i|^p - |\sum_{i=1}^{\Theta} x_i|^p}{\sum_{i,j=1}^{\Theta} |x_i - x_j|^p} = 0.$$

Hence, in these cases, we only obtain one bound, namely

$$c_{p,\Theta} h^p \leq \Theta^{p-1} - \widetilde{\lambda}_{N-1}(\widetilde{\Delta}_p) \text{ if } p > 2,$$

and

$$\Theta^{p-1} - \widetilde{\lambda}_{N-1}(\widetilde{\Delta}_p) \leq C_{p,\Theta} h \text{ if } 1 \leq p < 2.$$

Similarly to Theorem 8.4.3, we can prove the following

Theorem 8.4.4. *Let Γ be a hypergraph such that $2 \leq |e| \leq \Theta$, for each edge e. Then, given $p \in (1, +\infty)$,*

$$c_{p,\Theta} h^p \leq \Theta^{p-1} - \widetilde{\lambda}_{N-1}(\widetilde{\Delta}_p) \text{ when } p \geq 2,$$

$$\Theta^{p-1} - \widetilde{\lambda}_{N-1}(\widetilde{\Delta}_p) \leq C_{p,\Theta} h \text{ when } 1 < p \leq 2.$$

where the universal positive constants $c_{p,\Theta}$ and $C_{p,\Theta}$ only depend on p and Θ. In the particular case of $p = 2$, the constants satisfy $C_{2,\Theta} \leq 2$ and $c_{2,\Theta} \geq 1/2(\Theta - 1)$.

We conclude this chapter by presenting Cheeger inequalities for an oriented hypergraph $\Gamma = (V, E, \varphi)$. Given two disjoint sets $A, B \subseteq V$, let

$$|E_+(A,B)| := \sum_{v \in A, \, v' \in B} \max \left\{ \sum_{e \in E} o_e(v, v'), 0 \right\},$$

$$|E_-(A)| := \sum_{\{v,v'\} \subset A} \max \left\{ -\sum_{e \in E} o_e(v, v'), 0 \right\},$$

$$|\partial(A \cup B)| := \sum_{v \in A \cup B, \, v' \notin A \cup B} \left| \sum_{e \in E} o_e(v, v') \right|,$$

and

$$\text{vol}_w(A \cup B) := \sum_{v \in A \cup B} \left(\sum_{e \ni v} (|e| - 1) - \sum_{v' \in V} \left| \sum_{e \in E} o_e(v, v') \right| \right),$$

where

$$o_e(v, v') := \begin{cases} 1, & \text{if } v, v' \in e, v \text{ and } v' \text{ are co-oriented in } e \\ -1, & \text{if } v, v' \in e, v \text{ and } v' \text{ are anti-oriented in } e \\ 0, & \text{otherwise.} \end{cases}$$

We define

$$h_B := \min_{A \cap B = \emptyset \neq A \cup B} \frac{2(|E_+(A, B)| + |E_-(A)| + |E_-(B)|) + |\partial(A \cup B)| + \text{vol}_w(A \cup B)}{\text{vol}(A \cup B)}.$$

Before proving the next two theorems, we need the following preliminary results.

Proposition 8.4.1. *For any* $\mathbf{x} \neq 0$, *there exists* $s \geq 0$ *such that*

$$\frac{\sum_{vv'} \left(w_{vv'}^+ |\hat{x}_v^s + \hat{x}_{v'}^s| + w_{vv'}^- |\hat{x}_v^s - \hat{x}_{v'}^s| \right)}{\sum_{v \in V} \deg(v) |\hat{x}_v^s|}$$

$$\leq \frac{p \sum_{vv'} \left(w_{vv'}^+ |x_v + x_{v'}| + w_{vv'}^- |x_v - x_{v'}| \right) \left(|x_v|^p + |x_{v'}|^p \right)^{\frac{1}{p^*}}}{2^{1/p^*} \sum_v \deg(v) |x_v|^p},$$

where

$$\hat{x}^s := \mathbf{1}_{\{v \in V : x_v > s^{1/p}\}} - \mathbf{1}_{\{v \in V : x_v < -s^{1/p}\}} \in \{-1, 0, 1\}^N.$$

Proof First, one can prove that

$$\int_0^\infty |\hat{x}_v^t| \, dt = |x_v|^p,$$

$$\int_0^\infty |\hat{x}_v^t + \hat{x}_{v'}^t| \, dt = \begin{cases} |x_v|^p + |x_{v'}|^p, & \text{if } x_v x_{v'} \geq 0, \\ ||x_v|^p - |x_{v'}|^p|, & \text{if } x_v x_{v'} \leq 0, \end{cases}$$

$$\int_0^\infty |\hat{x}_v^t - \hat{x}_{v'}^t| \, dt = \begin{cases} |x_v|^p + |x_{v'}|^p, & \text{if } x_v x_{v'} \leq 0, \\ ||x_v|^p - |x_{v'}|^p|, & \text{if } x_v x_{v'} \geq 0, \end{cases}$$

$$|a|^p + |b|^p \leq \begin{cases} |a + b|(|a|^p + |b|^p)^{1/p^*} & \text{if } ab \geq 0, \\ |a - b|(|a|^p + |b|^p)^{1/p^*} & \text{if } ab \leq 0, \end{cases}$$

and

$$||a|^p - |b|^p| \le \begin{cases} \frac{p}{2^{1/p^*}}|a - b|(|a|^p + |b|^p)^{1/p^*} & \text{if } ab \ge 0, \\ \frac{p}{2^{1/p^*}}|a + b|(|a|^p + |b|^p)^{1/p^*} & \text{if } ab \le 0, \end{cases}$$

where the last inequality has been used in [5]. Since $p/2^{1/p^*} \ge 1$ for $p \ge 1$, we have that

$$\int_0^\infty |\hat{x}_v^t \pm \hat{x}_{v'}^t| dt \le \frac{p}{2^{1/p^*}}|x_v \pm x_{v'}|(|x_v|^p + |x_{v'}|^p)^{1/p^*}.$$

Also, one can show that there exists $s \ge 0$ such that

$$\frac{\sum\limits_{vv'}\left(w_{vv'}^+|\hat{x}_v^s + \hat{x}_{v'}^s| + w_{vv'}^-|\hat{x}_v^s - \hat{x}_{v'}^s|\right)}{\sum\limits_{v \in V} \deg(v)|\hat{x}_v^s|}$$

$$\le \frac{\int_0^\infty \sum\limits_{vv'}\left(w_{vv'}^+|\hat{x}_v^t + \hat{x}_{v'}^t| + w_{vv'}^-|\hat{x}_v^t - \hat{x}_{v'}^t|\right)dt}{\int_0^\infty \sum\limits_{v \in V} \deg(v)|\hat{x}_v^t|dt}$$

$$\le \frac{p\sum\limits_{vv'}\left(w_{vv'}^+|x_v + x_{v'}| + w_{vv'}^-|x_v - x_{v'}|\right)\left(|x_v|^p + |x_{v'}|^p\right)^{\frac{1}{p^*}}}{2^{1/p^*}\sum\limits_{v \in V} \deg(v)|x_v|^p}.$$

$\square$

An immediate consequence is the following

Corollary 8.4.1. *For any* $\mathbf{x} \ne \mathbf{0}$*, there exists* $t \ge 0$ *such that*

$$\frac{\sum\limits_{vv'} w_{vv'}|\hat{x}_v^t - s_{vv'}\hat{x}_{v'}^t| + \sum\limits_{v \in V} w_v|\hat{x}_v^t|}{\sum\limits_{v \in V} \deg(v)|\hat{x}_v^t|}$$

$$\le \frac{\sum\limits_{vv'} w_{vv'}|x_v - s_{vv'}x_{v'}|(|x_v| + |x_{v'}|) + \sum\limits_{v} w_v x_v^2}{\sum\limits_{v} \deg(v)x_v^2},$$

where

$$\hat{x}^t := \mathbf{1}_{V_+^t(\mathbf{x})} - \mathbf{1}_{V_-^t(\mathbf{x})}$$

and $V_\pm^t(\mathbf{x}) := \{v \in V : \pm x_v > t\}$.

Theorem 8.4.5. *Let* $\Gamma = (V, E, \varphi)$ *be the connected,* Θ*-uniform oriented hypergraph. Then,*

$$\frac{h_B^2}{2(\Theta - 1)} \le \Theta - \lambda_N(\tilde{\Delta}_2) \le 2h_B.$$

Proof Let G be a signed graph that has vertex set V, edge signature

$$s_{vv'} := \text{sign}\left(\sum_{e \in E} \varphi(e,v)\varphi(e,v')\right),$$

and edge weight

$$w_{vv'} := \left|\sum_{e \in E} \varphi(e,v)\varphi(e,v')\right|$$

$$= |\#\{e : \varphi(e,v)\varphi(e,v') = 1\} - \#\{e : \varphi(e,v)\varphi(e,v') = -1\}|.$$

Then, it is easy to see that h_B also represents the Cheeger constant of G, and

$$\Theta \sum_{v \in V} \deg(v)x_v^2 - \sum_{e \in E}\left(\sum_{v \in V} \varphi(e,v)x_v\right)^2 = \sum_{vv'} w_{vv'}(x_v - s_{vv'}x_{v'})^2 + \sum_{v \in V} w_v x_v^2,$$

where

$$\deg(v) = \sum_{e \in E} \varphi(e,v)^2,$$

and

$$w_v := (\Theta - 1)\deg(v) - \sum_{v'} w_{vv'}.$$

Therefore,

$$\Theta - \lambda_N = \Theta - \max_{\mathbf{x} \neq 0} \frac{\sum_{e \in E}(\sum_{v \in V}\varphi(e,v)x_v)^2}{\sum_{v \in V}\deg(v)x_v^2}$$

$$= \min_{\mathbf{x} \neq 0} \frac{\sum_{v \sim v'} w_{vv'}(x_v - s_{vv'}x_{v'})^2 + \sum_{v \in V} w_v x_v^2}{\sum_{v \in V}\deg(v)x_v^2}$$

$$\leq \min_{\mathbf{x} \in \{-1,0,1\}^N} \frac{\sum_{v \sim v'} w_{vv'}(x_v - s_{vv'}x_{v'})^2 + \sum_{v \in V} w_v x_v^2}{\sum_{v \in V}\deg(v)x_v^2}$$

$$\leq 2 \min_{\mathbf{x} \in \{-1,0,1\}^N} \frac{\sum_{v \sim v'} w_{vv'}|x_v - s_{vv'}x_{v'}| + \sum_{v \in V} w_v|x_v|}{\sum_{v \in V}\deg(v)|x_v|}$$

$$= 2h_B.$$

Moreover,

$$h_B = \min_{\mathbf{x} \in \{-1,0,1\}^N} \frac{\sum_{v \sim v'} w_{vv'}|x_v - s_{vv'}x_{v'}| + \sum_{v \in V} w_v|x_v|}{\sum_{v \in V}\deg(v)|x_v|}$$

$$\leq \min_{\mathbf{x} \neq 0} \frac{\sum_{v \sim v} w_{vv'}|x_v - s_{vv'}x_{v'}|(|x_v| + |x_{v'}|) + \sum_v w_v x_v^2}{\sum_v \deg(v)x_v^2} \tag{8.4.4}$$

$$\leq \min_{\mathbf{x} \neq 0} \frac{\sqrt{\sum_{v \sim v'} w_{vv'}|x_v - s_{vv'}x_{v'}|^2 + \sum_v w_v x_v^2}\sqrt{\sum_{v \sim v'} w_{vv'}(|x_v| + |x_{v'}|)^2 + \sum_v w_v x_v^2}}{\sum_v \deg(v)x_v^2}$$

$$\le \min_{\mathbf{x}\neq\mathbf{0}} \sqrt{C \frac{\sum_{v\sim v'} w_{vv'}|x_v - s_{vv'}x_{v'}|^2 + \sum_v w_v x_v^2}{\sum_v \deg(v)x_v^2}}$$

$$= \sqrt{C(\Theta - \lambda_N)},$$

where

$$C := \max_{v\in V} \frac{2\sum_{v'} w_{vv'} + w_v}{\deg(v)} \le 2(\Theta - 1),$$

and the inequality (8.4.4) follows from Corollary 8.4.1. The claim follows. □

Finally, we now consider the Cheeger constant

$$h := \min_{A\cap B=\emptyset\neq A\cup B} \frac{2(|E_+(A,B)| + |E_-(A)| + |E_-(B)|) + |\partial(A\cup B)|}{\mathrm{vol}\,(A\cup B)}.$$

Theorem 8.4.6. *Let* $\Gamma = (V,E,\varphi)$ *be a connected,* Θ-*uniform oriented hypergraph. For any* $p \in (1,+\infty)$*, there exist universal constants* $C_{p,\Theta} \ge c_{p,\Theta} > 0$ *such that*

$$c_{p,\Theta}h^p \le \Theta^{p-1} - \widetilde{\lambda}_{N-1}(\tilde{\Delta}_p) \text{ when } p \ge 2,$$

$$\Theta^{p-1} - \widetilde{\lambda}_{N-1}(\tilde{\Delta}_p) \le C_{p,\Theta}h \text{ when } 1 < p \le 2.$$

Proof The technique is a combination of the proofs of Theorems 8.4.3 and 8.4.5. We have that

$$\Theta^{p-1} - \lambda_N(\Delta_p) = \Theta^{p-1} - \max_{\mathbf{x}\neq\mathbf{0}} \frac{\sum_{e\in E}|\sum_{v\in V}\varphi(e,v)x_v|^p}{\sum_{v\in V}\deg(v)|x_v|^p}$$

$$= \min_{\mathbf{x}\neq\mathbf{0}} \frac{\sum_{e\in E}(|e|^{p-1}\sum_{v\in e}|x_v|^p - |\sum_{v\in e}\varphi(e,v)x_v|^p)}{\sum_{v\in V}\deg(v)|x_v|^p}$$

$$\le M_{p,\Theta} \min_{\mathbf{x}\neq\mathbf{0}} \frac{\sum_{vv'}(w_{vv'}^+|x_v + x_{v'}|^p + w_{vv'}^-|x_v - x_{v'}|^p)}{\sum_{v\in V}\deg(v)|x_v|^p}$$

$$\le M_{p,\Theta} \min_{\mathbf{x}\in\{-1,0,1\}^N} \frac{\sum_{vv'}(w_{vv'}^+|x_v + x_{v'}|^p + w_{vv'}^-|x_v - x_{v'}|^p)}{\sum_{v\in V}\deg(v)|x_v|^p}$$

$$\le 2^{p-1}M_{p,\Theta} \min_{\mathbf{x}\in\{-1,0,1\}^N} \frac{\sum_{vv'}(w_{vv'}^+|x_v + x_{v'}| + w_{vv'}^-|x_v - x_{v'}|)}{\sum_{v\in V}\deg(v)|x_v|}$$

$$= 2^{p-1}M_{p,\Theta}h.$$

Moreover,

$$
\begin{aligned}
h &= \min_{\mathbf{x}\in\{-1,0,1\}^N} \frac{\sum_{vv'}(w^+_{vv'}|x_v + x_{v'}| + w^-_{vv'}|x_v - x_{v'}|)}{\sum_{v\in V}\deg(v)|x_v|} \\[2mm]
&\leq \frac{p}{2^{1/p^*}}\min_{\mathbf{x}\neq\mathbf{0}} \frac{\sum_{vv'}(w^+_{vv'}|x_v + x_{v'}| + w^-_{vv'}|x_v - x_{v'}|)(|x_v|^p + |x_{v'}|^p)^{1/p^*}}{\sum_v \deg(v)|x_v|^p} \\[2mm]
&\leq \frac{p}{2^{1/p^*}}\min_{\mathbf{x}\neq\mathbf{0}}
\end{aligned}
$$

$$
\frac{\left(\sum_{vv'}(w^+_{vv'}|x_v + x_{v'}|^p + w^-_{vv'}|x_v - x_{v'}|^p)\right)^{\frac{1}{p}}\left(\sum_{vv'}(w^+_{vv'} + w^-_{vv'})(|x_v|^p + |x_{v'}|^p)\right)^{\frac{1}{p^*}}}{\sum_v \deg(v)|x_v|^p}
$$

$$
\begin{aligned}
&\leq \frac{Cp}{2^{1/p^*}}\min_{\mathbf{x}\neq\mathbf{0}} \frac{\left(\sum_{vv'}(w^+_{vv'}|x_v + x_{v'}|^p + w^-_{vv'}|x_v - x_{v'}|^p)\right)^{\frac{1}{p}}\left(\sum_v \deg(v)|x_v|^p\right)^{\frac{1}{p^*}}}{\sum_v \deg(v)|x_v|^p} \\[3mm]
&\leq C\min_{\mathbf{x}\neq\mathbf{0}}\left(\frac{\sum_{vv'}(w^+_{vv'}|x_v + x_{v'}|^p + w^-_{vv'}|x_v - x_{v'}|^p)}{\sum_v \deg(v)|x_v|^p}\right)^{\frac{1}{p}} \\[3mm]
&\leq C\min_{\mathbf{x}\neq\mathbf{0}}\left(\frac{1}{m_{p,\Theta}}\frac{\sum_{e\in E}(|e|^{p-1}\sum_{v\in e}|x_v|^p - |\sum_{v\in e}\varphi(e,v)x_v|^p)}{\sum_{v\in V}\deg(v)|x_v|^p}\right)^{\frac{1}{p}} \\[3mm]
&= \frac{C}{m_{p,\Theta}^{1/p}}\left(\Theta^{p-1} - \max_{\mathbf{x}\neq\mathbf{0}}\frac{\sum_{e\in E}|\sum_{v\in V}\varphi(e,v)x_v|^p}{\sum_{v\in V}\deg(v)|x_v|^p}\right)^{\frac{1}{p}} \\[3mm]
&= \tilde{C}\left(\Theta^{p-1} - \lambda_N(\Delta_p)\right)^{\frac{1}{p}},
\end{aligned}
$$

where the first inequality follows from Proposition 8.4.1. This proves the claim. $\qquad\square$

We conclude by observing that, by Theorem 8.4.5 and by taking $p = 2$ in Theorem 8.4.6, we can also infer that

$$
\frac{h^2}{2(\Theta - 1)} \leq \frac{h_B^2}{2(\Theta - 1)} \leq \Theta - \lambda_N(\Delta_2) \leq 2h \leq 2h_B.
$$

This improves Theorem 1 in [214].

9

Nodal Domains

In this chapter, we derive general nodal domain theorems of Courant type for Laplacians and p-Laplacians on graphs and hypergraphs.

In Chapters 4 and 8, we have systematically investigated the relation between Laplacian eigenvalues and Cheeger-type constants, that is, the ease or efficiency with which a connected geometric structure can be cut into two pieces. The Cheeger-type constants are defined as the extremal values, the minima, of a corresponding functional. But, we may then also naturally ask for the domains realizing such a minimum, and we may call such a pair of domains and their joint border a *Cheeger cut*. And we can then also ask whether the Laplace operator can help us with that. Naturally, associated with the eigenvalues, we have the eigenfunctions, and these possess *nodal domains*, that is, connected regions on which they do not change sign. One might then suspect that the first nonzero eigenvalue has just two such nodal domains and that they then represent, or at least approximate, a Cheeger cut. While this often is a good intuition, it is not always true for the standard Laplacian Δ, but as we shall see, and perhaps already expect from the fact that for the 1-Laplacian Δ_1, the Cheeger inequality becomes an equality, it turns out to work for that operator.

But the geometry of nodal domains is of geometric interest in any case, and it can also draw upon an independent history, which we shall now briefly sketch.

The Sturm–Liouville oscillation theory determines the number of zeros of eigenfunctions of certain eigenvalue problems for ordinary differential equations. The basic result is that the zeros of the kth eigenfunction of a vibrating string divide the string into exactly k subintervals. As a natural generalization of Sturm–Liouville oscillation theory to higher-dimensional cases, Courant [69] proved a result regarding the zeros of eigenfunctions of certain elliptic differential operators, which is now referred to as Courant's nodal domain theorem. A (strong) *nodal domain* is a maximal connected region on which an eigenfunction is either positive or negative. In particular, the eigenfunction must not have zeros on such a nodal domain. Courant then showed that the number of nodal domains for the Dirichlet eigenvalue problem on a connected domain, as described in Section 3.1, is bounded by the degree of the eigenvalue. In particular, an eigenfunction for the first eigenvalue λ_1 has one nodal domain, as it does not change sign, and an eigenfunction for λ_2 has precisely two nodal domains, one on which it is positive and the other on which it is negative.

Analogous results hold on Riemannian manifolds. In particular, often the two nodal domains of the first eigenfunction provide good approximations of the Cheeger cut of the manifold into two large components separated by a small hypersurface. However, this is not always true. In the graph case, sometimes, the two nodal domains of the first eigenfunction provide a cut that is very different from the Cheeger cut. An instructive example where the nodal domain approximation of the Cheeger cut fails is provided by Guattery and Miller [120], and we will present more examples in the first section. Abundant extensions of Courant's nodal domain theorem have been obtained, including the general nodal domain theories on p-Laplacians [71], Riemannian manifolds [58], matrices, and graphs [74].

In this chapter, we investigate various discrete versions of nodal domain theorems, which help us understand the eigenfunctions for discrete structures. Following [106, 210], we shall work within the framework of signed graphs. Theorem 9.2.2 is very general and includes various approaches presented in the literature as special cases. It first appeared in [107], but we present a new proof here. Moreover, Theorem 9.3.1 also unifies many known results, and it includes new statements that are proved here for the first time. Its proof is based on the idea of constructing m-dimensional spaces of functions for comparison in the min–max inequalities, which characterize eigenvalues by Rayleigh quotients. These spaces are obtained by restricting an eigenfunction with m nodal domains to these domains, thereby creating m pairwise orthogonal functions.

9.1 Nodal Domain Theorems on Graphs and Matrices

The fundamental discrete nodal domain theorems for eigenfunctions of *generalized Laplacians*, that is, symmetric matrices with nonpositive off-diagonal entries, were established by Davies, Gladwell, Leydold, and Stadler in 2001 [74]. For detailed historical reviews and further advances in this topic, we refer to [44–46].

We start by recalling the definition of strong and weak nodal domains for functions on graphs from [74, Definitions 1 and 2]:

Definition 9.1.1 ([74]). Let $\Gamma = (V, E)$ be a graph and let $f: V \to \mathbb{R}$ be a function. A positive (respectively, negative) *strong nodal domain* of f is a maximal connected induced subgraph of Γ on vertices $v \in V$ with $f(v) > 0$ (respectively, $f(v) < 0$).

A positive (respectively, negative) *weak nodal domain* of f is a maximal connected induced subgraph of Γ on vertices $v \in V$ with $f(v) \geq 0$ (respectively, $f(v) \leq 0$) that contains at least one nonzero vertex.

The main theorem in [74] shows that, on a connected graph, any eigenfunction corresponding to the kth eigenvalue λ_k of the graph Laplacian has at most $k + r - 1$ strong nodal domains, where r is the multiplicity of λ_k and at most k weak nodal domains. Also, their nodal domain theorem was established for general Schrödinger matrices, which include the Laplacian matrices as special cases and are defined as follows. Given an $N \times N$ symmetric matrix $M = (M_{ij})$, the *induced graph* $\Gamma = (V, E)$ is given by

$$V := \{v_1, \dots, v_N\} \quad \text{and} \quad E := \{\{v, v'\}: M_{vv'} \neq 0 \text{ and } v \neq v'\},$$

and M is called a *compatible matrix* on Γ. A *Schrödinger matrix* on Γ is a compatible matrix with nonpositive off-diagonal entries.

Theorem 9.1.1 ([74]). *Let M be a Schrödinger matrix on a simple graph Γ. Let λ_k be the kth eigenvalue of M, and let r be the multiplicity of λ_k. Then, any eigenvector corresponding to λ_k has at most $k + r - 1$ strong nodal domains.*

Moreover, if Γ is connected, then any eigenvector corresponding to λ_k has at most k weak nodal domains.

It is natural to ask if there are nodal domain theorems for compatible matrices on graphs that do not necessarily satisfy the Schrödinger condition. This question was first systematically studied by Mohammadian [210], who proposed the following definitions.

Definition 9.1.2 ([210]). Let M be a symmetric matrix, let $\Gamma = (V, E)$ be the corresponding induced graph, and let $f: V \to \mathbb{R}$ be a function. Let also $\Gamma^<_{M,f}$ be the subgraph of Γ that has vertex set $\Omega := \{v \in V: f(v) \neq 0\}$ and edge set

$$E^< := \{\{v, v'\} \in E: f(v)M_{vv'}f(v') < 0\}.$$

Similarly, let $\Gamma^{\leq}_{M,f}$ be the graph that has vertex set V and edge set

$$E^{\leq} := \{\{v, v'\} \in E: f(v)M_{vv'}f(v') \leq 0\}.$$

A *strong (respectively, weak) nodal domain* of f is a connected component of $\Gamma^<_{M,f}$ (respectively, $\Gamma^{\le}_{M,f}$).

Mohammadian also proved the following nodal domain theorem:

Theorem 9.1.2 ([210]). *Let M be a symmetric matrix such that its induced graph Γ is connected. Then, for any kth eigenfunction f_k of M,*

$$c(\Gamma^{\le}_{M,f_k}) \le k, \quad \text{and} \quad c(\Gamma^<_{M,f_k}) \le k + r - 1,$$

where $c(\cdot)$ is the number of connected components, and r is the multiplicity of the kth eigenvalue.

However, Mohammadian's definition of weak nodal domains differs from Definition 9.1.1. Recently, Ge and Liu [106] provided a new definition of weak nodal domain that is compatible with the classical case:

Definition 9.1.3 ([106]). *Let $\Gamma = (V, E, s)$ be a signed graph, and let $f : V \to \mathbb{R}$ be a function. A sequence $\{v_k\}_{k=1}^n$ of vertices is called a *strong nodal domain path* of f, or simply an S-path of f, if*

$$v_k \sim v_{k+1} \quad \text{and} \quad f(v_k)s(v_k v_{k+1})f(v_{k+1}) > 0,$$

for each $k = 1, 2, \ldots, n - 1$.

*Similarly, a sequence $\{v_k\}_{k=1}^n$ of vertices, for $n \ge 2$, is called a *weak nodal domain path* of f, or simply a W-path of f, if for any two vertices v_i and v_j such that $f(v_i) \ne 0$, $f(v_j) \ne 0$, and $f(v_\ell) = 0$ for any $i < \ell < j$, we have that*

$$f(v_i)s(v_i v_{i+1}) \cdots s(v_{j-1}v_j)f(v_j) > 0.$$

In particular, according to this definition, every path that contains at most one vertex on which f is nonzero is a weak nodal domain path.

Now, fix a signed graph $\Gamma = (V, E, s)$, let $f : V \to \mathbb{R}$ be a function, and let $\Omega := \{v \in V : f(v) \ne 0\}$. We define equivalence relations $\overset{S}{\sim}$ and $\overset{W}{\sim}$ on Ω, as follows. Given $v, v' \in \Omega$, let $v \overset{S}{\sim} v'$ (respectively, $v \overset{W}{\sim} v'$) if $v = v'$ or there exists an S-path (respectively, a W-path) connecting v and v'. Let also $\{S_i\}_{i=1}^p$ (respectively, $\{W_i\}_{i=1}^q$) be the equivalence classes of the relation $\overset{S}{\sim}$ (respectively, $\overset{W}{\sim}$) on Ω. The induced subgraph of an equivalence class S_i is called a *strong nodal domain* of f. Similarly, the induced subgraph of a set

$$W_i^0 := W_i \cup \{v \in V : \text{there exists a W-path from } v \text{ to some vertex in } W_i\}$$

is called a *weak nodal domain* of f. We also let $\mathfrak{S}(f)$ (respectively, $\mathfrak{W}(f)$) be the number of strong (respectively, weak) nodal domains of f.

Theorem 9.1.3 ([106]). *Let M be a symmetric matrix, let $\Gamma = (V, E)$ be its induced signed graph, and let λ_k be its kth eigenvalue. If f_k is an eigenfunction corresponding to λ_k, then*

$$\mathfrak{S}(f_k) \le k + r - 1, \text{ and } \mathfrak{W}(f_k) \le k + c - 1, \qquad (9.1.1)$$

where r is the multiplicity of λ_k and c is the number of connected components of Γ. In particular, if Γ is connected, then $\mathfrak{W}(f_k) \le k$.

The *proof* will be contained in that of Theorem 9.2.2, which is a more general result.

We conclude this section with some examples that show how different the nodal domains of the second eigenfunction and the Cheeger cuts can be from each other. In particular, on the following two pages, we use a gray line for representing the Cheeger cut, and we use a gray dotted line for the cut that is determined by the two nodal domains of the first nontrivial eigenfunction.

The *ladder graph* on $2n$ vertices consists of two paths with vertices $v_1, \ldots, v_n$ and $v_{2n}, \ldots, v_{n+1}$, respectively, together with an edge for each pair (v_j, v_{2n+1-j}). As a generalization, we define the *broken ladder graph $BL(n, k)$* as the graph obtained from the ladder graph on $2n$ vertices by removing k rungs from one side, for a given $k \in \{0, \ldots, n - 1\}$. In particular, $BL(n, 0)$ is the original ladder graph, while for $k > 0$, the edges between the first k pairs (v_j, v_{2n+1-j}) are removed. (See also illustrations below.) Let $\lambda_2(BL(n, k))$ be the second eigenvalue (therefore, the first nonzero eigenvalue) of the normalized Laplacian of $BL(n, k)$, and let $h(BL(n, k))$ be the Cheeger constant of $BL(n, k)$. For $n = 8$ and $k = 0, 1, \ldots, 6$, we have

$$\lambda_2(BL(8,0)) = 0.060171,$$
$$\lambda_2(BL(8,1)) = 0.0660208,$$
$$\lambda_2(BL(8,2)) = 0.0720744,$$
$$\lambda_2(BL(8,3)) = 0.0764965,$$
$$\lambda_2(BL(8,4)) = 0.0637421,$$
$$\lambda_2(BL(8,5)) = 0.0424294,$$
$$\lambda_2(BL(8,6)) = 0.0301588,$$

and these are all simple eigenvalues. Moreover,

$$h(BL(8,0)) = 1/11,$$
$$h(BL(8,1)) = 1/10,$$
$$h(BL(8,2)) = 1/9,$$
$$h(BL(8,3)) = 1/8,$$

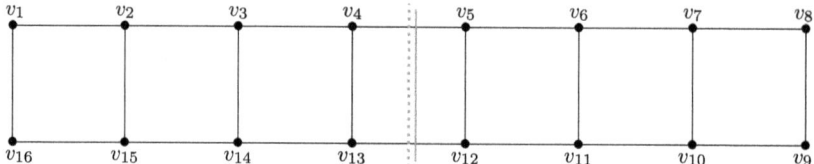

Figure 9.1 Ladder graph $BL(8,0)$.

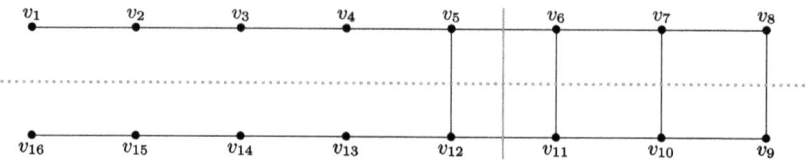

Figure 9.2 Ladder graph $BL(8,4)$.

$$h(BL(8,4)) = 1/8,$$
$$h(BL(8,5)) = 1/9,$$
$$h(BL(8,6)) = 1/11.$$

Of course, they all fall into the range determined in Theorem 4.3.1. For the ladder graph $BL(8,0)$ on 16 nodes shown in Figure 9.1, one eigenvector of $\lambda_2(BL(8,0))$, with entries rounded to two decimal entries, is

$$(0.30, 0.32, 0.22, 0.08, -0.08, -0.22, -0.32, -0.30,$$
$$- 0.30, -0.32, -0.22, -0.08, 0.08, 0.22, 0.32, 0.30).$$

Its two nodal domains,

$$\{v_1, v_2, v_3, v_4, v_{13}, v_{14}, v_{15}, v_{16}\} \quad \text{and} \quad \{v_5, v_6, v_7, v_8, v_9, v_{10}, v_{11}, v_{12}\},$$

give the exact Cheeger cut.

The same holds for $BL(8,1)$, $BL(8,2)$, and $BL(8,3)$. However, things start to change from the ladder graph $BL(8,4)$ shown in Figure 9.2. In fact, an eigenvector corresponding to $\lambda_2(BL(8,4))$, with entries rounded to three decimal entries, is

$$(0.337, 0.446, 0.359, 0.226, 0.078, 0.022, 0.006, 0.001,$$
$$- 0.001, -0.006, -0.022, -0.078, -0.226, -0.359, -0.446, -0.337).$$

Its two nodal domains, $\{v_1, v_2, v_3, v_4, v_5, v_6, v_7, v_8\}$ and $\{v_9, v_{10}, v_{11}, v_{12}, v_{13}, v_{14}, v_{15}, v_{16}\}$, yield a cut that is far from the Cheeger cut, which is given by

$$\{v_1, v_2, v_3, v_4, v_5, v_{12}, v_{13}, v_{14}, v_{15}, v_{16}\} \quad \text{and} \quad \{v_6, v_7, v_8, v_9, v_{10}, v_{11}\}$$

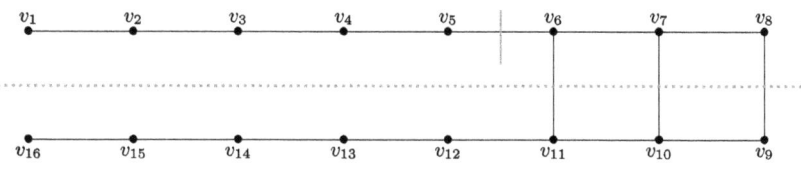

Figure 9.3 Ladder graph $BL(8,5)$.

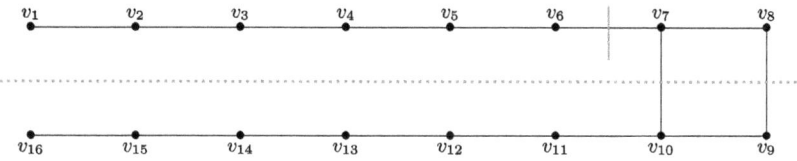

Figure 9.4 Ladder graph $BL(8,6)$.

Moreover, an eigenvector corresponding to $\lambda_2(BL(8,5))$ on the graph of Figure 9.3, with entries rounded to three decimal entries, is

$$(0.305, 0.412, 0.359, 0.275, 0.168, 0.057, 0.016, 0.004,$$
$$-0.004, -0.016, -0.057, -0.168, -0.275, -0.359, -0.412, -0.305).$$

Thus, the two nodal domains, $\{v_1, v_2, v_3, v_4, v_5, v_6, v_7, v_8\}$ and $\{v_9, v_{10}, v_{11}, v_{12}, v_{13}, v_{14}, v_{15}, v_{16}\}$, do not yield the Cheeger cut, which is given by

$$\{v_1, v_2, v_3, v_4, v_5\} \quad \text{and} \quad \{v_6, v_7, v_8, v_9, v_{10}, v_{11}, v_{12}, v_{13}, v_{14}, v_{15}, v_{16}\}.$$

Similarly, an eigenvector corresponding to $\lambda_2(BL(8,6))$ on the graph of Figure 9.4, with entries rounded to two decimal entries, is

$$(0.28, 0.38, 0.35, 0.29, 0.22, 0.13, 0.04, 0.01,$$
$$-0.01, -0.04, -0.13, -0.22, -0.29, -0.35, -0.38, \quad 0.28).$$

Its two nodal domains, $\{v_1, v_2, v_3, v_4, v_5, v_6, v_7, v_8\}$ and $\{v_9, v_{10}, v_{11}, v_{12}, v_{13}, v_{14}, v_{15}, v_{16}\}$, again do not yield the right Cheeger cut, given by

$$\{v_1, v_2, v_3, v_4, v_5, v_6\} \quad \text{and} \quad \{v_7, v_8, v_9, v_{10}, v_{11}, v_{12}, v_{13}, v_{14}, v_{15}, v_{16}\}.$$

The broken ladder graph $BL(2k, k)$ was introduced by Guattery and Miller in [120] to show that the first nontrivial eigenfunction of the unnormalized Laplacian does not approximate the unnormalized Cheeger cut. Our preceding computations demonstrate that the same phenomenon still occurs for the normalized Laplacian and the standard Cheeger cut. For the 1-Laplacian, in contrast, the cut obtained from the nodal domains of the second eigenfunction coincides with the Cheeger cut, as we shall see in Theorem 9.2.1.

9.2 Nonlinear Nodal Domain Theorems on Graphs

In order to derive nodal domain properties for p-Laplacians on graphs or signed graphs, we need to generalize the concepts from Sections 7.1 and 7.2.

Let $\Gamma = (V, E)$ be a graph with the following additional structure:

- Two positive measures $w\colon E \to \mathbb{R}^+$ and $\mu\colon V \to \mathbb{R}^+$;
- Real coefficients $\kappa\colon V \to \mathbb{R}$; and
- A sign function $s\colon E \to \{-1, 1\}$.

For $p > 1$, the *signed p-Laplacian* $\Delta_p^s\colon C(V) \to C(V)$ is defined by

$$\Delta_p^s f(v) := \sum_{v' \in V} w_{vv'} |f(v) - s(vv')f(v')|^{p-2} (f(v) - s(vv')f(v'))$$
$$+ \kappa_v |f(v)|^{p-2} f(v),$$

for $v \in V$ and $f \in C(V)$. Moreover, the *eigenvalue problem* for Δ_p^s, when $p > 1$, consists in finding $\lambda \in \mathbb{R}$ and $f \neq 0$ such that

$$\Delta_p^s f(v) = \lambda \mu_v |f(v)|^{p-2} f(v), \quad \forall v \in V.$$

The *signed 1-Laplacian*[1] Δ_1^s is the set-valued map defined by

$$\Delta_1^s f(v) := \left\{ \sum_{v' \in V} w_{vv'} z_{vv'} + \kappa_v z_v \,\middle|\, \begin{array}{l} z_{vv'} \in \mathrm{Sgn}(f(v) - s(vv')f(v')), \\ z_{vv'} = -s(vv')z_{v'v}, \\ z_v \in \mathrm{Sgn}(f(v)) \end{array} \right\},$$

where

$$\mathrm{Sgn}(t) := \begin{cases} \{1\} & \text{if } t > 0, \\ [-1, 1] & \text{if } t = 0, \\ \{-1\} & \text{if } t < 0. \end{cases}$$

The *eigenvalue problem* for Δ_1^s consists in finding $\lambda \in \mathbb{R}$ and $f \neq 0$ such that

$$\Delta_1^s f(v) \bigcap \lambda \mu_v \mathrm{Sgn}(f(v)) \neq \emptyset, \quad \forall v \in V.$$

By applying the Lusternik–Schnirelman theory, we can define a sequence of min–max eigenvalues for Δ_p^s as follows:

$$\lambda_k(\Delta_p^s) := \inf_{\mathrm{genus}(S) \geq k} \sup_{f \in S} \frac{\sum_{\{v, v'\} \in E} w_{vv'} |f(v) - s(vv')f(v')|^p + \kappa_v |f(v)|^p}{\sum_{v \in V} \mu_v |f(v)|^p},$$

---

[1] The idea to consider a signed version of the 1-Laplacian comes from a short discussion with Delio Mugnolo at the workshop [131]. Mugnolo asked the third author: Can the graph 1-Laplacian characterize the bipartiteness of a graph? Now we know the answer is yes, but we need to consider the signed version of the 1-Laplacian with a negative signature.

for $k \geq 1$, where genus(S) denotes the Krasnoselskii genus of the centrally symmetric compact subset S in $C(V) \cong \mathbb{R}^V$ (see Definition 2.6.2). We also denote by Gen_k the collection of all centrally symmetric compact subsets with Krasnoselskii genus larger than or equal to k.

Remark 9.2.1. Any symmetric matrix can be represented by Δ_2^s, for some suitable parameters $w_{vv'}$, $s(vv')$, and κ_v. Also, if $\kappa \equiv 0$, then Δ_p^s is the classical p-Laplacian on signed graphs. If $s \equiv +1$, then Δ_p^s is the p-Schrödinger operator on graphs [76]. Moreover, if we let $\kappa \equiv 0$ and $w \equiv 1$, and we let μ be the weighted degree function on V, then we obtain an equivalent formulation of the normalized p-Laplacian that has been discussed in Chapter 7.

We now prove some preliminary results on positive and negative nodal domains for the p-Laplacians on simple graphs (i.e., graphs with $s \equiv +1$).

Proposition 9.2.1. *Let Γ be a connected (simple) graph, and let $p > 1$. Then, any eigenfunction corresponding to the second eigenvalue $\lambda_2(\Delta_p)$ of the p-Laplacian on Γ has exactly one positive weak nodal domain and one negative weak nodal domain.*

Proof By Proposition 7.1.2, we have that

$$\sum_{v \in V} \deg v \cdot \phi_p(f(v)) = 0.$$

This implies that $\{v \in V : f(v) > 0\} \neq \emptyset$ and $\{v \in V : f(v) < 0\} \neq \emptyset$. Thus, f has at least one positive and one negative weak nodal domain. But as we shall see in Theorem 9.2.2, any eigenfunction corresponding to $\lambda_2(\Delta_p)$ has at most two weak nodal domains. Hence, the claim follows.   □

Example 9.2.1. In the graph with 10 vertices that is depicted in Figure 9.5, the function

$$f := \mathbf{1}_{\{5,6,7,8,9,10\}} - \mathbf{1}_{\{1,2,3,4\}}$$

is an eigenfunction corresponding to $\lambda_2(\Delta_1)$, and it has exactly three weak nodal domains.

Thus, Proposition 9.2.1 does not hold for $p = 1$. Instead, the following property holds.

Proposition 9.2.2. *Let Γ be a connected (simple) graph. Then, any eigenfunction f corresponding to $\lambda_2(\Delta_1)$ has exactly one positive weak nodal domain or exactly one negative weak nodal domain. Moreover, if f has at least two positive strong nodal domains and two negative strong nodal domains, then the number of weak nodal domains of f is exactly two.*

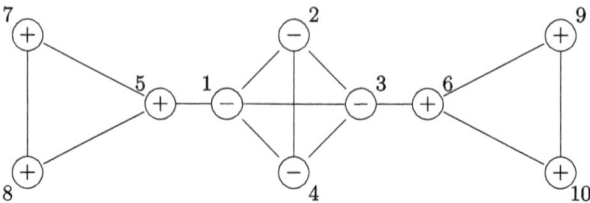

Figure 9.5 An example graph discussed in Example 9.2.1.

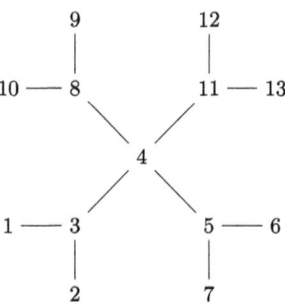

Figure 9.6 An example graph discussed in Example 9.2.2 where the second 1-Laplacian eigenvalue has variational multiplicity three and the hypotheses of Proposition 9.2.2 are satisfied.

Proof Assume without loss of generality, that $\{f > 0\} \neq \emptyset$. If $\{f < 0\} = \emptyset$, then f has exactly one positive weak nodal domain. If $\{f < 0\} \neq \emptyset$, then by Theorem 7.2.3, either $\{f \geq 0\}$ or $\{f \leq 0\}$ induces a connected subgraph on Γ, which implies that f has either exactly one positive weak nodal domain or exactly one negative weak nodal domain.

Now, assume that f has two positive strong nodal domains and two negative strong nodal domains. Then, both $\{f > 0\}$ and $\{f < 0\}$ induce disconnected subgraphs, and by Theorem 7.2.3, both $\{f \geq 0\}$ and $\{f \leq 0\}$ induce connected subgraphs, which implies that f has exactly one positive weak nodal domain and exactly one negative weak nodal domain.   □

Example 9.2.2. In the graph with 13 vertices that is depicted in Figure 9.6, the function

$$f := \mathbf{1}_{\{1,2,3,5,6,7\}} - \mathbf{1}_{\{8,9,10,11,12,13\}}$$

is an eigenfunction corresponding to $\lambda_2(\Delta_1)$. It has four strong nodal domains (two positive ones and two negative ones), and it has exactly two weak nodal domains.

While we have seen above that the nodal domains of $\lambda_2(\Delta)$ can be very different from a Cheeger cut, we shall now show that for Δ_1, the nodal domains of $\lambda_2(\Delta_1)$ do yield a Cheeger cut (with some technical provisos that will of course be spelled out in detail). This is of course in line with the fact that this eigenvalue equals the Cheeger constant itself.

We shall denote the complement of a vertex set A by A^c. Thus, for every nonempty $A \subset V$ with nonempty complement, (A, A^c) is a cut of Γ.

Theorem 9.2.1. *For any eigenfunction f of a (simple) graph Γ corresponding to $\lambda_2(\Delta_1)$, and for any strong nodal domain A of f, (A, A^c) is a Cheeger cut of Γ.*

Proof Let $\lambda_2 := \lambda_2(\Delta_1)$ and, without loss of generality, assume that A is a positive strong nodal domain of f. Then, by the definition of a nodal domain,

$$\mathrm{Sgn}(f(v) - f(v')) \subset \mathrm{Sgn}(\mathbf{1}_A(v) - \mathbf{1}_A(v')),$$

for any $\{v, v'\} \in E$, and $\mathrm{Sgn}(f(v)) \subset \mathrm{Sgn}(\mathbf{1}_A(v))$, for any $v \in V$.

Thus, $\Delta_1 f(v) \subset \Delta_1 \mathbf{1}_A(v)$ and therefore,

$$\Delta_1 \mathbf{1}_A(v) \bigcap \lambda_2 \deg v \cdot \mathrm{Sgn}(\mathbf{1}_A(v)) \supset \Delta_1 f(v) \bigcap \lambda_2 \deg v \cdot \mathrm{Sgn}(f(v)) \neq \emptyset,$$

for all $v \in V$, which implies that $\mathbf{1}_A$ is also an eigenfunction corresponding to λ_2. As a consequence, $R_1(\mathbf{1}_A) = R_1(f) = \lambda_2$. Moreover, by Theorem 7.2.1, $\lambda_2 = h$. Also, one can check that $R_1(\mathbf{1}_A) = |\partial A|/\mathrm{vol}\,(A)$. Next, we show that $\mathrm{vol}\,(A) \leq \mathrm{vol}\,(A^c)$. To prove this, we can assume that Γ is connected, as otherwise we can simply consider the connected component containing A. Since $(\lambda_2, \mathbf{1}_A)$ is an eigenpair, by the definition of the 1-Laplace eigen-equation, there exists $z_{vv'} \in \mathrm{Sgn}(f(v) - f(v'))$ with $z_{vv'} = -z_{v'v}$, such that

$$\begin{cases} \sum_{v' \sim v} z_{vv'} = \lambda_2 \deg v, & \forall v \in A, \\ |\sum_{v' \sim v} z_{vv'}| \leq \lambda_2 \deg v, & \forall v \in A^c. \end{cases}$$

Therefore,

$$\lambda_2 \mathrm{vol}\,(A) = \lambda_2 \sum_{v \in A} \deg v$$

$$= \sum_{v \in A} \sum_{v' \sim v} z_{vv'}$$

$$= \sum_{\{v, v'\} \in E} (z_{vv'} + z_{v'v}) - \sum_{v \in A^c} \sum_{v' \sim v} z_{vv'}$$

$$= -\sum_{v \in A^c} \sum_{v' \sim v} z_{vv'}$$

$$\leq \sum_{v \in A^c} \lambda_2 \deg v$$

$$= \lambda_2 \mathrm{vol}\,(A^c),$$

which implies that vol $(A) \leq$ vol (A^c), since $\lambda_2 > 0$ for any connected graph. Hence, we have that vol $(A) = \min\{\text{vol}(A), \text{vol}(A^c)\}$, and thus

$$\frac{|\partial A|}{\min\{\text{vol}(A), \text{vol}(A^c)\}} = \frac{|\partial A|}{\text{vol}(A)} = R_1(\mathbf{1}_A) = \lambda_2 = h,$$

which implies that (A, A^c) is a Cheeger cut of Γ. $\qquad\qquad\square$

By revisiting the proof of Theorem 9.2.1, we can also prove

Lemma 9.2.1. *Let f be an eigenfunction of a (simple) graph Γ corresponding to $\lambda_2(\Delta_1)$. Then,*

$$\text{vol}\,\{f > 0\} \leq \text{vol}\,\{f < 0\} + \text{vol}\,\{f = 0\}$$

and

$$\text{vol}\,\{f < 0\} \leq \text{vol}\,\{f > 0\} + \text{vol}\,\{f = 0\}.$$

Note that, in the statement of Lemma 9.2.1, we cannot replace $\{f > 0\}$ and $\{f < 0\}$ by a positive and a negative strong nodal domain, respectively. For example, the graph in Example 9.2.1 is such that $\{5,7,8\}$ and $\{6,9,10\}$ are the two positive strong nodal domains, while $\{1,2,3,4\}$ is the negative strong nodal domain of the eigenfunction

$$f = \mathbf{1}_{\{5,6,7,8,9,10\}} - \mathbf{1}_{\{1,2,3,4\}}$$

corresponding to $\lambda_2(\Delta_1)$. In this case, vol $\{5,7,8\} = $ vol $\{6,9,10\} < $ vol $\{1,2,3,4\}$.

The preceding suggests that when f not only has a positive and a negative domain but also a nontrivial zero set, that zero set could be assigned to either of these domains, leading to two different Cheeger cuts. In fact, again by the proof of Proposition 9.2.1, if $\{f > 0\} \neq \emptyset$, then $(\{f > 0\}, \{f \leq 0\})$ is a Cheeger cut, and similarly, when $\{f < 0\} \neq \emptyset$, then $(\{f < 0\}, \{f \geq 0\})$ is a Cheeger cut. We conclude

Corollary 9.2.1. *If an eigenfunction for $\lambda_2(\Delta_1)$ has zeros and it changes sign, then Γ has at least two different Cheeger cuts.*

However, we should also note that either $\{f > 0\}$ or $\{f < 0\}$ might be empty. For example, for the broken ladder $BL(8,4)$, the Cheeger cut

$$(\{1,2,3,4,5,12,13,14,15,16\}, \{6,7,8,9,10,11\})$$

is unique, while the eigenfunction

$$f = \mathbf{1}_{\{6,7,8,9,10,11\}}$$

corresponding to $\lambda_2(\Delta_1)$ has no points that take negative values.

We now turn to the nodal domains of eigenfunctions for higher-order eigenvalues. We need the following preliminary lemma (see [5, 157, 266]).

Lemma 9.2.2. *Let $p \geq 1$, and let $t, s, a, b \in \mathbb{R}$. Then,*

$$\begin{cases} |ta + sb|^p \geq (|t|^p a + |s|^p b)|a + b|^{p-2}(a + b), & \text{if } ab \leq 0, \\ |ta + sb|^p \leq (|t|^p a + |s|^p b)|a + b|^{p-2}(a + b), & \text{if } ab \geq 0. \end{cases} \tag{9.2.1}$$

For $p > 1$, the equality holds if and only if either $ab = 0$ or $t = s$.
In the particular case of $p = 1$, we further have that, for any $z \in \text{Sgn}(a + b)$,

$$\begin{cases} |ta + sb| \geq (|t|a + |s|b)z, & \text{if } ab \leq 0, \\ |ta + sb| \leq (|t|a + |s|b)z, & \text{if } ab \geq 0. \end{cases}$$

For a signed graph (Γ, s) and a function $f : V \to \mathbb{R}$, we let $\overline{\mathfrak{S}}(f)$ (respectively, $\overline{\mathfrak{W}}(f)$) denote the number of strong (respectively, weak) nodal domains of f in the *opposite signed graph* $(\Gamma, -s)$ (see Definition 4.6.2).

Theorem 9.2.2. *Let $\Gamma = (V, E)$ be a signed graph on N nodes. For any eigenfunction f corresponding to the kth min–max eigenvalue $\lambda_k(\Delta_p^s)$, we have*

$$\mathfrak{W}(f) \leq \mathfrak{S}(f) \leq k + r - 1 \quad \text{and} \quad \overline{\mathfrak{W}}(f) \leq \overline{\mathfrak{S}}(f) \leq N - k + r,$$

where r is the multiplicity of $\lambda_k(\Delta_p^s)$. If $p > 1$, we further have

$$\mathfrak{W}(f) \leq k + c - 1 \quad \text{and} \quad \overline{\mathfrak{W}}(f) \leq N - k + c,$$

where c is the number of connected components of Γ.

Proof We split the proof into multiple steps.

- Let $\lambda_k := \lambda_k(\Delta_p^s)$. Assume that f has m strong nodal domains and denote them by $V_1, \ldots, V_m$. The idea of this step is that we shall consider the m-dimensional function space spanned by the restrictions of f to $V_1, \ldots, V_m$, and we shall then use the variational properties of the eigenvalues. For this purpose, we consider the linear function space X spanned by $f|_{V_1}, \ldots, f|_{V_m}$, where

$$f|_{V_i}(v) := \begin{cases} f(v), & \text{if } v \in V_i, \\ 0, & \text{if } v \notin V_i. \end{cases}$$

Since $V_1, \ldots, V_m$ are pairwise disjoint, we have that $\dim X = m$. Moreover, each edge $\{v, v'\}$ intersects at most two nodal domains. Thus, by the definition of nodal domains, for $i \neq j$ and $\{v, v'\} \in E$ with $v \in V_i$ and $v' \in V_j$, we have

$$(f|_{V_i}(v) - s(vv')f|_{V_i}(v')) \cdot (f|_{V_j}(v) - s(vv')f|_{V_j}(v'))$$
$$= - f|_{V_i}(v)s(vv')f|_{V_j}(v')$$
$$\geq 0,$$

since f has opposite signs on adjacent nodal domains connected by positive edges and the same signs on adjacent nodal domains connected by negative edges.

For the case of $p > 1$, let $g := \sum_{i=1}^{m} t_i f|_{V_i}$ and consider the function ϕ_p introduced in (7.1.4). Then,

$$\sum_{\{v,v'\} \in E} w_{vv'} |g(v) - s(vv')g(v')|^p + \sum_{v \in V} \kappa_v |g(v)|^p$$

$$= \sum_{\{v,v'\} \in E} w_{vv'} \left| \sum_{i=1}^{m} t_i (f|_{V_i}(v) - s(vv')f|_{V_i}(v')) \right|^p + \sum_{i=1}^{m} |t_i|^p \sum_{v \in V_i} \kappa_v |f(v)|^p$$

$$\leq \sum_{v \sim v'} w_{vv'} \sum_{i=1}^{m} |t_i|^p (f|_{V_i}(v) - s(vv')f|_{V_i}(v')) \, \phi_p (f(v) - s(vv')f(v'))$$

$$+ \sum_{i=1}^{m} |t_i|^p \sum_{v \in V_i} \kappa_v |f(v)|^p$$

$$= \sum_{i=1}^{m} |t_i|^p \sum_{v \in V_i} f(v) \left(\sum_{v' \in V} w_{vv'} \phi_p (f(v) - s(vv')f(v')) \right) + \sum_{i=1}^{m} |t_i|^p \sum_{v \in V_i} \kappa_v |f(v)|^p$$

$$= \sum_{i=1}^{m} |t_i|^p \sum_{v \in V_i} f(v) \left(\sum_{v' \in V} w_{vv'} \phi_p (f(v) - s(vv')f(v')) + \kappa_v |f(v)|^{p-2} f(v) \right)$$

$$= \sum_{i=1}^{m} |t_i|^p \sum_{v \in V_i} f(v) \cdot \Delta_p^s f(v)$$

$$= \sum_{i=1}^{m} |t_i|^p \sum_{v \in V_i} f(v) \cdot \lambda_k \mu_v |f(v)|^{p-2} f(v)$$

$$= \lambda_k \sum_{i=1}^{m} |t_i|^p \sum_{v \in V_i} \mu_v |f(v)|^p$$

$$= \lambda_k \sum_{v \in V} \mu_v |g(v)|^p,$$

where the inequality follows by taking $a = f|_{V_i}(v) - s(vv')f|_{V_i}(v')$ and $b = f|_{V_j}(v) - s(vv')f|_{V_j}(v')$ in Lemma 9.2.2.

For $g \in X \setminus 0$, the preceding inequality implies that

$$R_p(g) := \frac{\sum_{\{v,v'\} \in E} w_{vv'} |g(x) - s(vv')g(v')|^p + \kappa_v |g(v)|^p}{\sum_{v \in V} \mu_v |g(v)|^p} \leq \lambda_k.$$

Moreover, by the definition of the variational eigenvalues, we have

$$\lambda_m = \inf_{X' \in \mathrm{Gen}_m} \sup_{g' \in X' \backslash 0} R_p(g') \le \sup_{g \in X \backslash 0} R_p(g) \le \lambda_k < \lambda_{k+r},$$

which implies that $m \le k + r - 1$.

For the case of $p = 1$, we have

$$\sum_{\{v,v'\} \in E} w_{vv'} |g(v) - s(vv')g(v')| + \sum_{v \in V} \kappa_v |g(v)|$$

$$= \sum_{\{v,v'\} \in E} w_{vv'} \left| \sum_{i=1}^{m} t_i (f|_{V_i}(v) - s(vv')f|_{V_i}(v')) \right| + \sum_{i=1}^{m} |t_i| \sum_{v \in V_i} \kappa_v |f(v)|$$

$$\le \sum_{\{v,v'\} \in E} w_{vv'} \sum_{i=1}^{m} |t_i| \, (f|_{V_i}(v) - s(vv')f|_{V_i}(v')) \, z_{vv'} + \sum_{i=1}^{m} |t_i| \sum_{v \in V_i} \kappa_v |f(v)|$$

$$= \sum_{i=1}^{m} |t_i| \sum_{v \in V_i} f(v) \left(\sum_{v' \in V} w_{vv'} z_{vv'} \right) + \sum_{i=1}^{m} |t_i| \sum_{v \in V_i} \kappa_v |f(v)|$$

$$= \sum_{i=1}^{m} |t_i| \sum_{v \in V_i} f(v) \left(\sum_{v' \in V} w_{vv'} z_{vv'} + \kappa_v z_v \right)$$

$$\in \sum_{i=1}^{m} |t_i| \sum_{v \in V_i} f(v) \cdot \Delta_1^s f(v) \cap \lambda_k \mu_v \mathrm{Sgn}(f(v))$$

$$\subset \sum_{i=1}^{m} |t_i| \sum_{v \in V_i} f(v) \cdot \lambda_k \mu_v \mathrm{Sgn}(f(v))$$

$$= \lambda_k \sum_{i=1}^{m} |t_i| \sum_{v \in V_i} \mu_v |f(v)|$$

$$= \lambda_k \sum_{v \in V} \mu_v |g(v)|,$$

where the inequality follows by taking

$$a = f|_{V_i}(v) - s(vv')f|_{V_i}(v') \quad \text{and} \quad b = f|_{V_j}(v) - s(vv')f|_{V_j}(v')$$

in Lemma 9.2.2, while the inclusion relation follows by the fact that

$$\sum_{v' \in V} w_{vv'} z_{vv'} + \kappa_v z_v \in \Delta_1^s f(v) \cap \lambda_k \mu_v \mathrm{Sgn}(f(v)),$$

$$z_{vv'} \in \mathrm{Sgn}(f(v) - s(vv')f(v')),$$

$$z_{vv'} = -s(vv')z_{v'v}, \text{ and}$$

$$z_v \in \mathrm{Sgn}(f(v)).$$

The rest of the proof is the same as above. This shows that $\mathfrak{S}(f) \le k + r - 1$.

- We shall now derive the corresponding inequality for $\overline{\mathfrak{S}}(f)$, which is the number of strong nodal domains of f in the opposite signed graph $(\Gamma, -s)$. Assume now that f has m strong nodal domains on $(\Gamma, -s)$, where m is not necessarily the same as in the previous step and denote them by $V_1, \ldots, V_m$. Let X now be the linear function space spanned by $f|_{V_1}, \ldots, f|_{V_m}$. Then, similar to the previous step, one can prove that $R_p(g) \geq \lambda_k$ for any $g \in X \setminus \{0\}$. By the intersection property of Krasnoselskii's $\mathbb{Z}_2$-genus (see Lemma 2.6.3), we have that

$$X' \cap X \setminus \{0\} \neq \emptyset, \quad \text{for any} \quad X' \in \mathrm{Gen}_{N-m+1}.$$

  Therefore,

$$
\begin{aligned}
\lambda_{N-m+1} &= \inf_{X' \in \mathrm{Gen}_{N-m+1}} \sup_{g' \in X' \setminus 0} R_p(g') \\
&\geq \inf_{X' \in \mathrm{Gen}_{N-m+1}} \inf_{g' \in X' \cap X \setminus 0} R_p(g') \\
&\geq \inf_{g \in X \setminus 0} R_p(g) \\
&\geq \lambda_k \\
&> \lambda_{k-r},
\end{aligned}
$$

  which implies that $N - m + 1 \geq k - r + 1$, or equivalently, $m \leq N - k + r$. This shows that $\overline{\mathfrak{S}}(f) \leq N - k + r$.
- The inequalities $\mathfrak{W}(f) \leq \mathfrak{S}(f)$ and $\overline{\mathfrak{W}}(f) \leq \overline{\mathfrak{S}}(f)$ trivially follow from the definition of strong and weak nodal domains.
- Assume now that f has m weak nodal domains and denote them by $U_1, \ldots, U_m$. Then, $m \leq k + r - 1$. Let also $V_1, \ldots, V_c$ denote the c connected components of the vertex set of the graph, that is, $\bigsqcup_{l=1}^{c} V_l = V$. Then, $w_{vv'} = 0$ whenever $v \in V_l$, $v' \in V_{l'}$ and $l \neq l'$. Therefore, for any $i \in \{1, \ldots, m\}$, there exists a unique $l \in \{1, \ldots, c\}$ such that $U_i \subset V_l$. For $l = 1, \ldots, c$, let now

$$I_l = \{i \in \{1, \ldots, m\} : U_i \subset V_l\}$$

  be the index set corresponding to V_l. Then, $\bigsqcup_{l=1}^{c} I_l = \{1, \ldots, m\}$. Assume, by contradiction, that $m \geq k + c$. Let X be the linear function space spanned by $f|_{U_1}, \ldots, f|_{U_m}$, and let Z be the linear function space spanned by

$$f|_{\cup_{i \in I_1} U_i}, \ldots, f|_{\cup_{i \in I_c} U_i}.$$

  Then, Z is a linear subspace of X, $\dim X = m$, and $\dim Z = c$. Moreover, by the above steps, $R_p(g') \leq \lambda_k$, for any $g' \in X \setminus 0$. By the definition of variational eigenvalues, it then follows that

$$\max_{g' \in Y \setminus \{0\}} R_p(g') = \lambda_k,$$

for any linear subspace $Y \subset X \setminus (Z \setminus 0)$ with $\dim Y \geq k$. Now, fix such a space Y. Then, the basic property of variational eigenvalues implies that there exists an eigenfunction

$$g = \sum_{i=1}^{m} t_i f|_{U_i} \in Y \setminus 0,$$

with $R_p(g) = \lambda_k$. This implies that, in this case, we have equality in Lemma 9.2.2. As a consequence, for any U_i and U_j that are connected by an edge, we have that one of the following holds:

(a) $t_i = t_j$, or
(b) There exist $v \in U_i$ and $v' \in U_j$ such that $f(v) = 0 \neq f(v')$ and $\{v, v'\} \in E$.

In this case, since f and g are eigenfunctions, we have that

$$\sum_{v' \in V} w_{vv'} |s(vv')f(v')|^{p-2}(s(vv')f(v')) = 0,$$

$$\sum_{v' \in V} w_{vv'} |s(vv')g(v')|^{p-2}(s(vv')g(v')) = 0.$$

Since every $v' \sim v$ has to satisfy $v' \in U_i$ or $v' \in U_j$, from these equalities, we obtain

$$\left(|t_i|^{p-2}t_i - |t_j|^{p-2}t_j\right) \sum_{v' \in U_j} w_{vv'} |f(v')|^{p-2}(s(vv')f(v'))$$

$$= |t_i|^{p-2}t_i \sum_{v' \in V} w_{vv'} \phi_p(s(vv')f(v')) - \sum_{v' \in V} w_{vv'} \phi_p(s(vv')g(v'))$$

$$= 0.$$

By the definition of a weak nodal domain path, we have that, for any $v', v'' \in U_j$,

$$(s(vv')f(v')) \cdot (s(vv'')f(v'')) = f(v')s(v'v)s(vv'')f(v'') \geq 0.$$

This implies that

$$\sum_{v' \in U_j} w_{vv'} |f(v')|^{p-2}(s(vv')f(v')) \neq 0.$$

Thus, $|t_i|^{p-2}t_i - |t_j|^{p-2}t_j = 0$, which yields $t_i = t_j$.

Hence, $t_i = t_j$ whenever U_i and U_j are connected by an edge. Therefore, in each connected component V_l, we have that $t_i = t_j$ whenever $i, j \in I_l$, for $l = 1, \ldots, m$. But this implies that $g \in Z \setminus 0$, which is a contradiction, since $g \in Y$. This proves that $\mathfrak{W}(f) \leq k + c - 1$.

- Assume now that f has m weak nodal domains $U_1, \ldots, U_m$ on $(\Gamma, -s)$. Assume also, by contradiction, that $m \geq N - k + c + 1$. Let X be the linear function space spanned by $f|_{U_1}, \ldots, f|_{U_m}$, and let Z be the linear function space spanned by

$$f|_{\bigcup_{i \in I_1} U_i}, \ldots, f|_{\bigcup_{i \in I_c} U_i}.$$

Then, Z is a linear subspace of X, $\dim X = m$, and $\dim Z = c$. Furthermore, as in the previous steps, we have that $R_p(g') \geq \lambda_k$, for any $g' \in X \setminus \{0\}$. The intersection property of Krasnoselskii's $\mathbb{Z}_2$-genus (Lemma 2.6.3) implies that

$$\min_{g' \in Y \setminus \{0\}} R_p(g') = \lambda_k,$$

for any linear subspace $Y \subset X \setminus (Z \setminus 0)$ such that $\dim Y \geq N - k + 1$. Fix such a space Y. Then the properties of variational eigenvalues imply that there exists an eigenfunction

$$g = \sum_{i=1}^{m} t_i f|_{U_i} \in Y \setminus 0,$$

with $R_p(g) = \lambda_k$. This implies that, also in this case, we must have equality in Lemma 9.2.2. The rest of the proof is the same as in the previous step. In particular, we arrive to a contradiction. This shows that $\overline{\mathfrak{W}}(f) \leq N - k + c$.

$\square$

Theorem 9.2.2 unifies many results that give upper bounds to the number of nodal domains for p-Laplacians on graphs and signed graphs, including Theorem 4.1 in [106], Theorem 5.4 in [157] for signed graphs, Theorems 3.4 and 3.5 in [266], as well as the inequality

$$\mathfrak{N}(f) \leq \min\{\mathfrak{S}(f), \overline{\mathfrak{S}}(f)\} \leq \min\{k + r - 1, N - k + r\},$$

where $\mathfrak{N}(f)$ is the number of connected components of the support of f (see Theorem 5.3 in [157] and Theorem 2.2 in [160] and Theorem 9.3.1). We would like to remark here that in [160], the authors also introduce the concept of up and down nodal domains for general nonlinear eigenproblems. In this general setting, the strong nodal count $\mathfrak{S}(f)$ (resp., $\overline{\mathfrak{S}}(f)$) actually equals the number of down (resp., up) nodal domains.

Remark 9.2.2. If f is an eigenfunction corresponding to an eigenvalue λ of Δ_1^s, then for any nodal domain A of f, both $f|_A$ and $\mathbf{1}_{A \cap \{f > 0\}} - \mathbf{1}_{A \cap \{f < 0\}}$ are also eigenfunctions corresponding to λ.

9.3 Nodal Domain Theorems on Hypergraphs

In this section, we work with concepts of positive and negative nodal domains similar to that in Definition 9.1.1, originally introduced for oriented hypergraphs in [219]. The results provided here are mainly taken from [157]. However, the first two statements in Theorem 9.3.1 are presented here for the first time. We give several preliminary definitions before presenting analogues of nodal domain theorems for general hypergraphs.

Definition 9.3.1 ([157, 219]). Let $\Gamma = (V, E, \varphi)$ be an oriented hypergraph. Given a function $f : V \to \mathbb{R}$, let $\text{supp}_\pm(f) := \{v \in V : \pm f(v) > 0\}$. A *positive strong nodal domain* of f is a connected component of the induced sub-hypergraph of Γ on $\text{supp}_+(f)$. A *negative strong nodal domain* of f is a connected component of the induced sub-hypergraph of Γ on $\text{supp}_-(f)$.

A positive (respectively, negative) *weak nodal domain* of f is a maximal connected induced sub-hypergraph of Γ on $\{v \in V : f(v) \geq 0\}$ (respectively, $\{v \in V : f(v) \leq 0\}$), that contains at least one nonzero vertex.

In the sequel, we shall need to work with functions that have the same sign on all the inputs of any oriented hyperedge and also the same sign on all outputs. For a simple graph, this is a trivial condition, as each oriented edge has only a single input and a single output, but for general hypergraphs, this is a nontrivial condition. We shall need this condition both for weak signs (≥ 0, ≤ 0) and strong signs (> 0, < 0).

Definition 9.3.2. A function $f : V \to \mathbb{R}$ is *sign-separable* if, for any $e \in E$, $f(v)f(v') \geq 0$ for all v and v' which are co-oriented in e. That is, the set e_{in} of inputs of e (respectively, the set e_{out} of outputs of e) is contained either in $\{v \in V : f(v) \geq 0\}$ or in $\{v \in V : f(v) \leq 0\}$.

Definition 9.3.3. A function $f : V \to \mathbb{R}$ is *strongly sign-separable* if, for any $e \in E$, the set e_{in} (respectively, the set e_{out}) is contained in either of $\{v \in V : f(v) > 0\}$, or $\{v \in V : f(v) < 0\}$, or $\{v \in V : f(v) = 0\}$.

Clearly, any strongly sign-separable function is also sign-separable. Moreover, if Γ is a simple graph, then any $f : V \to \mathbb{R}$ is strongly sign-separable.

Theorem 9.3.1. *Let* $\Gamma = (V, E, \varphi)$ *be an oriented hypergraph on N vertices, and let* $p \geq 1$. *Let* f *be an eigenfunction corresponding to the kth min–max eigenvalue* λ_k *of* Δ_p, *and let r be the multiplicity of* λ_k.

1. *If* f *is sign-separable, then* $\mathfrak{S}(f) \leq k + r - 1$.
2. *If* $p > 1$, Γ *is connected and* f *is strongly sign-separable, then* $\mathfrak{W}(f) \leq k$.
3. *If* Γ *has only inputs, then* $\mathfrak{S}(f) \leq N - k + r$.
4. *If* $p > 1$, Γ *is connected and has only inputs, then* $\mathfrak{W}(f) \leq N - k + 1$.
5. $\mathfrak{N}(f) \leq \min\{k + r - 1, N - k + r\}$.

Proof 1. Claim: If f is sign-separable, then $\mathfrak{S}(f) \leq k + r - 1$.

Assume the contrary, by contradiction. Then, f is a sign-separable eigenfunction of λ_k that has at least $k + r$ nodal domains, denoted by $V_1, \ldots, V_{k+r}$. Assume, without loss of generality, that

$$\lambda_k = \lambda_{k+1} = \ldots = \lambda_{k+r-1} < \lambda_{k+r}.$$

Moreover, let X be the space of linear functions spanned by $f|_{V_1}, \ldots, f|_{V_{k+r}}$. Since $V_1, \ldots, V_{k+r}$ are pairwise disjoint, this implies that $\dim X = k + r$. Hence, given $g \in X \setminus \{0\}$, there exists $(t_1, \ldots, t_{k+r}) \neq \mathbf{0}$ such that

$$g = \sum_{i=1}^{k+r} t_i f|_{V_i}.$$

By definition of positive and negative nodal domains, each hyperedge e intersects at most one positive nodal domain and at most one negative nodal domain. Moreover, since f is sign-separable, for $l \neq l'$ and $e \in E$,

$$\left(\delta f|_{V_l}(e)\right)\left(\delta f|_{V_{l'}}(e)\right) \geq 0.$$

For $p > 1$, we have

$$\sum_{e \in E}\left|\sum_{v \in e_{in}} g(v) - \sum_{v' \in e_{out}} g(v')\right|^p = \sum_{e \in E}\left|\sum_{v \in e_{in}}\sum_{l=1}^{k+r} t_l f|_{V_l}(v) - \sum_{v' \in e_{out}}\sum_{l=1}^{k+r} t_l f|_{V_l}(v')\right|^p$$

$$= \sum_{e \in E}\left|\sum_{l=1}^{k+r} t_l \delta f|_{V_l}(e)\right|^p \leq \sum_{e \in E}\sum_{l=1}^{k+r} |t_l|^p \delta f|_{V_l}(e)\,|\delta f(e)|^{p-2}\delta f(e)$$

$$= \sum_{l=1}^{k+r}|t_l|^p \sum_{v \in V_l} f(v)\sum_{e \ni v}|\delta f(e)|^{p-2}\left(\sum_{v' \in e\,:\,o_e(v,v')=-1} f(v') - \sum_{v' \in e\,:\,o_e(v,v')=1} f(v')\right)$$

$$= \sum_{l=1}^{k+r}|t_l|^p \sum_{v \in V_l} f(v)\lambda_k \deg v \cdot |f(v)|^{p-2}f(v)$$

$$= \lambda_k \sum_{l=1}^{k+r}|t_l|^p \sum_{v \in V_l} \deg v \cdot |f(v)|^p = \lambda_k \sum_{v \in V}\deg v \cdot |g(v)|^p,$$

where the inequality follows by taking $a = \sum_{v \in e} f|_{V_l}(v)$ and $b = \sum_{v \in e} f|_{V'_l}(v)$ in Lemma 9.2.2.

Similarly, for $p = 1$, we have

$$\sum_{e \in E} \left| \sum_{v \in e} g(v) \right| = \sum_{e \in E} \left| \sum_{l=1}^{k+r} t_l \delta f|_{V_l}(e) \right|$$

$$\leq \sum_{e \in E} \sum_{l=1}^{k+r} |t_l| \cdot \delta f|_{V_l}(e) z_e$$

$$= \sum_{l=1}^{k+r} |t_l| \sum_{v \in V_l} f(v) \left(\sum_{e_{in} \ni v} z_e - \sum_{e_{out} \ni v} z_e \right)$$

$$= \sum_{l=1}^{k+r} |t_l| \sum_{v \in V_l} f(v) \lambda_k \deg v \cdot z_v$$

$$= \lambda_k \sum_{l=1}^{k+r} |t_l| \sum_{v \in V_l} \deg v \cdot |f(v)|$$

$$= \lambda_k \sum_{v \in V} \deg v \cdot |g(v)|.$$

Then we have that $R_p(g) \leq \lambda_k$, where R_p is the Rayleigh quotient corresponding to Δ_p, which has been discussed, for oriented hypergraphs, in Section 7.4. By the min–max principle for Δ_p, we have

$$\lambda_{k+r} = \min_{X' \in \mathrm{Gen}_{k+r}} \max_{g' \in X' \setminus 0} R_p(g') \leq \max_{g \in X \setminus 0} R_p(g) = \lambda_k,$$

but this is a contradiction, since $\lambda_k < \lambda_{k+r}$. This proves the first claim.

2. Claim: If $p > 1$ and f is strongly sign-separable, then $\mathfrak{W}(f) \leq k$.

   If the multiplicity of λ_k is $r = 1$, then this claim follows directly from the first one. Therefore, it suffices to focus on the case where $r \geq 2$.

   Assume the contrary, by contradiction. Then, f is a strong sign-separable eigenfunction of λ_k, which has multiplicity $r \geq 2$, and f has at least $k + 1$ weak nodal domains, which we denote by $V_1, \ldots, V_{k+1}$. Assume that

$$\lambda_{k-1} < \lambda_k = \lambda_{k+1} = \ldots = \lambda_{k+r-1} < \lambda_{k+r},$$

and let X be the space of functions spanned by $f|_{V_1}, \ldots, f|_{V_{k+1}}$. As in the proof of the first claim, this implies that $R_p(g) \leq \lambda_k$, for all $g \in X \setminus \mathbf{0}$, and therefore

$$\max_{g \in Y \setminus \{0\}} R_p(g) = \lambda_k,$$

for any linear subspace $Y \subset X \setminus \{f\}$ such that $\dim Y \geq k$. Now, fix such a space Y. Then, there exists an eigenfunction

$$g = \sum_{i=1}^{k+1} t_i f|_{V_i} \in Y$$

such that $R_p(g) = \lambda_k$. By Lemma 9.2.2, for any V_l and $V_{l'}$ that are connected by a path, we have one of the following:

(a) $t_l = t_{l'}$, or
(b) There exist $v \in V_l$, $e \in E$, and $v' \in V_{l'}$ such that $f(v) = 0 \neq f(v')$ and $e \supset \{v, v'\}$.

In this case, since f and g are eigenfunctions, we have that

$$\sum_{e \ni v} |\delta f(v, e)|^{p-2} \delta f(v, e) = 0 \quad \text{and} \quad \sum_{e \ni v} |\delta g(v, e)|^{p-2} \delta g(v, e) = 0,$$

where

$$\delta f(v, e) := \sum_{v' \in e \,:\, o_e(v, v') = -1} f(v') - \sum_{v' \in e \,:\, o_e(v, v') = 1} f(v').$$

Since every e such that $v \in e$ has to satisfy $e \subset V_l$ or $e \subset V_{l'}$, from these equalities, we obtain

$$\left(|t_l|^{p-2} t_l - |t_{l'}|^{p-2} t_{l'} \right) \sum_{e \ni v, e \subset V_{l'}} \left| \delta f|_{V_{l'}}(v, e) \right|^{p-2} \delta f|_{V_{l'}}(v, e)$$

$$= \phi_p(t_l) \sum_{e \ni v} \phi_p(\delta f(v, e)) - \sum_{e \ni v} \phi_p(\delta g(v, e)) = 0.$$

The strongly sign-separable property implies that

$$\sum_{e \ni v, \, e \subset V_{l'}} \left| \delta f|_{V_{l'}}(v, e) \right|^{p-2} \delta f|_{V_{l'}}(v, e) \neq 0.$$

Thus, $|t_l|^{p-2} t_l - |t_{l'}|^{p-2} t_{l'} = 0$, which yields $t_l = t_{l'}$.

Hence, $t_l = t_{l'}$ whenever V_l and $V_{l'}$ are connected by a path. Since Γ is connected, $V_1, \ldots, V_{k+1}$ are all connected by a path. Therefore, $t_1 = \ldots = t_{k+1}$. This implies that $g \in \text{span}(f)$, which is a contradiction.

3. Claim: If Γ has only inputs, then $\mathfrak{S}(f) \leq N - k + r$.

   Assume the contrary, by contradiction. Then, f is an eigenfunction of λ_k, that has multiplicity r, and f has at least $N - k + r + 1$ nodal domains, which we denote by $V_1, \ldots, V_{N-k+r+1}$. Let X be the linear function space, which is spanned by

$$f|_{V_1}, \ldots, f|_{V_{N-k+r+1}}.$$

Then, since $V_1, \ldots, V_{N-k+r+1}$ are pairwise disjoint, $\dim X = N - k + r + 1$. Hence, given $g \in X \setminus 0$, there exists $(t_1, \ldots, t_{N-k+r+1}) \neq \mathbf{0}$ such that

$$g = \sum_{i=1}^{N-k+r+1} t_i f|_{V_i}.$$

By definition of positive and negative nodal domains, each hyperedge e intersects at most one positive nodal domain and at most one negative nodal domain. Thus, for $l \neq l'$ and $e \in E$, we have

$$\left(\sum_{v \in e_{in}} f|_{V_l}(v) \right) \cdot \left(\sum_{v \in e_{in}} f|_{V_{l'}}(v) \right) \leq 0.$$

Now, with a little abuse of notation, let $e = e_{in}$. For $p > 1$, we have that

$$\sum_{e \in E} \left| \sum_{v \in e} g(v) \right|^p = \sum_{e \in E} \left| \sum_{v \in e} \sum_{l=1}^{N-k+r+1} t_l f|_{V_l}(v) \right|^p$$

$$= \sum_{e \in E} \left| \sum_{l=1}^{N-k+r+1} t_l \left(\sum_{v \in e} f|_{V_l}(v) \right) \right|^p$$

$$\geq \sum_{e \in E} \sum_{l=1}^{N-k+r+1} |t_l|^p \left(\sum_{v \in e} f|_{V_l}(v) \right) \left| \sum_{v \in e} f(v) \right|^{p-2} \sum_{v \in e} f(v)$$

$$= \sum_{l=1}^{N-k+r+1} |t_l|^p \sum_{v \in V_l} f(v) \left(\sum_{e \in E :\, e \ni v} \left| \sum_{v' \in e} f(v') \right|^{p-2} \left(\sum_{v' \in e} f(v') \right) \right)$$

$$= \sum_{l=1}^{N-k+r+1} |t_l|^p \sum_{v \in V_l} f(v) \lambda_k \deg v \cdot |f(v)|^{p-2} f(v)$$

$$= \lambda_k \sum_{l=1}^{N-k+r+1} |t_l|^p \sum_{v \in V_l} \deg v \cdot |f(v)|^p$$

$$= \lambda_k \sum_{v \in V} \deg v \cdot |g(v)|^p,$$

where the inequality follows by taking $a = \sum_{v \in e} f|_{V_l}(v)$ and $b = \sum_{v \in e} f|_{V_{l'}}(v)$ in Lemma 9.2.2. Similarly, for $p = 1$, we have

$$\sum_{e \in E} \left| \sum_{v \in e} g(v) \right| = \sum_{e \in E} \left| \sum_{l=1}^{N-k+r+1} t_l \left(\sum_{v \in e} f|_{V_l}(v) \right) \right|$$

$$\geq \sum_{e \in E} \sum_{l=1}^{N-k+r+1} |t_l| \left(\sum_{v \in e} f|_{V_l}(v) \right) z_e$$

$$= \sum_{l=1}^{N-k+r+1} |t_l| \sum_{v \in V_l} f(v) \left(\sum_{e \in E : e \ni v} z_e \right)$$

$$= \sum_{l=1}^{N-k+r+1} |t_l| \sum_{v \in V_l} f(v) \lambda_k \deg v \cdot z_v$$

$$= \lambda_k \sum_{l=1}^{N-k+r+1} |t_l| \sum_{v \in V_l} \deg v \cdot |f(v)|$$

$$= \lambda_k \sum_{v \in V} \deg v \cdot |g(v)|,$$

where $z_e \in \mathrm{Sgn}(\sum_{v \in e} f(v))$ and $z_v \in \mathrm{Sgn}(f(v))$. By Lemma 9.2.2, it follows that $R_p(g) \geq \lambda_k$.

By the intersection property of the $\mathbb{Z}_2$-genus (Lemma 2.6.3), $X' \cap X \setminus \{0\} \neq \emptyset$, for any $X' \in \mathrm{Gen}_{k-r}$. Therefore,

$$\lambda_{k-r} = \inf_{X' \in \mathrm{Gen}_{k-r}} \sup_{g' \in X' \setminus 0} R_p(g')$$

$$\geq \inf_{X' \in \mathrm{Gen}_{k-r}} \sup_{g' \in X' \cap X \setminus 0} R_p(g')$$

$$\geq \inf_{X' \in \mathrm{Gen}_{k-r}} \inf_{g' \in X' \cap X \setminus 0} R_p(g')$$

$$\geq \inf_{X' \in \mathrm{Gen}_{k-r}} \inf_{g' \in X \setminus 0} R_p(g')$$

$$= \inf_{g \in X \setminus 0} R_p(g)$$

$$\geq \lambda_k.$$

Together with $\lambda_{k-r} \leq \ldots \leq \lambda_{k-1} \leq \lambda_k$, this implies that

$$\lambda_{k-r} = \ldots = \lambda_{k-1} = \lambda_k,$$

therefore, the multiplicity of λ_k is at least $r + 1$. This is a contradiction.

4. Claim: If $p > 1$, Γ is connected and has only inputs, then $\mathfrak{W}(f) \leq N - k + 1$.

Assume the contrary, by contradiction. Then, f is an eigenfunction of λ_k that has multiplicity $r \geq 2$, and f has at least $N - k + 2$ nodal domains, which we denote by $V_1, \ldots, V_{N-k+2}$. Let X be the linear function space spanned by $f|_{V_1}, \ldots, f|_{V_{N-k+2}}$. Since $V_1, \ldots, V_{N-k+2}$ are pairwise disjoint, we have that $\dim X = N - k + 2$. Also, for any linear subspace $Y \subset X \setminus \{f\}$ such that $\dim Y = N - k + 1$, as in the proof of the third claim, one can show that

$$\min_{g \in Y \setminus 0} R_p(g) = \lambda_k.$$

Now, given $g \in Y \setminus \{0\}$, there exists $(t_1, \ldots, t_{N-k+2}) \neq \mathbf{0}$ such that

$$g = \sum_{i=1}^{N-k+2} t_i f|_{V_i}.$$

By definition of nonpositive and non-negative nodal domains, each hyper-edge e intersects at most one nonpositive nodal domain and at most one non-negative nodal domain. By following the same reasoning as in the proof of the second claim, we can show that $t_1 = \ldots = t_{N-k+2}$. Hence, $g = t_1 f$, which leads to a contradiction, since $f \notin Y$ while $g \in Y$.

5. Claim: $\mathfrak{N}(f) \le \min\{k + r - 1, N - k + r\}$.

Assume that the support of f has m connected components and denote them by $V_1, \ldots, V_m$. Let then X be the linear function space spanned by $f|_{V_1}, \ldots, f|_{V_m}$. As for the first and third claims, one can check that $R_p(g) = \lambda_k$ for any $g \in X \setminus 0$. By the min–max principle for Δ_p,

$$\lambda_m = \inf_{X' \in \mathrm{Gen}_m} \sup_{g' \in X' \setminus 0} R_p(g') \le \sup_{g \in X \setminus 0} R_p(g) = \lambda_k \le \ldots \le \lambda_{k+r-1} < \lambda_{k+r},$$

which implies that $m \le k + r - 1$. Also, by the intersection property of Krasnoselskii's $\mathbb{Z}_2$-genus (Lemma 2.6.3), $X' \cap X \setminus \{0\} \neq \emptyset$ for any $X' \in \mathrm{Gen}_{N-m+1}$. Therefore,

$$\lambda_{N-m+1} = \inf_{X' \in \mathrm{Gen}_{N-m+1}} \sup_{g' \in X' \setminus 0} R_p(g')$$

$$\ge \inf_{X' \in \mathrm{Gen}_{N-m+1}} \inf_{g' \in X' \cap X \setminus 0} R_p(g')$$

$$\ge \inf_{g \in X \setminus 0} R_p(g)$$

$$\ge \lambda_k$$

$$\ge \ldots$$

$$\ge \lambda_{k-r+1}$$

$$> \lambda_{k-r},$$

which yields $N - m + 1 \ge k - r + 1$, or equivalently, $m \le N - k + r$.

$\square$

PART III

Additional Topics: Interlacing, Tensors, Nonbacktracking Laplacians, and Applications

10

Interlacing and Spectral Classes

We derive various interlacing results under particular manipulations of graphs and simplicial complexes, such as contraction, duplication, or graph doubling. We also define asymptotic spectral classes using the Wasserstein distance.

10.1 Interlacing

Interlacing theorems provide inequalities between the eigenvalues of graphs that are related by some geometric operations, such as adding or deleting an edge, or more generally, some subgraph.

Interlacing theorems originally arose in an algebraic, rather than a geometric context, concerned with the eigenvalues of two linear operators A, B (which are assumed symmetric to make their eigenvalues real) that are related by some operation. An interlacing theorem then states that for some integers n_1, n_2, the eigenvalues α_k of A and the eigenvalues β_k of B, both arranged in increasing order, satisfy an inequality of the form

$$\alpha_{k-n_1} \leq \beta_k \leq \alpha_{k+n_2} \quad \text{for all } k. \tag{10.1.1}$$

The original example, the *Cauchy interlacing theorem*, controls the eigenvalues of a symmetric matrix B obtained by deleting some rows and columns of a symmetric A in terms of those of A. Let us recall the result and its proof, as they will also guide our subsequent constructions.

Theorem 10.1.1. *Let A be a symmetric $(N \times N)$-matrix with eigenvalues $\alpha_1 \leq \ldots \leq \alpha_N$, and let $P \colon \mathbb{R}^N \to \mathbb{R}^M$ be an orthogonal projection. $B := PAP^\star$ then is a symmetric $(M \times M)$-matrix, and its eigenvalues $\beta_1 \leq \ldots \leq \beta_M$ satisfy*

$$\alpha_k \leq \beta_k \leq \alpha_{k+N-M}. \tag{10.1.2}$$

Proof We shall apply Theorem 2.6.1. Thus, take $H = \mathbb{R}^N$ with the Euclidean scalar product, let b_i be an eigenvector of B for the eigenvalue β_i, and let S_k be spanned by $b_1, \ldots, b_k$. Then, by (2.6.2), assuming $g \neq 0$ throughout,

$$
\begin{aligned}
\beta_k &= \max_{g \in S_k} \frac{(Bg, g)}{(g, g)} \\
&= \max_{g \in S_k} \frac{(PAP^\star g, g)}{(g, g)} \\
&> \min_{H_k \in \mathcal{H}_k} \max_{g \in H_k} \frac{(AP^\star g, P^\star g)}{(g, g)} \\
&= \alpha_k .
\end{aligned}
\tag{10.1.3}
$$

For the upper bound, let V_{M-k+1} be spanned by $b_k, \ldots, b_M$. By (2.6.2) again,

$$
\begin{aligned}
\beta_k &= \min_{g \in V_{M-k+1}} \frac{(Bg, g)}{(g, g)} \\
&= \min_{g \in V_{M-k+1}} \frac{(PAP^\star g, g)}{(g, g)} \\
&= \min_{g \in V_{M-k+1}} \frac{(AP^\star g, P^\star g)}{(g, g)} \\
&\leq \max_{H_{M-k+1} \in \mathcal{H}_{M-k+1}} \min_{g \in H_{M-k+1}} \frac{(AP^\star g, P^\star g)}{(g, g)} \\
&= \max_{H_{M-k+1} \in \mathcal{H}_{M-k+1}} \min_{g \in H_{M-k+1}} \frac{(AP^\star g, P^\star g)}{(P^\star g, P^\star g)} \frac{(P^\star g, P^\star g)}{(g, g)} \\
&\leq \alpha_{k+N-M},
\end{aligned}
\tag{10.1.4}
$$

since

$$
\frac{(P^\star g, P^\star g)}{(g, g)} \leq 1.
\tag{10.1.5}
$$

$\square$

In fact, one even has equality in the last line in (10.1.4), but the interpolation of $\frac{(P^\star g, P^\star g)}{(P^\star g, P^\star g)}$ and the use of (10.1.5) will be important for our subsequent reasoning.

We may, of course, directly apply Theorem 10.1.1 to the adjacency matrix of a graph and derive interpolation theorems for algebraic Laplacians. But, we want to develop a general scheme for interlacing inequalities for Laplacians of graphs and simplicial complexes. For that purpose, we shall systematically expand the scheme of the previous proof, following [137].

We can then systematically study the effects of various operations or modifications of simplicial complexes on Laplacian spectra, like the deletion, collapsing, and contracting of subcomplexes, simplicial maps, and coverings.

In the remainder of this chapter, we shall mainly derive interlacing equalities for up-Laplacians L_{up} of simplicial complexes. The same techniques can of course also yield such results for down-Laplacians. But we know from (4.8.6) that the nonzero spectrum of L_{down}^{q+1} equals that of L_{up}^q, and therefore, it suffices to treat the latter. Anyway, in the most important case, that of graphs, the vertex Laplacian is an up-Laplacian.

10.1.1 Deletion of Subcomplexes

The main result of this section is

Theorem 10.1.2. *Let (Σ, w_Σ) and $(\Sigma', w_{\Sigma'})$ be $(q + 1)$-dimensional, weighted simplicial complexes for which their proper difference (Definition 2.5.11) (Θ, w_Θ) exists. Let $\lambda_1 \leq \lambda_2 \leq \ldots \leq \lambda_N$ and $\theta_1 \leq \theta_2 \leq \ldots \leq \theta_N$ be the eigenvalues of the Laplace operators $L_{up}^q(\Sigma)$ and $L_{up}^q(\Theta)$, respectively. Then, we have*

$$\theta_k \leq \lambda_{k+D_{\Sigma'}} \tag{10.1.6}$$

for $k = 1, 2, \ldots, N - D_{\Sigma'}$, and for $k = 1, 2, \ldots, N$

$$\lambda_{k-D_W} \leq \theta_k, \tag{10.1.7}$$

where $D_W = \dim C^{q+1}(\Sigma') - \dim H^{q+1}(\Sigma')$, $D_{\Sigma'} = \dim C^q(\Sigma')$, and where we put $\lambda_j = 0$ whenever $j \leq 0$.

In fact, we also get an upper bound (10.1.6) for all k, if we define λ_j for $j > N$ appropriately as the general upper bound for the eigenvalues. For the normalized Laplacian, we have

$$\lambda_j = N + 2 \quad \text{for } j > N, \tag{10.1.8}$$

while for the combinatorial Laplacian,

$$\lambda_j = |V| \text{ (the cardinality of the vertex set)} \quad \text{for } j > N. \tag{10.1.9}$$

According to our general convention, we shall denote the set of q-simplices of a simplicial complex Λ by Λ_q.

Before entering the proof, we need some technical constructions, using the setting of Section 2.5.2.

For a weighted simplicial complex (Σ, w_Σ), a subcomplex $(\Sigma', w_{\Sigma'})$, and their proper difference (Θ, w_Θ), we let $i_q \colon C^q(\Theta) \to C^q(\Sigma)$ be the inclusion map (i.e., $(i_q g)[F] = g([F])$, for every $F \in S_q(\Sigma)$, $g \in C^q(\Theta)$) and let $\pi_{q+1} \colon C^{q+1}(\Sigma) \to C^{q+1}(\Theta)$ be the projection map (i.e., $(\pi_{q+1} \bar{f})[\bar{G}] = \bar{f}([\bar{G}])$, for all $\bar{G} \in S_{q+1}(\Theta)$ and $\bar{f} \in C^{q+1}(\Sigma))$. Their (formal, because the scalar product may be degenerate) adjoints $i_q^* : C^q(\Sigma) \to C^q(\Theta)$ and $\pi_{q+1}^* : C^{q+1}(\Theta) \to C^{q+1}(\Sigma)$ of the inclusion and projection maps satisfy

$$i_q^* f([G]) = \begin{cases} \frac{w_\Sigma(G)}{w_\Theta(G)} f(i_q[G]) = \frac{w_\Sigma(G)}{w_\Theta(G)} f(i_q[G]) & \text{if } w_\Theta(G) > 0, \\ 0 & \text{otherwise,} \end{cases}$$

(10.1.10)

and

$$\pi_{q+1}^* \bar{g}([\bar{F}]) = \frac{w_\Theta(\bar{F})}{w_\Sigma(\bar{F})} \bar{g}(\pi_{q+1}[\bar{F}]) = \frac{w_\Theta(\bar{F})}{w_\Sigma(F)} \bar{g}([\bar{F}]).$$

(10.1.11)

Lemma 10.1.1. *For* $0 \le q < \dim \Sigma,$

$$\delta_\Theta^q = \pi_{q+1} \delta_\Sigma^q i_q$$

(10.1.12)

$$(\delta_\Theta^q)^\star = i_q^\star (\delta_\Sigma^q)^\star \pi_{q+1}^\star.$$

(10.1.13)

Moreover,

$$i_q^* i_q g([G]) = \begin{cases} \frac{w_\Sigma(G)}{w_\Theta(G)} g([G]) & \text{for } w_\Theta(G) > 0, \\ 0 & \text{otherwise} \end{cases}$$

(10.1.14)

and

$$\pi_{q+1}^* \pi_{q+1} \bar{f}([\bar{F}]) = \frac{w_\Theta(\bar{F})}{w_\Sigma(\bar{F})} \bar{f}([\bar{F}]).$$

(10.1.15)

(In fact, i_q^* is not quite the adjoint of i_q, but what we need is (10.1.14) whose factor interpolates between the scalar products on $C^q(\Sigma)$ and $C^q(\Theta)$ on all q-simplices that have a positive weight in Θ.)

Proof (10.1.12) and (10.1.13) follow from the commutativity of the diagrams

$$
\begin{array}{ccc}
C^{q+1}(\Sigma) & \xleftarrow{\ \delta_\Sigma\ } & C^q(\Sigma) \\
\downarrow{\scriptstyle \pi_{q+1}} & & \uparrow{\scriptstyle i_q} \\
C^{q+1}(\Theta) & \xleftarrow{\ \delta_\Theta\ } & C^q(\Theta)
\end{array}
$$

(10.1.16)

$$
\begin{array}{ccc}
C^{q+1}(\Sigma) & \xrightarrow{\ \delta_\Sigma^*\ } & C^q(\Sigma) \\
\uparrow{\scriptstyle \pi_{q+1}^*} & & \downarrow{\scriptstyle i_q^*} \\
C^{q+1}(\Theta) & \xrightarrow{\ \delta_\Theta^*\ } & C^q(\Theta)
\end{array}
$$

(10.1.17)

Of course, the second is a consequence of the first, since π_{q+1}^* is the adjoint of π_{q+1}, and i_q^* satisfies this role for i_q on the image of δ_Σ^*.

(10.1.15) and (10.1.14) follow from (10.1.11) and (10.1.10).   □

We shall now use the diagrams (10.1.16) and (10.1.17) to factorize the Rayleigh quotient of $L_{up}^q(\Theta)$. For simplicity of notation, we omit the indices q of i and L and $q + 1$ of π. We have, recalling (2.6.6) and the convention that $R(g) = 0$ if $g = 0$,

$$R_{L_{up}(\Theta)}(g) = \frac{(\delta_\Theta^* \delta_\Theta g, g)}{(g,g)} = \frac{(i^* \delta_\Sigma^* \pi^* \pi \delta_\Sigma i g, g)}{(g,g)}$$

$$= \frac{(\pi^* \pi \delta_\Sigma i g, \delta_\Sigma i g)}{(\delta_\Sigma i g, \delta_\Sigma i g)} \frac{(\delta_\Sigma i g, \delta_\Sigma i g)}{(i g, i g)} \frac{(i g, i g)}{(g, g)}$$

$$= R_{\pi^* \pi}(\delta_\Sigma i g) R_{L_{up}(\Sigma)}(i g) R_{i^* i}(g). \qquad (10.1.18)$$

We now look at the three Rayleigh quotients in (10.1.18) in turn.

We start with the operator $\pi^* \pi$.

Lemma 10.1.2. *The eigenvalues of the operator $\pi^* \pi \colon C^{q+1}(\Sigma) \to C^{q+1}(\Sigma)$ are*

$$\left\{ \frac{w_\Theta(\bar{F})}{w_\Sigma(\bar{F})} \mid \bar{F} \in \Sigma_{q+1} \right\},$$

hence $R_{\pi^* \pi}(f) \le 1$. Moreover,

$$R_{\pi^* \pi}(\delta_\Sigma i g) = 1 \text{ if and only if } g_{|C^q(\Sigma')} \in \ker \delta_{\Sigma'}. \qquad (10.1.19)$$

These are the functions $g \in C^n(\Theta)$ that are orthogonal to $W := \operatorname{coker} \delta_{\Sigma'}$, and

$$D_W := \dim W = \dim C^{q+1}(\Sigma') - \dim H^{q+1}(\Sigma'). \qquad (10.1.20)$$

Proof The first part follows from (10.1.18).

By (10.1.15) $\bar{f} := \delta_\Sigma i g$ satisfies

$$R_{\pi^* \pi}(\bar{f}) = \frac{\sum_{\bar{F} \in \Theta_{q+1}} w_\Theta(\bar{F}) \bar{f}([\bar{F}])^2}{\sum_{\bar{F} \in \Sigma_{q+1}} w_\Sigma(\bar{F}) \bar{f}([\bar{F}])^2}. \qquad (10.1.21)$$

This is equal to 1 precisely if $w_{\Sigma'}(\bar{F}) \bar{f}([\bar{F}])^2 = (w_\Sigma(\bar{F}) - w_\Theta(\bar{F})) \bar{f}([\bar{F}])^2 = 0$ for every $\bar{F} \in \Sigma_{q+1}$, that is, if $\bar{f}$ is identically equal to zero on Σ'. This means that the restriction of $g \in C^q(\Theta)$ on $C^q(\Sigma')$ is in the kernel of $\delta_{\Sigma'}$. Therefore, the function g is orthogonal to $W := \operatorname{coker} \delta_{\Sigma'}$. Now

$$\dim \operatorname{coker} \delta^q = \dim \operatorname{im} \delta^q = \dim \ker \delta^{q+1} - \dim H^{q+1}, \qquad (10.1.22)$$

and since Σ' and K are $(q + 1)$-dimensional, $\ker \delta^{q+1} = C^{q+1}(\Sigma')$ and

$$\dim \operatorname{im} \delta_{\Sigma'} = \dim C^{n+1}(\Sigma') - \dim H^{n+1}(\Sigma') \qquad (10.1.23)$$

and we get (10.1.20). $\qquad \square$

We now turn to the eigenvalues of $i^* i$.

Lemma 10.1.3. *Let $g \in C^q(\Theta)$. Then,*

(i) $(g,g) > 0 \Rightarrow R_{i^*i}(g) \geq 1$.
(ii) $R_{i^*i}(g) = 1$ if and only if $g \perp C^q(\Sigma')$.

Proof From (10.1.14),

$$\frac{\sum_{G \in \Theta_q} w_K(G)g([G])^2}{\sum_{G \in \Theta_q} w_\Theta(G)g([G])^2}. \tag{10.1.24}$$

Since $w_K(G) \geq w_\Theta(G)$, (i) holds. By (10.1.24), $R_{i^*i}(g) = 1$ if and only if $g([G]) = 0$ for every $G \in \Sigma'_q$, yielding (ii). □

Proof of Theorem 10.1.2 We shall systematically use the fact that maximization (minimization) over a smaller set increases the maximum (decreases the minimum).

Let $\mathcal{H}_k$ be the set of k-dimensional subspaces of $C^q(\Sigma)$ and with $N = \dim C^q(\Sigma)$, let $\lambda_1 \leq \lambda_2 \leq \ldots \leq \lambda_N$ and $\theta_1 \leq \theta_2 \leq \ldots \leq \theta_N$ be the eigenvalues of the operators $L_{up}^q(\Sigma)$ and $L_{up}^q(\Theta)$, respectively. Then, recalling $W := \operatorname{coker} \delta_{\Sigma'}$ with $D_W = \dim W = \dim C^{q+1}(\Sigma') - \dim H^{q+1}(\Sigma')$ from (10.1.20),

$$
\begin{aligned}
\theta_k &= \min_{\mathcal{H}_k \in \mathcal{H}_k} \max_{g \in H_k} R_{L_{up}^q(\Theta)}(g) \\
&= \min_{\mathcal{H}_k \in \mathcal{H}_k} \max_{g \in H_k} R_{\pi^*\pi}(\delta_\Sigma ig) R_{L_{up}^q(\Sigma)}(ig) R_{i^*i}(g) \quad \text{by (10.1.18)} \\
&\geq \min_{\mathcal{H}_k \in \mathcal{H}_k} \max_{g \in H_k} R_{\pi^*\pi}(\delta_\Sigma ig) R_{L_q^{up}(\Sigma)}(ig) \quad \text{by Lemma 10.1.3} \\
&\geq \min_{\mathcal{H}_k} \max_{g \in H_k, g \perp W} R_{L_{up}^q(\Sigma)}(ig) \quad \text{by Lemma 10.1.2} \\
&\geq \min_{\mathcal{H}_k} \max_{g \in H_{k-D_W}} R_{L_{up}^q(\Sigma)}(ig) \\
&= \min_{\mathcal{H}_{k-D_W}} \max_{g \in H_{k-D_W}} R_{L_{up}^q(\Sigma)}(ig) \\
&= \lambda_{k-D_W}.
\end{aligned}
$$

This proves (10.1.7).

For (10.1.6), we use (10.1.18) and Theorem 2.6.1 again, getting

$$
\begin{aligned}
\theta_k &= \max_{\mathcal{H}_{N-k+1}} \min_{f \in H_{N-k+1}} R_{L_{up}^q(\Theta)}(f) \\
&= \max_{\mathcal{H}_{N-k+1}} \min_{f \in H_{N-k+1}} R_{\pi^*\pi}(\delta_\Sigma if) R_{L_{up}^q(\Sigma)}(if) R_{i^*i}(f) \\
&\leq \max_{\mathcal{H}_{N-k+1}} \min_{g \in H_{N-k+1}} R_{\Delta_\Sigma}(ig) R_{i^*i}(g) \quad \text{by Lemma 10.1.2} \\
&\leq \max_{\mathcal{H}_{N-k+1}} \min_{g \in H_{N-k+1}, g \perp C^q(\Sigma')} R_{\Delta_\Sigma}(ig) \quad \text{by Lemma 10.1.3}
\end{aligned}
$$

$$\leq \max_{\mathcal{H}_{N-k+1}} \min_{g \in H_{N-k+1-D_{\Sigma'}}} R_{\Delta_{\Sigma}}(ig) \tag{10.1.25}$$

$$= \max_{\mathcal{H}_{N-k+1-D_{\Sigma'}}} \min_{g \in H_{N-k+1-D_{\Sigma'}}} R_{\Delta_{\Sigma}}(ig)$$

$$= \lambda_{k+D_{\Sigma'}}. \tag{10.1.26}$$

□

Since we are particularly interested in the normalized graph Laplacian, we formulate the version of Theorem 10.1.2 for graphs explicitly.

Corollary 10.1.1. *Let* Γ' *be a weighted subgraph of the weighted graph* Γ *with* N *vertices and eigenvalues* $\lambda_1 \leq \ldots \leq \lambda_N$ *of the normalized Laplacian, and let* Θ *be their proper difference, with eigenvalues* $\theta_1 \leq \ldots \leq \theta_N$ *of its normalized Laplacian. Assume that the positive weight part of* Θ *is a (possibly disconnected) graph. Suppose that* Γ' *has* N' *vertices and* M' *connected components. Then,*

$$\lambda_{k-N'+M'} \leq \theta_k \leq \lambda_{k+N'} \quad \text{for } 1 \leq k \leq N, \tag{10.1.27}$$

if we put $\lambda_j = 0$ *for* $j \leq 0$ *and* $\lambda_j = 2$ *for* $j \geq N + 1$.

Proof Using the Euler formula $\sum_{i=0,1}(-1)^i \dim C^i(G) = \sum_i (-1)^i \dim H^i(G)$ for every graph, we get $D_W = \dim C^1(\Gamma') - \dim H^1(\Gamma') = \dim C^0(\Gamma') - \dim H^0(\Gamma') = N' - M'$, and we also have $D_{\Gamma'} = \dim C^0(\Gamma') = N'$. The result thus follows from Theorem 10.1.2, recalling (10.1.8). □

The result for the combinatorial Laplacian is almost the same, except that the largest eigenvalue is now bounded by N instead of 2.

Corollary 10.1.2. *Let* Γ' *be a weighted subgraph of the weighted graph* Γ *with* N *vertices and eigenvalues* $\lambda_1 \leq \ldots \leq \lambda_N$ *of the combinatorial Laplacian, and let* Θ *be their proper difference, with eigenvalues* $\theta_1 \leq \ldots \leq \theta_N$ *of its normalized Laplacian. Assume that the positive weight part of* Θ *is a (possibly disconnected) graph. Suppose that* Γ' *has* N' *vertices and* M' *connected components. Then,*

$$\lambda_{k-N'+M'} \leq \theta_k \leq \lambda_{k+N'} \quad \text{for } 1 \leq k \leq N, \tag{10.1.28}$$

if we put $\lambda_j = 0$ *for* $j \leq 0$ *and* $\lambda_j = N$ *for* $j \geq N + 1$. □

Example: Suppose that $\Gamma' = (v, v')$ is a single edge, where v is a pending vertex of Γ, that is, its only neighbor is v'. Assume $w_{\Gamma'}(v, v') = w_{\Gamma}(v, v')$. Then, the proper difference Θ, obtained by removing the edge (v, v'), has v as an isolated vertex, and $\deg_{\Theta} v' = \deg_{\Gamma} v' - 1$. Then,

$$\theta_k \geq \lambda_{k-1} \quad \text{for all } k. \tag{10.1.29}$$

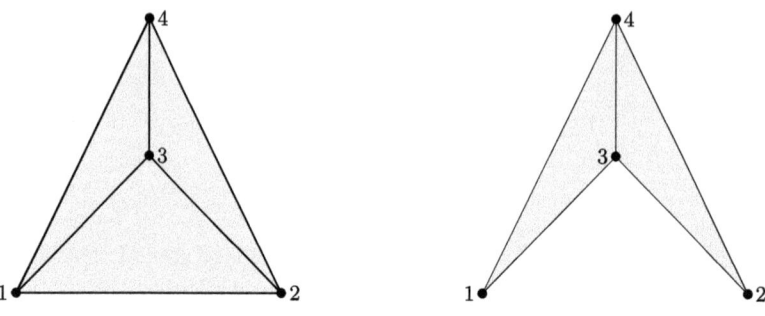

Figure 10.1 A collapse.

In particular, $\theta_2 = \theta_1 = 0$, but $\theta_3 \geq \lambda_2$.

We now present another operation, which, however, can also be interpreted as the deletion of a subcomplex.

Definition 10.1.1. Let Σ be a simplicial complex and $(\bar{F}, F)$ a pair of simplices with $F \in \partial\bar{F}$. The face $\bar{F}$ is called a *free face*, if it is not a facet of any other simplex in Σ. A simplicial complex Σ' obtained from Σ by deleting $\bar{F}$ and F and the simplicial map $\varphi \colon \Sigma \to \Sigma'$ that represents such an operation yields an *elementary collapse* of Σ. A sequence of elementary collapses is called a *collapse*.

In Figure 10.1, the edge $[1,2]$ and the simplex $[1,2,3]$ are collapsed.

An elementary collapse can also be seen as the deletion of a subcomplex, according to Definition 2.5.11, at least when the weights are properly chosen. Therefore, interlacing inequalities for elementary collapses follow from Theorem 10.1.2 when the Laplacians are either combinatorial (all weights $\equiv 1$) or normalized (the weight of a q-simplex is the sum of the weights of the $(q + 1)$-simplices of which it is a facet).

Theorem 10.1.3. *Let $\varphi \colon \Sigma \to \Sigma'$ be an elementary collapse, and let $\lambda_1 \leq \ldots \leq \lambda_{N_\Sigma}$ and $\theta_1 \leq \ldots \theta_{N_{\Sigma'}}$ be the eigenvalues of the combinatorial or the normalized Laplacians on Σ and Σ'. Then,*

$$\lambda_k \leq \theta_k \leq \lambda_{k+q+3}, \qquad (10.1.30)$$

where $N = \lambda_{N_\Sigma+1} = \ldots = \lambda_{N_\Sigma+q+3}$, and N is the number of vertices of Σ.

$\square$

10.1.2 Simplicial Maps

Definition 10.1.2. A *simplicial map* $\varphi \colon \Sigma \to \Sigma'$ is a map from the vertices of the simplicial complex Σ to those of Σ' with the property that $\varphi(v_1), \ldots, \varphi(v_k)$ span a simplex whenever $v_1, \ldots, v_k$ do.

We consider such a simplicial map $\varphi \colon \Sigma \to \Sigma'$ and assume that we have a weight function w_Σ on Σ, and we then want to use φ to construct a weight function on Σ' so that the Laplacians on Σ and Σ' are naturally related, and we can prove interlacing inequalities for their eigenvalues. In fact, there are two natural possibilities to define $w_{\Sigma'}$. The first option is

$$w_{\Sigma'}(G) = \sum_{\substack{F \in \Sigma_q : \\ \varphi(F) = G}} w_\Sigma(F), \text{ for every } G \in \Sigma'_q. \tag{10.1.31}$$

The second option is

$$w_{\Sigma'}(G) = \sum_{\substack{F \in \Sigma_q : \\ \varphi(F) = G}} w_\Sigma(F) - \sum_{\bar{F} \in \Sigma_{q+1} : \varphi(\bar{F}) = G} w_\Sigma(\bar{F}), \text{ for } G \in \Sigma'_q. \tag{10.1.32}$$

In fact, it suffices for our purposes to consider q- and $(q + 1)$-simplices in Σ that are mapped to a q-simplex in Σ' because this suffices for determining the eigenvalues of L^q_{up}, which is the operator we want to study here.

The weight function $w_{\Sigma'}$ in (10.1.32) is then motivated by the following consideration. Let (Σ, w_Σ) be a complex with normalized weights, that is, $w_\Sigma(F) = \sum_{\bar{F} : F \in \bar{F}} w_\Sigma(\bar{F})$, and let $\varphi \colon \Sigma \to \Sigma'$ be a simplicial map. We consider the situation where $\bar{F}_1, \ldots, \bar{F}_k$ are $(q + 1)$-faces of Σ, but their φ-images are only q-dimensional, that is, $\varphi(\bar{F}_i) = G_i \in \Sigma'_q$. Σ' then is a complex in which $\varphi(\bar{F}_i)$ are loops; thus,

$$\deg G = \sum_{\substack{\bar{G} \in \Sigma'_{q+1} : \\ G \in \partial \bar{G}}} w_{\Sigma'}(\bar{G}) + \sum_{\substack{\bar{F}_i \in \bar{\Sigma}_{q+1} : \\ G \in \partial \varphi(\bar{F}_i)}} w_\Sigma(\bar{F}_i).$$

If now (10.1.31) holds, then $\deg G = \sum_{\substack{\bar{G} \in \Sigma'_{q+1} : \\ G \in \partial \bar{G}}} w_{\Sigma'}(\bar{G}) + \sum_{\substack{\bar{F}_i \in \bar{\Sigma}_{q+1} : \\ G \in \partial \varphi(\bar{F}_i)}} w_\Sigma(\bar{F}_i)$, and

$$w_{\Sigma'}(G_i) = \sum_{\substack{F \in \Sigma_q : \\ \varphi(F) = G}} w_\Sigma(F)$$

$$= \sum_{\substack{\bar{F} \in \Sigma_{q+1}, \bar{F} \neq \bar{F}_i : \\ F \in \partial \bar{F}}} w_\Sigma(\bar{F}) + 2 w_\Sigma(\bar{F}_i)$$

$$= \sum_{\substack{\bar{F} \in \Sigma_{q+1}: \\ F \in \partial \bar{F}}} w_\Sigma(\bar{F}) + w_\Sigma(\bar{F}_i)$$

$$= \deg G_i + w_\Sigma(\bar{F}_i).$$

Therefore the weight function $w_{\Sigma'}$ is not normalized. With definition (10.1.32) for $w_{\Sigma'}$, however, we get $w_{\Sigma'}(G) = \deg G$, and $w_{\Sigma'}$ is normalized. Consequently, with this choice of weights, if $L_{up}^q(\Sigma, w_\Sigma)$ is the normalized Laplacian of Σ, then $L_{up}^q(\Sigma', w_{\Sigma'})$ is the normalized Laplacian of Σ'.

Having defined the weights, we obtain the adjoints $(\varphi^*)^q$ of the cochain maps φ^q defined in Section 2.5.1.

Lemma 10.1.4. *When Σ' is equipped with the weight function (10.1.31) or (10.1.32), the diagram*

$$
\begin{array}{ccc}
C^q(\Sigma) & \xleftarrow{\;L_{up}(\Sigma)\;} & C^q(\Sigma) \\[2pt]
\Big\downarrow{\varphi^*} & & \Big\uparrow{\varphi} \\[6pt]
C^q(\Sigma') & \xleftarrow{\;L_{up}(\Sigma')\;} & C^q(\Sigma')
\end{array}
\qquad (10.1.33)
$$

commutes. In particular, for $G \in \Sigma_q'$, we have the relation

$$L_{up}^q(\Sigma')e_{[G]} = \varphi^* L_{up}^q(\Sigma)\varphi e_{[G]} = \sum_{\substack{F \in \Sigma_q: \\ \varphi(F)=G}} \sum_{\substack{F' \in \Sigma_q: \\ F, \Phi \in \partial \bar{F}}} \mathrm{sgn}([F],[G])\,\mathrm{sgn}([F],$$

$$\partial[\bar{F}])\,\mathrm{sgn}([\Phi],[\varphi(\Phi)])\,\mathrm{sgn}([\Phi],\partial[\bar{F}])\frac{w_\Sigma(\bar{F})}{w_{\Sigma'}(\varphi(\Phi))}e_{[\varphi(\Phi)]}. \qquad (10.1.34)$$

Proof We first assume that φ preserves dimensions of simplices, that is, all q-simplices are mapped to q-simplices. In other words, φ is injective on the vertices $v_0, \ldots, v_q$ of every q-simplex F. In that case, (10.1.31) and (10.1.32) agree.

Let $[\bar{F}] = [v_0, \ldots, v_{q+1}]$ be a positively oriented $(q+1)$-simplex in Σ, and

$$[F] = [v_0, \ldots, \hat{v}_i, \ldots v_{q+1}], \quad [\Phi] = [v_0, \ldots, \hat{v}_j, \ldots v_{q+1}]$$

its oriented facets. Let $\bar{G} = \{\varphi(v_0), \ldots, \varphi(v_{q+1})\}$, $G = \{\varphi(v_0), \ldots, \varphi(\hat{v}_i), \ldots, \varphi(v_{q+1})\}$, and $\varphi(\Phi) = \{\varphi(v_0), \ldots, \varphi(\hat{v}_j), \ldots, \varphi(v_{q+1})\}$ be the images of $\bar{F}, F$, and Φ, respectively. Assume $[\bar{G}], [G]$, and $[\varphi(\Phi)]$ are all positively oriented. Then,

$$[\bar{G}] = [\varphi(v_0), \ldots, \varphi(v_{q+1})]\,\mathrm{sgn}([\bar{F}],[\bar{G}]),$$

$$[G] = [\varphi(v_0), \ldots, \varphi(\hat{v}_i), \ldots, \varphi(v_{q+1})]\,\mathrm{sgn}([F],[G])$$

and

$$[\varphi(\Phi)] = [\varphi(v_0), \ldots, \varphi(\hat{v}_j), \ldots, \varphi(v_{q+1})]\, \mathrm{sgn}([F], [G]).$$

Therefore,

$$\mathrm{sgn}([G], \partial[\bar{G}]) = \mathrm{sgn}\big([\varphi(v_0), \ldots, \varphi(\hat{v}_i), \ldots, \varphi(v_{q+1})]\, \mathrm{sgn}([F], [G]),$$
$$[\varphi(v_0), \ldots, \varphi(v_{q+1})]\, \mathrm{sgn}([\bar{F}], [\bar{G}])\big)$$
$$= (-1)^i\, \mathrm{sgn}([F], [G])\, \mathrm{sgn}([\bar{F}], [\bar{G}])$$
$$= \mathrm{sgn}([F], \partial[\bar{F}])\, \mathrm{sgn}([F], [G])\, \mathrm{sgn}([\bar{F}], [\bar{G}]). \tag{10.1.35}$$

Therefore, (10.1.34) is

$$\varphi^* L_{up}^q(\Sigma)\varphi e_{[G]} =$$

$$= \sum_{\substack{F \in \Sigma_q: \\ \varphi(F) = G}} \sum_{\substack{\Phi \in \Sigma_q: \\ F, \Phi \in \partial \bar{F}}} \mathrm{sgn}([G], \partial[\bar{G}])\, \mathrm{sgn}([\varphi(\Phi)], \partial[\bar{G}]) \frac{w_\Sigma(\bar{F})}{w_{\Sigma'}(\varphi(\Phi))} e_{[\varphi(\Phi)]}$$

$$\tag{10.1.36}$$

$$= \sum_{\substack{F \in \Sigma_q: \\ \varphi(F) = G}} \sum_{\substack{\bar{F} \in \Sigma_{q+1}: \\ F \in \partial \bar{F}}} w_\Sigma(\bar{F}) \sum_{\substack{\Phi: \Phi \in \partial \bar{F}}} \mathrm{sgn}([G], \partial[\bar{G}])\, \mathrm{sgn}([\varphi(\Phi)], \partial[\bar{G}]) \frac{e_{[\varphi(\Phi)]}}{w_{\Sigma'}(\varphi(\Phi))}.$$

$$\tag{10.1.37}$$

Since by (10.1.31), $w_{\Sigma'}(G) = \sum_{\substack{F \in \Sigma_q: \\ \varphi(F) = G}} w_\Sigma(F)$, (10.1.36) yields

$$\varphi^* L_{up}^q(\Sigma)\varphi e_{[G]} = \sum_{\substack{\varphi(\Phi) \in \Sigma_q': \\ G, \varphi(\Phi) \in \partial \bar{G}}} \mathrm{sgn}([G], \partial[\bar{G}])\, \mathrm{sgn}([G]', \partial[\bar{G}]) \frac{w_{\Sigma'}(\bar{G})}{w_{\Sigma'}(\varphi(\Phi))} e_{[\varphi(\Phi)]}$$

$$= L_{up}^q e_{[G]}. \tag{10.1.38}$$

Thus, diagram (10.1.33) commutes.

We now consider the situation where there exists some q-simplex in Σ, whose image under map φ is m-dimensional, $m < q$. Then, we find $[\bar{F}] = [v_0, \ldots, v_{q+1}]$ with $\varphi(v_i) = \varphi(v_j)$ for some $i \neq j$, and so, $[F_i] = [v_0, \ldots, \hat{v}_i, \ldots v_{q+1}]$, $[F_j] = [v_0, \ldots, \hat{v}_j, \ldots v_{q+1}]$, and $\varphi(\bar{F}) = \varphi(F_i) = \varphi(F_j) = G$. Then,

$$\sum_{\substack{F \in \{F_i, F_j\} \\ F, \Phi \in \partial \bar{F}}} \sum_{\substack{\bar{F} \in \Sigma_{q+1}:}} \mathrm{sgn}([F], [G])\, \mathrm{sgn}([F], \partial[\bar{F}])[\Phi, \varphi(\Phi)]\, \mathrm{sgn}([F]',$$

$$\partial[\bar{F}]) \frac{w_\Sigma(\bar{F})}{w_{\Sigma'}(\varphi(\Phi))} e_{[\varphi(\Phi)]} \tag{10.1.39}$$

$$= \frac{w_\Sigma(\bar{F})}{w_{\Sigma'}(G)} \text{sgn}([F_i],[G])(-1)^i \text{sgn}([F_j],[G])(-1)^j e_{[G]}$$

$$+ \frac{w_\Sigma(\bar{F})}{w_{\Sigma'}(G)} \text{sgn}([F_i],[G])(-1)^i \text{sgn}([F_i],[G])(-1)^i e_{[G]}$$

$$+ \frac{w_\Sigma(\bar{F})}{w_{\Sigma'}(G)} \text{sgn}([F_i],[G])(-1)^i \text{sgn}([F_j],[G])(-1)^j e_{[G]}$$

$$+ \frac{w_\Sigma(\bar{F})}{w_{\Sigma'}(G)} \text{sgn}([F_j],[G])(-1)^j \text{sgn}([F_j],[G])(-1)^j e_{[G]}$$

$$= \frac{w_\Sigma(\bar{F})}{w_{\Sigma'}(G)} (\text{sgn}([F_i],[G])(-1)^i (-1)^j \text{sgn}([F_i],[G])(-1)^{j-i+1} e_{[G]} + e_{[G]})$$

$$+ \frac{w_\Sigma(\bar{F})}{w_{\Sigma'}(G)} (\text{sgn}([F_j],[G])(-1)^j (-1)^i \text{sgn}([F_j],[G])(-1)^{j-i+1} e_{[G]} + e_{[G]})$$

$$= 0. \tag{10.1.40}$$

Therefore, $L_{up}^q(\Sigma') e_{[G]} = \varphi^* L_{up}^q(\Sigma)\varphi e_{[G]}$, and the diagram (10.1.33) commutes for both choices (10.1.32) and (10.1.31) for the weight function $w_{\Sigma'}$. □

We can now derive interlacing results for simplicial maps. We first state

Theorem 10.1.4. *Let* Σ, Σ' *be simplicial complexes with* N_Σ *and* $N_{\Sigma'}$ *q-simplices, and let* Σ *have N vertices. Let* $\varphi\colon \Sigma \to \Sigma'$ *be a simplicial map, and let the weight functions* w_Σ *and* $w_{\Sigma'}$ *be related by (10.1.31). Let* $\lambda_1 \leq \ldots \leq \lambda_{N_\Sigma}$ *and* $\theta_1 \leq \ldots \leq \theta_{N_{\Sigma'}}$ *be the eigenvalues of* $L_{up}^q(\Sigma, w_\Sigma)$ *and* $L_{up}^q(\Sigma', w_{\Sigma'})$. *Then,*

$$\lambda_k \leq \theta_k \leq \lambda_{k+N_\Sigma-N_{\Sigma'}}, \tag{10.1.41}$$

with $\lambda_{N_\Sigma+1} = \ldots = \lambda_{2N_\Sigma-N_{\Sigma'}} = N$.

Proof For $g \in C^q(\Sigma)$, we have

$$(\varphi g, \varphi g) = \sum_{G \in \Sigma'_{q+1}} \sum_{\substack{F \in \Sigma_q: \\ \varphi(F)=G}} (\text{sgn}([F],[G]) g(\varphi F))^2 w_\Sigma(F) \tag{10.1.42}$$

$$= \sum_{G \in \Sigma'_q} g(G)^2 w_{\Sigma'}(G) = (g, g). \tag{10.1.43}$$

Here and in the sequel, $\mathcal{H}_\ell$ will be the collection of all ℓ-dimensional subspaces of $C^q(\Sigma')(\cong \mathbb{R}^{N_{\Sigma'}})$ and we shall take the maximum (or minimum) over all vector spaces $H_\ell \in \mathcal{H}_\ell$, as prescribed by Theorem 2.6.1.

By Theorem 2.6.1, (10.1.34), and (10.1.42), we then have

$$\theta_k = \max_{H_{k-1} \in \mathcal{H}_{k-1}} \min_{g \perp H_{k-1}} \frac{(L_{up}(\Sigma')g, g)}{(g, g)} \tag{10.1.44}$$

$$= \max_{H_{k-1}} \min_{g \perp H_{k-1}} \frac{(\varphi^* L_{up}(\Sigma)\varphi g, g)}{(g,g)} \tag{10.1.45}$$

$$= \max_{H_{k-1}} \min_{g \perp H_{k-1}} \frac{(L_{up}(\Sigma)\varphi g, \varphi g)}{(\varphi g, \varphi g)}. \tag{10.1.46}$$

Let $\bigcup_j F_{ji} = \varphi^{-1}(G_i)$ and $\bigcup_i G_i = \Sigma_q'$ and W be the vector space generated by

$$\mathrm{sgn}([F_{ji}],[G_i])e_{[F_{ji}]} - \mathrm{sgn}([F_{(i+1)j}],[G_i])e_{[F_{(i+1)j}]}.$$

The dimension of W is $N_\Sigma - N_{\Sigma'}$. We let $\mathcal{V}_\ell$ be the collection of all ℓ-dimensional subspaces of $C^q(\Sigma)(\cong \mathbb{R}^{N_\Sigma})$, and we shall take the maximum (or minimum) over all vector spaces $V_\ell \in \mathcal{V}_\ell$, again as prescribed by Theorem 2.6.1.

Let $\varphi g = f$. From (10.1.46), we get

$$\theta_k = \max_{V_{k-1} \in \mathcal{V}_{k-1}} \min_{f \perp V_{k-1}, f \perp W} \frac{(L_{up}(\Sigma)f, f)}{(f,f)} \tag{10.1.47}$$

$$\geq \max_{V_{k-1}} \min_{f \perp V_{k-1}} \frac{(L_{up}(\Sigma)f, f)}{(f,f)} \tag{10.1.48}$$

$$\geq \lambda_k. \tag{10.1.49}$$

Inequality (10.1.48) holds since we are minimizing over a larger set. From (10.1.47), we also obtain the upper interlacing inequality

$$\theta_k \leq \max_{H_{k-1+N_\Sigma - N_{\Sigma'}}} \min_{f \perp H_{k-1+N_\Sigma - N_{\Sigma'}}} \frac{(L_{up}(\Sigma)f, f)}{(f,f)} \tag{10.1.50}$$

$$= \lambda_{k+N_\Sigma - N_{\Sigma'}}. \tag{10.1.51}$$

Recalling (10.1.40), we also obtain the result when the dimension of simplices is not necessarily preserved. □

For the second choice of the weight function of Σ', we have

Theorem 10.1.5. *Let (Σ, w_Σ) and $(\Sigma', w_{\Sigma'})$ be weighted simplicial complexes with their weights related by (10.1.32), and $\varphi \colon \Sigma \to \Sigma'$ a simplicial map. Let Z be the vector space with basis $\{e_{[G]} \mid \exists \bar{F} \in \Sigma_{q+1} \colon \varphi(\bar{F}) = G\}$, let z be its dimension, and let W be the vector space spanned by*

$$\{\mathrm{sgn}([F_{ji}],[G_i])e_{[F_{ji}]} - \mathrm{sgn}([F_{(i+1)j}],[G_i])e_{[F_{(i+1)j}]} \mid \bigcup_j F_{ji}$$

$$= \varphi^{-1}(G_i), \text{ and} \bigcup_i G_i = \Sigma_q'\}.$$

Then, the eigenvalues $\lambda_1 \leq \lambda_2 \leq \ldots \leq \lambda_{N_\Sigma}$ *of* $L_{up}^q(\Sigma)$ *and the eigenvalues* $\theta_1 \leq \theta_2 \leq \ldots \leq \theta_{N_\Sigma}$ *of* $L_{up}^q(\Sigma')$ *satisfy*

$$\lambda_k \leq \theta_k \leq \lambda_{k+z+N_\Sigma - N_{\Sigma'}}. \tag{10.1.52}$$

Proof We shall state inequalities here that are analogous to those in Theorem 10.1.4, and therefore, we do not repeat all details.

Let W be, as before, a vector space spanned by

$$\{\mathrm{sgn}([F_{ji}],[G_i])e_{[F_{ji}]} - \mathrm{sgn}([F_{(i+1)j}],[G_i])e_{[F_{(i+1)j}]} \mid \bigcup_j F_{ji}$$

$$= \varphi^{-1}(G_i), \text{ and } \bigcup_i G_i = \Sigma'_q\}.$$

Let $\phi\colon C^q(\Sigma') \to C^q(\Sigma')$ be an operator with

$$\phi e_{[G]} = \sum_{\bar{F} \in \Sigma_{q+1}\,:\, \varphi(\bar{F})=G} \frac{w_\Sigma(\bar{F})}{w_{\Sigma'}(G)} e_{[G]},$$

and let Z be a vector space of dimension z spanned by vectors

$$\{e_{[G]} \mid G \in \Sigma'_q, \exists \bar{F} \in \Sigma_{q+1}\colon \varphi(\bar{F}) = G\}.$$

Assume that $\varphi\colon \Sigma \to \Sigma'$ is a simplicial map, and w_Σ and $w_{\Sigma'}$ are weight functions satisfying 10.1.32.

As before, $\mathcal{H}_\ell$ will be the collection of all ℓ-dimensional subspaces of $C^q(\Sigma')$, and we shall take the maximum (or minimum) over all vector spaces $H_\ell \in \mathcal{H}_\ell$, as prescribed by Theorem 2.6.1.

With these conventions, we have

$$\theta_k = \max_{H_{k-1} \in \mathcal{H}_{k-1}} \min_{g \perp H_{k-1}} \frac{(L_{up}(\Sigma')g, g)}{(g, g)} \tag{10.1.53}$$

$$= \max_{H_{k-1}} \min_{g \perp H_{k-1}} \frac{(\varphi^* L_{up}(\Sigma)\varphi g, g)}{(g, g)} \tag{10.1.54}$$

$$= \max_{H_{k-1}} \min_{g \perp H_{k-1}} \frac{(L_{up}(\Sigma)\varphi g, \varphi g)}{(\varphi g, \varphi g) - (\phi g, g)} \tag{10.1.55}$$

$$\geq \max_{H_{k-1}} \min_{g \perp H_{k-1}} \frac{(L_{up}(\Sigma)\varphi g, \varphi g)}{(\varphi g, \varphi g)} \tag{10.1.56}$$

$$= \max_{H_{k-1}} \min_{f \perp H_{k-1}, f \perp W} \frac{(L_{up}(\Sigma)f, f)}{(f, f)} \tag{10.1.57}$$

$$\geq \max_{H_{k-1}} \min_{g \perp H_{k-1}} \frac{(L_{up}(\Sigma)f, f)}{(f, f)} \tag{10.1.58}$$

$$\geq \lambda_k. \tag{10.1.59}$$

On the other hand, the upper interlacing inequality follows from (10.1.55). As before, $\mathcal{V}_\ell$ is the collection of all ℓ-dimensional subspaces of $C^q(\Sigma)(\cong \mathbb{R}^{N_\Sigma})$.

$$\theta_k \leq \max_{V_{k-1} \in \mathcal{V}_{k-1}} \min_{g \perp V_{k-1}} \frac{(L_{up}(\Sigma)\varphi g, \varphi g)}{(\varphi g, \varphi g)} \tag{10.1.60}$$

$$\leq \max_{V_{k-1}} \min_{g \perp V_{k-1}, g \perp Z} \frac{(L_{up}(\Sigma)\varphi g, \varphi g)}{(\varphi g, \varphi g)} \tag{10.1.61}$$

$$\leq \max_{V_{k+z-1}} \min_{g \perp V_{k+z-1},} \frac{(L_{up}(\Sigma)\varphi g, \varphi g)}{(\varphi g, \varphi g)} \tag{10.1.62}$$

$$= \max_{V_{k+z-1}} \min_{f \perp V_{k+z-1}, f \perp W} \frac{(L_{up}(\Sigma)f, f)}{(f, f)} \tag{10.1.63}$$

$$\leq \max_{V_{k+z+N_\Sigma-N_{\Sigma'}-1}} \min_{f \perp V_{k+z+N_\Sigma-N_{\Sigma'}-1}} \frac{(L_{up}(\Sigma)f, f)}{(f, f)} \tag{10.1.64}$$

$$\leq \lambda_{k+N_\Sigma-N_{\Sigma'}+z} \tag{10.1.65}$$

$\square$

10.1.3 Coverings

In a topological setting, a continuous map $f\colon M \to N$ between two compact spaces is a *covering* if every point in the image has a neighborhood V whose preimage $f^{-1}(V)$ consists of disjoint sets $U_1,\ldots,U_m$ that are all homeomorphically mapped to V by f. In particular, in the context of Riemannian manifolds, we require f to be an isometry, and every Laplacian eigenfunction of N can then be pulled back by f to a Laplacian eigenfunction of M. Consequently, the spectrum of N is contained in that of M. In order to get such a result in the simplicial context, we first need to clarify the notion of a covering between simplicial complexes. A naive version would simply require for a simplicial map $\varphi\colon \Sigma \to \Sigma'$ between simplicial complexes to be a covering such that the preimage of every simplex F is a disjoint union of simplices all mapped bijectively to F. That, however, does not suffice to let the spectrum of Σ' be contained in that of Σ. Simply take a path with one edge (i.e., a single edge with its two vertices) and a path with two edges and join them at a vertex to get a path with three edges.

The union of the two original paths then would be a covering in that sense of the path with three edges, but the latter has the eigenvalues $1/2$ and $3/2$, which do not belong to the spectra of the shorter paths. In fact, an even simpler

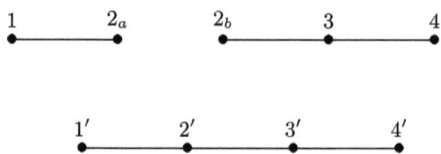

Figure 10.2 A covering, but not a strong covering.

example would join two paths of length 1 into a path of length 2. The latter has the eigenvalue 1, but the former does not.

The problem with that naive notion is that the boundaries of simplices may have different numbers of preimages than those simplices themselves. Since the Laplace operator involves boundary operations, of course, we can then not hope for strong results. Or putting it differently, the analogue of a neighborhood in a Riemannian manifold in the context of simplicial complexes should also involve the boundary facets of simplices. We thus formulate

Definition 10.1.3. A simplicial map $\varphi\colon \Sigma \to \Sigma'$ is a *strong covering* if for every q-simplex G in Σ' that is a facet of some $\bar{G}$ and every $F \in \varphi^{-1}(G)$, there exists $\bar{F} \in \Sigma_{q+1}$ such that F is a facet of $\bar{F}$ and $\varphi(\bar{F}) = \bar{G}$.

This property is not satisfied in (Figure 10.2), because there, the vertex $2'$ in the image is a facet of the edge $[2',3']$, but its preimage 2_a is not a facet of an edge that is mapped to the edge $[2',3']$.

Remark 10.1.1. Thus, under a strong covering for any two q-simplices G_1 and G_2 in Σ' that are $(q+1)$-up neighbors, and for every $F_1 \in \varphi^{-1}(G_1)$, there must exist $F_2 \in \Sigma_q$ such that F_1 and F_2 are $(q+1)$-up neighbors and $\varphi(F_2) = G_2$. Again, this is not satisfied in (Figure 10.2) because the vertices $2',3'$ are 1-up neighbors, but the preimage 2_a is not a 1-neighbor of any vertex that is mapped to $3'$.

Lemma 10.1.5. *Let* $\varphi\colon \Sigma \to \Sigma'$ *be a strong covering. Then, every* $F \in \Sigma_q$ *and every* $G \in \Sigma'_q$ *with* $\varphi(F) = G$ *are facets of the same number of* $(q+1)$-*simplices, that is,*

$$|\{\bar{F} \in \Sigma_{q+1} \mid F \in \partial\bar{F}\}|= |\{\bar{G} \in \Sigma'_{q+1} \mid G \in \partial\bar{G}\}|.$$

In particular, all preimages of a q-simplex in Σ' are facets of the same number of $(q+1)$-simplices. Thus, the local structure is the same.

Proof Assume that $F \in \partial\bar{F}_i$ for $\bar{F}_i \in \Sigma_{q+1}$, for $i \in \{1,\ldots,k\}$. Then, since φ is a (strong) covering, $\varphi(\bar{F}_i) \in \Sigma'_{q+1}$ and $\varphi(F) \in \partial\varphi(\bar{F}_i)$, and $\varphi(\bar{F}_i) \neq \varphi(\bar{F}_j)$ for $i \neq j$. Consider $\bar{G} \in \Sigma'_{q+1}$ with $G \in \partial\bar{G}$. Since φ is a strong covering, we can find $\bar{F} \in \Sigma_{q+1}$ with $F \in \partial\bar{F}$, such that $\varphi(\bar{F}) = \bar{G}$. Moreover, since φ is a simplicial map, if $G \in \partial\bar{G}, \partial\bar{G}'$ and $\bar{G} \neq \bar{G}'$, then $\bar{F} \neq \bar{F}'$. □

Lemma 10.1.6. *Let $\varphi\colon \Sigma \to \Sigma'$ be a strong covering. Assume that Σ' is $(q+1)$-up path connected for every q (i.e., for any two $G, G' \in \Sigma'_q$, there is a sequence $G = G_1, \bar{G}_1, G_2, \bar{G}_2, \ldots, \bar{G}_m, G_{m+1} = G'$ such that G_i and G_{i+1} are facets of the $(q+1)$-simplex $\bar{G}_i$ for $i = 1, \ldots, m$). Then, all simplices in Σ' have the same number of preimages, called the degree of the covering. That is*

$$|\{F \in \Sigma_q \mid F \in \varphi^{-1}(G), G \in \Sigma'_q\}| = c$$

for some positive integer c, for every $F \in \Sigma$.

Proof Let G and G' be $(q+1)$-up neighbors in Σ'. Then, there is some $\bar{G} \in \Sigma'_{q+1}$ with $G, G' \in \partial\bar{G}$. Since φ is a strong covering, for every $F \in \varphi^{-1}(G)$, there exists $\bar{F}$ with $\varphi(\bar{F}) = \bar{G}$. Thus, there exists $F' \in \partial\bar{F}$ in the preimage of G'. Therefore, if G and G' are $(q+1)$-up neighbors, then $|\{F \in \Sigma_q \mid F \in \varphi^{-1}(G)\}| = |\{F \in \Sigma_q \mid F' \in \varphi^{-1}(G')\}|$. Since Σ' is $(q+1)$-path connected, for q-faces G and G' of Σ', the conclusion follows. □

Theorem 10.1.6. *Let $\varphi\colon \Sigma \to \Sigma'$ be a strong covering where Σ' is $(q+1)$-up path connected, and let the weights in Σ and Σ' be related by $\frac{w_\Sigma(\bar{F})}{w_\Sigma(F)} = \frac{w_{\Sigma'}(\varphi(\bar{F}))}{w_{\Sigma'}(\varphi(F))}$, for every pair $\bar{F}, F$ with $F \in \partial\bar{F}$. Then, $\varphi L^q_{up}(\Sigma') = L^q_{up}(\Sigma)\varphi$, that is,*

$$\mathrm{Spec}(L^q_{up}(\Sigma')) \subset \mathrm{Spec}(L^q_{up}(\Sigma)\varphi). \tag{10.1.66}$$

Proof We have

$$\varphi L_{up}(\Sigma')e_{[G]} = \tag{10.1.67}$$

$$= \sum_{\substack{\bar{G} \in \Sigma'_{q+1}: \\ G', G \in \partial\bar{G}}} \sum_{\substack{F' \in \Sigma_q: \\ \varphi(F') = G'}} \mathrm{sgn}([F'], [G'])\, \mathrm{sgn}([G], \partial[\bar{G}])\, \mathrm{sgn}([G'], \partial[\bar{G}]) \frac{w_{\Sigma'}(\bar{G})}{w_{\Sigma'}(G')} e_{[F']}.$$

$$L_{up}(\Sigma)\varphi e_{[G]} = \tag{10.1.68}$$

$$= \sum_{\substack{F \in \Sigma_q: \\ \varphi(F) = G}} \sum_{\substack{\bar{F} \in \Sigma_{q+1}: \\ F, F' \in \partial\bar{F}}} \mathrm{sgn}([F], [G])\, \mathrm{sgn}([F], \partial[\bar{F}])\, \mathrm{sgn}([F'], \partial[\bar{F}]) \frac{w_\Sigma(\bar{F})}{w_\Sigma(F')} e_{[F']}.$$

We now claim

$$\{F' \in \Sigma_q \mid \exists \bar{G} \in \Sigma'_{q+1}, : G, \varphi(F') \in \partial\bar{G}\} =$$

$$\{F' \in \Sigma_q \mid \exists \bar{F} \in \Sigma_{q+1}, : F, F' \in \partial\bar{F}, \text{ where } \varphi(F) = G\}. \tag{10.1.69}$$

If a multiple of $e_{[F']}$ occurs in the sum in (10.1.68), then there exists $F \in \Sigma_q$ with $\varphi(F) = G$, and F and F' are $(q + 1)$-up neighbors. Therefore, $\varphi(F')$ and G are $(q + 1)$-up neighbors as well, and $e_{[F']}$ is a summand in (10.1.67).

On the other hand, if a multiple of $e_{[F']}$ appears in (10.1.67), then F' is a q-face of the simplicial complex Σ, such that $\varphi(F')$ and G are $(q + 1)$-up neighbors. Since φ is a strong covering, there must exist $F \in \Sigma_q$ with $\varphi(F) = G$, and F, F' are $(q + 1)$-up neighbors; hence, $e_{[F']}$ is a summand in (10.1.68). This proves (10.1.69).

By (10.1.35), we have

$$\text{sgn}([F], [G]) \, \text{sgn}([F], \partial[\bar{F}]) \, \text{sgn}([F'], \partial[\bar{F}])$$
$$= \text{sgn}([G], \partial[\bar{G}]) \, \text{sgn}([\bar{F}], [\bar{G}]) \, \text{sgn}([F'], \partial[\bar{F}])$$
$$= \text{sgn}([G], \partial[\bar{G}]) \, \text{sgn}([G'], \partial[\bar{G}]) \, \text{sgn}([F'], [G']),$$

which makes (10.1.67) and (10.1.68) equal.   □

In particular, for the normalized Laplace operators, we obtain an inclusion of spectra.

Corollary 10.1.3. *If $\varphi \colon \Sigma \to \Sigma'$ is a strong covering, and Σ' is $(q + 1)$-up path connected, then for the normalized Laplace operators Δ_{up}^q, we have*

$$\text{Spec}(\Delta_{up}^q(\Sigma')) \subset \text{Spec}(\Delta_{up}^q(\Sigma)). \qquad (10.1.70)$$

□

10.1.4 Contraction

In this section, we analyze the effect of another operation, namely contraction, on the eigenvalues of the Laplace operator.

Definition 10.1.4. Let Σ be a simplicial complex of dimension $q + 1$, and let $\bar{F}_1$ be a $(q + 1)$-simplex, for which precisely two of its facets F, F', are incident to other $(q + 1)$-simplices. An *elementary contraction* then is obtained by the construction of a simplicial complex Σ' and a simplicial map $\varphi \colon \Sigma \to \Sigma'$ in the following manner. We first delete the simplex $\bar{F}_1$ and identify F and F'. Thus, there exist vertices $v \in F$ and $v' \in F'$ with $\varphi(v) = \varphi(v')$, and φ is injective on all other vertices of Σ. When there exist $2m$ $(q + 1)$-simplices $\bar{F}_2$, $\bar{F}_2', \ldots, \bar{F}_{m+1}, \bar{F}_{m+1}'$ of Σ such that F is a facet of the $\bar{F}_i$, and F' is a facet of the $\bar{F}_i'$ and $\bar{F}_i$ and $\bar{F}_i'$ share another facet, we let φ also identify $\bar{F}_i$ and $\bar{F}_i'$, for $i = 2, \ldots, m$. When $m > 0$, we say that the resulting elementary contraction is of type (*i*). When there are no such pairs, we can let φ be injective on Σ_{q+1} and say that the resulting elementary contraction is of type (*ii*).

A sequence of elementary contractions is called a *contraction*.

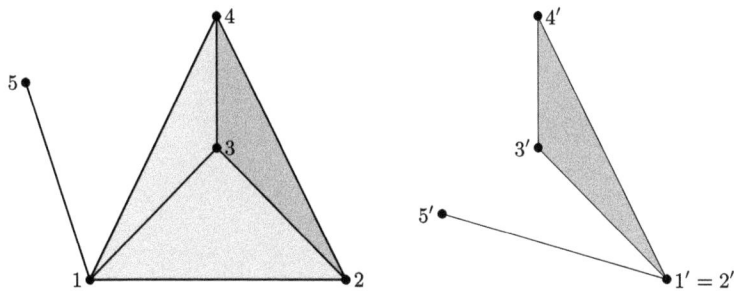

Figure 10.3 An elementary contraction of type (i).

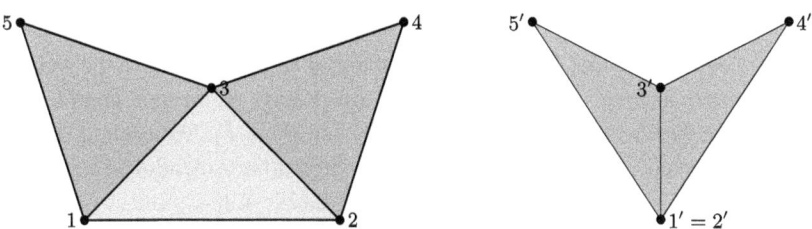

Figure 10.4 An elementary contraction of type (ii).

Remark 10.1.2. Assume that the elementary contraction $\varphi \colon \Sigma \to \Sigma'$ identifies the q-faces F and F' of the $(q + 1)$-simplex $\bar{F}$. If F' is not a facet of any other simplex, then this operation is a collapse, as in Definition 10.1.1, and therefore can be treated as such. Thus, to avoid redundancy, in the definition of an elementary contraction, we had required that both F and F' (but no further facets of $\bar{F}$) are incident to other $(q + 1)$-simplices.

Figures 10.3 and 10.4 represent two main types of elementary contractions, type *(i)* and type *(ii)*, respectively. In Figure 10.3, the simplex with vertices 1,2,3 is contracted, and then the simplex 1,3,4 has to be mapped onto the simplex $1' = 2',3',4'$. In contrast, in Figure 10.4, the simplex 1,2,3 is contracted without further ado by simply identifying the vertices 1 and 2.

Although collapses and contractions are simplicial maps, we have to recall that the interlacing theorems of Section 10.1.2 require specific choices of weights on Σ', as specified in (10.1.32) and (10.1.31).

Theorem 10.1.7. *Let $L_{up}^q(\Sigma)$ be either the combinatorial Laplacians $L_{up}^q(\Sigma, w_\Sigma)$ with $w_\Sigma \equiv 1$, or the normalized Laplacian $\Delta_{up}^q(\Sigma)$, and analogously for the Laplacian on Σ'. Let $\varphi \colon \Sigma \to \Sigma'$ be an elementary contraction, and let $\lambda_1 \leq \ldots \leq \lambda_{N_\Sigma}, \theta_1 \leq \ldots \leq \theta_{N_{\Sigma'}}$ be the eigenvalues of $L_{up}^q(\Sigma)$ and $L_{up}^q(\Sigma')$, then*

(i)

$$\lambda_{k-m(q+2)} \le \theta_k \le \lambda_{k+N_\Sigma-N_{\Sigma'}+m(q+2)}, \tag{10.1.71}$$

if the contraction φ is of type (i), or

(ii)

$$\lambda_k \le \theta_k \le \lambda_{k+q+2}, \tag{10.1.72}$$

if φ is of type (ii),

where $\lambda_{N_\Sigma+1} = \ldots = \lambda_{2N_\Sigma-N_{\Sigma'}+m(q+1)} = N$, $\lambda_0 = \lambda_{-1} = \ldots = \lambda_{-n-1} = 0$, and N is the number of vertices of Σ.

Proof Assume first that L_{up}^q is the combinatorial Laplacian, that is, the weight functions w_Σ and $w_{\Sigma'}$ are identically equal to 1. Let the elementary contraction φ be given by identification of the facets F_1 and F_1' of $\bar{F}_1$. We need to distinguish between the two types of elementary contractions φ mentioned earlier. If φ is of type *(i)*, let $\bar{F}_2, \bar{F}_2', \ldots, \bar{F}_{m+1}, \bar{F}_{m+1}'$ be the $(q+1)$-simplices of Σ that are identified by φ. Let $\psi : C^q(\Sigma') \to C^q(\Sigma')$ be a map with

$$\psi e_{[G]} = \sum_{G' \in \partial \bar{G}_i} w_\Sigma(\bar{F}_i) \, \mathrm{sgn}([G], \partial[\bar{G}_i]) \, \mathrm{sgn}([G'], \partial[\bar{G}_i]) \frac{1}{w_{\Sigma'}(G')} e_{[G']},$$

$$\tag{10.1.73}$$

if φ is of type *(i)*, $G \in \partial \varphi(\bar{F}_i)$, where $\bar{G}_i = \varphi(\bar{F}_i)$, for all $2 \le i \le m+1$, and

$$\psi e_{[G]} = 0, \tag{10.1.74}$$

for other choices of $G \in \Sigma'$, if φ is of type *(ii)*.

According to (10.1.37), by the arguments of Section 10.1.2, we have

$$\varphi^* L_{up}^q(\Sigma) \varphi e_G$$
$$= \sum_{\substack{\bar{F} \in \Sigma_{q+1}: \\ G \in \partial \varphi(\bar{F})}} w_\Sigma(\bar{F}) \sum_{G' \in \partial \bar{G}} \mathrm{sgn}([G], \partial[\bar{G}]) \, \mathrm{sgn}([G'], \partial[\bar{G}]) \frac{e_{[G']}}{w_{\Sigma'}(G')}.$$

Therefore,

$$\varphi^* L_{up}^q(\Sigma) \varphi e_{[G]} = L_{up}^q(\Sigma') e_{[G]}$$
$$+ w_\Sigma(\bar{F}_i) \sum_{G' \in \partial \bar{G}_i} \mathrm{sgn}([G], \partial[\bar{G}_i]) \, \mathrm{sgn}([G'], \partial[\bar{G}_i]) \frac{e_{[G']}}{w_{\Sigma'}(G')},$$

if φ is of type *(i)* and $G \in \partial \varphi(\bar{F}_i)$, $2 \le i \le m+1$, and

$$\varphi^* L_{up}^q(\Sigma) \varphi e_{[G]} = L_{up}^q(\Sigma') e_{[G]}, \tag{10.1.75}$$

otherwise. Thus, $L_{up}^q(\Sigma') = \varphi^* L_{up}^q(\Sigma)\varphi - \psi$. We then have the following equalities for $(\varphi g, \varphi g)$.

$$(\varphi g, \varphi g) = \sum_{G \in \Sigma_q'} \sum_{\substack{F \in \Sigma_q: \\ \varphi(F)=G}} (\text{sgn}([F],[G]) g(\varphi F))^2 w_\Sigma(F) \qquad (10.1.76)$$

$$= \sum_{G \in \Sigma_q'} g([G])^2 \sum_{\substack{F \in \Sigma_q: \\ \varphi(F)=G}} w_\Sigma(F) \qquad (10.1.77)$$

$$= (g, g) + (\phi_1 g, g) + (\phi_2 g, g), \qquad (10.1.78)$$

where

$$\phi_1 e_{[G]} = \begin{cases} w_\Sigma(F) e_{[G]} & \text{if } \varphi \text{ is of type } (i), \text{ and } \varphi(F) = G \in \partial \varphi(\bar{F}_i), 2 \le i \le m+1, \\ 0 & \text{otherwise,} \end{cases}$$

$$(10.1.79)$$

and

$$\phi_2 e_{[G]} = \begin{cases} w_\Sigma(F_1) e_{[G]} & \text{if } F_1 \in \varphi^{-1}(G), \text{ and } \varphi \text{ is of type } (ii), \\ 0 & \text{otherwise.} \end{cases} \qquad (10.1.80)$$

We then get the interlacing inequalities from the following sequence. Let $\mathcal{Y}$ denote the subspace of $C^q(\Sigma')$ on which $(\psi g, g) = 0$ and $y := \dim \mathcal{Y} \le (q + 2)m$ if φ is of type (i), and let $y = 0$ if φ is an elementary contraction of type (ii). Let also $\mathcal{W}$ be the vector space generated by

$$\{\text{sgn}([F_{ji}],[G_i]) e_{[F_{ji}]}$$
$$- \text{sgn}([F_{(i+1)j}],[G_i]) e_{[F_{(i+1)j}]} \mid \bigcup_j F_{ji} = \varphi^{-1}(G_i), \text{ and } \bigcup_i G_i = \Sigma_q'\},$$

which has dimension $\dim \mathcal{W} = N_\Sigma - N_{\Sigma'}$. Then,

$$\theta_k = \min_{\mathcal{V}_{N_{\Sigma'}}} \max_{k \in \mathcal{V}_{N_{\Sigma'}-k}} \max_{g \perp \mathcal{V}_{N_{\Sigma'}-k}} \frac{(L_{up}(\Sigma')g, g)}{(g, g)} \qquad (10.1.81)$$

$$= \min_{\mathcal{V}_{N_{\Sigma'}-k}} \max_{g \perp \mathcal{V}_{N_{\Sigma'}-k}} \frac{(\varphi^* L_{up}(\Sigma)\varphi g, g) - (\psi g, g)}{(g, g)} \qquad (10.1.82)$$

$$\ge \min_{\mathcal{V}_{N_{\Sigma'}-k}} \max_{g \perp \mathcal{V}_{N_{\Sigma'}-k}, g \perp \mathcal{Y}} \frac{(\varphi^* L_{up}(\Sigma)\varphi g, g)}{(g, g)} \qquad (10.1.83)$$

$$= \min_{\mathcal{V}_{N_{\Sigma'}-k}} \max_{g \perp \mathcal{V}_{N_{\Sigma'}-k}, g \perp \mathcal{Y}} \frac{(L_{up}(\Sigma)\varphi g, \varphi g)}{(\varphi g, \varphi g) - (\phi_1 g, g) - (\phi_2 g, g)} \qquad (10.1.84)$$

$$\ge \min_{\mathcal{V}_{N_{\Sigma'}-k}} \max_{g \perp \mathcal{V}_{N_{\Sigma'}-k}, g \perp \mathcal{Y}} \frac{(L_{up}(\Sigma)\varphi g, \varphi g)}{(\varphi g, \varphi g)} \qquad (10.1.85)$$

$$\ge \min_{\mathcal{V}_{N_{\Sigma'}-k+y}} \max_{g \perp \mathcal{V}_{N_{\Sigma'}-k+y}} \frac{(L_{up}(\Sigma)\varphi g, \varphi g)}{(\varphi g, \varphi g)} \qquad (10.1.86)$$

$$= \min_{\mathcal{V}_{N_{\Sigma'}-k+y}} \max_{f \perp \mathcal{V}_{N_{\Sigma'}-k+y}, f \perp W} \frac{(L_{up}(\Sigma)f,f)}{(f,f)} \tag{10.1.87}$$

$$\geq \min_{\mathcal{V}_{N_{\Sigma'}-k+y}} \max_{f \perp \mathcal{V}_{N_{\Sigma}-k+y}} \frac{(L_{up}(\Sigma)f,f)}{(f,f)} \tag{10.1.88}$$

$$\geq \min_{\mathcal{V}_{N_{\Sigma}-k+y}} \max_{f \perp \mathcal{V}_{N_{\Sigma}-k+y}} \frac{(L_{up}(\Sigma)f,f)}{(f,f)} \tag{10.1.89}$$

$$\geq \lambda_{k-y}. \tag{10.1.90}$$

We have used that $N_\Sigma - N_{\Sigma'} = q + 1$ if φ is an elementary collapse of type (*ii*). Similarly, for the upper interlacing inequality, we consider the vector space

$$\mathcal{Z} := \{g \in C^q(\Sigma') \mid (\phi_1 g, g) + (\phi_2 g, g) = 0\}$$

of dimension $z := \dim \mathcal{Z} = 1$ if φ is of type (*ii*) and $z = m(q + 2)$ for type (*i*). The upper interlacing inequality then follows from (10.1.55) via

$$\theta_k \leq \max_{\mathcal{V}_{k-1} \in \mathcal{V}_{k-1}} \min_{g \perp \mathcal{V}_{k-1}} \frac{(L_{up}(\Sigma)\varphi g, \varphi g) - (\psi g, g)}{(\varphi g, \varphi g) - (\phi_1 g, g) - (\phi_2 g, g)} \tag{10.1.91}$$

$$\leq \max_{\mathcal{V}_{k-1}} \min_{g \perp \mathcal{V}_{k-1}} \frac{(L_{up}(\Sigma)\varphi g, \varphi g)}{(\varphi g, \varphi g) - (\phi_1 g, g) - (\phi_2 g, g)} \tag{10.1.92}$$

$$\leq \max_{\mathcal{V}_{k-1}} \min_{g \perp \mathcal{V}_{k-1}, g \perp \mathcal{Z}} \frac{(L_{up}(\Sigma)\varphi g, \varphi g)}{(\varphi g, \varphi g)} \tag{10.1.93}$$

$$\leq \max_{\mathcal{V}_{k+z-1}} \min_{g \perp \mathcal{V}_{k+z-1},} \frac{(L_{up}(\Sigma)\varphi g, \varphi g)}{(\varphi g, \varphi g)} \tag{10.1.94}$$

$$= \max_{\mathcal{V}_{k+z-1}} \min_{f \perp \mathcal{V}_{k+z-1}, f \perp W} \frac{(L_{up}(\Sigma)f,f)}{(f,f)} \tag{10.1.95}$$

$$\leq \max_{\mathcal{V}_{k+z+N_\Sigma-N_L-1}} \min_{f \perp \mathcal{V}_{k+z+N_\Sigma-N_L-1}} \frac{(L_{up}(\Sigma)f,f)}{(f,f)} \tag{10.1.96}$$

$$\leq \lambda_{k+N_\Sigma-N_L+z}. \tag{10.1.97}$$

Thus, in case (*i*), we have

$$\lambda_{k-m(q+2)} \leq \theta_k \leq \lambda_{k+N_\Sigma-N_{\Sigma'}+m(q+2)}, \tag{10.1.98}$$

while in case (*ii*), we get

$$\lambda_k \leq \theta_k \leq \lambda_{k+N_\Sigma-N_{\Sigma'}+1}. \tag{10.1.99}$$

For the normalized Laplacian, the weight functions w_Σ and $w_{\Sigma'}$ have value 1 on all simplices of dimension $q + 1$, whereas $w_\Sigma(F) = \deg F$ and $w_{\Sigma'}(G) = \deg G$, for all q-faces F, G, of Σ and Σ', respectively. The maps $\phi_1, \phi_2 \colon C^q(\Sigma') \to C^q(\Sigma')$ are then given by

$$\phi_1 e_{[G]} = \begin{cases} w_\Sigma(\bar{F}_i) e_{[G]} & \text{if } \varphi \text{ is of type } (i), \text{ and } G \in \partial \bar{G}_i, 2 \le i \le m+1, \\ 0 & \text{otherwise,} \end{cases}$$

$$(10.1.100)$$

and

$$\phi_2 e_{[G]} = \begin{cases} 2 w_\Sigma(\bar{F}_1) e_{[G]} & \text{if } G = \varphi(F_1), \\ 0 & \text{otherwise.} \end{cases} \qquad (10.1.101)$$

The rest of the proof is the same as for the combinatorial Laplacian, and the same interlacing inequalities hold. $\qquad \square$

10.2 Spectral Classes

10.2.1 Asymptotic Classes of Graphs

In Section 4.1, we have already seen several families of graphs, like the complete graphs K_N, the complete bipartite graphs $K_{n,m}$ (including the star graphs $K_{1,n}$), the cycles C_N, the wheels W_N, the petals P_ℓ, and the books B_ℓ. They usually arise from the iteration of some construction, like duplicating a vertex or an edge, and they exhibit corresponding regularities of their spectra that lead to asymptotics as $N \to \infty$. As we have seen and as we want to explore in more detail, those properties can be quite similar for different classes.

For instance, the petals P_ℓ and the books B_ℓ both arise from the repeated duplication of an edge. In the case of the petals, one starts with the triangle $p_1 = K_3$, while for the books, B_1 is a line graph with four vertices.

In the case of the books, we duplicate the middle edge, thereby generating an eigenvalue $3/2$ according to (4.1.86). And since the books are bipartite, by Lemma 4.1.13, we then also get the eigenvalue $1/2$. In fact, the line with four vertices already has the eigenvalues $0, 1/2, 3/2, 2$, and so, by the edge duplications, we just increase the multiplicities of the eigenvalues $1/2$ and $3/2$.

In this context, let us consider another construction of repeated edge duplications, the water wheel graphs WW_ℓ. Here, we start with the square $WW_1 = C_4$ (Figure 10.5) and we repeatedly duplicate one of the edges.

These graphs WW_ℓ are again bipartite (Figure 10.6), and so we have the eigenvalues $0, 2$, and the edge duplications produce the eigenvalues $1/2$ and $3/2$. The remaining eigenvalues are $\lambda = \frac{1}{2} + \frac{\ell}{\ell+1}$ with an eigenfunction that is $\frac{2\ell}{\ell+1}$ at the central and -1 at the peripheral vertices, and $\lambda = \frac{3}{2} - \frac{\ell}{\ell+1}$ with an eigenfunction that is $\pm \frac{2\ell}{\ell+1}$ at the central and ± 1 at the peripheral vertices, according to Lemma 4.1.13. We notice that these eigenvalues converge to $3/2$

Figure 10.5 The square $WW_1 = C_4$.

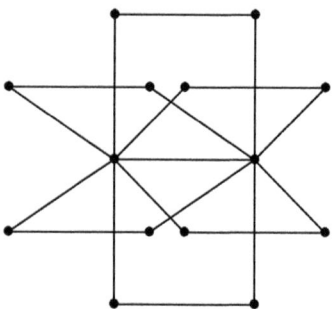

Figure 10.6 The water wheel graphs WW_ℓ.

and $1/2$ as $\ell \to \infty$. Thus, the spectrum of WW_ℓ approaches that of B_ℓ for $\ell \to \infty$. This is not surprising as the only difference between the two types of graph is the presence of the connection between the two central vertices in WW_ℓ and its absence in B_ℓ. Also, P_ℓ is similar, as it can be seen as the result of merging those two central vertices into a single one.

We now look at another construction, *graph doubling*. This is to be distinguished from the duplications of motifs that we have considered earlier. Here, we take a graph Γ_0 and an isomorphic copy Γ_0'. The vertex v of Γ_0 corresponds to the vertex v' of Γ_0' and the graph Γ then is formed from the union of Γ_0 and Γ_0' by also connecting each v with its twin v'. Thus, the degree $\deg_\Gamma v$ of each vertex w in Γ equals $\deg_{\Gamma_0} v + 1$ and the same of course for each v'. We now assume that Γ_0 is k-regular, that is

$$\deg_{\Gamma_0} v = k \quad \text{for all } v. \tag{10.2.1}$$

Let then λ be an eigenvalue of Γ_0 with eigenfunction u, that is,

$$u(v) - \frac{1}{k} \sum_{v' \sim_{\Gamma_0} v} u(v') = \lambda u(v). \tag{10.2.2}$$

We extend u to Γ by putting $u(v') = u(v)$ for all v. The corresponding quantity in Γ then is

$$u(v) - \frac{1}{k+1} \sum_{v' \sim_\Gamma v} u(v') = u(v) - \frac{1}{k+1} u(v) - \frac{1}{k+1} \sum_{v' \sim_{\Gamma_0} v} u(v') = \frac{k}{k+1} \lambda u(v)$$
(10.2.3)

by (10.2.2), and so, u then becomes an eigenfunction on Γ for the eigenvalue $\frac{k}{k+1}\lambda$.

If we instead extend u to Γ by putting $u(v') = -u(v)$ for all v, we obtain

$$u(v) - \frac{1}{k+1} \sum_{v' \sim_\Gamma v} u(v') = u(v) + \frac{1}{k+1} u(v) - \frac{1}{k+1} \sum_{v' \sim_{\Gamma_0} v} u(v')$$

$$= \left(\frac{k}{k+1} \lambda + \frac{2}{k+1} \right) u(v), \qquad (10.2.4)$$

that is, the eigenvalue $\frac{k}{k+1}\lambda + \frac{2}{k+1}$.

We record the preceding as

Lemma 10.2.1. *Let Γ be the graph obtained from doubling a k-regular graph Γ_0. Then, Γ is $(k+1)$-regular, and it has the eigenvalues*

$$\frac{k}{k+1} \lambda \quad and \quad \frac{k}{k+1} \lambda + \frac{2}{k+1}, \qquad (10.2.5)$$

where the λ's are the eigenvalues of Γ_0.

$\square$

In the same manner, we can then also triple, quadruple, ..., graphs, and if the starting graph is regular, determine their spectra.

From Lemma 10.2.1, we conclude

Corollary 10.2.1. *The cube graph Q_n on 2^n vertices has the eigenvalues $\frac{2j}{n}$ with multiplicities $\binom{n}{j}$ for $j = 0, \ldots, n$.*

Proof Q_n arises by iterated doubling of the line K_2, which has the eigenvalues 0 and 2. The next graph Q_2 (which is the cycle C_4) then acquires the eigenvalue 1 with multiplicity 2, once coming from the eigenvalue 0 via (10.2.4), and once coming from 2 via (10.2.3). At the next step, from (10.2.3) and (10.2.4), we get the eigenvalues 2/3 and 4/3 with multiplicity 3, with two of each coming from the two eigenvalues 1. Iterating this procedure proves the result. $\square$

We note that the spectra of the cubes Q_n are quite different from, and in particular behave asymptotically very different than, those of the other graph families we have discussed. In the next section, we want to develop an approach that can quantify that difference.

There are also systematic constructions of classes of graphs that are not regular. The most important ones are the *random graphs* introduced by Erdős and Rényi [91]. Here, one has a collection of N vertices and some $0 \le p \le 1$, and one connects any two vertices with probability p. These connections, that is, the edges of the graph, are randomly drawn independently of each other. For $p = 0$, we get of course a completely disconnected graph without any edges, whereas for $p = 1$, we get the complete graph K_N. For other values of p, the graph is usually not connected, but when p is above a certain threshold (in fact, the threshold is $\frac{1}{N-1}$, as already shown by Erdős and Rényi), then with probability 1, there is a large component containing most of the vertices.

10.2.2 Spectral Asymptotics

We now want to compare the spectra of different graphs of possibly different size, and we want to define a distance between any two such spectra. For that purpose, we shall describe the constructions of [118, 119]. Let $\sigma(\Gamma) = \{\lambda_i, i = 1, \dots, N\}$ denote the spectrum, that is, the collection of eigenvalues of the (possibly weighted, but finite and simple) graph Γ with N nodes. We consider here the vertex Laplacian $\Delta = L^0$, but obviously, an analogous theory can be developed for any (normalized) Laplacian on a simplicial complex or a hypergraph. And, one can also develop a corresponding theory for possibly infinite graphs, simplicial complexes and hypergraphs, but this is not our concern.

We define the measure

$$\mu_{\sigma(\Gamma)} := \frac{1}{N} \sum_{i=1}^{N} \delta(\lambda_i), \tag{10.2.6}$$

where $\delta(\lambda)$ denotes the Dirac measure supported at λ. Thus, for any function $\phi \colon [0,2] \to \mathbb{R}$,

$$\mu_{\sigma(\Gamma)}(\phi) = \frac{1}{N} \sum_{i=1}^{N} \phi(\lambda_i). \tag{10.2.7}$$

Because of the normalizing factor $\frac{1}{N}$ and since the eigenvalues satisfy $0 \le \lambda_i \le 2$, $\mu_{\sigma(\Gamma)}$ is a probability measure on $[0,2]$. We can therefore invoke the theory of probability measures on compact metric spaces (X, d). On the space $\mathcal{P}(X)$ of probability measures on X, we have the p-Wasserstein distance defined by

$$d_p(\mu, \nu) := \left(\inf_{\pi \in \Pi(\mu, \nu)} \int_{X \times X} d(x, y)^p \, d\pi(x, y) \right)^{1/p}, \tag{10.2.8}$$

where $\Pi(\mu, \nu)$ is the collection of all measures on $X \times X$ whose first marginal is μ and whose second marginal is ν. That means that for all Borel subsets A, B

of X, $\pi(A \times X) = \mu(A), \pi(X \times B) = \nu(B)$. This amounts to a *coupling*, and we shall describe it here for the case where μ and ν, as in (10.2.6), are sums of rescaled Dirac measures. Thus, let

$$\mu = \sum_{i=1}^{N} a_i \delta(\lambda_i), \quad \nu = \sum_{j=1}^{M} b_j \delta(\sigma_j) \tag{10.2.9}$$

with $0 \le \lambda_i, \sigma_j \le 2, a_i, b_j \ge 0$ and $\sum a_i = 1 = \sum b_j$. Then, π should be a measure of the form

$$\pi = \sum c_{ij} \delta(\lambda_i, \sigma_j) \tag{10.2.10}$$

with

$$c_{ij} \ge 0, \sum_{i,j} c_{ij} = 1, \sum_{k} c_{ik} = a_i, \sum_{\ell} c_{\ell j} = b_j \quad \text{for all } i, j.$$

Thus, from the perspective of μ, by (10.2.9), each λ_i is assigned a couple of σ_j's so that the total weight remains a_i. The relation (10.2.8) then becomes

$$d_p(\mu, \nu) = \left(\inf_{(c_{ij})} \sum c_{ij} |\lambda_i - \sigma_j|^p \right)^{1/p}. \tag{10.2.11}$$

For $1 \le p < \infty$, d_p defines a metric on $\mathcal{P}(X)$, turning it into a compact metric space and inducing the topology of weak convergence of measures. Every $\mu \in \mathcal{P}(X)$ is a convex combination of Dirac measures supported at points of X. Therefore, in particular, the diameter of $(\mathcal{P}(X), d_p)$ equals the diameter of (X, d). In our case, $X = [0, 2]$ with the Euclidean metric $(d(x, y) = |x - y|)$, and so the diameter of $(\mathcal{P}[0, 2], d_p)$ is 2. We note, however,

Lemma 10.2.2. *When the average of μ is $= 1$, that is, $\int_{[0,2]} x d\mu(x) = 1$, then*

$$d_1(\mu, \nu) \le 1 \tag{10.2.12}$$

for any $\nu \in \mathcal{P}[0, 2]$.

Proof Since any probability measure is a convex combination of Dirac measures, it can be approximated by finite such combinations, and so, we may assume that μ, ν are as in (10.2.9). The condition on the average of μ then becomes

$$\sum_{i=1}^{N} a_i \lambda_i = 1. \tag{10.2.13}$$

(10.2.13) and $\sum a_i = 1$ imply that when $0 \le \sigma \le 2$,

$$\sum_{i=1}^{N} a_i |\lambda_i - \sigma| \le 1. \tag{10.2.14}$$

Since also $\sum b_j = 1$, therefore also

$$\sum_{i,j} a_i b_j |\lambda_i - \sigma_j| \le 1. \qquad (10.2.15)$$

Thus, putting $c_{ij} = a_i b_j$, (10.2.12) follows from (10.2.11). $\qquad \square$

As in [118],

Definition 10.2.1. The *spectral p-distance* between two graphs Γ_1, Γ_2 is

$$d_p(\Gamma_1, \Gamma_2) := d_p(\mu_{\sigma(\Gamma_1)}, \mu_{\sigma(\Gamma_2)}). \qquad (10.2.16)$$

We shall usually take $p = 1$ and in this case simply speak of the *spectral distance* between Γ_1 and Γ_2.

We note that this does not define a metric on the set of finite graphs, but only a pseudo-metric, because isospectral graphs have vanishing distance with respect to d_p.

We then have (with a simpler proof) an inequality of [118].

Lemma 10.2.3. *For any two (possibly weighted) graphs Γ_1, Γ_2,*

$$d_1(\Gamma_1, \Gamma_2) \le 1. \qquad (10.2.17)$$

Proof When both graphs consist of isolated vertices only, with no edges, then their spectral measures both equal $\delta(0)$, and so, their spectral distance vanishes. Slightly expanding this argument, we may assume that one of them is connected and has more than one vertex. But then, its eigenvalues satisfy

$$\sum_{i=1}^{N} \lambda_i = N, \qquad (10.2.18)$$

and hence, the spectral measure (10.2.6) satisfies (10.2.13) (with $a_i = \frac{1}{N}$ for all i). Lemma 10.2.2 then implies the result. $\qquad \square$

We note that the result is not completely trivial because the diameter of $\mathcal{P}[0,2]$ is 2 and not 1. Therefore, we need (10.2.18).

There is a trivial example showing that the bound (10.2.17) is sharp. We let Γ_1 consist of an isolated vertex, hence $\mu_{\sigma(\Gamma_1)} = \delta(0)$, and Γ_2 has two vertices connected by an edge, hence $\mu_{\sigma(\Gamma_2)} = \frac{1}{2}(\delta(0) + \delta(2))$, and the spectral distance is 1. For a slightly less trivial example, let Γ_1 be a complete graph K_N or a complete bipartite graph $K_{m,n}$. For $N \to \infty$ or $m + n \to \infty$, the spectral measure of Γ_1 then converges to $\delta(1)$, and the spectral distance from the two vertex graph Γ_2 converges to 1. In fact, inspection of the proof of Lemma 10.2.3 shows that such an extreme case can occur only if one of the two spectral measures is supported at 0 and 2, and this is possible only for graphs that have vertices only of degree 0 or 1. We can thus formulate

Corollary 10.2.2. *When one of the finite graphs is connected and has at least three vertices, then*

$$d_1(\Gamma_1, \Gamma_2) < 1. \tag{10.2.19}$$

□

We may now introduce a concept.

Definition 10.2.2. Let (Γ_N) be some family of graphs, with N denoting the number of vertices. When

$$\lim_{N \to \infty} \mu_{\sigma(\Gamma_N)} =: \mu_{\sigma(\Gamma_N)N \to \infty} \tag{10.2.20}$$

exists as a weak limit of measures, we call it the *spectral class* of the family.

Examples can be obtained from the families discussed in Section 10.2.1:

- The families K_N of complete and $K_{m,n}$ ($N = m + n$) of complete bipartite graphs both belong to the spectral class $\delta(1)$.
- The petals P_ℓ, books B_ℓ, and water wheels WW_ℓ ($\ell \to \infty$) belong to the spectral class $\frac{1}{2}\left(\delta\left(\frac{1}{2}\right) + \delta\left(\frac{3}{2}\right)\right)$.
- In contrast, by Corollary 10.2.1, the spectral class of the cubes Q_ℓ is not given by a finite sum of delta measures. The same holds for paths, cycles, or wheels whose spectra we have analyzed in Section 4.1.

We can also define and determine spectral classes for all geometric objects that possess Laplacian spectra, for instance, for families of oriented hypergraphs. For an oriented hypergraph Γ on N nodes, the corresponding *spectral measure* can be defined by (10.2.6). While for graphs, this is a probability measure on $[0, 2]$, in the more general case, it is a probability measure on $[0, N]$. Similarly, also the *spectral class* of a family of hypergraphs can be defined by (10.2.20), as investigated in [215].

For example, if we consider the signless Laplacian Δ^+ from Definition 4.9.5,

- For any fixed $c \geq 2$, the family of c-complete hypergraphs belongs to the spectral class $\delta(1)$, as a consequence of Lemma 4.9.3.
- For any fixed $t, M \in \mathbb{N}$, the family Γ_N of hyperflowers with t twins on N nodes and M edges belongs to the spectral class $\delta(0)$, as a consequence of Proposition 4.9.2.
- Instead of considering a family of hyperflowers in which the number of central nodes grows, as in the previous case, we can also consider a family of hyperflowers in which the number of hyperedges (and therefore the number

of peripheral vertices) grows. In particular, for any fixed $t, c \in \mathbb{N}$, we can consider the family Γ_{c+tM} of hyperflowers with t twins on $c + tM$ nodes and M edges. This belongs to the spectral class $\frac{t-1}{t} \cdot \delta(0) + \frac{1}{t} \cdot \delta(t)$, again by Proposition 4.9.2.

Remark 10.2.1. Several general notions of the convergence of families of graphs have been developed in the literature, and several of these also imply convergence properties of spectra. In this book, however, we do not go into this subject; relevant references include [2, 190, 280] and many more.

11

Spectral Theory of Weighted Hypergraphs via Tensors

This chapter generalizes spectral theory to hypergraphs via tensor notions.

The spectral theory of graphs, as presented in Chapters 4 and 5, was based on matrices, and therefore, tools from linear algebra could be employed. We have also extended this approach to simplicial complexes in Section 4.8 and to oriented hypergraphs in Section 4.9. A problem with the latter is, however, that there no longer is a 1 : 1 correspondence between hypergraphs and their associated matrices. Therefore, the spectrum obtained through such a matrix approach may fail to capture some important properties of the underlying hypergraph. Therefore, in this chapter, we pursue a different generalization. We associate a tensor to a hypergraph. The order of this tensor equals the maximal size of a hyperedge. Thus, for an ordinary graph, the tensor is simply a matrix, but for hypergraphs with larger hyperedges, we obtain higher-order tensors.

Now, for tensors, we can also formulate eigenvalue problems. When the order is larger than two, however, these problems become nonlinear. In fact, these problems then become NP-hard [135]. This is the price we have to pay for letting more information about the hypergraph enter the spectral theory. Nevertheless, in this chapter, we study the spectral theory of weighted hypergraphs via tensors, and we mainly follow [105]. We do not develop the underlying spectral theory of tensors here but rather refer to the monograph [233]. That theory can provide the basis for the tensor spectral theory of uniform, unweighted hypergraphs. We also refer to [15, 16, 21, 85, 222, 223] for a rich literature on the nonuniform, unweighted case.

11.1 Tensors

We start by recalling several definitions and properties of tensors. Given $n \in \mathbb{N}$, we denote by $[n]$ the set $\{1,\ldots,n\}$. We indicate a vector in $\mathbb{C}^n$ by $\mathbf{x} =$

$(x_1, \ldots, x_n)$.[1] We write $\mathbf{x} \geq 0$ if $x_i \geq 0$ for every $i \in [n]$.

Let $k, n \geq 2$. A kth order n-dimensional *tensor* T consists of n^k complex entries

$$T_{i_1, \ldots, i_k} \in \mathbb{C},$$

where $i_1, \ldots, i_k \in [n]$. The tensor T is *symmetric* if its entries are invariant under any permutation of their indices.

An $n \times n$ matrix is a second-order n-dimensional tensor. One of the many differences between the spectral theory of matrices and that of tensors is that the eigenvalue equations (11.1.1) for higher-order tensors are nonlinear, and therefore, the eigenvalues of a real symmetric tensor are not necessarily real.

Given a vector $\mathbf{x} \in \mathbb{C}^n$, define $T\mathbf{x}^{k-1} \in \mathbb{C}^n$ by

$$(T\mathbf{x}^{k-1})_i := \sum_{i_2, \ldots, i_k = 1}^{n} T_{i, i_2, \ldots, i_k} x_{i_2} \cdots x_{i_k}.$$

Let $\mathbf{x}^{[k-1]} \in \mathbb{C}^n$ be the vector with entries $x_i^{[k-1]} := x_i^{k-1}$. If

$$T\mathbf{x}^{k-1} = \lambda \mathbf{x}^{[k-1]} \tag{11.1.1}$$

for some $\lambda \in \mathbb{C}$ and some nonzero vector $\mathbf{x} \in \mathbb{C}^n$, then we say that $(\lambda, \mathbf{x})$ is an *eigenpair* for T. The number λ is an *eigenvalue* of T, and $\mathbf{x}$ is an *eigenvector*. The *spectral radius* of T, denoted $\rho(T)$, is the largest modulus of the eigenvalues of T.

Let Id denote the kth order n-dimensional *unit tensor*, with entries

$$\mathrm{Id}_{i_1, \ldots, i_k} := \begin{cases} 1 & \text{if } i_1 = \ldots = i_k \\ 0 & \text{otherwise.} \end{cases}$$

Just like matrices, tensors have a determinant. As illustrated in [233, Section 2.1.3], the determinant of a tensor T is the resultant of the system of equations $T\mathbf{x}^{k-1} = 0$. It is a polynomial in the entries of T that vanishes if and only if the system has a nonzero solution. As shown in [233, Theorem 2.12], the eigenvalues of T are the roots of the *characteristic polynomial*

$$\phi_T(\lambda) := \det(T - \lambda \, \mathrm{Id})$$

---

[1] We point out that this convention is different from that employed in earlier chapters where, for consistency of the transformation rules for covariant and contravariant tensors in Riemannian geometry, we had used upper indices for the components of vectors. In the present chapter; however, this would be unwieldy, as we frequently need to take powers of components of tensors.

of T. The *spectrum* of T, denoted $\text{Spec}(T)$, is the multiset of its eigenvalues, counted with multiplicity as roots of the characteristic polynomial. An eigen-pair $(\lambda, \mathbf{x})$ is given by an *H-eigenvalue* and an *H-eigenvector*, respectively, if $\lambda \in \mathbb{R}$ and $\mathbf{x} \in \mathbb{R}^n$. We let $\text{Hspec}(T)$ denote the set of distinct H-eigenvalues of T.

The tensor T is *non-negative* if all its entries are non-negative. The tensor T is *weakly irreducible* if, for any nonempty proper index subset J of $[n]$, there is at least one entry

$$T_{i_1,\dots,i_k} \neq 0,$$

where $i_1 \in J$ and at least one index $i_j \in [n] \setminus J$, for $j = 2, \dots, k$.

For example, the adjacency matrix of a graph is weakly irreducible if and only if the graph is connected, and as we shall see in Theorem 11.3.2, this property generalizes to the tensors associated with hypergraphs.

Similarly, T is *reducible* if there exists a nonempty proper set J of $[n]$ such that

$$T_{i_1,\dots,i_k} = 0,$$

for each $i_1 \in J$, $i_2, \dots, i_k \in [n] \setminus J$. It is *irreducible* if it is not reducible.

In [233, Theorems 3.25 and 3.26], the Perron–Frobenius theorem for non-negative matrices has been generalized to weakly irreducible non-negative tensors.

Theorem 11.1.1. *If T is non-negative and weakly irreducible, then*

$$\rho(T) = \max_{\substack{\mathbf{x} \geq 0 \\ \mathbf{x} \neq 0}} \min_{x_i > 0} \frac{(T\mathbf{x}^{k-1})_i}{x_i^{k-1}}$$

is a positive H-eigenvalue, with positive H-eigenvector $\mathbf{x}$. Furthermore, $\rho(T)$ is the unique H-eigenvalue of T with a positive H-eigenvector, and $\mathbf{x}$ is the unique positive H-eigenvector associated with $\rho(T)$ up to a multiplicative constant.

Now, T is *diagonally dominated* if, for each $i \in [n]$,

$$T_{i,\dots,i} \geq \sum_{\substack{i_2,\dots,i_k \in [n] \\ \text{not all equal to } i}} |T_{i,i_2,\dots,i_k}|. \tag{11.1.2}$$

By [234, Theorem 2], if T is any tensor and λ is an eigenvalue for T, then there exists $i \in [n]$ such that

$$|\lambda - T_{i,\dots,i}| \leq \sum_{\substack{i_2,\dots,i_k \in [n] \\ \text{not all equal to } i}} |T_{i,i_2,\dots,i_k}|. \tag{11.1.3}$$

An immediate consequence is the following

Corollary 11.1.1. *If T is a diagonally dominated tensor, then all its H-eigenvalues are non-negative.*

For a given tensor, the *algebraic multiplicity* of an eigenvalue λ, denoted $am(\lambda)$, is its multiplicity as a root of the characteristic polynomial. The *geometric multiplicity* of λ, denoted $gm(\lambda)$, is the dimension of its *eigenvariety*, that is, the variety of its eigenvectors. The *span multiplicity* of λ, denoted $sm(\lambda)$, is the dimension of the vector space

$$\text{span}(\{\mathbf{x} \in \mathbb{C}^n : \mathbf{x} \text{ eigenvector for } \lambda\}).$$

Moreover, if λ is a real eigenvalue, its *Hspan multiplicity*, denoted $Hsm(\lambda)$, is the dimension of the vector space

$$\text{span}(\{\mathbf{x} \in \mathbb{R}^n : \mathbf{x} \text{ eigenvector for } \lambda\}).$$

In the case of symmetric matrices, we know that every eigenvalue λ is an H-eigenvalue and

$$am(\lambda) = sm(\lambda) = Hsm(\lambda) = gm(\lambda).$$

For symmetric tensors of higher order, this is not always the case.

Clearly, $sm(\lambda) \geq gm(\lambda)$, but since the eigenvalue equation (11.1.1) is non-linear, linear combinations of eigenvectors in general are not eigenvectors, and so, we could have strict inequality here. In [139, Conjecture 1.1], it is conjectured that, for any λ,

$$am(\lambda) \geq gm(\lambda)(k-1)^{gm(\lambda)-1}.$$

Moreover, in [139, Conjecture 1.1], it is conjectured that, for any λ,

$$am(\lambda) \geq sm(\lambda).$$

We need one more preliminary definition before introducing tensors attached to hypergraphs, namely, a normalization constant.

Given a collection of indices $i_1, \ldots, i_k$, let $\{i_1, \ldots, i_k\}$ be the set obtained from $i_1, \ldots, i_k$ by not accounting for multiplicity. Given a kth-order n-dimensional tensor T and an index set $J \subseteq [n]$ of cardinality r, we say that an entry $T_{i_1, \ldots, i_k}$ of T *corresponds to* J if $\{i_1, \ldots, i_k\} = J$. We let,

$$\mathcal{N}(r, k) := |\text{entries of } T \text{ corresponding to } J|.$$

Given $j \in J$, we define its *jth row* as the $(k-1)$-order n-dimensional tensor T_j obtained by setting the first index of T equal to j. We also set

$$N(r, k) := \frac{\mathcal{N}(r, k)}{r} = |\text{entries of the row } T_j \text{ corresponding to } J|.$$

We have that $\mathcal{N}(r,k) = \left\{{k \atop r}\right\}r!$, where

$$\left\{{k \atop r}\right\} := \frac{1}{r!}\sum_{j=0}^{r}(-1)^j\binom{r}{j}(r-j)^k$$

is the *Stirling number of the second kind*. This can be seen from the fact that $\left\{{k \atop r}\right\}$ counts the number of r-partitions of a set of cardinality k, while $r!$ counts the number of permutations of r objects. Hence,

$$N(r,k) = \left\{{k \atop r}\right\}(r-1)!\,.$$

In particular, $N(k,k) = (k-1)!$.

11.2 Hypergraph Tensors

We fix a weighted hypergraph $\Gamma = (V, E, w)$ with vertex set $V = \{v_1,\ldots,v_N\}$, hyperedge set $E = \{e_1,\ldots,e_M\}$ and weight function $w\colon E \to \mathbb{R}_{>0}$. Throughout the chapter, we assume that each hyperedge contains at least two vertices and that $e_i \neq e_j$ if $i \neq j$.

We let

$$\underline{d} := \min_{v\in V} \deg v, \quad \overline{d} := \max_{v\in V} \deg v \quad \text{and} \quad \Theta := \max_{e\in E}|e|.$$

For $r \leq \Theta$, we let

$$E_r := \{e \in E\colon |e| = r\}.$$

The *adjacency tensor* of Γ is the Θ-th order N-dimensional tensor $A = A(\Gamma)$ with entries

$$A_{i_1,\ldots,i_\Theta} := \begin{cases} 0 & \text{if } \{v_{i_1},\ldots,v_{i_\Theta}\} \notin E \\ \frac{w(e)}{N(r,\Theta)} & \text{if } \{v_{i_1},\ldots,v_{i_\Theta}\} = e \in E_r. \end{cases}$$

The *Kirchhoff Laplacian tensor* of Γ is the Θ-th order N-dimensional tensor $K = K(\Gamma)$ with entries

$$K_{i_1,\ldots,i_\Theta} := \begin{cases} \deg v_{i_1} & \text{if } i_1 = \ldots = i_\Theta \\ -A_{i_1,\ldots,i_\Theta} & \text{otherwise.} \end{cases}$$

The *Chung Laplacian tensor* of Γ is the Θ-th order N-dimensional tensor $\mathcal{L} = \mathcal{L}(\Gamma)$ with entries

$$\mathcal{L}_{i_1,\ldots,i_\Theta} := \begin{cases} 1 & \text{if } i_1 = \ldots = i_\Theta \\ -A_{i_1,\ldots,i_\Theta} \cdot \prod_{j\in\{i_1,\ldots,i_\Theta\}} \frac{1}{\Theta\sqrt{\deg v_j}} & \text{otherwise.} \end{cases}$$

The *normalized Laplacian tensor* of Γ is the Θ-th order N-dimensional tensor $\Delta = \Delta(\Gamma)$ with entries

$$\Delta_{i_1,\dots,i_\Theta} := \begin{cases} 1 & \text{if } i_1 = \dots = i_\Theta \\ -\dfrac{A_{i_1,\dots,i_\Theta}}{\deg v_{i_1}} & \text{otherwise.} \end{cases}$$

In the case of simple graphs, the adjacency, Kirchhoff Laplacian, Chung Laplacian, and normalized Laplacian tensors coincide with the adjacency, Kirchhoff Laplacian, Chung Laplacian, and normalized Laplacian matrices, respectively.

The tensors defined earlier also have signless versions, simply deleting the minus signs in their definitions. Thus, the *signless Kirchhoff Laplacian tensor* of Γ is the Θ-th order N-dimensional tensor $K^+ = K^+(\Gamma)$ with entries

$$K^+_{i_1,\dots,i_\Theta} := \begin{cases} \deg v_{i_1} & \text{if } i_1 = \dots = i_\Theta \\ A_{i_1,\dots,i_\Theta} & \text{otherwise.} \end{cases}$$

The *signless Chung Laplacian tensor* of Γ is the Θ-th order N-dimensional tensor $\mathcal{L}^+ = \mathcal{L}^+(\Gamma)$ with entries

$$\mathcal{L}^+_{i_1,\dots,i_\Theta} := \begin{cases} 1 & \text{if } i_1 = \dots = i_\Theta \\ A_{i_1,\dots,i_\Theta} \cdot \prod_{j\in\{i_1,\dots,i_\Theta\}} \dfrac{1}{\sqrt[\Theta]{\deg v_j}} & \text{otherwise.} \end{cases}$$

The *signless normalized Laplacian tensor* of Γ is the Θ-th order N-dimensional tensor $\Delta^+ = \Delta^+(\Gamma)$ with entries

$$\Delta^+_{i_1,\dots,i_\Theta} := \begin{cases} 1 & \text{if } i_1 = \dots = i_\Theta \\ \dfrac{A_{i_1,\dots,i_\Theta}}{\deg v_{i_1}} & \text{otherwise.} \end{cases}$$

Remark 11.2.1. From the definition, it is apparent that the tensors A, K, K^+, $\mathcal{L}$, and $\mathcal{L}^+$ are symmetric and that the tensors A, K^+, $\mathcal{L}^+$, and Δ^+ are non-negative.

Example 11.2.1. Let $\Gamma = (V, E, w)$ be the weighted hypergraph with vertex set $V = \{v_1, v_2, v_3\}$, hyperedge set $E = \{\{v_1, v_2\}, \{v_1, v_2, v_3\}\}$, and weights $w(\{v_1, v_2\}) = 1$ and $w(\{v_1, v_2, v_3\}) = 2$ (Figure 11.1).

In this case, $N = \Theta = 3$. The nonzero entries of the adjacency tensor are

$$A_{112} = A_{121} = A_{211} = A_{122} = A_{212} = A_{221} = \frac{w(\{v_1, v_2\})}{N(2,3)} = \frac{1}{3}$$

and

$$A_{123} = A_{132} = A_{213} = A_{231} = A_{312} = A_{321} = \frac{w(\{v_1, v_2, v_3\})}{N(3,3)} = 1.$$

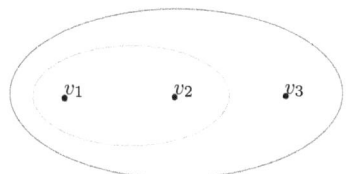

Figure 11.1 The hypergraph in Example 11.2.1.

Also, since deg v_1 = deg v_2 = 3 and deg v_3 = 2, the nonzero entries of K are

$$K_{112} = K_{121} = K_{211} = K_{122} = K_{212} = K_{221} = -\frac{1}{3},$$

$$K_{123} = K_{132} = K_{213} = K_{231} = K_{312} = K_{321} = -1,$$

$$K_{111} = K_{222} = 3, \qquad K_{333} = 2.$$

Similarly, since $\sqrt[9]{\deg v_1} = \sqrt[9]{\deg v_2} = \sqrt[3]{3}$ and $\sqrt[9]{\deg v_3} = \sqrt[3]{2}$, the nonzero entries of $\mathcal{L}$ are

$$\mathcal{L}_{112} = \mathcal{L}_{121} = \mathcal{L}_{211} = \mathcal{L}_{122} = \mathcal{L}_{212} = \mathcal{L}_{221} = -\frac{1}{9},$$

$$\mathcal{L}_{123} = \mathcal{L}_{132} = \mathcal{L}_{213} = \mathcal{L}_{231} = \mathcal{L}_{312} = \mathcal{L}_{321} = -\frac{1}{\sqrt[3]{18}},$$

$$\mathcal{L}_{111} = \mathcal{L}_{222} = \mathcal{L}_{333} = 1,$$

while the nonzero entries of Δ are

$$\Delta_{112} = \Delta_{121} = \Delta_{211} = \Delta_{122} = \Delta_{212} = \Delta_{221} = -\frac{1}{9},$$

$$\Delta_{123} = \Delta_{132} = \Delta_{213} = \Delta_{231} = -\frac{1}{3},$$

$$\Delta_{312} = \Delta_{321} - -\frac{1}{2}$$

$$\Delta_{111} = \Delta_{222} = \Delta_{333} = 1.$$

11.3 First Spectral Properties

In this section, we prove the first spectral properties of the hypergraph tensors that we introduced in Section 11.2.

We start by proving that the Chung Laplacian of a weighted hypergraph has the same spectrum as the normalized Laplacian tensor. The same holds for the signless versions.

Proposition 11.3.1. *The tensors $\mathcal{L}$ and Δ have the same eigenvalues, counted with algebraic multiplicity. Moreover,*

$$(\lambda, \mathbf{x}) \text{ is an eigenpair for } \mathcal{L} \iff (\lambda, \mathbf{y}) \text{ is an eigenpair for } \Delta,$$

where $\mathbf{y} \in \mathbb{C}^N$ is the vector with entries

$$y_j = \frac{x_j}{\sqrt[\Theta]{\deg v_j}}.$$

Hence, the eigenvalues of $\mathcal{L}$ and Δ have also the same geometric and span multiplicities, and $\mathrm{Hspec}(\mathcal{L}) = \mathrm{Hspec}(\Delta)$. The same holds for $\mathcal{L}^+$ and Δ^+.

Proof For the algebraic multiplicities, it suffices to prove the claim for $\mathcal{L}$ and Δ. The other case then follows from the observation that

$$\mathcal{L}^+ = 2 \cdot \mathrm{Id} - \mathcal{L} \quad \text{and} \quad \Delta^+ = 2 \cdot \mathrm{Id} - \Delta. \tag{11.3.1}$$

As recalled, for instance, in [92] and in [247, Theorem 2.3], if $d_1, \ldots, d_N \in \mathbb{R}$, then $\mathcal{L}$ has the same eigenvalues, counted with algebraic multiplicity, as the Θ-th order N-dimensional tensor $\mathcal{L}'(d_1, \ldots, d_N)$ with entries

$$\mathcal{L}'(d_1, \ldots, d_N)_{i_1, \ldots, i_\Theta} = \frac{1}{d_{i_1}^{\Theta-1}} \cdot \mathcal{L}_{i_1, \ldots, i_\Theta} \cdot d_{i_2} \ldots d_{i_\Theta}.$$

By taking $d_j = \sqrt[\Theta]{\deg v_j}$ for each $j \in [N]$, we have that $\mathcal{L}'(d_1, \ldots, d_N) = \Delta$. This proves the claim for the algebraic multiplicities.

The claim for the other multiplicities follows directly from Theorem 2.5 in [247]. $\qquad\square$

Proposition 11.3.2. *Each of the tensors $A, K, K^+, \mathcal{L}, \mathcal{L}^+, \Delta,$ and Δ^+ has $(\Theta - 1)^{N-1} \cdot N$ eigenvalues, counted with algebraic multiplicity, whose sum is:*

- 0, *for A;*
- $(\Theta - 1)^{N-1} \cdot \left(\sum_{i=1}^{N} \deg v_i \right)$, *for K and K^+; and*
- $(\Theta - 1)^{N-1} \cdot N$, *for $\mathcal{L}, \mathcal{L}^+, \Delta,$ and Δ^+.*

Proof As shown in [232, Theorem 1], a symmetric tensor has $(\Theta - 1)^{N-1} \cdot N$ eigenvalues, counted with algebraic multiplicity, whose sum is $(\Theta - 1)^{N-1}$ times the sum of its diagonal elements. Hence, the claim for $A, K, K^+, \mathcal{L}$, and $\mathcal{L}^+$ follows. The claim for Δ, and Δ^+ follows by Proposition 11.3.1. $\qquad\square$

Proposition 11.3.1 tells us that, as in the case of graphs, the Laplacians $\mathcal{L}$ and Δ are the same, from the spectral viewpoint, and the same holds for their signless versions. Remarks 11.3.1 and 11.3.2 are a further step in this direction. They show that the spectra of all our tensors are closely related.

Remark 11.3.1. From (11.3.1),

$$(\lambda,\mathbf{x}) \text{ is an eigenpair for } \mathcal{L} \iff (2-\lambda,\mathbf{x}) \text{ is an eigenpair for } \mathcal{L}^+$$

and

$$(\lambda,\mathbf{x}) \text{ is an eigenpair for } \Delta \iff (2-\lambda,\mathbf{x}) \text{ is an eigenpair for } \Delta^+.$$

Moreover, since

$$K_{i_1,\ldots,i_\Theta} = (\deg v_{i_1})\Delta_{i_1,\ldots,i_\Theta},$$

we have that

$$(0,\mathbf{x}) \text{ is an eigenpair for } K \iff (0,\mathbf{x}) \text{ is an eigenpair for } \Delta$$

and similarly, $(0,\mathbf{x})$ is an eigenpair for $K^+ \iff (0,\mathbf{x})$ is an eigenpair for Δ^+.
Finally, $(0,\mathbf{x})$ is an eigenpair for $A \iff (1,\mathbf{x})$ is an eigenpair for Δ and Δ^+.

Remark 11.3.2. If Γ is d-regular, then

$$K = d \cdot \mathrm{Id} - A, \quad \mathcal{L} = \Delta = \frac{1}{d} \cdot K \quad \text{and} \quad K^+ = 2d \cdot \mathrm{Id} - K.$$

Hence, in this case, it is easy to see that

$$\lambda \text{ is an eigenvalue for } K \iff d - \lambda \text{ is an eigenvalue for } A$$

$$\iff \frac{\lambda}{d} \text{ is an eigenvalue for } \mathcal{L} = \Delta$$

$$\iff 2 - \frac{\lambda}{d} \text{ is an eigenvalue for } \mathcal{L}^+ = \Delta^+$$

$$\iff 2d - \lambda \text{ is an eigenvalue for } K^+,$$

with the same multiplicities. In particular, the spectral theories of the different tensors are equivalent to each other for regular weighted hypergraphs.

Such observations allow us to expand on the previous knowledge on their eigenvalues. For instance, we are in position to discuss the existence of H-eigenvectors and H-eigenvalues.

Proposition 11.3.3. *1. The tensors A, K^+, $\mathcal{L}^+$, and Δ^+ have at least one H-eigenvalue. Their largest H-eigenvalue equals their spectral radius and has a non-negative H-eigenvector.*
2. The tensors $\mathcal{L}$ and Δ have at least one H-eigenvalue. Their smallest H-eigenvalue equals $2 - \rho(\mathcal{L})$ and has a non-negative H-eigenvector.

Proof The first claim follows from [233, Theorem 2.4], which applies to non-negative tensors. The second claim follows from the first one, together with Proposition 11.3.1 and Remark 11.3.1. □

Now, we want to prove that our hypergraph Laplacian tensors are diagonally dominated. For this purpose, we compute the sums of their rows. This will also allow us to bound their H-eigenvalues.

Lemma 11.3.1. *Given* $i_1 \in [N]$,

$$\sum_{i_2,\ldots,i_\Theta \in [N]} A_{i_1,i_2,\ldots,i_\Theta} = \deg v_{i_1},$$

$$\sum_{i_2,\ldots,i_\Theta \in [N]} K_{i_1,i_2,\ldots,i_\Theta} = \sum_{i_2,\ldots,i_\Theta \in [N]} \Delta_{i_1,i_2,\ldots,i_\Theta} = 0,$$

$$\sum_{i_2,\ldots,i_\Theta \in [N]} K^+_{i_1,i_2,\ldots,i_\Theta} = 2 \cdot \deg v_{i_1},$$

$$\sum_{i_2,\ldots,i_\Theta \in [N]} \Delta^+_{i_1,i_2,\ldots,i_\Theta} = 2.$$

Proof Given $i_1 \in [N]$,

$$\sum_{i_2,\ldots,i_\Theta \in [N]} A_{i_1,i_2,\ldots,i_\Theta}$$

$$= \sum_{r=2}^{\Theta} \left(\sum_{e \in E_r \,:\, v_{i_1} \in e} |\text{entries of the row } A_{i_1} \text{ corresponding to } e| \cdot \frac{w(e)}{N(r,\Theta)} \right)$$

$$= \sum_{r=2}^{\Theta} \sum_{e \in E_r \,:\, v_{i_1} \in e} w(e) = \deg v_{i_1}.$$

This implies that

$$\sum_{i_2,\ldots,i_\Theta \in [N]} K_{i_1,i_2,\ldots,i_\Theta} = \deg v_{i_1} - \sum_{i_2,\ldots,i_\Theta \in [N]} A_{i_1,i_2,\ldots,i_\Theta} = 0,$$

while

$$\sum_{i_2,\ldots,i_\Theta \in [N]} \Delta_{i_1,i_2,\ldots,i_\Theta} = \frac{1}{\deg v_{i_1}} \left(\sum_{i_2,\ldots,i_\Theta \in [N]} K_{i_1,i_2,\ldots,i_\Theta} \right) = 0.$$

Similarly,

$$\sum_{i_2,\ldots,i_\Theta \in [N]} K^+_{i_1,i_2,\ldots,i_\Theta} = \deg v_{i_1} + \sum_{i_2,\ldots,i_\Theta \in [N]} A_{i_1,i_2,\ldots,i_\Theta} = 2 \cdot \deg v_{i_1}$$

and

$$\sum_{i_2,\ldots,i_\Theta \in [N]} \Delta^+_{i_1,i_2,\ldots,i_\Theta} = \frac{1}{\deg v_{i_1}} \left(\sum_{i_2,\ldots,i_\Theta \in [N]} K^+_{i_1,i_2,\ldots,i_\Theta} \right)$$

$$= \frac{1}{\deg v_{i_1}} (2 \cdot \deg v_{i_1}) = 2. \qquad \square$$

Now we show that hypergraph Laplacian tensors are diagonally dominated, and we prove some bounds for the eigenvalues of all hypergraph tensors eigenvalues.

Theorem 11.3.1. *The tensors K, Δ, K^+, and Δ^+ are diagonally dominated. Moreover, if λ is an eigenvalue for K or K^+, then*

$$|\lambda - \overline{d}| \leq \overline{d}.$$

If μ is an eigenvalue for Δ or Δ^+, then

$$|\mu - 1| \leq 1.$$

If ν is an eigenvalue for A, then

$$|\nu| \leq \overline{d}.$$

In particular, all the real eigenvalues of K and K^+ are in $[0, 2\overline{d}]$, all the real eigenvalues of Δ and Δ^+ (equivalently, $\mathcal{L}$ and $\mathcal{L}^+$) are in $[0, 2]$, and all the real eigenvalues of A are in $[-\overline{d}, \overline{d}]$.

Proof By Lemma 11.3.1, for each $i \in [N]$, we have

$$K_{i,\ldots,i} = \deg v_i = \sum_{\substack{i_2,\ldots,i_\Theta \in [N] \\ \text{not all equal to } i}} |K_{i,i_2,\ldots,i_\Theta}| ;$$

$$K^+_{i,\ldots,i} = \deg v_i = \sum_{\substack{i_2,\ldots,i_\Theta \in [N] \\ \text{not all equal to } i}} \left|K^+_{i,i_2,\ldots,i_\Theta}\right| ;$$

$$\Delta_{i,\ldots,i} = 1 = \sum_{\substack{i_2,\ldots,i_\Theta \in [N] \\ \text{not all equal to } i}} |\Delta_{i,i_2,\ldots,i_\Theta}| ;$$

$$\Delta^+_{i,\ldots,i} = 1 = \sum_{\substack{i_2,\ldots,i_\Theta \in [N] \\ \text{not all equal to } i}} \left|\Delta^+_{i,i_2,\ldots,i_\Theta}\right| .$$

Hence, K, Δ, K^+, and Δ^+ satisfy (11.1.2), implying that they are diagonally dominated.

The second claim follows from Equation (11.1.3). $\qquad \square$

Now we move to irreducibility, and we give a necessary and sufficient condition for our tensors to be weakly irreducible.

Theorem 11.3.2. *A, K, K^+, $\mathcal{L}$, $\mathcal{L}^+$, Δ, and Δ^+ are weakly irreducible tensors if and only if Γ is connected.*

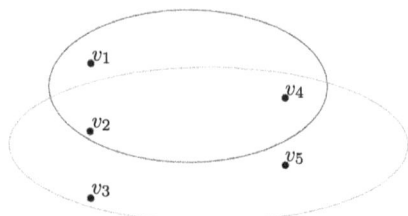

Figure 11.2 A reducible hypergraph. This can be seen by taking $V_1 = \{v_1, v_2, v_3\}$ and $V_2 = \{v_4, v_5\}$.

Proof Without loss of generality, we only prove the claim for A. By definition, the tensor A is weakly irreducible if and only if for any nonempty proper index subset J of $[N]$, there is at least one entry

$$A_{i_1,\dots,i_\Theta} \neq 0,$$

where $i_1 \in J$ and at least one index $i_j \in [N] \setminus J$, for $j = 2, \dots, \Theta$. By definition of A, this happens if and only if, for each nonempty proper subset J of $[N]$, there exist $i_1 \in J$ and $i_j \in [N] \setminus J$ such that v_{i_1} and v_{i_j} share a common hyperedge. Hence, A is weakly irreducible if and only if the hypergraph is connected. □

Now that weak irreducibility is settled, we address irreducibility. The following combinatorial property of the hypergraph will allow us to characterize irreducible hypergraph tensors.

Definition 11.3.1. A hypergraph $\Gamma = (V, E, w)$ is *reducible* (Figure 11.2) if one can decompose the vertex set as a disjoint union $V = V_1 \sqcup V_2$ such that V_1 and V_2 are both nonempty and, for each hyperedge e,

$$e \cap V_1 \neq \emptyset \Rightarrow |e \cap V_1| \geq 2.$$

A hypergraph is *irreducible* if it is not reducible.

Let us consider some examples:

- Every disconnected hypergraph is reducible because of our general assumption that each hyperedge contains at least two vertices.
- If Γ has one vertex v that is only contained in hyperedges of cardinality ≥ 3, then, by setting $V_1 = V \setminus \{v\}$ and $V_2 = \{v\}$, it is clear that Γ is reducible. This implies, in particular, that the majority of connected hypergraphs are reducible.

- If $\Gamma = (V, E)$ and there exists $E' \subseteq E$ such that $\Gamma' = (V, E')$ is a connected graph, then Γ is irreducible. In particular, every connected graph is irreducible.

The following result motivates the terminology of Definition 11.3.1.

Theorem 11.3.3. *A, K, K^+, $\mathcal{L}$, $\mathcal{L}^+$, Δ, and Δ^+ are irreducible tensors if and only if Γ is an irreducible hypergraph.*

Proof If Γ is reducible, let $V = V_1 \sqcup V_2$ be a decomposition of the vertex set as in Definition 11.3.1. Then, setting $J = \{i \in [N]: v_i \in V_1\}$ shows that the tensors associated with Γ are reducible. Vice versa, if $J \subset [N]$ shows that the tensors associated with Γ are reducible, then setting $V_1 = \{v_i \in V : i \in J\}$ and $V_2 = V \setminus V_1$ shows that Γ is a reducible hypergraph. $\square$

Another important property that we want to understand better is the spectral radius.

Proposition 11.3.4. *The spectral radii of A and K^+ satisfy*

$$\underline{d} \le \rho(A) \le \overline{d} \quad and \quad 2\underline{d} \le \rho(K^+) \le 2\overline{d}.$$

In particular, if Γ is d-regular, then

$$\rho(A) = d \quad and \quad \rho(K^+) = 2d.$$

Moreover, if $\Gamma' = (V, E', w')$ is another weighted hypergraph on N vertices and maximum hyperedge cardinality Θ, and it is obtained from Γ by removing hyperedges or by decreasing some hyperedge weights, then

$$\rho(A(\Gamma')) \le \rho(A(\Gamma)) \quad and \quad \rho(K(\Gamma')) \le \rho(K(\Gamma)).$$

Proof If T is a non-negative tensor, then by [233, Lemma 3.20],

$$\min_{i_1 \in [N]} \sum_{i_2, \dots, i_\Theta \in [N]} T_{i_1, i_2, \dots, i_\Theta} \le \rho(T) \le \max_{i_1 \in [N]} \sum_{i_2, \dots, i_\Theta \in [N]} T_{i_1, i_2, \dots, i_\Theta}.$$

By Lemma 11.3.1, this implies that $\underline{d} \le \rho(A) \le \overline{d}$ and $2\underline{d} \le \rho(K^+) \le 2\overline{d}$. Moreover, if $\Gamma' = (V, E', w')$ is another weighted hypergraph on N vertices and maximum hyperedge cardinality Θ, and it is obtained from Γ by removing hyperedges or by decreasing some hyperedge weights, then

$$A(\Gamma')_{i_1, \dots, i_\Theta} \le A_{i_1, \dots, i_\Theta} \quad and \quad K^+(\Gamma')_{i_1, \dots, i_\Theta} \le K^+_{i_1, \dots, i_\Theta},$$

for each $i_j \in [N]$ and $j \in [\Theta]$. By [197, Lemma 2.3], this implies that

$$\rho(A(\Gamma')) \le \rho(A) \quad and \quad \rho(K(\Gamma')) \le \rho(K). \qquad \square$$

Proposition 11.3.5. *The spectral radius of $\mathcal{L}^+$ and Δ^+ is $\rho(\mathcal{L}^+) = \rho(\Delta^+) = 2$.*

Proof Since $\mathcal{L}^+$ and Δ^+ are isospectral, they have the same spectral radius. Now, since Δ is a non-negative tensor, by [233, Lemma 3.20],

$$\min_{i_1 \in [N]} \sum_{i_2,\ldots,i_\Theta \in [N]} \Delta^+_{i_1,i_2,\ldots,i_\Theta} \leq \rho(\Delta^+) \leq \max_{i_1 \in [N]} \sum_{i_2,\ldots,i_\Theta \in [N]} \Delta^+_{i_1,i_2,\ldots,i_\Theta}.$$

By Lemma 11.3.1, the claim follows.   □

Thanks to Proposition 11.3.5, we can derive more information on the H-eigenvalues and H-eigenvectors.

Corollary 11.3.1. *If Γ is connected, then 2 is an eigenvalue for $\mathcal{L}^+$ and Λ^+, and $(1,\ldots,1) \in \mathbb{R}^N$ is the unique positive H-eigenvector of Δ^+ associated with 2, up to a multiplicative constant.*

Proof The fact that 2 is always an eigenvalue for $\mathcal{L}^+$ and Δ^+ follows by Propositions 11.3.3 and 11.3.5. Moreover, it is easy to check that a corresponding eigenvector for Δ^+ is $(1,\ldots,1) \in \mathbb{R}^N$. By Theorem 11.1.1, the claim follows.   □

Corollary 11.3.2. *If Γ is connected, then 0 is an eigenvalue for K, $\mathcal{L}$, and Δ, and $(1,\ldots,1) \in \mathbb{R}^N$ is the unique positive H-eigenvector of Δ associated with 0, up to a multiplicative constant.*

Proof It follows from Remark 11.3.1 and Corollary 11.3.1.   □

The last two results concerned connected hypergraphs. From our viewpoint, it is not very restrictive to assume that the hypergraph is indeed connected. If this is not the case, we can study the spectrum of a hypergraph from the spectra of its connected components.

Theorem 11.3.4. *Given two hypergraphs $\Gamma_1 = (V_1, E_1, w_1)$ and $\Gamma_2 = (V_2, E_2, w_2)$, let $\Gamma := \Gamma_1 \sqcup \Gamma_2 = (V_1 \sqcup V_2, E_1 \sqcup E_2, w)$, where $w|_{E_i} := w_i$ for $i = 1, 2$. Let also $T \in \{A, K, K^+, \mathcal{L}, \mathcal{L}^+, \Delta, \Delta^+\}$. Then, the eigenvalues of $T(\Gamma)$ are precisely the eigenvalues of $T(\Gamma_1)$ together with the eigenvalues of $T(\Gamma_2)$. Moreover, an eigenvalue that has algebraic multiplicity m for $T(\Gamma_1)$ has algebraic multiplicity $m(\Theta - 1)^{|V_2|}$ for $T(\Gamma)$.*

Proof This is an immediate consequence of [249, Corollary 4.2].   □

We can already make some observations on the comparison between the spectral theory of hypergraphs via matrices and the one via tensors. Recall, from Chapter 4.1, that for a simple connected graph on N nodes, its normalized Laplacian Δ has N real, non-negative eigenvalues. These are contained in the interval $[0,2]$, and their sum is N. Also, 0 in this case is an eigenvalue with multiplicity 1, and the constant vectors are the corresponding eigenvectors.

These properties generalize in a different way when considering the normalized Laplacian matrix for (oriented) hypergraphs or the normalized Laplacian tensor for (weighted) hypergraphs. In fact,

- As shown in Chapter 4.9, the normalized Laplacian matrix for a connected (oriented) hypergraph on N nodes has still N real, non-negative eigenvalues whose sum is N. But in this case, the eigenvalues are contained in the interval $[0, \Theta]$, 0 is not necessarily an eigenvalue, and, if it is an eigenvalue, the corresponding eigenvectors do not need to be constant.
- As we have seen in this chapter, the normalized Laplacian tensor a for (weighted) connected hypergraph on N nodes has $(\Theta - 1)^{N-1} \cdot N$ eigenvalues whose sum is $(\Theta - 1)^{N-1} \cdot N$. These do not need to be real, but the real ones are always contained in $[0,2]$, and, as in the graph case, 0 is an eigenvalue and the constant vectors are the unique corresponding real eigenvectors.

11.4 Θ-Duplicate Vertices

For graphs, in Chapter 4.1, we have discussed *duplicate vertices*, that is, vertices that do not share common edges but are structurally equivalent. Such vertices are important in applied network theory because their presence is important for the study of the *network redundancy* and *robustness*, as discussed, for instance, in [17, 198]. We have also seen, in Chapter 4.9, how the notion of duplicate vertices can be generalized to hypergraphs using the hypergraph adjacency matrix. In this section, we study an alternative generalization of this concept to the case of hypergraphs, which is based on the adjacency tensor.

Definition 11.4.1. Two vertices v_i and v_j are Θ-*duplicates* if they do not share common hyperedges and the corresponding rows of the adjacency tensor are the same, that is,

$$A_{i,i_2,\ldots,i_\Theta} = A_{j,i_2,\ldots,i_\Theta} \text{ for every } i_2,\ldots,i_\Theta \in [N].$$

In particular, Θ-duplicates have the same degree. It also follows from the definition that if v_i is a vertex that has a Θ-duplicate v_j, then v_i is not contained in any hyperedge of cardinality smaller than Θ. Indeed, if $\Theta = 2$, this is trivially true. If $\Theta > 3$, the existence of such a hyperedge would imply that $A_{i,i,i_3,\ldots,i_\Theta} \neq 0$, for some $\{i_3,\ldots,i_\Theta\} \subset [N]$. But, as v_j is a Θ-duplicate, then $A_{j,i,i_3,\ldots,i_\Theta} \neq 0$, which contradicts the assumption that v_i and v_j are not contained in a common hyperedge.

Theorem 11.4.1. *Assume that there are n vertices that are Θ-duplicates of each other.*

1. *If $\Theta = 2$, then*

   - *0 is an eigenvalue for A with multiplicity at least $n - 1$;*
   - *1 is an eigenvalue for $\mathcal{L}$, $\mathcal{L}^+$, Δ, and Δ^+, with multiplicity at least $n - 1$;*
   - *The degree d of the Θ-duplicates is an eigenvalue for K and K^+ with multiplicity at least $n - 1$.*

2. *If $\Theta > 2$, then,*

   - *0 is an eigenvalue for A with Hspan multiplicity at least n and geometric multiplicity at least n;*
   - *1 is an eigenvalue for $\mathcal{L}$, $\mathcal{L}^+$, Δ, and Δ^+, with Hspan multiplicity at least n and geometric multiplicity at least n;*
   - *d is an eigenvalue for K and K^+ with Hspan multiplicity at least n and geometric multiplicity at least n.*

Proof By Remark 11.3.1, it suffices to prove each claim for one operator. The first one has been proved in Corollary 4.9.2 for Δ. For the second one, we focus on A and we assume that $v_1, \ldots, v_n$ are Θ-duplicate vertices.

If $\Theta > 2$, we have already observed that v_i is not contained in any hyperedge of cardinality smaller than Θ. This implies that every monomial in the equations $(A\mathbf{x}^{\Theta-1})_i$ with $i > n$ is divisible by at least one variable x_j, with $j > n$. This means that the eigenvariety contains the n-dimensional linear space defined by the equations

$$x_{n+1} = \ldots = x_N = 0,$$

therefore, gm(0) $\geq n$. In particular, $\mathbf{e}_1, \ldots, \mathbf{e}_n \in \mathbb{R}^N$, that is, the first n vectors of the canonical basis of $\mathbb{R}^N$, are eigenvectors with eigenvalue 0, so Hsm(0) $\geq n$.   □

Proposition 11.4.1. *Let v_i and v_j be Θ-duplicate vertices.*

- *If $(\lambda, \mathbf{x})$ is an eigenpair for A and $\lambda \neq 0$, then $x_i^{\Theta-1} = x_j^{\Theta-1}$.*
- *If $(\mu, \mathbf{y})$ is an eigenpair for $\mathcal{L}$, $\mathcal{L}^+$, Δ, or Δ^+ and $\mu \neq 1$, then $y_i^{\Theta-1} = y_j^{\Theta-1}$.*
- *If $(\nu, \mathbf{z})$ is an eigenpair for K or K^+ and $\nu \neq \deg v_i$, then $z_i^{\Theta-1} = z_j^{\Theta-1}$.*

Proof We only prove the claim for A, the other cases being similar. Since $(\lambda, \mathbf{x})$ is an eigenpair for A,

$$\sum_{i_2,\ldots,i_\Theta \in [N]} A_{i,i_2,\ldots,i_\Theta} x_{i_2} \ldots x_{i_\Theta} = \lambda x_i^{\Theta-1} \qquad (11.4.1)$$

and

$$\sum_{i_2,\dots,i_\Theta \in [N]} A_{j,i_2,\dots,i_\Theta} x_{i_2} \cdots x_{i_\Theta} = \lambda x_j^{\Theta-1}. \tag{11.4.2}$$

Since v_i and v_j are Θ-duplicates, the left-hand sides in (11.4.1) and (11.4.2) coincide. Thus,

$$\lambda x_i^{\Theta-1} = \lambda x_j^{\Theta-1}.$$

Since $\lambda \neq 0$, this implies that $x_i^{\Theta-1} = x_j^{\Theta-1}$. $\qquad\square$

11.5 The Hyperflower

In Chapter 4.9, we defined hyperflowers. As a particular case, we consider here Θ-hyperflowers. In the notation of Definition 4.9.9, these are hyperflowers with $t = 1$ and a core consisting of $\Theta - 1$ vertices.

The Θ-*hyperflower* is the unweighted, Θ-uniform hypergraph $\Gamma = (V, E)$ on N nodes and $M = N - \Theta + 1$ hyperedges, such that

- $V = \{v_1, \dots, v_N\}$
- $E = \{\ell_\Theta, \dots, \ell_N\}$
- $\ell_j = \{v_1, \dots, v_{\Theta-1}, v_j\}$ for every $j \in \{\Theta, \dots, N\}$.

We say that the vertices $v_1, \dots, v_{\Theta-1}$ are the *central vertices* of Γ, while $v_\Theta, \dots, v_N$ are its *peripheral vertices*.

If Γ is a Θ-hyperflower, all its central vertices belong to all hyperedges; hence, they all have degree M. Moreover, the M peripheral vertices of Γ have degree 1. They are Θ-duplicates of each other, and therefore, by Theorem 11.4.1, if $\Theta > 2$, then 0 is an eigenvalue for A with Hspan multiplicity at least M, while 1 is an eigenvalue for $\mathcal{L}$, $\mathcal{L}^+$, Δ, and Δ^+, with Hspan multiplicity at least M. In Proposition 11.5.1, we improve this result for hypergraphs that have no edges of size 2, as, for instance, the Θ-hyperflower for $\Theta \geq 3$.

Proposition 11.5.1. *Let Γ be a hypergraph on N vertices such that every hyperedge contains at least three vertices. Then,*

- *0 is an eigenvalue for A, with Hspan multiplicity equal to N, and*
- *1 is an eigenvalue for $\mathcal{L}$, $\mathcal{L}^+$, Δ, and Δ^+, with Hspan multiplicity equal to N.*

In particular, this holds for the Θ-hyperflower whenever $\Theta \geq 3$.

Proof We only prove the claim for A. The other claims then follow from Remark 11.3.1. By hypothesis, an entry $A_{i_1,\ldots,i_\triangledown}$ of A is nonzero only if $|\{i_1,\ldots,i_\triangledown\}| \geq 3$. Therefore, given $i \in [N]$, each monomial of the polynomial

$$\sum_{i_2,\ldots,i_k \in [N]} A_{i,i_2,\ldots,i_\triangledown} x_{i_2} \cdots x_{i_\triangledown}$$

involves at least two different variables. This implies that such polynomial vanishes on every vector of the canonical basis of $\mathbb{R}^N$, thus $0 \in \operatorname{Spec}(A)$ and $\operatorname{Hsm}(0) = N$. □

Theorem 11.5.1. *Let Γ be a Θ-hyperflower. Given $\lambda \in \mathbb{C}\backslash\{0\}$ and $\mathbf{x} \in \mathbb{C}^N\backslash\{0\}$, we have that $(\lambda,\mathbf{x})$ is an eigenpair for A if and only if, up to multiplying $\mathbf{x}$ by a nonzero constant, all the following conditions hold:*

1. *$x_j^{\Theta-1} = 1$ for each $j \in \{\Theta,\ldots,N\}$;*
2. *$x_i^\Theta = \sum_{j=\Theta}^N x_j$ for each $i \in [\Theta - 1]$; and*
3. *$\lambda = x_1 \cdots x_{\Theta-1}$.*

Proof By Proposition 11.4.1, if $(\lambda,\mathbf{x})$ is an eigenpair for A and $\lambda \neq 0$, then $x_j^{\Theta-1}$ must be constant for $j \in \{\Theta,\ldots,N\}$. Assume first that $x_j = 0$ for all $j \in \{\Theta,\ldots,N\}$. Then, for $i \in \{1,\ldots,\Theta - 1\}$, the left-hand side of

$$\sum_{i_2,\ldots,i_\Theta \in [N]} A_{i,i_2,\ldots,i_\Theta} x_{i_2} \cdots x_{i_\Theta} = \lambda x_i^{\Theta-1}$$

vanishes. This implies that $\lambda x_i^{\Theta-1} = 0$; therefore, $x_i = 0$ for each $i \in [\Theta - 1]$, but this is a contradiction since $\mathbf{x} \neq \mathbf{0}$. Hence, $x_j^{\Theta-1}$ is constant and nonzero for $j \in \{\Theta,\ldots,N\}$. Up to multiplying $\mathbf{x}$ by a nonzero constant, we can assume that

$$x_j^{\Theta-1} = 1 \text{ for each } j \in \{\Theta,\ldots,N\}. \tag{11.5.1}$$

Now, since Γ is Θ-uniform, $(\lambda,\mathbf{x})$ is an eigenpair for A if and only if

$$\sum_{(v_k,v_{i_2},\ldots,v_{i_\Theta}) \in E} x_{i_2} \cdots x_{i_\Theta} = \lambda x_k^{\Theta-1} \tag{11.5.2}$$

for each $k \in [N]$. If v_j is a peripheral vertex, then by (11.5.1), (11.5.2) becomes

$$x_1 \cdots x_{\Theta-1} = \lambda. \tag{11.5.3}$$

Hence, since $\lambda \neq 0$, we have that $x_i \neq 0$ for each $i \in [\Theta - 1]$.

If v_i is a central vertex, that is, $i \in [\Theta - 1]$, then (11.5.2) becomes

$$\sum_{j=\Theta}^N x_j \cdot \frac{x_1 \cdots x_{\Theta-1}}{x_i} = \lambda x_i^{\Theta-1}.$$

By (11.5.3), the latter equality is equivalent to

$$\sum_{j=\Theta}^{N} x_j = x_i^{\Theta}, \qquad (11.5.4)$$

for each $i \in [\Theta - 1]$. This proves the claim. $\qquad\square$

Remark 11.5.1. If $(\lambda, \mathbf{x})$ satisfies Theorem 11.5.1, then x_i^{Θ} is constant for all $i \in [\Theta - 1]$ and

$$\left| x_i^{\Theta} \right| = \left| \sum_{j=\Theta}^{N} x_j \right| \leq \sum_{j=\Theta}^{N} \left| x_j \right| = M, \qquad (11.5.5)$$

implying that $|x_i| \leq \sqrt[\Theta]{M}$. Hence,

$$|\lambda| = |x_1 \cdots x_{\Theta-1}| \leq \sqrt[\Theta]{M^{\Theta-1}}.$$

Corollary 11.5.1. *If Γ is a Θ-hyperflower with M hyperedges and ω is a Θ-th root of 1, then*

$$\sqrt[\Theta]{M^{\Theta-1}}, \quad \omega \sqrt[\Theta]{M^{\Theta-1}}, \quad \ldots, \quad \omega^{\Theta-1} \sqrt[\Theta]{M^{\Theta-1}}$$

are eigenvalues of $A(\Gamma)$. If, furthermore, $M = n(\Theta - 1) + 1$ for some positive integer n, then $1, \omega, \ldots, \omega^{\Theta-1}$ are also eigenvalues of $A(\Gamma)$.

Proof If ω is a Θ-th root of 1 and β is a $(\Theta - 1)$-th root of ω, let

- $x_j := \beta^{\Theta}$ for each $j \in \{\Theta, \ldots, N\}$ and
- $x_i := \beta \sqrt[\Theta]{M}$ for each $i \in [\Theta - 1]$.

By Theorem 11.5.1, $\mathbf{x}$ is an eigenvector for the eigenvalue

$$x_i^{\Theta-1} = \omega \sqrt[\Theta]{M^{\Theta-1}}.$$

Hence,

$$\sqrt[\Theta]{M^{\Theta-1}}, \quad \omega \sqrt[\Theta]{M^{\Theta-1}}, \quad \ldots, \quad \omega^{\Theta-1} \sqrt[\Theta]{M^{\Theta-1}}$$

are eigenvalues of Γ.

Now, assume that $M = n(\Theta - 1) + 1$ for some positive integer n. Let again ω be a Θ-th root of 1. Let α be a $(\Theta - 1)$-th root of ω and let $z := \alpha^{\Theta}$, so that z is a $(\Theta - 1)$-th root of 1. Assume that the M elements $x_{\Theta}, \ldots, x_N$ are given by

$$z \ (n+1 \text{ times}), \quad z^2 \ (n \text{ times}), \quad \ldots, \quad 1 \ (n \text{ times}).$$

Then, since $\sum_{k=1}^{\Theta-1} z^k = 0$, by Theorem 11.5.1, we must have

$$\sum_{j=\Theta}^{N} x_j = z = x_i^{\Theta},$$

if we want $\mathbf{x}$ to be an eigenvector. If, in particular, $x_i := \alpha$ for each $i \in [\Theta - 1]$, then the earlier condition is satisfied, and, by Theorem 11.5.1, $\mathbf{x}$ is an eigenvector with eigenvalue

$$x_i^{\Theta-1} = \alpha^{\Theta-1} = \omega.$$

Hence, $1, \omega, \ldots, \omega^{\Theta-1}$ are eigenvalues of $A(\Gamma)$.   □

For simplicity, in Theorem 11.5.1 and Corollary 11.5.1, we focused on the adjacency tensor of a Θ-hyperflower Γ. However, similar results can be shown also for the other tensors associated with Γ. The following two theorems for $K(\Gamma)$ and $\Delta(\Gamma)$, respectively, can be proved as Theorem 11.5.1.

Theorem 11.5.2. *Let Γ be a Θ-hyperflower with M hyperedges. Given $\lambda \in \mathbb{C} \setminus \{1, M\}$ and $\mathbf{x} \in \mathbb{C}^N \setminus \{0\}$, we have that $(\lambda, \mathbf{x})$ is an eigenpair for K if and only if, up to multiplying $\mathbf{x}$ by a nonzero constant, all the following conditions hold:*

1. $x_j^{\Theta-1} = 1$ *for each $j \in \{\Theta, \ldots, N\}$;*
2. $x_i^{\Theta} = \frac{(1-\lambda)}{(M-\lambda)} \cdot \left(\sum_{j=\Theta}^{N} x_j\right)$ *for each $i \in [\Theta - 1]$;*
3. $\lambda = 1 - x_1 \cdots x_{\Theta-1}.$

Theorem 11.5.3. *Let Γ be a Θ-hyperflower with M hyperedges. Given $\lambda \in \mathbb{C} \setminus \{1\}$ and $\mathbf{x} \in \mathbb{C}^N \setminus \{0\}$, we have that $(\lambda, \mathbf{x})$ is an eigenpair for Δ if and only if, up to multiplying $\mathbf{x}$ by a nonzero constant, all the following conditions hold:*

1. $x_j^{\Theta-1} = 1$ *for each $j \in \{\Theta, \ldots, N\}$;*
2. $x_i^{\Theta} = \frac{1}{M} \cdot \left(\sum_{j=\Theta}^{N} x_j\right)$ *for each $i \in [\Theta - 1]$;*
3. $\lambda = 1 - x_1 \cdots x_{\Theta-1}.$

We now present the full spectrum of the various Laplacian tensors for hyperflowers.

Example 11.5.1. Let $\Gamma = (V, E)$ be the 3-hyperflower with $V = \{v_1, v_2, v_3, v_4\}$ and $E = \{\{v_1, v_2, v_3\}, \{v_1, v_2, v_4\}\}$ (Figure 11.3).

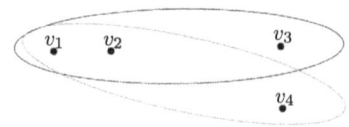

Figure 11.3 The hyperflower in Example 11.5.1.

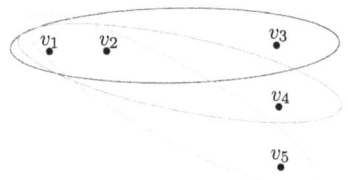

Figure 11.4 The hyperflower in Example 11.5.2.

The characteristic polynomial of A is $(\lambda^3 - 4)^3 \lambda^{23}$. Hence, the eigenvalues of A are

0 with multiplicity 23 $\qquad\qquad$ $\sqrt[3]{4}$ with multiplicity 3

$\sqrt[3]{4}\omega$ with multiplicity 3 $\qquad\qquad$ $\sqrt[3]{4}\omega^2$ with multiplicity 3,

where ω is a third root of 1. In particular, the distinct eigenvalues are exactly the ones in Proposition 11.5.1 and Corollary 11.5.1.

The characteristic polynomial of K is $(\lambda^2 - 5\lambda + 8)^3 (\lambda - 1)^{13} (\lambda - 2)^{10} \lambda^3$. Its roots are

0 with multiplicity 3 $\qquad\qquad$ $\dfrac{5}{2} + \dfrac{\sqrt{7}}{2}\mathbf{i}$ with multiplicity 3

$\dfrac{5}{2} - \dfrac{\sqrt{7}}{2}\mathbf{i}$ with multiplicity 3 $\qquad\qquad$ 1 with multiplicity 13

2 with multiplicity 10.

The characteristic polynomial of Δ is $(\lambda^2 - 3\lambda + 3)^3 (\lambda - 1)^{23} \lambda^3$. Thus, its eigenvalues are

0 with multiplicity 3 $\qquad\qquad$ 1 with multiplicity 23

$2 + \omega$ with multiplicity 3 $\qquad\qquad$ $2 + \omega^2$ with multiplicity 3.

Example 11.5.2. Let $\Gamma = (V, E)$ be the 3-hyperflower with $V = \{v_1, v_2, v_3, v_4, v_5\}$ and $E = \{\{v_1, v_2, v_3\}, \{v_1, v_2, v_4\}, \{v_1, v_2, v_5\}\}$ (Figure 11.4).

The characteristic polynomial of A is $(\lambda^3 - 9)^3 (\lambda^2 + \lambda + 1)^9 (\lambda - 1)^9 \lambda^{44}$; therefore, its eigenvalues are

0 with multiplicity 44 $\qquad\qquad$ 1 with multiplicity 9

$\sqrt[3]{9}$ with multiplicity 3 $\qquad\qquad$ $\sqrt[3]{9}\omega$ with multiplicity 3

$\sqrt[3]{9}\omega^2$ with multiplicity 3 $\qquad\qquad$ ω with multiplicity 9

ω^2 with multiplicity 9.

Also in this case, the distinct eigenvalues of A are exactly the ones in Proposition 11.5.1 and Corollary 11.5.1. Now, the characteristic polynomial of K is

$$(\lambda^3 - 7\lambda^2 + 15\lambda - 8)^9(\lambda^2 - 7\lambda + 15)^3(\lambda - 1)^{36}(\lambda - 3)^8\lambda^3.$$

The eigenvalues of K are

   0 with multiplicity 3

   0.7944305695994095 with multiplicity 9

   1 with multiplicity 36

   3 with multiplicity 8

   3.1027847152002956 + 0.6654569511528129i with multiplicity 9

   3.1027847152002956 − 0.6654569511528129i with multiplicity 9

   $\dfrac{7}{2} + \dfrac{\sqrt{11}}{2}$i, with multiplicity 3

   $\dfrac{7}{2} - \dfrac{\sqrt{11}}{2}$i with multiplicity 3.

The characteristic polynomial of Δ is

$$\frac{(9\lambda^3 - 27\lambda^2 + 27\lambda - 8)^9(\lambda^2 - 3\lambda + 3)^3(\lambda - 1)^{44}\lambda^3}{387420489}.$$

Its eigenvalues are

   0, with multiplicity 3

   0.519250143230864, with multiplicity 9

   1, with multiplicity 44

   1.240374928384567 + 0.4163415888278001i, with multiplicity 9

   1.240374928384567 − 0.4163415888278001i, with multiplicity 9

   $2 + \omega$, with multiplicity 3

   $2 + \omega^2$, with multiplicity 3.

11.6 Spectral Symmetries

We now discuss some spectral symmetries. First, we recall [92, Theorem 3.12] and we apply it to the hypergraph tensors.

Definition 11.6.1. Let T be a tensor and let ℓ be a positive integer. The tensor T is *spectral ℓ-symmetric* if

$$\mathrm{Spec}(T) = e^{\frac{2\pi i}{\ell}}\, \mathrm{Spec}(T).$$

Definition 11.6.2. Let $k \geq 2$ and $\ell \geq 2$ such that $\ell \mid k$. A kth order n-dimensional tensor T is *(k,ℓ)-colorable* if there exists a map $\phi \colon [n] \rightarrow [k]$ such that, if $T_{i_1,\ldots,i_k} \neq 0$, then

$$\phi(i_1) + \ldots + \phi(i_k) \equiv \frac{k}{\ell} \quad \mathrm{mod}\ k.$$

Such ϕ is an *(k,ℓ)-coloring* of T.

Definition 11.6.3. The hypergraph Γ is *(Θ,ℓ)-colorable*, for some $\ell \geq 2$ such that $\ell \mid \Theta$, if there exists a map $\phi \colon V \rightarrow [\Theta]$ such that, if $\{v_{i_1},\ldots,v_{i_\Theta}\} \in E$, then

$$\phi(v_{i_1}) + \ldots + \phi(v_{i_\Theta}) \equiv \frac{\Theta}{\ell} \quad \mathrm{mod}\ \Theta.$$

Remark 11.6.1. Clearly, the hypergraph Γ is (Θ,ℓ)-colorable if and only if its associated tensors are (Θ,ℓ)-colorable.

Theorem 11.6.1 ([92]). *Let T be a symmetric weakly irreducible non-negative tensor of order k. Then, T is spectral ℓ-symmetric if and only if T is (k,ℓ)-colorable.*

As an immediate consequence of Theorem 11.6.1, we obtain the following

Corollary 11.6.1. *A connected hypergraph Γ is (Θ,ℓ)-colorable if and only if one of A, K^+, $\mathcal{L}^+$, and Δ^+ is spectral ℓ-symmetric, if and only if all of them are spectral ℓ-symmetric.*

Proof The claim for A, K^+, and $\mathcal{L}^+$ follows directly from Theorem 11.6.1, since these are all symmetric non negative tensors, and, by Theorem 11.3.2, they are also weakly irreducible, as we are assuming that Γ is connected. The claim for Δ^+ then follows by Proposition 11.3.1. $\qquad\square$

Example 11.6.1. The Θ-hyperflowers are (Θ,Θ)-colorable. This can be seen by coloring the central vertices with Θ and the peripheral vertices with 1. By Corollary 11.6.1, this implies that A is spectral Θ-symmetric, as we can observe in Examples 11.5.1 and 11.5.2.

We now discuss another kind of spectral symmetry for the hypergraph tensors. For a simple graph Γ, the following are equivalent:

1. Γ is bipartite
2. $\lambda \in \mathrm{Spec}(A) \iff -\lambda \in \mathrm{Spec}(A)$, with the same multiplicity

3. $\text{Spec}(K) = \text{Spec}(K^+)$
4. $\lambda \in \text{Spec}(\Delta) \iff 2 - \lambda \in \text{Spec}(\Delta)$, with the same multiplicity.

In fact, we have seen that this is true for Δ in Lemma 4.1.13, and the proof for the other operators is analogous.

There are various ways of generalizing the notion of bipartite graph to the case of hypergraphs, see for instance, the *balanced hypergraphs* in [191, Section 3] or the *bipartite hypergraphs* in Chapter 4.9. We now consider *odd-bipartite hypergraphs*.

Definition 11.6.4. The hypergraph Γ is *odd-bipartite* if Θ is even and one can decompose the vertex set as a disjoint union $V = V_1 \sqcup V_2$ such that, if $i_1, \ldots, i_\Theta \in [N]$, then

$$\{v_{i_1}, \ldots, v_{i_\Theta}\} \in E \Rightarrow \text{There is an odd number of vertices of}$$
$$V_1 \text{ among } v_{i_1}, \ldots, v_{i_\Theta},$$

where the vertices $v_{i_1}, \ldots, v_{i_\Theta}$ are counted with repetitions.

Clearly, if $\Theta = 2$, Γ is a bipartite graph if and only if Γ is odd-bipartite.

We refer to [105, Theorem 6.11] and [248, Theorem 2.2 and Theorem 2.3] for the proof of the following result.

Theorem 11.6.2. *If Γ is a connected, weighted hypergraph, the following conditions are equivalent:*

1. *Θ is even and Γ is odd-bipartite.*
2. $\text{Spec}(A) = -\text{Spec}(A)$ *and* $\text{Hspec}(A) = -\text{Hspec}(A)$.
3. $\text{Hspec}(A) = -\text{Hspec}(A)$.
4. $\text{Spec}(K) = \text{Spec}(K^+)$ *and* $\text{Hspec}(K) = \text{Hspec}(K^+)$.
5. $\text{Hspec}(K) = \text{Hspec}(K^+)$.
6. $\text{Spec}(\Delta) = \text{Spec}(\Delta^+)$ *and* $\text{Hspec}(\Delta) = \text{Hspec}(\Delta^+)$.
7. $\text{Hspec}(\Delta) = \text{Hspec}(\Delta^+)$.

Remark 11.6.2. The second condition in Theorem 11.6.2 means that

$$\lambda \in \text{Spec}(A) \iff -\lambda \in \text{Spec}(A), \text{ with the same multiplicity,}$$

and $\lambda \in \text{Hspec}(A) \iff -\lambda \in \text{Hspec}(A)$. Similarly, by Remark 11.3.1, the sixth condition in Theorem 11.6.2 means that

$$\lambda \in \text{Spec}(\Delta) \iff 2 - \lambda \in \text{Spec}(\Delta), \text{ with the same multiplicity,}$$

and $\lambda \in \text{Hspec}(\Delta) \iff 2 - \lambda \in \text{Hspec}(\Delta)$. Moreover, by Proposition 11.3.1, the claim for Δ and Δ^+ also holds for $\mathcal{L}$ and $\mathcal{L}^+$.

12

The Nonbacktracking Laplacian

A nonbacktracking random walk, that is, where the random walker is not allowed to travel an edge back and forth, leads to a new Laplace-type operator with novel properties. It encodes structural features of the underlying graph in a manner different from other Laplace operators.

Spectral graph theory is typically based on the study of the spectrum of symmetric operators associated with graphs. As discussed in Chapters 4, 5, and 8, the main advantage is that such spectra can be easily computed and are relatively easy to interpret in terms of geometric properties of the graph. The price to pay is the presence of several families of isospectral graphs, that is, nonisomorphic graphs that share the same spectrum and therefore cannot be uniquely distinguished by their eigenvalues.

An alternative to the spectral theory of graphs via symmetric operators, which partially overcomes this problem and also comes with its own interesting aspects, is the spectral theory of *nonbacktracking operators*. The idea behind it is the following. Consider a simple graph $G = (V, E)$ and a random walker on V who moves, with uniform probability, from one vertex v to any of its neighbors, except for the one that she visited immediately prior to v. This process is called a *nonbacktracking random walk* since the walker does not go back to the vertex she just visited, and it is clearly non-Markovian. However, it is possible to construct a directed graph $\mathcal{G} = (\mathcal{V}, \mathcal{E})$ called the *nonbacktracking graph* or the *Hashimoto graph* of G, with the property that a nonbacktracking random walk on V can be equivalently seen as a simple random walk on $\mathcal{V}$, and this process is Markovian on $\mathcal{V}$. We shall see how to construct $\mathcal{G}$ in the first section of this chapter.

The reader may have already noticed that we are deviating here from our general convention to denote a graph by Γ and call it G instead. The reason is that our notation in this chapter is set up so as to utilize a systematic correspondence between ordinary and calligraphic letters. The ordinary letters, like G, will be used for quantities defined on or by the original graph, and the calligraphic letters, $\mathcal{G}$ in this case, for the corresponding quantities on the nonbacktracking graph constructed from it. So, for instance, V is the vertex set of G and $\mathcal{V}$ that of $\mathcal{G}$. Also, in this chapter, it will be natural to work with *directed* graphs.

The adjacency matrix $\mathcal{A}$ and the normalized Laplacian $\mathcal{L}$ of the new directed graph $\mathcal{G}$ are nonsymmetric operators, and, as we shall see, their spectra can be considered in order to investigate properties of the original graph G. We recall the discussion of general non symmetric Laplacians in Section 4.7.

The transpose of $\mathcal{A}$ is called the *nonbacktracking matrix* of G, and it was introduced by Ki-ichiro Hashimoto in 1989 in the context of graph zeta functions [127]. Later on, many properties of this matrix have been investigated in the area of algebraic graph theory [22, 66, 257], and it has then been shown that it is very powerful in spectral graph theory, as well as in its applications to network analysis [8, 36, 67, 111, 116, 177, 202, 226, 259–262]. Moreover, Backhausz, Szegedy, and Virág applied nonbacktracking random walks to the study of Ramanujan graphings [13], while Shrestha, Scarpino, and Moore in 2015 [250], as well as Castellano and Pastor-Satorras in 2018 [48], applied the study of nonbacktracking paths to epidemic spreading on networks.

The matrix $\mathcal{L}$ is called the *nonbacktracking Laplacian* of G. It was introduced in [156] (on which this chapter is mainly based) and further investigated in [220].

Before proceeding, we should warn the reader that, once one learns about the beauty of the nonbacktracking matrices, there is no going back to only working with symmetric operators.

12.1 Basic Definitions

We now give a formal and detailed description of the concepts that we introduced at the beginning of this chapter.

We fix a simple graph $G = (V, E)$ on N nodes and M edges and (for reasons that will soon be clear) we assume that it has minimum degree ≥ 2. For each edge, we fix an *orientation*; that is, we let one of its endpoints be its *input* and the other one be its *output*. Moreover, we let $e = [v, w]$ denote the oriented edge whose input is v and whose output is w. We also write $\text{in}(e) := v$ and $\text{out}(e) := w$, and we let $e^{-1} := [w, v]$. We also let $e_1, \ldots, e_M$ denote the edges of G with the orientation that has been (arbitrarily) fixed, and we let

$$e_{M+1} := e_1^{-1}, \ldots, e_{2M} := e_M^{-1}$$

denote the edges with the inverse orientation.

Earlier, we described a nonbacktracking random walk as a non-Markovian process on the vertex set of G. We can equivalently define it as a Markovian process on the set of oriented edges, as follows.

Definition 12.1.1. A *nonbacktracking random walk* on G is a discrete-time Markov process on the oriented edges such that the probability of going from e_i to e_j is

$$\mathbb{P}(e_i \rightarrow e_j) = \begin{cases} \frac{1}{\deg(\mathrm{out}(e_i))-1} & \text{if } \mathrm{out}(e_i) = \mathrm{in}(e_j) \text{ and } \mathrm{in}(e_i) \neq \mathrm{out}(e_j) \\ 0 & \text{otherwise.} \end{cases}$$

Moreover, the nonbacktracking matrix of G can be defined directly, without first considering the nonbacktracking graph, in the following way.

Definition 12.1.2. The matrix $B := B(G)$ is the $2M \times 2M$ matrix with $(0,1)$–entries such that

$$B_{ij} = 1 \iff \mathrm{out}(e_i) = \mathrm{in}(e_j) \text{ and } \mathrm{in}(e_i) \neq \mathrm{out}(e_j).$$

The *nonbacktracking matrix* of G is $B^\top$, the transpose matrix of B.

Now, we fix a directed graph $\mathcal{G} = (\mathcal{V}, \mathcal{E})$ on $\mathcal{N}$ nodes and $\mathcal{M}$ edges. If $\mathcal{G}$ has an edge from a vertex v to a vertex w, we write $v \rightarrow w$ and denote such an edge by $(v \rightarrow w)$. Given a vertex $v \in \mathcal{V}$, we define its *outdegree* or simply its *degree* as

$$\deg_{\mathcal{G}} v := \deg v := |\{w \in \mathcal{V} : (v \rightarrow w) \in \mathcal{E}\}|.$$

We assume that $\mathcal{G}$ has minimum degree ≥ 1, and, analogously to the undirected case, we define the following operators.

We now recall the definitions from Section 4.7.

Definition 12.1.3. The *degree matrix* of $\mathcal{G}$ is the $\mathcal{N} \times \mathcal{N}$ diagonal matrix $\mathcal{D} := \mathcal{D}(\mathcal{G}) := (\mathcal{D}_{vw})_{v,w \in \mathcal{V}}$ whose diagonal entries are

$$\mathcal{D}_{vv} := \deg v.$$

The *adjacency matrix* of $\mathcal{G}$ is the $\mathcal{N} \times \mathcal{N}$ matrix $\mathcal{A} := \mathcal{A}(\mathcal{G}) := (\mathcal{A}_{vw})_{v,w \in \mathcal{V}}$ defined by

$$\mathcal{A}_{vw} := \begin{cases} 1 & \text{if } v \rightarrow w \\ 0 & \text{otherwise.} \end{cases} \tag{12.1.1}$$

The *normalized Laplacian* of $\mathcal{G}$ is the $\mathcal{N} \times \mathcal{N}$ matrix

$$\mathcal{L}(\mathcal{G}) := \mathrm{Id} - \mathcal{D}^{-1}\mathcal{A}. \tag{12.1.2}$$

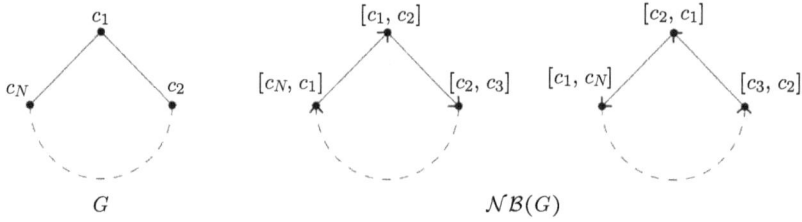

Figure 12.1 The cycle graph G and its nonbacktracking graph $\mathcal{N}B(G)$.

Remark 12.1.1. Similarly to the undirected case, the normalized Laplacian $\mathcal{L}$ of the directed graph $\mathcal{G}$ can be equivalently seen as an operator such that, given $f : \mathcal{V} \to \mathbb{C}$ and $\omega \in \mathcal{V}$,

$$\mathcal{L}f(\omega) = f(\omega) - \frac{1}{\deg \omega}\left(\sum_{\omega \to \tau} f(\tau)\right).$$

We can now define the nonbacktracking graph and the nonbacktracking Laplacian of G.

Definition 12.1.4. The *nonbacktracking graph* of G is the directed graph $\mathcal{N}B(G)$ on vertices $e_1, \ldots, e_{2M}$, that has B as its adjacency matrix.

The *nonbacktracking Laplacian* of G, denoted by $\mathcal{L} := \mathcal{L}(G)$, is the normalized Laplacian $\mathcal{L}(\mathcal{N}B(G))$ of $\mathcal{N}B(G)$.

Example 12.1.1. If G is the cycle graph on N nodes, then $\mathcal{N}B(G)$ is given by two disconnected directed cycles on N nodes (Figure 12.1).

Example 12.1.2. In this example, we consider a tripod (Figure 12.2). The tripod has pending vertices, that is, vertices of degree 1, and hence violates our general assumption that all vertices have degree at least 2, and therefore, the associated nonbacktracking graph is not strongly connected. The vertices corresponding to the oriented edges going out of the central node in the tripod have only incoming arrows but no outgoing ones, and conversely, the vertices corresponding to oriented edges into the central node have only outgoing arrows.

Example 12.1.3. We next consider a graph that has more than one cycle (Figure 12.3). If we want to traverse the oriented edge from v to w in the backward direction, we can simply go around the other cycle, return to w, and then go back to v. This is a nonbacktracking path.

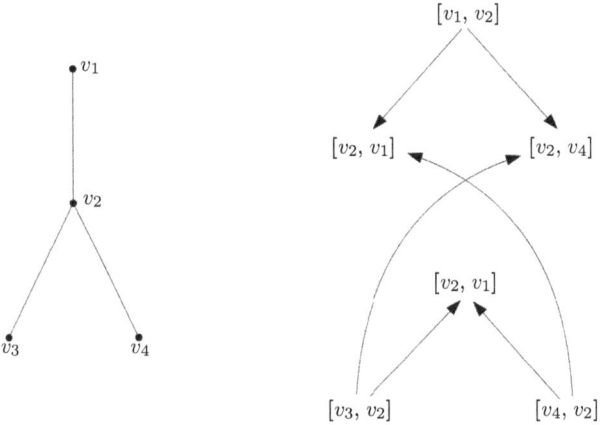

Figure 12.2 The tripod and its nonbacktracking graph.

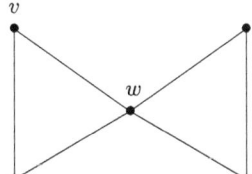

Figure 12.3 A graph that has more than one cycle.

In order to employ a systematic notation, we shall denote the nonbacktracking graph of $G = (V, E)$ by $\mathcal{G} = (\mathcal{V}, \mathcal{E})$. Clearly, if G has N nodes and M edges, then $\mathcal{G}$ has $2M$ nodes. Moreover, for each oriented edge e_i,

$$\deg_{\mathcal{G}} e_i = \deg_G(\text{out}(e_i)) - 1. \qquad (12.1.3)$$

Hence, the assumption that G has minimum degree ≥ 2 implies that $\mathcal{G}$ has minimum degree ≥ 1. In particular, this assumption makes the nonbacktracking Laplacian well-defined.

Remark 12.1.2. The off-diagonal entries of the nonbacktracking Laplacian of G are given by

$$\mathcal{L}_{ij} = -\frac{B_{ij}}{\deg_{\mathcal{G}}(e_i)} = -\frac{B_{ij}}{\deg_G(\text{out}(e_i)) - 1} = -\mathbb{P}(e_i \to e_j),$$

for $i \neq j$. Therefore, the nonbacktracking Laplacian encodes the probabilities of nonbacktracking random walks on G.

Remark 12.1.3. The *line-digraph* constructed by Harary and Norman [126] is given by the nonbacktracking graph with the addition of all directed edges of the form $[v, w] \to [w, v]$.

12.2 The Nonbacktracking Graph

We now investigate some properties of nonbacktracking graphs. We fix again a simple graph $G = (V, E)$ on N nodes and M edges that has minimum degree ≥ 2, and we let $\mathcal{G} = (\mathcal{V}, \mathcal{E})$ be its nonbacktracking graph. According to our construction, $\mathcal{G}$ has $2M$ nodes. Another straightforward property is that, by (12.1.3), G is $(k + 1)$–regular if and only if $\mathcal{G}$ is k–regular. Similarly, G is bipartite if and only if $\mathcal{G}$ is bipartite, since G has odd-length cycles if and only if $\mathcal{G}$ has odd-length cycles. Moreover, as shown in [66] and [220], every simple graph can be uniquely reconstructed from its nonbacktracking graph.

Proposition 12.2.1 gives the number of (directed) edges of $\mathcal{G}$.

Proposition 12.2.1. *Let $G = (V, E)$ be a simple graph on N nodes and M edges, with minimum degree ≥ 2, and let $\mathcal{G} = (\mathcal{V}, \mathcal{E})$ be its nonbacktracking graph. Then, $\mathcal{G}$ has $\sum_{v \in V} (\deg_G v)^2 - 2M$ edges.*

Proof By definitions of vertex degrees for G and $\mathcal{G}$,

$$|\mathcal{E}| = \sum_{e \in \mathcal{V}} \deg_{\mathcal{G}} e, \text{ while } M = \frac{\sum_{v \in V} \deg_G v}{2}.$$

Therefore,

$$|\mathcal{E}| = \sum_{e \in \mathcal{V}} \deg_{\mathcal{G}} e = \sum_{v \in V} \deg_G v \cdot (\deg_G v - 1) = \sum_{v \in V} (\deg_G v)^2 - 2M,$$

where the second equality holds because, for each $v \in V$, there are $\deg_G v$ oriented edges of the form $[w, v]$ and for each of these oriented edges, there are $\deg_G v - 1$ oriented edges of the form $[v, z]$ with $z \neq w$.   □

Now, in Example 12.1.1, we saw that if G is a cycle graph, then its nonbacktracking graph is given by two disconnected cycles. Theorem 12.2.1 shows that this is the only case in which a connected simple graph has a disconnected nonbacktracking graph.

Theorem 12.2.1. *Let $G = (V, E)$ be a simple connected graph on N nodes and M edges, with minimum degree ≥ 2, and let $\mathcal{G} = (\mathcal{V}, \mathcal{E})$ be its nonbacktracking graph. Then, the following are equivalent:*

1. G is not the cycle graph;

2. *G has at least two cycles;*
3. *$\mathcal{G}$ is weakly connected; and*
4. *$\mathcal{G}$ is strongly connected.*

Proof Clearly, since the minimum degree in G is ≥ 2, the first two conditions are equivalent to each other. Moreover, 4 clearly implies 3, and, by Example 12.1.1, 3 implies 1. Hence, if we prove that 1 implies 4, we are done.

An essential observation in the proof to follow is that if v, v' are two neighboring vertices and if we go from v to v', we can return to v in a nonbacktracking manner by inserting a nonbacktracking path starting at and returning to v' but not passing through v. That path need not be a cycle itself but will of course contain some cycle. Figure 12.3 illustrates this.

In fact, we just need to show the following two properties because their combination will allow us to go from any oriented edge to any other oriented edge.

1. If (v_0, v_1) and (v'_0, v'_1) are edges, then for some orientations, say $[v_0, v_1]$ and $[v'_0, v'_1]$, we can go from one to the other in $\mathcal{G}$.
2. If $[v_0, v_1]$ is an oriented edge, we can go from it to the oppositely oriented edge $[v_1, v_0]$ in $\mathcal{G}$.

Now 1 directly follows from the connectedness of G. For 2, we start at $[v_0, v_1]$ and follow any nonbacktracking path. We never get stuck because we assume that G has no pending vertices and because all vertices have a degree of at least 2. G is finite, and so, such a path π_1 either returns onto itself before reaching v_0 again. In that case, by the earlier observation, we can use it to reverse direction without backtracking and return to v_1 and hence the oppositely oriented edge $[v_1, v_0]$. Or π_1 eventually returns to v_0. But since we assume that G is not a cycle, there must be some vertex (of degree at least 3) on that path π_1 where we can branch off into a different path π_2. That path again either returns onto itself before reaching v_0 again, in which case we can proceed as before, or it brings us back to π_1 at some other vertex of degree at least 3, which will allow us to reverse directions on π_1 and return to v_1 without going through v_0 first. Thus, in any case, we obtain 2. $\qquad\square$

12.3 First Spectral Properties of the Nonbacktracking Laplacian

We first consider a directed graph $\mathcal{G} = (\mathcal{V}, \mathcal{E})$ on $\mathcal{N}$ nodes and minimum degree ≥ 1, which need not be the nonbacktracking graph associated with a simple graph. Since its normalized Laplacian $\mathcal{L}$ is an $\mathcal{N} \times \mathcal{N}$ matrix of trace $\mathcal{N}$, it has $\mathcal{N}$ eigenvalues (counted with algebraic multiplicity) that sum to $\mathcal{N}$. Moreover, by Proposition 3.2 in [23], the spectrum of $\mathcal{L}$ is contained in the complex disc $D(1,1)$, and therefore, the real eigenvalues are contained in $[0,2]$. Also, by Proposition 3.1 in [23], 0 is an eigenvalue for $\mathcal{L}$, and the constant functions $f: \mathcal{V} \to \mathbb{C}$ are the corresponding eigenfunctions. As a consequence, from the spectrum of $\mathcal{L}$, we can derive the number of connected components of $\mathcal{G}$. Notably, this does not hold for the adjacency matrix $\mathcal{A}$ of $\mathcal{G}$.

Another property of $\mathcal{L}$, which is the same as in the undirected case, is that if $\mathcal{G}$ is a directed bipartite graph, then the spectra of both $\mathcal{L}$ and $\mathcal{A}$ are symmetric. Therefore, in particular, 2 is an eigenvalue for $\mathcal{L}$ [156]. However, the other direction does not necessarily hold in the directed case: As shown in [156], there exist directed graphs that are not bipartite but have 2 as an eigenvalue of their normalized Laplacian.

Furthermore, it follows from (12.1.2) that the normalized Laplacian $\mathcal{L}$ and the adjacency matrix $\mathcal{A}$ of $\mathcal{G}$ satisfy:

(i) f is an eigenfunction for $\mathcal{A}$ with eigenvalue 0 if and only if f is an eigenfunction for $\mathcal{L}$ with eigenvalue 1;

(ii) If $\mathcal{G}$ is k–regular, then (λ, f) is an eigenpair for $\mathcal{A}$ if and only if $(1 - \frac{\lambda}{k}, f)$ is an eigenpair for $\mathcal{L}$.

All the earlier properties hold for any directed graph $\mathcal{G}$. From here on, we shall focus on the case of nonbacktracking graphs. In particular, we fix again a simple $G = (V, E)$ on N nodes and M edges that have minimum degree ≥ 2, and we let $\mathcal{G} = (\mathcal{V}, \mathcal{E})$ be its nonbacktracking graph. Thus, $\mathcal{A}$ as defined in (12.1.1) is the adjacency matrix of $\mathcal{G}$ (equivalently, the transpose of the nonbacktracking matrix of G) and $\mathcal{L}$ from (12.1.2) is the normalized Laplacian of $\mathcal{G}$ (equivalently, the nonbacktracking Laplacian of G). Also, as in the previous chapters, we let A and L denote the adjacency matrix and the normalized Laplacian of G, respectively.

The first important consequence of the earlier observations for the case of nonbacktracking graphs is the following.

Remark 12.3.1. As observed in Section 12.2, G is regular if and only if $\mathcal{G}$ is regular. In this case, we know that the spectral properties of L and A are equivalent to each other, and by property (ii) above, also the spectral properties of $\mathcal{L}$ and $\mathcal{A}$ are equivalent to each other. But since it is known that, in the regular case, the eigenvalues of $\mathcal{A}$ can be recovered from those of A [172, 175], it follows that in this case also the spectral theory of $\mathcal{L}$ can be recovered from that of A or, equivalently, of L.

Now, although the nonbacktracking Laplacian is not symmetric, as we shall see in Theorem 12.3.1, it satisfies the so-called *PT-symmetry* (where PT stands for parity-time), following the terminology in [36]. Before stating it, we define the $2M \times 2M$ matrix

$$P := \begin{pmatrix} 0 & \mathrm{Id} \\ \mathrm{Id} & 0 \end{pmatrix}$$

It satisfies

$$P^2 = \mathrm{Id}, \qquad P^\top = P = P^{-1}, \tag{12.3.1}$$

and its eigenvalues are

$$\pm 1, \quad \text{each with multiplicity } M. \tag{12.3.2}$$

Moreover, for $\mathbf{x}, \mathbf{y} \in \mathbb{C}^{2M}$, we let $\langle \mathbf{x}, \mathbf{y} \rangle := \overline{\mathbf{x}}^\top \mathbf{y}$ be the usual complex inner product, and we define their *P-product* as

$$(\mathbf{x}, \mathbf{y})_P := \langle \mathbf{x}, P\mathbf{y} \rangle = \overline{\mathbf{x}}^\top P \mathbf{y}.$$

From (12.1.1), we observe

Lemma 12.3.1.

$$\left(\mathcal{A}^\top P \right)_{i,j} = \begin{cases} 1, & \textit{if } \mathrm{in}(e_i) = \mathrm{in}(e_j), \ \mathrm{out}(e_i) \neq \mathrm{out}(e_j), \\ 0, & \textit{otherwise.} \end{cases} \tag{12.3.3}$$

$\square$

Next,

Theorem 12.3.1. *The nonbacktracking adjacency matrix $\mathcal{A}$ and the nonbacktracking Laplacian $\mathcal{L}$ of a graph satisfy*

1. *$\mathcal{A}^\top = P\mathcal{A}P$ and $\mathcal{L}^\top = P\mathcal{L}P$;*
2. *$\mathcal{A}$ and $\mathcal{L}$ are self-adjoint with respect to the P-product;*
3. *If $\mathbf{x}$ is an eigenvector of $\mathcal{A}$ or $\mathcal{L}$ for an eigenvalue $\lambda \in \mathbb{C}$ is not real, then*

$$\sum_{[v,w]} \overline{x_{[v,w]}} \cdot x_{[w,v]} = \sum_{i=1}^{M} (\overline{x_i} \cdot x_{i+M} + \overline{x_{i+M}} \cdot x_i) = 0.$$

Proof 1. Since for a nonbacktracking graph $\mathcal{G}$, we can pass from the oriented edge e_i to the oriented edge e_j precisely if we can pass from $-e_j$ to $-e_i$ (the oppositely oriented edges),

$$\mathcal{A}_{ij} = \mathcal{A}_{j+M, i+M}.$$

Similarly,

$$\mathcal{L}_{ij} = -\mathbb{P}(e_i \to e_j) = -\mathbb{P}(e_j^{-1} \to e_i^{-1}) = \mathcal{L}_{j+M, i+M}.$$

for $i \neq j, i, j \leq M$. These relations imply

$$\mathcal{A}^\top = P\mathcal{A}P, \quad \mathcal{L}^\top = P\mathcal{L}P. \tag{12.3.4}$$

2. With (12.3.4) and $P^\top = P$, we compute

$$(\mathbf{x}, \mathcal{A}\mathbf{y})_P = \langle \mathbf{x}, P\mathcal{A}\mathbf{y} \rangle = \langle P\mathbf{x}, \mathcal{A}\mathbf{y} \rangle = \langle \mathcal{A}^\top P\mathbf{x}, \mathbf{y} \rangle$$
$$= \langle P\mathcal{A}P^2\mathbf{x}, \mathbf{y} \rangle = \langle P\mathcal{A}\mathbf{x}, \mathbf{y} \rangle = \langle \mathcal{A}\mathbf{x}, P\mathbf{y} \rangle = (\mathcal{A}\mathbf{x}, \mathbf{y})_P.$$

and analogously

$$(\mathbf{x}, \mathcal{L}\mathbf{y})_P = (\mathcal{L}\mathbf{x}, \mathbf{y})_P.$$

Therefore, $\mathcal{A}$ and $\mathcal{L}$ are self-adjoint with respect to the P-product.
3. The second claim implies that, if $(\lambda, \mathbf{x})$ is an eigenpair for $\mathcal{L}$, then

$$\lambda(\mathbf{x}, \mathbf{x})_P = (\mathbf{x}, \lambda\mathbf{x})_P = (\mathbf{x}, \mathcal{L}\mathbf{x})_P = (\mathcal{L}\mathbf{x}, \mathbf{x})_P = (\lambda\mathbf{x}, \mathbf{x})_P = \overline{\lambda}(\mathbf{x}, \mathbf{x})_P$$

and, of course, the analogous result holds for $\mathcal{A}$.

Hence, if $\lambda \neq \overline{\lambda}$, that is, λ is not real, then $(\mathbf{x}, \mathbf{x})_P = 0$, that is, $\overline{\mathbf{x}}^\top P\mathbf{x} = 0$, which can be rewritten as

$$\sum_{[v,w]} \overline{x_{[v,w]}} \cdot x_{[w,v]} == \sum_{i=1}^{m} \left(\overline{x_i} \cdot x_{i+M} + \overline{x_{i+M}} \cdot x_i \right) = 0.$$

$\square$

12.4 Spectral Gap from 1

In Chapter 4, we have considered the spectral gap from the value 1 for the spectrum of the normalized Laplacian of a simple graph. We have seen that there exist simple graphs that have 1 as an eigenvalue; therefore, the spectral gap from 1 can be 0, and we have also seen (in Theorem 4.1.2) that this gap cannot be larger than $1/2$. In this section, we shall see that, in the case of the nonbacktracking Laplacian, things change.

Again, let $G = (V, E)$ be a simple graph on N nodes and M edges with minimum degree ≥ 2, let $\mathcal{G} = (\mathcal{V}, \mathcal{E})$ be its nonbacktracking graph, and let $\mathcal{A}$ and $\mathcal{L}$ be its nonbacktracking matrix and its nonbacktracking Laplacian, respectively. Given an operator $\mathcal{O}$, let also $\sigma(\mathcal{O})$ denote its spectrum, and let $\rho(\mathcal{O})$ denote its spectral radius, that is, the largest modulus of its eigenvalues.

It is known that, by the assumption that G has minimum degree ≥ 2, 0 is not in the spectrum of $\mathcal{A}$ [261]. This implies that 1 is not in the spectrum of $\mathcal{L}$ and, therefore, the spectral gap from 1 for $\mathcal{L}$,

$$\varepsilon := \min_{\lambda \in \sigma(\mathcal{L})} |1 - \lambda| = \min_{\mu \in \sigma(\mathcal{D}^{-1}\mathcal{A})} |\mu|, \tag{12.4.1}$$

is nonzero. In Theorem, 12.4.1 we give a lower bound for ε, and we prove that the bound is sharp. We also refer to [220] for the study of a sharp upper bound. Before stating the theorem, we prove Lemma 12.4.1. It uses the matrix P defined in Section 12.3, which satisfies $P\mathcal{A}P = \mathcal{A}^*$ and $P\mathcal{L}P = \mathcal{L}^*$.

Lemma 12.4.1. *The spectrum of $\mathcal{A}P$ is given by $\{-1\} \cup \{\deg v - 1\}_{v \in V}$, where each positive eigenvalue d has multiplicity equal to the number of nodes v with $\deg v - 1 = d$, and the multiplicity of -1 equals $2M - N$.*

Proof Note that the spectrum of $\mathcal{A}P$ equals the spectrum of $P^{\mathsf{T}}\mathcal{A}^{\mathsf{T}} = P\mathcal{A}^{\mathsf{T}}$, which equals the spectrum of $\mathcal{A}^{\mathsf{T}}P$, since 0 is not an eigenvalue.

With (12.3.3), the eigenproblem $\mathcal{A}^{\mathsf{T}}Pf = \lambda f$ simplifies to the system of equations

$$(\lambda + 1) f(e) = \sum_{\text{in}(e')=\text{in}(e)} f(e'), \quad \text{for each } e \in V. \tag{12.4.2}$$

First fix a node $v \in V$ with degree d and define the function $f_v \colon V \to \mathbb{R}$ by

$$f_v(e) := \begin{cases} d, & \text{if in}(e) = v \\ 0, & \text{otherwise.} \end{cases}$$

Then, f_v satisfies Equation (12.4.2) with eigenvalue $\lambda = d - 1$, and the set $\{f_v\}_{v \in V}$ is linearly independent. In other words, $\lambda = \deg v - 1$ is an eigenvalue of $\mathcal{A}^{\mathsf{T}}P$, for each node v, and each such eigenvalue has multiplicity equal to the number of nodes v with the corresponding degree.

Now fix an oriented edge $[v, w]$ with $v, w \in V$ and suppose v has degree d. Define the function $g_{[v,w]} \colon V \to \mathbb{R}$ by

$$g_{[v,w]}(e) := \begin{cases} -1, & \text{if in}(e) = v, \text{out}(e) = w \\ \frac{1}{d-1}, & \text{if in}(e) = v, \text{out}(e) \neq w \\ 0, & \text{otherwise.} \end{cases}$$

Then, $g_{[v,w]}$ satisfies equations (12.4.2) for the eigenvalue $\lambda = -1$. Now fix $v \in V$ and let $w_1, \ldots, w_d$ be its neighbors. It is clear that the matrix

$$[g_{[v,w_1]} | \cdots | g_{[v,w_d]}]$$

has rank $d - 1$. Furthermore, for two different nodes v and v', the vectors $g_{[v,w]}$ and $g_{[v',w']}$ are linearly independent. We have shown that the set $\{g_{[v,w]}\}_{v \sim w \in E}$, which contains $2M$ elements, contains exactly $\sum_v (\deg v - 1) = 2M - N$ linearly independent eigenfunctions of $\lambda = -1$, and we are done.   □

Corollary 12.4.1. *$\mathcal{A}$ has both $+1$ and -1 as eigenvalues.*

Proof By (12.3.2), P has ± 1 as the only eigenvalues, both with multiplicity M. We also have $M \geq N$ because we assume that every vertex has degree ≥ 2. And, $M = N$ then only occurs for a cycle graph, for which we shall give the argument later. Thus, by Lemma 12.4.1, we may assume that the multiplicity of the eigenvalue -1 of $\mathcal{A}P$ is $> M$. But then, this eigenspace must contain eigenvectors of P for both eigenvalues $+1$ and -1, and since $\mathcal{A} = (\mathcal{A}P)P$, we then get both eigenvalues $+1$ and -1 for $\mathcal{A}$ as well.

In the case of the cycle graph, all vertices have degree 2, and so, by Lemma 12.4.1 again, $\mathcal{A}P$ has the two eigenvalues ± 1 with multiplicity $M = N$, and we can apply a similar argument. □

As a consequence of Lemma 12.4.1,

$$\min_{\lambda \in \sigma(\mathcal{A}P)} |\lambda|^2 = 1.$$

Theorem 12.4.1. *Let G be a simple graph with maximum vertex degree Δ. Then, the spectral gap (12.4.1) from 1 for the nonbacktracking Laplacian $\mathcal{L}$ of G satisfies*

$$\varepsilon \geq \frac{1}{\Delta - 1}.$$

This bound is sharp.

Proof Since $0 \notin \sigma(\mathcal{A})$, the matrix $\mathcal{A}$ is invertible. Therefore, we can write

$$\varepsilon^{-1} = \max_{\lambda \in \sigma(\mathcal{D}\mathcal{A}^{-1})} |\lambda| = \rho(\mathcal{D}\mathcal{A}^{-1}).$$

Further, the spectral norm $\| \cdot \|_2$

$$\rho\left(\mathcal{D}\mathcal{A}^{-1}\right) \leq \|\mathcal{D}\mathcal{A}^{-1}\| \leq \|\mathcal{D}\|\|\mathcal{A}^{-1}\|.$$

The nonbacktracking diagonal matrix $\mathcal{D}$ satisfies $\|\mathcal{D}\|_2 = \Delta - 1$. And $\|\mathcal{A}^{-1}\|_2^2$ equals the largest norm among the eigenvalues of $\left(\mathcal{A}^{-1}\right)^* \mathcal{A}^{-1}$. But, by (12.3.4), $\mathcal{A}^\top = P\mathcal{A}P$, and thus,

$$\|\mathcal{A}^{-1}\|_2^2 = \max_{\mu \in \sigma(P\mathcal{A}^{-1}P\mathcal{A}^{-1})} |\mu| = \max_{\nu \in \sigma(P\mathcal{A}^{-1})} |\nu|^2 = \frac{1}{\min_{\lambda \in \sigma(\mathcal{A}P)} |\lambda|^2} = 1,$$

since the smallest norm of the eigenvalues of $\mathcal{A}P$ is 1, as shown in Lemma 12.4.1. Therefore,

$$\varepsilon^{-1} = \rho\left(\mathcal{D}\mathcal{A}^{-1}\right) \leq \|\mathcal{D}\|_2 \|\mathcal{A}^{-1}\|_2 = \Delta - 1.$$

Thus, the spectral gap ε is at least $(\Delta - 1)^{-1}$.

To prove that the lower bound is sharp, we use the fact that ± 1 are always eigenvalues of $\mathcal{A}$ by Corollary 12.4.1. It follows that, if G is Δ-regular, then $1 \pm \frac{1}{\Delta-1}$ are eigenvalues of $\mathcal{L}$. Hence, in particular, $\varepsilon = \frac{1}{\Delta-1}$ in this case. □

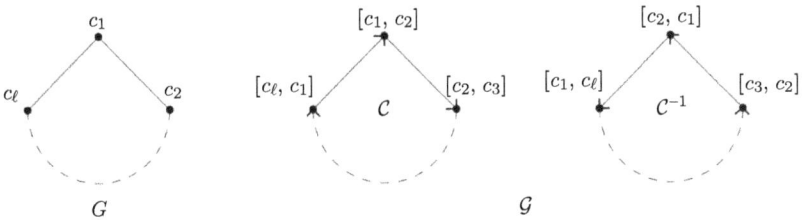

Figure 12.4 The setting of Section 12.5.

In the proof of Theorem 12.4.1, we have shown that, if G is Δ–regular, then $1 \pm \frac{1}{\Delta-1}$ are two real eigenvalues of $\mathcal{L}$. A natural question is whether regular graphs are the only graphs for which $\varepsilon = \frac{1}{\Delta-1}$, but the answer is no. In the next section, we show that, for instance, the presence of a Δ–regular cycle in the graph G produces the eigenvalues $1 \pm \frac{1}{\Delta-1}$ for $\mathcal{L}$.

12.5 Cycles

This section is dedicated to the signature that cycles can leave in the spectrum of the nonbacktracking Laplacian. As before, we consider a simple graph $G = (V, E)$ on N nodes and M edges that has minimum degree ≥ 2, and we let $\mathcal{G} = (\mathcal{V}, \mathcal{E})$ be its nonbacktracking graph. We start with the following observation.

Remark 12.5.1. We say that a cycle is *chordless* if no two vertices of the cycle are connected by an edge that does not belong to the cycle. Given $c_1, \ldots, c_\ell \in V$, then, as already observed in Example 12.1.1 and illustrated in Figure 12.4,

$\{c_1, \ldots, c_\ell\}$ is a simple chordless cycle in $G \iff C := \{[c_1, c_2], \ldots, [c_\ell, c_1]\}$
is a simple chordless cycle in $\mathcal{G}$

$\iff$ both C and $C^{-1} := \{[c_1, c_\ell], \ldots, [c_2, c_1]\}$

are simple chordless cycles in $\mathcal{G}$.

Moreover, $C \cap C^{-1} = \emptyset$ and there are no edges between C and C^{-1} in $\mathcal{G}$.

From here on, we also fix C and C^{-1} as mentioned earlier. Lemma 12.5.1 concerns a structural property of $\mathcal{G}$ related to the vertices outside $C \sqcup C^{-1}$:

Lemma 12.5.1. *If $e \in \mathcal{V} \setminus \left(C \sqcup C^{-1} \right)$, then exactly one of the following holds:*

1. *There are no edges between e and $C \sqcup C^{-1}$ in $\mathcal{G}$;*
2. *There exists $i \in \{1, \ldots, \ell\}$ such that the only edge between e and C is $(e \to [c_i, c_{i+1}])$, while the only edge between e and C^{-1} is $(e \to [c_i, c_{i-1}])$ (Figure 12.5); and*

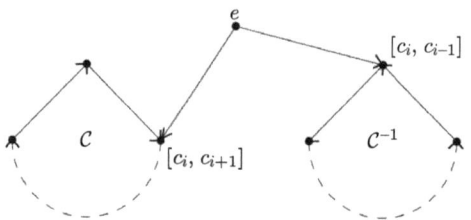

Figure 12.5 An illustration of Lemma 12.5.1.

3. *There exists* $i \in \{1,\ldots,\ell\}$ *such that the only edge between* e *and* C *is* $([c_{i-1},c_i] \to e)$, *while the only edge between* e *and* C^{-1} *is* $([c_{i+1},c_i] \to e)$.

Proof This essentially follows from the fact that each directed edge has a unique beginning and end and obvious symmetry considerations. Here are the details:

Assume that there exists a directed edge going from e to $C \sqcup C^{-1}$ in $\mathcal{G}$ and assume, without loss of generality, that this directed edge goes to C. Then, there exists $i \in \{1,\ldots,\ell\}$ such that $(e \to [c_i,c_{i+1}]) \in \mathcal{E}$. This implies that $\mathrm{out}(e) = c_i$ and $\mathrm{in}(e) \neq c_{i-1}, c_{i+1}$, since we are assuming that $e \notin C \sqcup C^{-1}$. Therefore, $(e \to [c_i,c_{i-1}]) \in \mathcal{E}$. Moreover, since $\mathrm{out}(e) = c_i$ and $\mathrm{in}(e) \neq c_{i-1}, c_{i+1}$, it is clear that there cannot be other edges between e and $C \sqcup C^{-1}$, other than $(e \to [c_i,c_{i+1}])$ and $(e \to [c_i,c_{i-1}])$.

In the same way, one can show that if there exists a directed edge going from $C \sqcup C^{-1}$ to e in $\mathcal{G}$, then there exists $i \in \{1,\ldots,\ell\}$ such that the only edge between e and C is $([c_{i-1},c_i] \to e)$, while the only edge between e and C^{-1} is $([c_{i+1},c_i] \to e)$. This proves the claim. □

Lemmas 12.5.2 and 12.3.1 will allow us to prove Theorem 12.5.1, which shows that the presence of a d–regular cycle in the graph G produces the eigenvalues $1 \pm \frac{1}{d-1}$ for $\mathcal{L}$, as anticipated in Section 12.4.

Lemma 12.5.2. *Let* $\mu \in \mathbb{R} \setminus \{0\}$ *and let* $f \colon V \to \mathbb{R}$, $f \neq 0$ *be a function whose support is contained in* $C \sqcup C^{-1}$. *Then,* $(1 - \frac{1}{\mu}, f)$ *is an eigenpair for* $\mathcal{L}$ *if and only if the following conditions hold:*

1.

$$f([c_{i-1},c_i]) = \frac{\mu}{\deg c_i - 1} \cdot f([c_i,c_{i+1}]), \ \forall i \in \{1,\ldots,\ell\};$$

2.

$$f([c_{i+1},c_i]) = \frac{\mu}{\deg c_i - 1} \cdot f([c_i,c_{i-1}]), \ \forall i \in \{1,\ldots,\ell\}; \text{ and}$$

3.

$$f([c_i, c_{i+1}]) + f([c_i, c_{i-1}]) = 0, \quad \forall i \in \{1,\ldots,\ell\}: \deg c_i > 2.$$

Proof $(1 - \frac{1}{\mu}, f)$ is an eigenpair for $\mathcal{L}$ if and only if

$$f([v,w]) = \frac{\mu}{\deg w - 1} \cdot \left(\sum_{[w,z],\, z \neq v} f([w,z]) \right), \quad \forall [v,w]. \qquad (12.5.1)$$

Since, by assumption, the support of f is contained in $\mathcal{C} \sqcup \mathcal{C}^{-1}$, (12.5.1) applied to the vertices in $\mathcal{C} \sqcup \mathcal{C}^{-1}$ gives the first two claims.

Now, let $e = [v,w] \in \mathcal{V} \setminus (\mathcal{C} \sqcup \mathcal{C}^{-1})$. If there are no directed edges from e to $\mathcal{C} \sqcup \mathcal{C}^{-1}$ in $\mathcal{G}$, then (12.5.1) is trivially satisfied. Otherwise, by Lemma 12.5.1, there exists $i \in \{1,\ldots,\ell\}$ such that the only edge from e to $\mathcal{C}$ is $(e \to [c_i, c_{i+1}])$, while the only edge from e to $\mathcal{C}^{-1}$ is $(e \to [c_i, c_{i-1}])$. This happens if and only if $\mathrm{out}(e) = c_i$, and, in this case, (12.5.1) becomes

$$f([c_i, c_{i+1}]) + f([c_i, c_{i-1}]) = 0.$$

Since an element $e \in \mathcal{V} \setminus (\mathcal{C} \sqcup \mathcal{C}^{-1})$ with $\mathrm{out}(e) = c_i$ exists if and only if $\deg c_i > 2$, this proves the last claim.

$\square$

Corollary 12.5.1. *Let $\mu \in \mathbb{R}_+$ and let $f: \mathcal{V} \to \mathbb{R}$, $f \neq 0$ be a function whose support is contained in $\mathcal{C} \sqcup \mathcal{C}^{-1}$, where $\mathcal{C}$ has even length. If $(1 - \frac{1}{\mu}, f)$ is an eigenpair for $\mathcal{L}$, then also $1 + \frac{1}{\mu}$ is an eigenvalue for $\mathcal{L}$ and it has an eigenfunction whose support is contained in $\mathcal{C} \sqcup \mathcal{C}^{-1}$.*

Proof Let $g: \mathcal{V} \to \mathbb{R}$ be defined by $g(e) := 0$ if $e \in \mathcal{V} \setminus (\mathcal{C} \sqcup \mathcal{C}^{-1})$ and

$$g([c_i, v]) := \begin{cases} f([c_i, v]), & \text{if } i \text{ is even} \\ -f([c_i, v]), & \text{if } i \text{ is odd}, \end{cases}$$

for $i = 1,\ldots,\ell$. Then, g satisfies the conditions in Lemma 12.5.2 for $1 + \frac{1}{\mu}$. $\square$

We can now identify conditions for cycles to generate eigenpairs. We first prove the following result on regular cycles:

Theorem 12.5.1. *Let $d > 1$. If there exists a simple chordless cycle in G whose vertices have constant degree d, then $1 - \frac{1}{d-1}$ is an eigenvalue for $\mathcal{L}$. If, additionally, such a cycle is even, then also $1 + \frac{1}{d-1}$ is an eigenvalue for $\mathcal{L}$.*

Moreover, the geometric multiplicity of $1 - \frac{1}{d-1}$ for $\mathcal{L}$ is larger than or equal to the number of d-regular simple chordless cycles in G, while the geometric multiplicity of $1 + \frac{1}{d-1}$ for $\mathcal{L}$ is larger than or equal to the number of d-regular even simple chordless cycles in G.

Proof Given a cycle whose vertices have constant degree d, let $f: V \to \mathbb{R}$ be defined by

$$f(e) := \begin{cases} 1 & \text{if } e \in \mathcal{C} \\ -1 & \text{if } e \in \mathcal{C}^{-1} \\ 0 & \text{otherwise.} \end{cases}$$

Then, f satisfies the conditions in Lemma 12.5.2 for the eigenvalue $1 - \frac{1}{d-1}$, implying that $(1 - \frac{1}{d-1}, f)$ is an eigenpair for $\mathcal{L}$. This proves the first claim for $1 - \frac{1}{d-1}$, and the first claim for $1 + \frac{1}{d-1}$ then follows from Corollary 12.5.1.

The second claim follows from the fact that the preceding functions are linearly independent if they are defined for distinct cycles. $\square$

The next result is proved in a similar way.

Theorem 12.5.2. *Let $d > 2$. If there exists a simple chordless cycle of length ℓ in G such that one vertex has degree d while all other vertices have degree 2, then $1 - \frac{1}{\sqrt[\ell]{d-1}}$ is an eigenvalue for $\mathcal{L}$. If, additionally, such a cycle is even, then also $1 + \frac{1}{\sqrt[\ell]{d-1}}$ is an eigenvalue for $\mathcal{L}$.*

Moreover, the geometric multiplicity of $1 - \frac{1}{\sqrt[\ell]{d-1}}$ for $\mathcal{L}$ is larger than or equal to the number of such cycles in G, while the geometric multiplicity of $1 + \frac{1}{\sqrt[\ell]{d-1}}$ for $\mathcal{L}$ is larger than or equal to the number of such even cycles in G.

Proof Fix a cycle $\{c_1, \ldots, c_\ell\}$ in G and let

$$\mathcal{C} := \{[c_1, c_2], \ldots, [c_\ell, c_1]\}, \quad \mathcal{C}^{-1} := \{[c_1, c_\ell], \ldots, [c_2, c_1]\}$$

be the corresponding cycles in $\mathcal{G}$. Let $c_0 := c_\ell$ and $c_{\ell+1} := c_1$, and assume that $\deg c_1 = d > 2$, while $\deg c_i = 2$ for $i \in \{2, \ldots, \ell\}$. The proof is similar to the one of Theorem 12.5.1. In this case, we can apply Lemma 12.5.2 for $\mu = \sqrt[\ell]{d-1}$ if we find a nonzero function $f: V \to \mathbb{R}$ whose support is contained in $\mathcal{C} \sqcup \mathcal{C}^{-1}$ and such that the following conditions hold:

1.

$$f([c_\ell, c_1]) = \frac{\sqrt[\ell]{d-1}}{d-1} \cdot f([c_1, c_2]);$$

2.

$$f([c_{i-1}, c_i]) = \sqrt[\ell]{d-1} \cdot f([c_i, c_{i+1}]), \quad \forall i \in \{2, \ldots, \ell\};$$

3.

$$f([c_2, c_1]) = \frac{\sqrt[\ell]{d-1}}{d-1} \cdot f([c_1, c_\ell]);$$

4.
$$f([c_{i+1}, c_i]) = \sqrt[\ell]{d-1} \cdot f([c_i, c_{i-1}]), \quad \forall i \in \{2, \ldots, \ell\};$$

5.
$$f([c_1, c_2]) + f([c_1, c_\ell]) = 0.$$

By letting

$$f([c_1, c_2]) := -f([c_1, c_\ell]) := 1,$$

$$f([c_\ell, c_1]) := -f([c_2, c_1]) := \frac{\sqrt[\ell]{d-1}}{d-1},$$

$$f([c_{\ell-1}, c_\ell]) := \sqrt[\ell]{d-1} \cdot f([c_\ell, c_1]),$$

$$\vdots$$

$$f([c_2, c_3]) := \sqrt[\ell]{d-1} \cdot f([c_3, c_4]),$$

$$f([c_3, c_2]) := \sqrt[\ell]{d-1} \cdot f([c_2, c_1]),$$

$$\vdots$$

$$f([c_\ell, c_{\ell-1}]) := \sqrt[\ell]{d-1} \cdot f([c_{\ell-1}, c_{\ell-2}]),$$

then clearly conditions 1, 2, and 5 above are satisfied, as well as condition 2 and for $i \in \{3, \ldots, \ell\}$ and condition 4 for $i \in \{2, \ldots, \ell-1\}$. Moreover, since

$$\sqrt[\ell]{d-1} \cdot f([c_2, c_3]) = \sqrt[\ell]{d-1} \cdot \sqrt[\ell]{d-1} \cdot f([c_3, c_4])$$

$$= (\sqrt[\ell]{d-1})^{\ell-1} \cdot f([c_\ell, c_1])$$

$$= (\sqrt[\ell]{d-1})^{\ell-1} \cdot \frac{\sqrt[\ell]{d-1}}{d-1}$$

$$= 1 - f([c_1, c_2]),$$

the second condition is satisfied also for $i = 2$. Similarly, since

$$\sqrt[\ell]{d-1} \cdot f([c_\ell, c_{\ell-1}]) = (\sqrt[\ell]{d-1})^{\ell-1} \cdot f([c_2, c_1]) = -1 = f([c_1, c_\ell]),$$

the fourth condition is satisfied also for $i = \ell$. By using the same method as in the proof of Theorem 12.5.1, this proves the claim. □

The previous results show that certain chordless cycles $\mathcal{C}$ in $\mathcal{G}$ produce eigenvalues of the form $1 - \frac{1}{\mu}$ for $\mathcal{L}$, where μ is a positive real number that depends on the cycle structure. The corresponding eigenfunctions are supported in $\mathcal{C} \sqcup \mathcal{C}^{-1}$. But this does not generalize to all cycles. In fact, Theorem 12.5.3 completely characterizes those cycles for which it is possible.

Theorem 12.5.3. *Assume that G is not a cycle graph. Let* $\{c_1,\ldots,c_\ell\}$ *be a simple chordless cycle in G and let*

$$\mathcal{C} := \{[c_1,c_2],\ldots,[c_\ell,c_1]\}, \quad \mathcal{C}^{-1} := \{[c_1,c_\ell],\ldots,[c_2,c_1]\}$$

be the corresponding cycles in $\mathcal{G}$*. Then,* $\mathcal{L}$ *admits an eigenpair of the form* $(1 - \frac{1}{\mu}, f)$*, where* $\mu \in \mathbb{R}_+$ *and* $f: \mathcal{V} \to \mathbb{R}$ *is a nonzero function with support in* $\mathcal{C} \sqcup \mathcal{C}^{-1}$*, if and only if*

$$\mu = \sqrt[\ell]{(\deg c_1 - 1) \cdot (\deg c_2 - 1) \cdots (\deg c_\ell - 1)}$$

and, after cyclically or anticyclically relabeling the vertices of the cycle, $\deg c_1 > 2$ *and, for all* $i \in \{2,\ldots,\ell\}$ *such that* $\deg c_i > 2$,

$$\frac{\mu}{\deg c_{i+1} - 1} \cdot \frac{\mu}{\deg c_{i+2} - 1} \cdots \frac{\mu}{\deg c_\ell - 1} \cdot \frac{\mu}{\deg c_1 - 1}$$
$$= \frac{\mu}{\deg c_{i-1} - 1} \cdot \frac{\mu}{\deg c_{i-2} - 1} \cdots \frac{\mu}{\deg c_2 - 1} \cdot \frac{\mu}{\deg c_1 - 1}. \tag{12.5.2}$$

Remark 12.5.2. In (12.5.2), there is a factor $\frac{\mu}{\deg c_1 - 1}$ on both sides. It cancels out, but writing it down helps understand the condition for $i = 2$.

Proof of Theorem 12.5.3 Since G is not a cycle graph, then after relabeling the vertices of the cycle, we can assume that $\deg c_1 > 2$. In this case, $(1 - \frac{1}{\mu}, f)$ is an eigenpair for $\mathcal{L}$ if and only if they satisfy the conditions of Lemma 12.5.2. In particular, $f([c_1,c_2]) \neq 0$, and we can assume that

$$f([c_1,c_2]) = 1.$$

The third condition of Lemma 12.5.2 applied to $i = 1$ then gives

$$f([c_1,c_\ell]) = -1,$$

while the first condition applied to $i \neq 2$ gives

$$f([c_\ell,c_1]) = \frac{\mu}{\deg c_1 - 1} \cdot f([c_1,c_2]) = \frac{\mu}{\deg c_1 - 1},$$
$$f([c_{\ell-1},c_\ell]) = \frac{\mu}{\deg c_\ell - 1} \cdot f([c_\ell,c_1]) = \frac{\mu}{\deg c_\ell - 1} \cdot \frac{\mu}{\deg c_1 - 1},$$
$$\vdots$$
$$f([c_2,c_3]) = \frac{\mu}{\deg c_3 - 1} \cdot f([c_3,c_4]) = \frac{\mu}{\deg c_3 - 1} \cdots \frac{\mu}{\deg c_\ell - 1} \cdot \frac{\mu}{\deg c_1 - 1},$$

and the second condition of Lemma 12.5.2 applied to $i \neq \ell$ gives

$$f([c_2,c_1]) = \frac{\mu}{\deg c_1 - 1} \cdot f([c_1,c_\ell]) = -\frac{\mu}{\deg c_1 - 1},$$

$$f([c_3, c_2]) = \frac{\mu}{\deg c_2 - 1} \cdot f([c_2, c_1]) = -\frac{\mu}{\deg c_2 - 1} \cdot \frac{\mu}{\deg c_1 - 1},$$

$$\vdots$$

$$f([c_\ell, c_{\ell-1}]) = \frac{\mu}{\deg c_{\ell-1} - 1} \cdot f([c_{\ell-1}, c_{\ell-2}])$$

$$= -\frac{\mu}{\deg c_{\ell-1} - 1} \cdot \frac{\mu}{\deg c_{\ell-2} - 1} \cdots \frac{\mu}{\deg c_1 - 1}.$$

In particular, the earlier conditions give a complete description of f, given μ. We now check the other conditions of Lemma 12.5.2. We observe that:

1. The first condition of Lemma 12.5.2 is satisfied for all $i \in \{1, \dots, \ell\}$ if and only if

$$1 = f([c_1, c_2]) = \frac{\mu}{\deg c_2 - 1} \cdot f([c_2, c_3])$$

$$= \frac{\mu}{\deg c_2 - 1} \cdot \frac{\mu}{\deg c_3 - 1} \cdots \frac{\mu}{\deg c_\ell - 1} \cdot \frac{\mu}{\deg c_1 - 1},$$

that is, if and only if

$$\mu = \sqrt[\ell]{(\deg c_1 - 1) \cdot (\deg c_2 - 1) \cdots (\deg c_\ell - 1)};$$

2. The second condition of Lemma 12.5.2 is satisfied for all $i \in \{1, \dots, \ell\}$ if and only if

$$-1 = f([c_1, c_\ell]) = \frac{\mu}{\deg c_\ell - 1} \cdot f([c_\ell, c_{\ell-1}])$$

$$= -\frac{\mu}{\deg c_\ell - 1} \cdot \frac{\mu}{\deg c_{\ell-1} - 1} \cdot \frac{\mu}{\deg c_{\ell-2} - 1} \cdots \frac{\mu}{\deg c_1 - 1},$$

that is, as before, if and only if

$$\mu = \sqrt[\ell]{(\deg c_1 - 1) \cdot (\deg c_2 - 1) \cdots (\deg c_\ell - 1)};$$

3. The third condition of Lemma 12.5.2 is satisfied if and only if, for all $i \in \{2, \dots, \ell\}$ such that $\deg c_i > 2$,

$$f([c_i, c_{i+1}]) = -f([c_i, c_{i-1}]), \qquad (12.5.3)$$

where

$$f([c_i, c_{i+1}]) = \frac{\mu}{\deg c_{i+1} - 1} \cdot \frac{\mu}{\deg c_{i+2} - 1} \cdots \frac{\mu}{\deg c_\ell - 1} \cdot \frac{\mu}{\deg c_1 - 1}$$

and

$$-f([c_i, c_{i-1}]) = \frac{\mu}{\deg c_{i-1} - 1} \cdot \frac{\mu}{\deg c_{i-2} - 1} \cdots \frac{\mu}{\deg c_1 - 1},$$

therefore (12.5.3) can be rewritten as

$$\frac{\mu}{\deg c_{i+1} - 1} \cdot \frac{\mu}{\deg c_{i+2} - 1} \cdots \frac{\mu}{\deg c_\ell - 1}$$
$$= \frac{\mu}{\deg c_{i-1} - 1} \frac{\mu}{\deg c_{i-2} - 1} \cdots \frac{\mu}{\deg c_2 - 1}.$$

Putting everything together, we have that $1 - \frac{1}{\mu}$ is an eigenvalue with an eigenfunction whose support is contained in $\mathcal{C} \sqcup \mathcal{C}^{-1}$ if and only if

$$\mu = \sqrt[\ell]{(\deg c_1 - 1) \cdot (\deg c_2 - 1) \cdots (\deg c_\ell - 1)}$$

and (12.5.2) is satisfied for all $i \in \{2, \ldots, \ell\}$ such that $\deg c_i > 2$.   □

12.6 Isospectrality

Recall that two graphs are said to be *isospectral* with respect to a given operator if they are nonisomorphic but have the same spectrum with respect to that operator. Hence, the smaller the number of isospectral graphs, the better the operator is at distinguishing nonisomorphic graphs. We recall that in Corollary 4.1.11, we have already seen some isospectral graphs, complete bipartite graphs with the same number of vertices, and for systematic constructions of isospectral graphs, we refer to [112, 258].

While we do not intend to present a systematic account of isospectral graphs in this book, we conclude this chapter by showing two isospectrality tables, which suggest that the nonbacktracking Laplacian has nicer isospectrality properties than the adjacency matrix, the normalized Laplacian, and the nonbacktracking matrix of a graph. We refer to [156] for the computational details.

In Tables 12.1 and 12.2, we show the number of simple graphs with minimum degree ≥ 2 not determined by their spectrum with respect to the

Table 12.1 *Graphs with minimum degree ≥ 2 not determined by their spectrum, by number of nodes N.*

N	#graphs	A	L	$\mathcal{A}$	$\mathcal{L}$
≤ 6	76	0	2	0	0
7	510	26	4	0	0
8	7 459	744	11	2	0
9	197 867	32 713	243	6	0
10	9 808 968	1 976 884	16 114	10 130	156
total	10 014 880	2 010 367	16 374	10 138	156

Table 12.2 *Graphs with minimum degree* ≥ 2 *and* $4 \leq N \leq 10$ *not determined by their spectrum, by number of edges* M.

M	#graphs	$\mathcal{A}$	$\mathcal{L}$
0	0	0	0
1	0	0	0
2	0	0	0
3	0	0	0
4	1	0	0
5	2	0	0
6	6	0	0
7	10	0	0
8	25	0	0
9	68	0	0
10	182	0	0
11	532	0	0
12	1 679	0	0
13	5 218	4	0
14	15 437	14	0
15	41 126	26	0
16	96 274	62	0
17	197 433	162	0
18	355 986	364	4
19	567 827	634	8
20	807 284	983	16
21	1 029 639	1 329	24
22	1 184 688	1 492	26
23	1 235 599	1 490	26
24	1 172 658	1 333	24
25	1 015 663	989	16
26	804 863	628	8
27	584 762	368	4
28	390 136	166	0
29	239 514	60	0
30	135 636	26	0
31	71 025	8	0
32	34 559	0	0
33	15 734	0	0
34	6 745	0	0
35	2 764	0	0
36	1 101	0	0
≥ 37	704	0	0
total	10 014 880	10 138	156

adjacency matrix A, to the normalized Laplacian L, to the nonbacktracking matrix $\mathcal{A}$ and to the nonbacktracking Laplacian $\mathcal{L}$. In Table 12.1, such graphs are partitioned by number of nodes N. In Table 12.2, graphs are partitioned by number of edges M. Interestingly, from Table 12.2, we can observe that the number of isospectral graphs with respect to $\mathcal{L}$ as a function of the number of edges M is in progression:

$$4, 8, 16, 24, 26, 26, 14, 16, 8, 4,$$

and in [156], it is conjectured that this is part of a larger pattern.

In Figures 12.6 and 12.7, we also illustrate the two smallest pairs of isospectral graphs with respect to the nonbacktracking Laplacian.

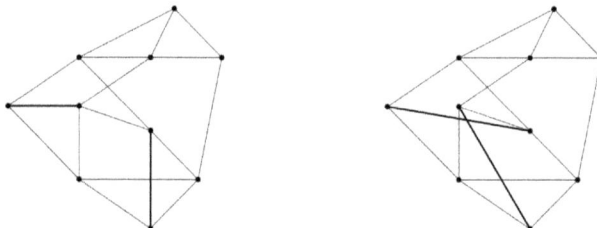

Figure 12.6 The first of the two smallest pairs of isospectral graphs with respect to $\mathcal{L}$.

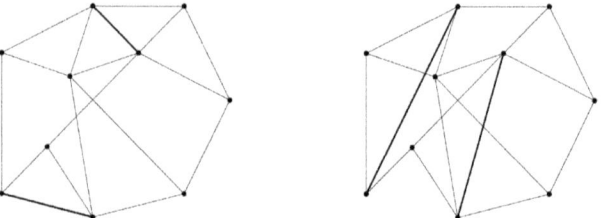

Figure 12.7 The second of the two smallest pairs of isospectral graphs with respect to $\mathcal{L}$.

13

Applications

Laplace operators and their spectral properties are powerful tools for the analysis of networks in the social and the biological sciences and in other domains. In computer science, the theory of families of expander graphs is particularly important. Eigenvalues are also a key for quantifying synchronization and other features of nonlinear dynamics.

13.1 Social Sciences

Sociology is about the structure and the dynamics of relations between individuals or groups, and therefore, it is natural for the field to invoke network concepts. In fact, sociology pioneered the analysis of empirical network data a long time before other disciplines took this up. It started with the work of Moreno and Jennings [212, 213] who explicitly proposed to analyze the formal structure of networks of interpersonal relations to find explanations for individual behavior. Harary [123–125] then introduced and investigated the structure of signed graphs, and we have made substantial use of that concept at various places of this monograph; see in particular Section 4.6. Of course, the sign of an edge is interpreted as the positive or negative nature of a relation between persons. And, when persons switch their attitudes, as formalized in Definition 4.6.4, the signs of all their relations change.

Other important developments in social network analysis, although not directly related to spectral theory, include the discovery of the small world phenomenon and the role of weak ties. The small world phenomenon means that in the modern world, any two individuals v, v' are separated only by very few intermediates; that is, in the network formed by personal contacts between people, one finds individuals $v = v_1 \sim v_2 \ldots v_{k-1} \sim v_k = v'$ for a rather small value of k. That is, the diameter of the network of personal contacts is quite small. This was convincingly demonstrated by the experiments of Milgram in the 1960s.

The importance of weak ties was emphasized by Granovetter [115]. It means that for many purposes, like getting new information or spotting new opportunities, strong ties between your circle of close friends are often less helpful than the much weaker ties with loose acquaintances.

But there are also concepts in social network analysis that depend on eigenvalues and eigenvectors of operators. Let us describe some.

13.1.1 Eigenvector Centrality

In a network, for instance, a social one, it is natural to try to quantify the importance of various nodes in the network. A first attempt might be to call a node important if it is connected to many others, that is, if it has a high degree. A better option might be to call a node important if it is connected to other important nodes. That now is a recursive definition. It was first formalized by Bonacich [33]. We let $A = (a_{vv'})$ be the adjacency matrix of the network. (For simplicity, we consider an unweighted and undirected network here, but the construction can easily be extended to the weighted case. The directed case is more complicated.) Then, if x_v is the importance of node v, it should be proportional to $\sum_{v'} a_{vv'} x_{v'}$, that is,

$$x_v = \lambda \sum_{v'} a_{vv'} x_{v'} \tag{13.1.1}$$

for some positive λ, which then is an eigenvalue of A. Also, all the values of x_v should be positive themselves. Now it follows from the Perron–Frobenius Theorem that, when the non-negative matrix A is irreducible, the largest eigenvalue of A is simple and has an eigenvector with positive entries. Thus, such an eigenvector solves the problem, and the value x_v for a suitably normalized such eigenvector then is called the *eigenvector centrality* of the node v.

In fact, these properties can be derived in a very simple manner. First, irreducibility of A means that whenever the vertex set V is decomposed into two nonempty subsets V_1, V_2, there exists at least one pair $v_1 \in V_1, v_2 \in V_2$ with

$$a_{v_1 v_2} \neq 0. \tag{13.1.2}$$

But this simply means that the underlying graph $\Gamma = (V, E)$ is connected.

By Theorem 2.6.1, the largest eigenvalue of A can be obtained as

$$\lambda = \max_{x \neq 0} \frac{\langle \mathbf{x}, A\mathbf{x} \rangle}{\langle \mathbf{x}, \mathbf{x} \rangle} \tag{13.1.3}$$

as the adjacency matrix of an undirected graph is symmetric; here $\langle\cdot,\cdot\rangle$ is the Euclidean product in $\mathbb{R}^N$ where N is the number of nodes of Γ. When for some $\mathbf{x} \in \mathbb{R}^N$, we replace every negative entry $x_v < 0$ by $|x_v|$, we do not change the denominator in (13.1.3) and do not decrease the numerator in (13.1.3). Therefore, we may assume that all $x_v \geq 0$ for a maximizer of (13.1.3). In particular, the maximum is positive, and therefore, we obtain a solution of (13.1.1) with some $\lambda > 0$. We want to exclude that some x_v vanish. But since not all x_v can be zero, for some such $x_v = 0$, we find some v' with $a_{vv'} \neq 0$ and $x_{v'} > 0$, since A is irreducible. But since all $x_{v'} \geq 0$, then by (13.1.1), also $x_v > 0$, a contradiction. Therefore, λ has an eigenvector with positive entries. Thus, from any maximizer of (13.1.3), by changing signs of the negative entries, if any, we have produced a maximizer with only positive entries. But then, when the original had both positive and negative entries, since A is irreducible, there would be some $a_{vv'} \neq 0$ connecting some v with $x_v > 0$ and some v' with $x_{v'} < 0$. But then, the replacement would have strictly increased the value of (13.1.3), contradicting the maximizing property. This argument shows that all eigenvectors are either positive or negative everywhere. Therefore, in particular, eigenvectors for this largest eigenvalue λ cannot be orthogonal to each other, and the eigenspace for λ is one-dimensional, indeed.

Thus, a positive eigenvector for the largest eigenvalue of A provides the centrality measure.

While such centrality measures were first conceived in mathematical sociology, they have found widespread applications also in other fields. In particular, Google's page rank algorithm is based on some such centrality measure. See, for instance, [34].

13.1.2 Cohesion of Networks

Personal ties can break in a social group, and this can lead to a split into subgroups that are no longer connected with each other. Or, in order to prevent the spreading of an epidemic, contacts between different groups should be reduced. Thus, depending on the interpretation of the edges, for instance, whether they correspond to the spreading of information, opinions, or diseases, it can be desirable or undesirable to break up a network into disconnected components. In any case, this raises the question how easy or difficult that is. This issue has been formally addressed in Section 4.3, and we have described there how this is controlled by the first nonzero eigenvalue λ_2 of the Laplacian. When that eigenvalue is small, it is easy to break up the network into large disconnected components by cutting a small number of edges. And the nodal domains of the corresponding eigenfunction yield an approximation of those components. See Chapter 9.

13.2 Biological Networks

There is an abundance of network-type data in the biological sciences, ranging from the species to the molecular level (see, for instance, [148]). The descendence relations between species, derived from paleontological data, and the genetic relationships, derived from the comparison of gene sequences, can ideally be organized in phylogenetic trees (see, for instance, [245] for the mathematics of such trees), although in practice processes such as hybridization, incipient speciations, or horizontal gene transfer may necessitate deviations from such a simple arrangement (see [14] for a mathematical theory that can incorporate that). Also foodwebs, that is, trophic interactions between species in ecosystems, are arranged in directed graphs – who eats whom. Interactions between individuals in animal groups can also be represented by directed or undirected graphs.

At the molecular level, networks can be constructed from correlation patterns, for instance, between gene expression in cells, chemical interactions as in metabolic networks, and physical interactions, for instance, between proteins but also by the geometric proximity of molecules or their parts in cells.

For an application of spectral analysis, let us consider the example of protein–protein interaction (PPI) networks, following [17]. Such a PPI network is modeled by a graph whose nodes stand for the proteins and edges indicate interactions. We plot the spectrum of the PPI networks of different species with a smoothing kernel (Figure 13.1). More precisely, the spectral density is represented as a sum of Lorentz distributions, $\rho(\lambda) = \sum_{k=2}^{N} \frac{\gamma}{(\lambda_k - \lambda)^2 + \gamma^2}$ with width $\gamma = .08$ where $\lambda_2, \ldots, \lambda_N$ are the positive eigenvalues. In Figure 13.2, we see the prominent peak near 1, indicating a high multiplicity of the eigenvalue 1.

This should be compared with the analogous representation in Figure 13.2 of the spectrum of a simulated network with repeated node duplication and subsequent random mutations. In fact, we have seen in Section 4.1 that repeated node duplications lead to high multiplicity of the eigenvalue 1. The simulated network therefore suggests a plausible evolutionary process behind the spectral structure of PPI networks. The mechanism should work through repeated duplications of the genes coding for the proteins in question, with subsequent mutations so as to diversify their function.

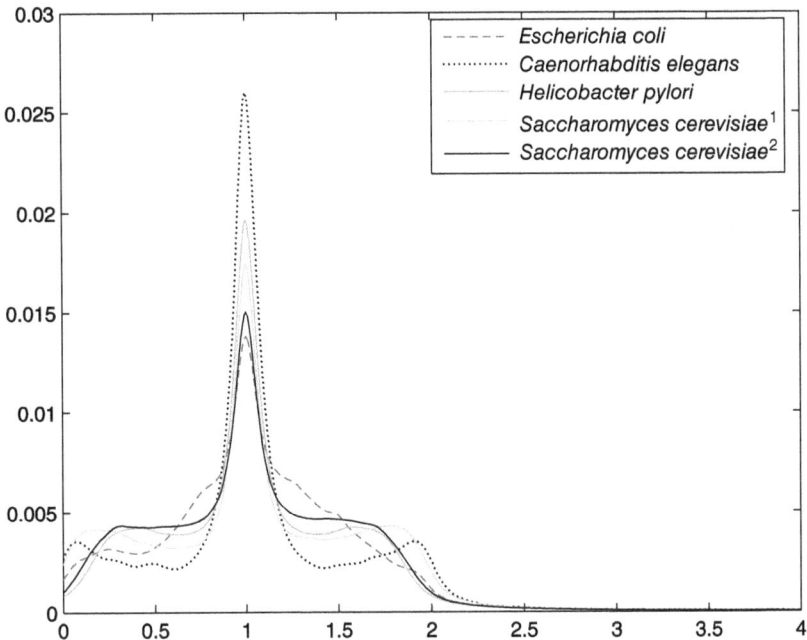

Figure 13.1 Spectra of PPI networks of different species, convolved with a Lorentz kernel.

In fact, in [18, 20], a systematic program has been proposed to classify biological and other networks according to their large-scale spectral properties. The approach of Section 10.2 may provide the appropriate mathematical structure for that program.

Another example of spectral analysis for biological networks is given in [216]. Here, networks of genetic expressions are modeled by *hypergraphs with real coefficients* (Definition 2.4.5), as follows. Each vertex v represents a cell, each edge e represents a gene, and the coefficient $\varphi(v, e)$ represents the fraction of transcripts in cell v mapping to gene e. It is assumed that the coefficients are normalized with respect to cells, so that $\sum_e \varphi(v, e) = 1$ for each v, as such cell-wise normalization is the norm in RNA-sequence analysis. Therefore, $\varphi(v, e)$ encodes the relative expression of gene e in cell v. The corresponding Laplace operator, which was introduced in [154] and generalizes the Laplacian for oriented hypergraphs in Chapter 4.9, is then defined as $L = D^{-1}\mathcal{I}\mathcal{I}^{\top}$, where D is the diagonal degree matrix and $\mathcal{I}$ is the incidence matrix of the hypergraph. For a data set of gene expression with N cells, the Laplacian spectrum consists of N real, non-negative eigenvalues, $\lambda_1 \leq \ldots \leq \lambda_N$, and the largest of these

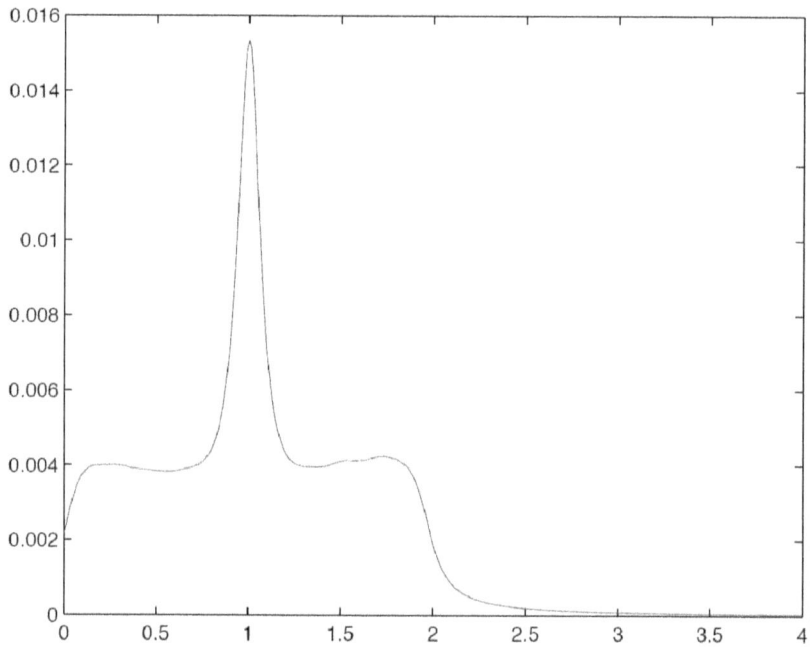

Figure 13.2 Spectrum of a model network with iterated node duplications and random mutations, again convolved with a Lorentz kernel.

eigenvalues is shown to encode *cellular redundancy*. In fact, as proved in [216], $1 \leq \lambda_N \leq N$, and λ_N is equal to 1 if and only if each edge has cardinality 1 (that is, each gene is concentrated in one single cell), while it is equal to N if and only if each gene is uniformly distributed among all cells (that is, the network has cellular redundancy). Moreover, the normalized quantity λ_N / N is chosen as a measure of cellular redundancy that is independent of the number of cells, and its accuracy is shown through an analysis of simulated and real data sets of gene expression.

13.3 Computer Science

In order to link the main applications of spectral theory in computer science with the perspective of our book, we should first clarify a mathematical issue. As we have seen in Section 4.1, there are different linear operators on the vertex functions of a graph whose spectra encode information about the underlying graph and that have been investigated in the literature in that regard. Here, we have mainly considered the (4.1.33), defined by

$$Lf(v) = f(v) - \frac{1}{\deg v} \sum_{v' \sim v} f(v') \qquad (13.3.1)$$

for $f: V \to \mathbb{R}$ and $v \in V$. The algebraic graph Laplacian (1.2.1) is

$$Kf(v) = \deg v\, f(v) - \sum_{v' \sim v} f(v'). \qquad (13.3.2)$$

Finally, the adjacency matrix operates on functions via

$$Af(v) = \sum_{v' \sim v} f(v'). \qquad (13.3.3)$$

Clearly, for a general graph, the spectra of these operators are different, and there is no obvious and simple relation between them. When, however, the graph is d-regular, that is, all vertices have the same degree d,

$$\deg v = d \qquad \text{for all } v \in V, \qquad (13.3.4)$$

the spectra are simply related. Whenever λ is an eigenvalue of L, then $\sigma = d\lambda$ is an eigenvalue of K, and $\mu = d - d\lambda$ is an eigenvalue of A, and conversely. Also, for a d-regular graph, the Cheeger constants of L and K are simply related to each other. In order to conform to the interpretation in the applications we have in mind, we use the notation

$$|\partial S| := |E(S, \overline{S})|, \qquad (13.3.5)$$

the size of the *boundary*[1] of the vertex set $S \subset V$, where we recall that $|E(S, \overline{S})|$ is the number of edges between S and its complement. The (Pólya)–Cheeger constant for L (4.3.12), (4.3.13) is

$$h_L = \min_{S \subset V} \frac{|\partial S|}{\min(\text{vol}(S), \text{vol}(\overline{S}))}, \qquad (13.3.6)$$

where, as we also recall,

$$\text{vol}(S) = \sum_v \deg v. \qquad (13.3.7)$$

The analogous constant for K is

$$h_K = \min_{S \subset V} \frac{|\partial S|}{\min(|S|, |\overline{S}|)}, \qquad (13.3.8)$$

---

[1] This is the edge boundary of S. The vertex boundary would instead consider those vertices in the complement of S that are connected to some vertex in S. Since several edges from S may lead to the same vertex, the vertex boundary may have fewer elements than the edge boundary. But when the graph is d-regular, the discrepancy could be at most by a factor d. – On the other hand, as in (13.3.7), we could also count those vertices with their degrees.

where now (not quite consistent with our earlier notation),

$$|S| = \text{number of vertices in } S. \tag{13.3.9}$$

Again, when the graph is d-regular, we have the simple relation $\text{vol}(S) = d|S|$, and so

$$h_K = d\, h_L. \tag{13.3.10}$$

In particular, the Cheeger inequality of Theorem 4.3.1 then is the same for K as for L, except that (13.3.10) leads to a different scaling on the left-hand side for h_K. Thus,

$$\frac{h_K^2}{2d} \leq \sigma_2 \leq 2h_K, \tag{13.3.11}$$

or equivalently

$$\frac{\sigma_2}{2} \leq h_K \leq \sqrt{2d\sigma_2}, \tag{13.3.12}$$

where now σ_2 is the second smallest eigenvalue of K, that is, for a connected graph, its smallest nonvanishing eigenvalue. And if $\mu_1 > \mu_2 \geq \cdots \geq \mu_N$ are the eigenvalues of A ($\mu_1 = d$, $\mu_1 > \mu_2$ as we assume that the graph is connected, and $\mu_N = -d$ precisely if the graph is bipartite), the Cheeger inequality (13.3.12) becomes

$$\frac{d - \mu_2}{2} \leq h_K \leq \sqrt{2d(d - \mu_2)}. \tag{13.3.13}$$

Therefore, while the notion of an expander graph, which we shall discuss in this section, is usually formulated in terms of the Cheeger inequality for A, we can as well work with our preferred operator L, as long as only d-regular graphs are considered, which in fact is the setting of expander theory.

It follows from a result of Alon and Boppana [3] (see, for instance, [136] for a proof) that for a d-regular connected graph with N vertices

$$\mu_2 \geq 2\sqrt{d-1}\left(1 - O\left(\frac{1}{\log^2 N}\right)\right). \tag{13.3.14}$$

This motivates the definition of a *Ramanujan graph* as such a graph with

$$\mu_2 \leq 2\sqrt{d-1}, \tag{13.3.15}$$

that is, with the largest possible spectral gap within this class. Families of Ramanujan graphs were constructed in [195, 201]; see also [61, 136].

We shall now provide the basic definition.

Definition 13.3.1. A family $\Gamma_m = (V_m, E_m)$ of d-regular graphs with $|V_m| \to \infty$ as $m \to \infty$ is called an *expander family* if

$$h_K(\Gamma_m) \geq \varepsilon \quad \text{for all } m \text{ for some } \varepsilon > 0. \qquad (13.3.16)$$

The idea is to have families of graphs with two seemingly contradictory properties, sparsity, and high connectivity. Of course, every individual connected graphs satisfies (13.3.16) for some $\varepsilon > 0$, but the point of the definition is to require this uniformly on families of graphs whose size goes to infinity.

Pinsker [229] showed with probabilistic arguments that almost all graphs satisfy some expansion property. Margulis [199] constructed the first explicit expander family. In his construction, $d = 8$ and $V_m = \mathbb{Z}_m \times \mathbb{Z}_m$. With everything understood mod m, the neighbors of the vertex (x, y) are $(x + y, y)$, $(x - y, y)$, $(x, y + x)$, $(x, y - x)$, $(x + y + 1, y)$, $(x - y + 1, y)$, $(x, y + x + 1)$, and $(x, y - x + 1)$. The explicit eigenvalue bound (13.3.15) was derived in [102].

This example, however, does not represent just some isolated discovery but emerges from a wider context opened by Kazhdan [170]. For the purposes here, it suffices to consider a discrete group G with a finite symmetric set Σ of generators.

Definition 13.3.2. The discrete group G possesses *Kazhdan's property (T)* if for some symmetric set Σ of generators, there exists some $\delta > 0$ with the property that for every unitary representation ρ of G in some Hilbert space H that does not fix any nonzero element of H, for every $v \in H, v \neq 0$, there exists some $s \in \Sigma$ with

$$\frac{\|\rho(s)v - v\|}{\|v\|} \geq \delta. \qquad (13.3.17)$$

(In fact, when (13.3.17) holds for all $v \neq 0$, then Σ has to generate G.)

This may sound like a rather special property, but Kazhdan discovered that it is possessed by all lattices in simple Lie groups H of $\mathbb{R}$-rank ≥ 2, like $\text{Sl}(n, \mathbb{R})$ for $n \geq 3$. A lattice is a discrete subgroup Γ for which the quotient $\Gamma \backslash (H/K)$ (where K is a maximal compact subgroup of H) has finite volume (induced by the natural Riemannian metric on H/K); an example is $\text{Sl}(n, \mathbb{Z})$ for $n \geq 3$. $\text{Sl}(2, \mathbb{Z})$, however, does not possess property (T).

Definition 13.3.3. Let Σ be a finite symmetric set of generators of the discrete group G. The *Cayley graph* $C(G, \Sigma)$ has G as its vertex set and connects $g, h \in G$ when there exists some $s \in \Sigma$ with $h = gs$.

We observe that the symmetry of Σ makes the Cayley graph undirected, even if G is not commutative.

Margulis [199, 200] then found that Kazhdan's property (T) of G is equivalent to the expansion property of the Cayley graphs of its quotients by finite index normal subgroups and applied this to the construction of expanders. Examples then are the Cayley graphs of $Sl(n, \mathbb{Z}/p\mathbb{Z})$ for $n \geq 3$ and primes p.

This already suggests that the theory of expander graphs has surprisingly rich and diverse connections with other areas of mathematics, such as number theory or representation theory; see, for instance, the survey in [193].

For practical purposes, since computing h_K is an **NP**-hard problem, one usually requires an eigenvalue bound, more precisely, a lower bound on the spectral gap $d - \mu_2$, as for instance in (13.3.15). By now, many other expander families are known, such as the Ramanujan graphs already mentioned. See, for instance, [61, 136] for details.

There are other variants of the expansion property; for instance, one considers a bipartite graph $\Gamma = (V, E)$, with V consisting of the two disjoint vertex classes L, R, with $|L| = n$. Every vertex in L should have d neighbors, which by bipartiteness are all in R. The set $N(S) \subset R$ of neighbors of any $S \subset L$ with $|S| \leq \frac{n_1}{2}$ should satisfy

$$|N(S)| \geq |S|. \tag{13.3.18}$$

In (13.3.18), the expansion factor is 1. But if S consists of a single vertex, $|S| = 1$, then $|N(S)| = d|S|$ by assumption, and so, for small $|S|$, one may require an expansion factor even larger than 1, and in fact even larger than $\frac{d}{2}$. [136] speaks of a *magical graph* in such a case.

We now introduce another important concept.

Definition 13.3.4. For a (directed) graph $\Gamma = (V, E)$, let $I, O \subset V$, called the input and the output set, both have n vertices. Γ is called a *super concentrator* if for every k and every $S \subset I$ and $T \subset O$ with $|S| = |T| = k$, there exist k vertex disjoint paths in Γ from S to T.

Super concentrators find many applications for the efficient design of networks. Super concentrators can be constructed from graphs with suitable expansion properties like (13.3.16) or (13.3.18), see, for instance, [61, 136, 199].

Another important application of expander graphs in computer science concerns error-correcting codes. The theoretical framework had been developed by Claude Shannon [246]. One has a number of messages, for simplicity, since we are using binary codes, 2^k for some $k \in \mathbb{N}$. To make them better distinguishable, one encodes them using $n > k$ binary symbols. Thus, every message is uniquely represented by a string of n 0s and 1s. One wants to set up the code in such a way that different code words differ in as many symbols as possible while not making n too large. The Hamming distance d_H between two such strings is defined as the number of positions where the two have different symbols. For example, $d_H(0101, 0011) = 2$ because the two strings differ at the second and at the third positions. When C is the set of code words, by assumption of cardinality 2^k, one wants to control the

$$\text{rate } R := \frac{\log_2 |C|}{n} = \frac{k}{n} \tag{13.3.19}$$

and the

$$\text{(normalized) distance } \delta := \frac{\min_{c \neq c' \in C} d_H(c, c')}{n} \tag{13.3.20}$$

simultaneously from below. Thus, the code words should not be too long but sufficiently different from each other. When the Hamming distance between any two code words is at least d_0, then any code word can be uniquely recovered even if transmitted with errors, as long as less than $\frac{d_0}{2}$ symbols are incorrectly transmitted. Of course, the receiver needs to know the code words, and when she receives a message, she simply looks up the code word with the smallest Hamming distance to that message.

In a line of research that started with [103] and was advanced by the discoveries in [199, 229], it was found by Sipser and Spielman [252] that the simultaneous control of the rate and the distance between code words can be achieved efficiently with the help of expander graphs, and we shall now describe this using the presentation in [136] and the concept of a magical graph described earlier.

But let us first explain the principle. The strings with entries 0 and 1 of length n are naturally considered as elements of the vector space $\mathbb{Z}_2^n$ and therefore amenable to operations from linear algebra over the field $\mathbb{Z}_2$. One can seek 2^k code words as those elements that are in the kernel of some matrix $A = (A_{ij})_{i,j=1,...n}$ of rank $n - k$. That matrix A then is represented as the adjacency matrix of a directed bipartite graph with one vertex class L consisting of n vertices $v_1, \ldots, v_n$. The subsets S of L then represent strings by putting a 1 in the ith position if $v_i \in S$. In order to get a good code with (13.3.20) controlled, one then shows that whenever the size of S is sufficiently small, we can find some $w_j \in N(S)$, the neighborhood of S (which by bipartiteness is contained in the other vertex class R, consisting of those vertices w_j where $A_{ij} = 1$ for some i), that has a unique neighbor $v_i \in S$. But since the corresponding element $A_{ij} = 1$ and for all other k, A_{kj} is applied to 0, and the jth component of the image of the string encoded by S is 1. This implies that this string is not in the kernel of A and hence cannot be a code word.

If the graph is magical, then a small $S \subset L$ has a set of neighbors $N(S) \subset R$ of size larger than $\frac{d}{2}|S|$. But since, by construction, every element of S has precisely d neighbors in $N(S)$, there has to exist some $w \in N(S)$ that has only one neighbor $v \in S$, which then makes the preceding argument work. If $|R| \le \frac{3n}{4}$, then the dimension of ker A is $\ge \frac{n}{4}$, and so, we achieve a rate of 4, that is, the code words are at most four times longer than the number of bits they encode. Since the code is linear, the minimal distance between any two code words equals the minimal distance between a code word and $0 \in$ ker A. But the earlier argument shows that vectors with only a few 1s cannot be code words, and therefore, we also obtain a lower bound on the Hamming distance between code words. Thus, an efficient error-correcting code is constructed from the magical graph.

In the literature, for instance [64, 194], notions of hypergraph (in fact, simplicial complex) expanders have also been proposed, and such expanders have been shown to exist and applied to various problems in computer science. Again, these objects are required to be highly regular. A hypergraph is called r-uniform if every hyperedge connects exactly r vertices, and such a hypergraph then is d-regular, if every subset of $r - 1$ vertices of such a hyperedge is contained in exactly d hyperedges. As before, one may require lower bounds for the eigenvalues of Laplace operators, now also in higher dimensions. A topological version of the expansion property was proposed by Gromov [117]. Here, a simplicial complex or a hypergraph is considered as a topological space. Gromov's condition then requires that for any continuous map from a pure r-dimensional simplicial complex (where *pure* means that every simplex is contained in some r-dimensional one) to $\mathbb{R}^{r-1}$, there exists some point p whose preimage is contained in an ε-fraction of the r-simplices. For some existence results of such families, see [100]. In the graph case, when the Cheeger inequality holds, if we consider a function f from the geometric realization of a graph to $\mathbb{R}$ and take some suitable $z \in \mathbb{R}$, the sets $\{f(v) > z\}$ and $\{f(v) < z\}$ are connected by many edges and so by the intermediate value theorem, the images of many edges have to contain z.

In any case, such higher-dimensional expanders are currently finding applications in general questions around error-correcting codes; for instance, in the quantum version, see, for example, the discussion at the end of [194].

13.4 Nonlinear Dynamics

Nonlinear dynamics is a branch of physics that overlaps with the mathematical theory of dynamical systems. One of its aims is to understand the formation of structures and patterns through general mechanisms of self-organization. These mechanisms include the interaction of simple units that follow some nonlinear dynamical rule in networks. The coupling between those units is typically of the diffusive type and therefore can be formalized by a Laplace operator. The spectrum of that operator and in particular its second (i.e., its first nonvanishing) and its largest eigenvalue then control properties of the network diffusion. When we have a normalized Laplace operator, the second eigenvalue should be large in order to facilitate the uniform spreading of activity through the network and to prevent the formation of local clusters that behave differently from the rest. The largest eigenvalue should often be bounded away from 2, in order to prevent period-2 oscillations of the network dynamics.

In order to illustrate these principles, let us consider a concrete setting, that of the synchronization of coupled chaotic oscillators (see [228] and the references therein). We shall follow [151] in linking this to an eigenvalue problem of the Laplacian of a connected graph Γ with vertex set V consisting of N nodes. One starts with a differentiable function $\Psi: \mathbb{R} \to \mathbb{R}$, often chosen to be the quadratic (logistic) map in the literature. The idea is that Ψ should generate chaotic dynamics when operating on the state of an individual node. (Here, we take a positive Lyapunov exponent of Ψ [see later] as an indication of chaotic dynamics.)

The nodes then are coupled, and one studies the system

$$u(v,n+1) = \epsilon \frac{1}{\deg v} \sum_{\substack{v' \\ v' \sim v}} \Psi(u(v',n)) + (1-\epsilon)\Psi(u(v,n)) \qquad (13.4.1)$$

for $n \in \mathbb{N}$, with some initialization $u(v,0)$. We can also rewrite this as

$$u(v,n+1) = \epsilon \left(\frac{1}{\deg v} \sum_{\substack{v' \\ v' \sim v}} \Psi(u(v',n)) - \Psi(u(v,n)) \right)$$
$$+ \Psi(u(v,n)) = \epsilon \Delta \Psi(u(v,n)) + \Psi(u(v,n)). \qquad (13.4.2)$$

That is, the dynamics at the individual nodes are diffusively coupled via the (normalized) graph Laplacian Δ. Thus, while in isolation, the dynamics at the nodes might display chaotic behavior; it was a somewhat surprising finding [165] that these individual chaotic dynamics can get synchronized when diffusively coupled. The dynamics is synchronized if

$$u(v,n) = \bar{u}(n) \qquad (13.4.3)$$

for some function $\bar{u}$ that is independent of the individual nodes. This function then needs to solve the uncoupled equation

$$\bar{u}(n + 1) = \Psi(\bar{u}(n)). \tag{13.4.4}$$

We then want to check the condition for linear stability of such a synchronized state. We shall use an orthonormal basis of $(u_k)_{k=1,...,N}$ of eigenvectors of Δ, and we recall our convention that we order the eigenvalues as $\lambda_1 = 0 < \lambda_2 \leq \ldots \leq \lambda_N \leq 2$. We consider a perturbation

$$u(v,n) = \bar{u}(n) + \delta\alpha_k(n)u_k(v), \tag{13.4.5}$$

by an eigenmode u_k for small enough δ and some $k \geq 2$. Here, we only need to consider the case $k \geq 2$ because the eigenfunction u_1 for the eigenvalue $\lambda_1 = 0$ is constant and therefore does not desynchronize a state. The question of linear stability analysis then is whether $\alpha_k(n)$ goes to 0 for $n \to \infty$ if $u(v,n)$ solves (13.4.1). Inserting (13.4.5) into (13.4.2) and expanding about $\delta = 0$ yields

$$\alpha_k(n + 1) = (1 - \epsilon\lambda_k)\Psi'(\bar{u}(n))\alpha_k(n), \tag{13.4.6}$$

where Ψ' denotes the derivative of Ψ. A sufficient local stability condition is

$$\lim_{n\to\infty} \frac{1}{n} \log \frac{\alpha_k(n)}{\alpha_k(0)} = \lim_{n\to\infty} \frac{1}{n} \log \prod_{v=0}^{n-1} \frac{\alpha_k(v + 1)}{\alpha_k(v)} < 0 . \tag{13.4.7}$$

Here, this is

$$\log|1 - \epsilon\lambda_k| + \lim_{n\to\infty} \frac{1}{n} \sum_{v=0}^{n-1} \log |\Psi'(\bar{u}(v))| < 0 , \tag{13.4.8}$$

with the Lyapunov exponent

$$\mu_0 = \lim_{n\to\infty} \frac{1}{n} \sum_{v=0}^{n-1} \log |\Psi'(\bar{u}(n))|$$

of Ψ. Therefore, the stability condition (13.4.8) is

$$|e^{\mu_0}(1 - \epsilon\lambda_k)| < 1. \tag{13.4.9}$$

The important point is that we may have temporal instability of the uncoupled dynamics, as witnessed by a positive Lyapunov exponent, that is,

$$\mu_0 > 0, \tag{13.4.10}$$

but still, (13.4.9) may hold for all $k \geq 2$, that is, the synchronized state may be linearly stable. We thus need to identify the conditions for 13.4.9 to hold for all

$k \geq 2$. This happens if

$$\frac{1 - e^{-\mu_0}}{\lambda_2} < \epsilon < \frac{1 + e^{-\mu_0}}{\lambda_N}. \qquad (13.4.11)$$

In order to satisfy (13.4.11), we need

$$\frac{\lambda_N}{\lambda_2} < \frac{e^{\mu_0} + 1}{e^{\mu_0} - 1}. \qquad (13.4.12)$$

While this condition is rather tight (and in fact, synchronization can be observed in numerical simulations for a much wider range of parameters), the important point is that it can still hold in situations where $\mu_0 > 1$, that is, where Ψ generates chaotic dynamics. And it is easier to satisfy the larger λ_2 is, that is, recalling the discussion in Section 4.3, the more difficult it is to cut the network into independent pieces.

While the phenomenon that chaotic dynamics can synchronize was a surprising discovery, a synchronized dynamic is still a somewhat boring situation – everybody does the same. But we have recently found [31, 218] that when we diffusively couple chaotic dynamics on hypergraphs via the hypergraph Laplacian (2.5.39), much richer dynamical patterns can emerge.

Another surprising phenomenon [9] is that transmission delays, that is, replacing (13.4.1) by

$$u(v, n + 1) = \epsilon \frac{1}{\deg v} \sum_{\substack{v' \\ v' \sim v}} \Psi(u(v', n - n_0)) + (1 - \epsilon)\Psi(u(v, n)) \qquad (13.4.13)$$

for some $n_0 > 0$ can facilitate synchronization. This is surprising because each node only is influenced by past states of its neighbors but still coordinates its dynamic with their current states. It remains to study, for instance, what will happen on hypergraphs when delays are introduced.

Remarks on Notation

We use the symbol Δ for our main Laplace-type operators, with the sign convention of geometers rather than of analysts. This makes our operators positive. Thus, the Euclidean Laplace operator on $\mathbb{R}^n$ is

$$\Delta = -\sum_{i=1}^{n} \frac{\partial^2}{\partial(x^i)^2}.$$

This is a special case of the Hodge Laplacian of Riemannian geometry,

$$\Delta = dd^* + d^*d,$$

operating on differential forms, where d is the exterior derivative and d^* its L^2-adjoint with respect to the product defined by the Riemannian metric.

In discrete geometry, however, there are various variants of the Laplace operator, and we denote those that are not our main ones by other symbols, such as $L, \mathcal{L}, A, \ldots$.

The degree of a vertex i of a (hyper)graph is denoted by $\deg i$; for the maximal (minimal) degree of the vertices in a (hyper)graph, we use the symbol $\overline{d}$ ($\underline{d}$). The cardinality of a hyperedge e is denoted by $|e|$. The cardinality is the number of vertices contained in e. Thus, the cardinality of an edge in a graph is always 2 (except for self-edges [which we usually do not allow], which would have cardinality 1). The maximal cardinality of hyperedges then is either $\overline{c}$ or Θ.

References

[1] R. Abelson and M. Rosenberg. Symbolic psycho-logic: A model of attitudinal cognition. *Behavioral Science*, 3(1):1–13, 1958.

[2] M. Abért, A. Thom, and B. Virág. Benjamini-Schramm convergence and pointwise convergence of the spectral measure. *preprint*, 2013. https://tu-dresden.de/mn/math/geometrie/thom/ressourcen/dateien/forschung/preprint_app?lang=en.

[3] N. Alon. Eigenvalues and expanders. *Combinatorica*, 6(2):83–96, 1986.

[4] N. Alon and V. Milman. λ_1, isoperimetric inequalities for graphs, and superconcentrators. *Journal of Combinatorial Theory, Series B*, 38(1):73–88, 1985.

[5] S. Amghibech. Eigenvalues of the discrete p-Laplacian for graphs. *Ars Combinatoria*, 67:283–302, 2003.

[6] A. Amit and N. Linial. Random graph coverings I: General theory and graph connectivity. *Combinatorica*, 22(1):1–18, 2002.

[7] G. Aronsson. Extension of functions satisfying Lipschitz conditions. *Arkiv för Matematik*, 6(6):551–561, 1967.

[8] F. Arrigo, P. Grindrod, D. J. Higham, and V. Noferini. Non-backtracking walk centrality for directed networks. *Journal of Complex Networks*, 6(1):54–78, 2018.

[9] F. Atay, J. Jost, and A. Wende. Delays, connection topology, and synchronization of coupled chaotic maps. *Physical Review Letters*, 92(14):144101, 2004.

[10] F. M. Atay and H. Tuncel. On the spectrum of the normalized Laplacian for signed graphs: Interlacing, contraction, and replication. *Linear Algebra and Its Applications*, 442:165–177, 2014.

[11] F. M. Atay and S. Liu. Cheeger constants, structural balance, and spectral clustering analysis for signed graphs. *Discrete Mathematics*, 343(1):111616, 2020.

[12] G. Aubert and P. Kornprobst. *Mathematical problems in image processing*. Springer, 2nd ed., 2006.

[13] Á. Backhausz, B. Szegedy, and B. Virág. Ramanujan graphings and correlation decay in local algorithms. *Random Structures & Algorithms*, 47(3):424–435, 2015.

[14] H.-J. Bandelt and A. Dress. Split decomposition: A new and useful approach to phylogenetic analysis of distance data. *Molecular Phylogenetics and Evolution*, 1(3):242–252, 1992.

[15] A. Banerjee and A. Char. On the spectrum of directed uniform and non-uniform hypergraphs. arXiv:1710.06367.

[16] A. Banerjee, A. Char, and B. Mondal. Spectra of general hypergraphs. *Linear Algebra and Its Applications*, 518:14–30, 2017.

[17] A. Banerjee and J. Jost. Laplacian spectrum and protein-protein interaction networks. arXiv:0705.3373.

[18] A. Banerjee and J. Jost. Spectral plots and the representation and interpretation of biological data. *Theory in Biosciences*, 126(1):15–21, 2007.

[19] A. Banerjee and J. Jost. On the spectrum of the normalized graph Laplacian. *Linear Algebra and Its Applications*, 428(11–12):3015–3022, 2008.

[20] A. Banerjee and J. Jost. Spectral plot properties: Towards a qualitative classification of networks. *Networks and Heterogeneous Media*, 3(2):395–411, 2008.

[21] A. Banerjee and S. Parui. On synchronization in coupled dynamical systems on hypergraphs. arXiv:2008.00469, 2020.

[22] H. Bass. The Ihara-Selberg zeta function of a tree lattice. *International Journal of Mathematics*, 3(06):717–797, 1992.

[23] F. Bauer. Normalized graph Laplacians for directed graphs. *Linear Algebra and Its Applications*, 436:4193–4222, 2012.

[24] F. Bauer and J. Jost. Bipartite and neighborhood graphs and the spectrum of the normalized graph Laplacian. *Communications in Analysis and Geometry*, 21:787–845, 2013.

[25] F. Bauer, F. Atay, and J. Jost. Synchronization in discrete-time networks with general pairwise coupling. *Nonlinearity*, 22(10):2333, 2009.

[26] F. Bauer, F. Atay, and J. Jost. Synchronized chaos in networks of simple units. *Europhysics Letters*, 89(2):20002, 2010.

[27] M. Belloni and B. Kawohl. The pseudo-p-Laplace eigenvalue problem and viscosity solutions as $p \to \infty$. *ESAIM: Control, Optimisation and Calculus of Variations*, 10(1):28–52, 2004.

[28] M. Berger, P. Gauduchon, and E. Mazet. *Le spectre d'une variété riemannienne*. Springer, Lecture Notes in Mathematics 194, 1974.

[29] N. Biggs. *Algebraic graph theory*. Cambridge University Press, 67, 1993.

[30] Y. Bilu and N. Linial. Lifts, discrepancy and nearly optimal spectral gap. *Combinatorica*, 26(5):495–519, 2006.

[31] T. Böhle, C. Kuehn, R. Mulas, and J. Jost. Coupled hypergraph maps and chaotic cluster synchronization. *EPL*, 136:40005, 2021.

[32] B. Bollobás. *Random graphs*. Springer, 1998.

[33] P. Bonacich. Factoring and weighting approaches to status scores and clique identification. *Journal of Mathematical Sociology*, 2(1):113–120, 1972.

[34] A. Bonato. *A course on the web graph*, volume 89. American Mathematical Society, 2008.

[35] E. M. Borba and U. Schwerdtfeger. Eigenvalue bounds for the signless p-Laplacian. *The Electronic Journal of Combinatorics*, 25(2), 2018.

[36] C. Bordenave, M. Lelarge, and L. Massoulié. Nonbacktracking spectrum of random graphs: Community detection and nonregular Ramanujan graphs. *Annals of Probability*, 46(1):1–71, 2018.

[37] K. Borsuk. Drei sätze über die n-dimensionale euklidische sphäre. *Fundamenta Mathematicae*, 20(1):177–190, 1933.

[38] A. Brouwer and W. Haemers. *Spectra of graphs*. Springer Science & Business Media, 2011.

[39] T. Bühler and M. Hein. Spectral clustering based on the graph p-Laplacian. In *Proceedings of the 26th Annual International Conference on Machine Learning*, pages 81–88, 2009.

[40] L. Bungert, M. Burger, Y. Korolev, T. Roith, and D. Tenbrinck. Structural analysis of an l-infinity variational problem and relations to distance functions. Report.

[41] P. Buser. On Cheeger's Inequality $\lambda_1 \geq h^2/4$. In *Geometry of the Laplace Operator (Proceedings of Symposia in Pure Mathematics)*, volume 36, pages 29–77, 1980.

[42] P. Buser. A note on the isoperimetric constant. *Annales scientifiques de l'École Normale Supérieure*, Ser. 4, 15(2):213–230, 1982.

[43] P. Buser. *Geometry and spectra of compact Riemann surfaces*, volume 106 of *Progress in mathematics*. Birkhäuser Boston, Inc., Boston, MA, 1992.

[44] T. Bıyıkoğlu. A discrete nodal domain theorem for trees. *Linear Algebra and Its Applications*, 360:197–205, 2003.

[45] T. Bıyıkoğlu, W. Hordijk, J. Leydold, T. Pisanski, and P. F. Stadler. Graph Laplacians, nodal domains, and hyperplane arrangements. *Linear Algebra and Its Applications*, 390:155–174, 2004.

[46] T. Bıyıkoğlu, J. Leydold, and P. F. Stadler. *Laplacian eigenvectors of graphs, Perron-Frobenius and Faber-Krahn type theorems*. Springer, 2007.

[47] Y. Canzani. *Analysis on manifolds via the Laplacian*. Notes, Harvard University, 2013.

[48] C. Castellano and R. Pastor-Satorras. Relevance of backtracking paths in recurrent-state epidemic spreading on networks. *Physical Review E*, 98(5):052313, 2018.

[49] A.-C. Castro, B. Pelletier, and P. Pudlo. The normalized graph cut and Cheeger constant: From discrete to continuous. *Advances in Applied Probability*, 44(4):907–993, 2012.

[50] K. C. Chang, S. Shao, D. Zhang, and W. Zhang. Lovász extension and graph cut. *Communications in Mathematical Sciences*, 19:761–786, 2021.

[51] K. C. Chang. The spectrum of the 1–Laplace operator. *Communications in Contemporary Mathematics*, 11:865–894, 2009.

[52] K. C. Chang. Spectrum of the 1-Laplacian and Cheeger's constant on graphs. *Journal of Graph Theory*, 81:167–207, 2016.

[53] K. C. Chang, S. Shao, and D. Zhang. The 1-Laplacian Cheeger cut: Theory and algorithms. *Journal of Computational Mathematics*, 33(5):443–467, 2015.

[54] K. C. Chang, S. Shao, and D. Zhang. Nodal domains of eigenvectors for 1-Laplacian on graphs. *Advances in Mathematics*, 308:529–574, 2017.

[55] I. Chavel. *Eigenvalues in Riemannian geometry*. Academic Press, 1984.

[56] J. Cheeger. A lower bound for the smallest eigenvalue of the Laplacian. In *Problems in Analysis*, pages 195–199. Princeton University Press, 1970.

[57] B. Cheng and B. L. Liu. On the nullity of graphs. *The Electronic Journal of Linear Algebra*, 16:60–67, 2007.

[58] S.-Y. Cheng. Eigenfunctions and nodal sets. *Commentarii Mathematici Helvetici*, 51:43–55, 1976.

[59] G. Choquet. Theory of capacities. *Annales de l'Institut Fourier*, 5:131–295, 1954.

[60] M. Chudnovsky, N. Robertson, P. Seymour, and R. Thomas. The strong perfect graph theorem. *Annals of Mathematics (2)*, 164(1):51–229, 2006.

[61] F. Chung. *Spectral graph theory*. American Mathematical Society, 1997.

[62] S. M. Cioaba, W. H. Haemers, J. R. Vermette, and W. Wong. The graphs with all but two eigenvalues equal to ±1. *Journal of Algebraic Combinatorics*, 41:887–897, 2015.

[63] F. Clarke. Generalized gradients and applications. *Transactions of the American Mathematical Society*, 205:247–262, 1975.

[64] D. Conlon, J. Tidor, and Y. Zhao. Hypergraph expanders of all uniformities from Cayley graphs. *Proceedings of the London Mathematical Society*, 121(5):1311–1336, 2020.

[65] S. Cook. The complexity of theorem-proving procedures, stoc'71: Proceedings of the third annual acm symposium on theory of computing, 1971.

[66] Y. Cooper. Properties determined by the Ihara zeta function of a graph. *The Electronic Journal of Combinatorics*, 16(1):R84, 2009.

[67] S. Coste and Y. Zhu. Eigenvalues of the non-backtracking operator detached from the bulk. *Random Matrices: Theory and Applications*, 10(03):2150028, 2021.

[68] R. Courant, K. O. Friedrichs, and H. Lewy. Über die partiellen Differential-gleichungen der mathematischen Physik. *Mathematische Annalen*, 100:32–74, 1928.

[69] R. Courant and D. Hilbert. *Methoden der Mathematischen Physik I*. Springer, 1924, ^{3}1968.

[70] R. Courant and D. Hilbert. *Methoden der Mathematischen Physik II*. Springer, 1937, ^{3}1967.

[71] M. Cuesta, D. G. De Figueiredo, and J.-P. Gossez. A nodal domain property for the p-Laplacian. *Comptes Rendus de l'Académie des Sciences – Series I – Mathematics*, 330:669–673, 2000.

[72] D. Cvetković, M. Doob, and H. Sachs. Spectra of graphs: Theory and applications. *(No Title)*, 1980.

[73] K. Das and S. Sun. Extremal graph on normalized Laplacian spectral radius and energy. *Electronic Journal of Linear Algebra*, 29(1):237–253, 2016.

[74] B. E. Davies, G. L. Gladwell, J. Leydold, and P. F. Stadler. Discrete nodal domain theorems. *Linear Algebra and Its Applications*, 336(1–3):51–60, 2001.

[75] M. Degiovanni and M. Marzocchi. A critical point theory for nonsmooth functionals. *Annali di Matematica Pura ed Applicata*, 167:73–100, 1994.

[76] P. Deidda, M. Putti, and F. Tudisco. Nodal domain count for the generalized graph p-Laplacian. *Applied and Computational Harmonic Analysis*, 64:1–32, 2023.

[77] P. Deidda. *The graph p-Laplacian eigenvalue problem*. PhD thesis, Università degli studi di Padova, 2023.

[78] M. Desai and V. Rao. A characterization of the smallest eigenvalue of a graph. *Journal of Graph Theory*, 18(2):181–194, 1994.

[79] P. Diaconis and D. Stroock. Geometric bounds for eigenvalues of Markov chains. *Annals of Applied Probability*, 1:36–61, 1991.

[80] J. Dieudonné, F. Ellison, W. Ellison, and P. Dugac. *Abrégé d'histoire des mathématiques 1700–1900*. Hermann, 1978.

[81] J. Dodziuk. Finite-difference approach to the Hodge theory of harmonic forms. *American Journal of Mathematics*, 98:79–104, 1976.

[82] J. Dodziuk. Difference equations, isoperimetric inequality and transience of certain random walks. *Transactions of the American Mathematical Society*, 284(2):787–794, 1984.

[83] J. Dodziuk and V. K. Patodi. Riemannian structures and triangulations of manifolds. *Journal of the Indian Mathematical Society*, 40:1–52, 1976.

[84] D. Dotterer and M. Kahle. Coboundary expanders. *Journal of Topology and Analysis*, 4:499–514, 2012.

[85] C. Duan, L. Wang, and X. Li. Some properties of the signless Laplacian and normalized Laplacian tensors of general hypergraphs. *Taiwanese Journal of Mathematics*, 24(2):265–281, 2020.

[86] B. Eckmann. Harmonische Funktionen und Randwertaufgaben in einem Komplex. *Commentarii Mathematici Helvetici*, 17(1):240–255, December 1944.

[87] J. Eichhorn. *Global analysis on open manifolds*. Nova Publishers, 2007.

[88] I. Ekeland and R. Temam. *Convex analysis and variational problems*. SIAM, 1999.

[89] A. Elmoataz, M. Toutain, and D. Tenbrinck. On the p-Laplacian and ∞-Laplacian on graphs with applications in image and data processing. *SIAM Journal on Imaging Sciences*, 8:2412–2451, 2015.

[90] P. Erdös, A. Rényi, and T. Sós. On a problem of graph theory. *Studia Scientiarum Mathematicarum Hungarica*, 1:215–235, 1966.

[91] P. Erdős and A. Rényi. On random graphs. *Publicationes Mathematicae*, 6(290), 1959.

[92] Y.-Z. Fan, T. Huang, Y.-H. Bao, C.-L. Zhuan-Sun, and Y.-P. Li. The spectral symmetry of weakly irreducible nonnegative tensors and connected hypergraphs. *Transactions of the American Mathematical Society*, 372(3):2213–2233, 2019.

[93] H. Federer. *Geometric measure theory*. Springer, 1979.

[94] H. Federer and W. Fleming. Normal and integral currents. *Annals of Mathematics*, pages 458–520, 1960.

[95] M. Fiedler. Algebraic connectivity of graphs. *Czechoslovak Mathematical Journal*, 23(2):298–305, 1973.

[96] M. Fiedler. A property of eigenvectors of nonnegative symmetric matrices and its application to graph theory. *Czechoslovak Mathematical Journal*, 25(4):619–633, 1975.

[97] A. Floer. Witten's complex and infinite dimensional Morse theory. *Journal of Differential Geometry*, 30:207–221, 1989.

[98] S. Fomin, M. Shapiro, and D. Thurston. Cluster algebras and triangulated surfaces. Part I: Cluster complexes. *Acta Mathematica*, 201:83–146, 2008.

[99] S. Foucart and H. Rauhut. *A mathematical introduction to compressive sensing*. Springer, 2013.

[100] J. Fox, M. Gromov, V. Lafforgue, A. Naor, and J. Pach. Overlap properties of geometric expanders. *Journal für die reine und angewandte Mathematik (Crelles Journal)*, 2012(671):49–83, 2012.

[101] S. Fujishige. *Submodular functions and optimization. Second edition*. Elsevier B. V., Amsterdam, 2005.

[102] O. Gabber and Z. Galil. Explicit constructions of linear-sized superconcentrators. *Journal of Computer and System Sciences*, 22(3):407–420, 1981.

[103] R. Gallager. Low density parity check codes. *Cambridge: M*, 1:1–73, 1963.

[104] G. Gallo, G. Longo, S. Pallottino, and S. Nguyen. Directed hypergraphs and applications. *Discrete Applied Mathematics*, 42:177–201, 1993.

[105] F. Galuppi, R. Mulas, and L. Venturello. Spectral theory of weighted hypergraphs via tensors. *Linear and Multilinear Algebra*, 71(3): 317–347, 2022.

[106] C. Ge and S. Liu. Symmetric matrices, signed graphs, and nodal domain theorems. *Calculus of Variations and Partial Differential Equations*, 62:Article 137, 2023.

[107] C. Ge, S. Liu, and D. Zhang. Nodal domain theorems for p-Laplacians on signed graphs. *Journal of Spectral Theory*, 13(3):937–989, 2023.

[108] S. A. Gershgorin. Über die Abgrenzung der Eigenwerte einer Matrix. *Izvestiya Akademii Nauk SSSR, Otdelenie Fiziko-Matematicheskikh Nauk*, (6):749–754, 1931.

[109] G. Gilboa and S. Osher. Nonlocal linear image regularization and supervised segmentation, *Multiscale Modeling & Simulation*, 6(2):595–630, 2007.

[110] E. Giusti. *Direct methods in the calculus of variations*. World Scientific, 2003.

[111] C. Glover and M. Kempton. Some spectral properties of the non-backtracking matrix of a graph. *Linear Algebra and Its Applications*, 618:37–57, 2021.

[112] C. Godsil and B. McKay. Constructing cospectral graphs. *Aequationes Mathematicae*, 25(1):257–268, 1982.

[113] C. Godsil and G. Royle. *Algebraic graph theory*, volume 207. Springer Science & Business Media, 2001.

[114] C. Gordon, D. L. Webb, and S. Wolpert. One cannot hear the shape of a drum. *Bulletin of the American Mathematical Society*, 27(1), 1992.

[115] M. S. Granovetter. The strength of weak ties. *American Journal of Sociology*, 78(1360), 1973.

[116] P. Grindrod, D. J. Higham, and V. Noferini. The deformed graph Laplacian and its applications to network centrality analysis. *SIAM Journal on Matrix Analysis and Applications*, 39(1):310–341, 2018.

[117] M. Gromov. Singularities, expanders and topology of maps. Part 2: From combinatorics to topology via algebraic isoperimetry. *Geometric and Functional Analysis*, 20:416–526, 2010.

[118] J. Gu, B. B. Hua, and S. P. Liu. Spectral distances on graphs. *Discrete Applied Mathematics*, 190:56–74, 2015.

[119] J. Gu, J. Jost, S. P. Liu, and P. Stadler. Spectral classes of regular, random, and empirical graphs. *Linear Algebra and Its Applications*, 489:30–49, 2016.

[120] S. Guattery and G. L. Miller. On the quality of spectral separators. *SIAM Journal on Matrix Analysis and Applications*, 3:701–719, 1998.

[121] A. Gundert and M. Szedlak. Higher dimensional discrete Cheeger inequalities. *Proceedings of the Annual Symposium on Computational Geometry*, 01 2014.

[122] A. Gundert and U. Wagner. On eigenvalues of random complexes. *Proceedings of the Annual Symposium on Computational Geometry*, 216:545–582, 2016.

[123] F. Harary. Structural duality. *Behavioral Science*, 2:255–265, 1957.

[124] F. Harary. On the notion of balance of a signed graph. *Michigan Mathematical Journal*, 2:143–146, 1955.

[125] F. Harary. On the measurement of structural balance. *Behavioral Science*, 4:316–323, 1959.

[126] F. Harary and R. Z. Norman. Some properties of line digraphs. *Rendiconti del Circolo Matematico di Palermo Series 2*, 9:161–168, 1960.

[127] K. Hashimoto. Zeta functions of finite graphs and representations of *p*-adic groups. In *Automorphic forms and geometry of arithmetic varieties*, volume 15 of *Advanced studies in pure mathematics*, pages 211–280. Elsevier, 1989.

[128] A. Hatcher. *Algebraic topology*. Cambridge University Press, 2002.

[129] A. S. Hathaway. Early history of the potential. *Bulletin of the American Mathematical Society*, 1:66–74, 1891.

[130] M. Hein and T. Bühler. An inverse power method for nonlinear eigenproblems with applications in 1-spectral clustering and sparse PCA. In *Advances in Neural Information Processing Systems (NIPS)*, pages 847–855, 2010.

[131] M. Hein, D. Lenz, and D. Mugnolo. Mini-workshop: Discrete *p*-Laplacians: Spectral theory and variational methods in mathematics and computer science. *Oberwolfach Reports*, 12(1):399–447, 2015. Abstracts from the mini-workshop held February 8–14, 2015, Organized by Matthias Hein, Daniel Lenz and Delio Mugnolo.

[132] M. Hein and S. Setzer. Beyond spectral clustering – Tight relaxations of balanced graph cuts. *Advances in Neural Information Processing Systems*, 24:2366–2374, 2011.

[133] M. Hein, S. Setzer, L. Jost, and S. S. Rangapuram. The total variation on hypergraphs-learning on hypergraphs revisited. In *Proceedings of the 26th International Conference on Neural Information Processing Systems-Volume 2*, pages 2427–2435, 2013.

[134] D. Hilbert. Grundzüge einer allgemeinen Theorie der linearen Integralgleichungen. (Erste Mitteilung). *Nachrichten der Gesellschaft der Wissenschaften zu Göttingen, Mathematisch-Physikalische Klasse*, 1904:49–91, 1904.

[135] C. J. Hillar and L.-H. Lim. Most tensor problems are NP-hard. *Journal of the ACM*, 60(6), 2013.

[136] S. Hoory, N. Linial, and A. Wigderson. Expander graphs and their applications. *Bulletin of the American Mathematical Society*, 43(4):439–561, 2006.

[137] D. Horak and J. Jost. Interlacing inequalities for eigenvalues of discrete Laplace operators. *Annals of Global Analysis and Geometry*, 43(2):177–207, 2013.

[138] D. Horak and J. Jost. Spectra of combinatorial Laplace operators on simplicial complexes. *Advances in Mathematics*, 244:303–336, 2013.

[139] S. Hu and Y. Ke. Multiplicities of tensor eigenvalues. *Communications in Mathematical Sciences*, 14:1049–1071, 2016.

[140] D. Janežič, A. Miličević, S. Nikolić, and N. Trinajstić. *Graph theoretical matrices in chemistry*. Mathematical Chemistry Monographs, 2007.

[141] Y. Jin, J. Jost, and G. Wang. A new nonlocal variational setting for image processing. *Inverse Problems and Imaging*, 9(2):415–430, 2015.

[142] P. Joharinad and J. Jost. *Mathematical principles of geometric and topological data analysis*. Springer, 2023.

[143] J. Jost. Equilibrium maps between metric spaces. *Calculus of Variations*, 2:173–204, 1994.

[144] J. Jost. *Nonpositive curvature: Geometric and analytic aspects*. Birkhäuser, 1997.

[145] J. Jost. Dynamical networks. In J. F. Feng, J. Jost, and M. Ping Qian, editors, *Networks : From biology to theory*, pages 35–62. Springer, 2007.

[146] J. Jost. *Mathematical methods in biology and neurobiology.* Universitext. Springer, 2014.

[147] J. Jost. *Mathematical concepts.* Springer, 2015.

[148] J. Jost. *Biologie und Mathematik.* Springer Spektrum, 2019.

[149] J. Jost. *Postmodern analysis.* Springer, 3rd edition, 2005.

[150] J. Jost. *Riemannian geometry and geometric analysis.* Springer, 7th edition, 2017.

[151] J. Jost and M. P. Joy. Spectral properties and synchronization in coupled map lattices. *Physical Review E,* 65(1, pt. 2):016201, 2002.

[152] J. Jost and X. Li-Jost. *Calculus of variations.* Cambridge University Press, 1998.

[153] J. Jost and R. Mulas. Hypergraph Laplace operators for chemical reaction networks. *Advances in Mathematics,* 351:870–896, 2019.

[154] J. Jost and R. Mulas. Normalized Laplace operators for hypergraphs with real coefficients. *Journal of Complex Networks,* 9(1):cnab009, 2021.

[155] J. Jost, R. Mulas, and F. Münch. Spectral gap of the largest eigenvalue of the normalized graph Laplacian. *Communications in Mathematics and Statistics,* 10(1), 2021.

[156] J. Jost, R. Mulas, and L. Torres. Spectral theory of the non-backtracking Laplacian for graphs. *Discrete Mathematics,* 346(10):113536, 2023.

[157] J. Jost, R. Mulas, and D. Zhang. p-Laplace operators for chemical hypergraphs. *Vietnam Journal of Mathematics,* 50(2):323–358, 2022.

[158] J. Jost, R. Mulas, and D. Zhang. Petals and books: The largest Laplacian spectral gap from 1. *Journal of Graph Theory,* 104(4):727–756, 2023.

[159] J. Jost and D. Zhang. Discrete-to-Continuous Extensions: Lovász extension, optimizations and eigenvalue problems. arXiv:2106.03189.

[160] J. Jost and D. Zhang. Discrete-to-Continuous Extensions: Piecewise multilinear extension, min-max theory and spectral theory. arXiv:2106.04116.

[161] J. Jost and D. Zhang. Cheeger inequalities on simplicial complexes. *Annali della Scuola Normale Superiore di Pisa. Classe di Scienze* (5), 2024.

[162] P. Juutinen and P. Lindqvist. On the higher eigenvalues for the ∞-eigenvalue problem. *Calculus of Variations and Partial Differential Equations,* 23(2):169–192, 2005.

[163] P. Juutinen, P. Lindqvist, and J. Manfredi. The ∞-eigenvalue problem. *Archive for Rational Mechanics and Analysis,* 148(2):89–105, 1999.

[164] M. Kac. Can one hear the shape of a drum? *The American Mathematical Monthly,* 73(4):1–23, 1966.

[165] K. Kaneko. Period-doubling of kink-antikink patterns, quasiperiodicity in antiferro-like structures and spatial intermittency in coupled logistic lattice: Towards a prelude of a "field theory of chaos". *Progress of Theoretical Physics,* 72(3):480–486, 1984.

[166] B. Kawohl and V. Fridman. Isoperimetric estimates for the first eigenvalue of the p-Laplace operator and the Cheeger constant. *Commentationes Mathematicae Universitatis Carolinae,* 44(4):659–667, 2003.

[167] B. Kawohl and J. Horák. On the geometry of the p-Laplacian operator. *Discrete & Continuous Dynamical Systems-S,* 10(4):799, 2017.

[168] B. Kawohl and F. Schuricht. Dirichlet problems for the 1-Laplace operator, including the eigenvalue problem. *Communications in Contemporary Mathematics,* 9(04):515–543, 2007.

[169] B. Kawohl and F. Schuricht. First eigenfunctions of the 1-Laplacian are viscosity solutions. *Communications on Pure and Applied Analysis*, 14:329–339, 2015.

[170] D. Kazhdan. On the connection of the dual space of a group with the structure of its closed subgroups (in Russian). *Funktsional'nyi Analiz i ego Prilozheniya*, 1(1):71–74, 1967.

[171] M. Keller, D. Lenz, and R. Wojciechowski. *Graphs and discrete Dirichlet spaces*, volume 358. Springer, 2021.

[172] M. Kempton. Non-backtracking random walks and a weighted Ihara's theorem. *Open Journal of Discrete Mathematics*, 6(4):207–226, 2016.

[173] G. Kirchhoff. Ueber die Auflösung der Gleichungen, auf welche man bei der Untersuchung der linearen Vertheilung galvanischer Ströme geführt wird. *Annual Review of Physical Chemistry*, 72, 1847.

[174] A. Kostenko and N. Nicolussi. *Laplacians on infinite graphs*. EMS Press, 2023.

[175] M. Kotani and T. Sunada. Zeta functions of finite graphs. *Journal of Mathematical Sciences-University of Tokyo*, 7(1):7–26, 2000.

[176] M. Krebs and A. Shaheen. *Expander families and Cayley graphs: A beginner's guide*. Oxford University Press, 2011.

[177] F. Krzakala, C. Moore, E. Mossel, J. Neeman, A. Sly, L. Zdeborová, and P. Zhang. Spectral redemption in clustering sparse networks. *Proceedings of the National Academy of Sciences*, 110(52):20935–20940, 2013.

[178] P. Kurasov. *Spectral geometry of graphs*. Springer Nature, 2024.

[179] C. Lange, S. P. Liu, N. Peyerimhoff, and O. Post. Frustration index and Cheeger inequalities for discrete and continuous magnetic Laplacians. *Calculus of Variations and Partial Differential Equations*, 54:4165–4196, 2015.

[180] J. R. Lee, S. O. Gharan, and L. Trevisan. Multiway spectral partitioning and higher-order Cheeger inequalities. *Journal of the ACM*, 61(6):1–30, 2014.

[181] P. Li. *Geometric analysis*, volume 134. Cambridge University Press, 2012.

[182] P. Li and S. T. Yau. Estimates of eigenvalues of a compact Riemannian manifold. *AMS Proceedings of Symposia in Pure Mathematics*, 36:205–240, 1980.

[183] A. Lichnerowicz. *Géometrie des groupes de transformations*. Dunod, Paris, 1958.

[184] N. Linial and R. Meshulam. Homological connectivity of random 2-complexes. *Combinatorica*, 26(4):475–487, 2006.

[185] S. Littig and F. Schuricht. Convergence of the eigenvalues of the p-Laplace operator as p goes to 1. *Calculus of Variations and Partial Differential Equations*, 49(1–2):707–727, 2014.

[186] Q. Liu and A. Mitsuishi. Principal eigenvalue problem for infinity Laplacian in metric spaces. *Advanced Nonlinear Studies*, 22(1):548–573, 2022.

[187] S. P. Liu. Multi-way dual Cheeger constants and spectral bounds of graphs. *Advances in Mathematics*, 268:306–338, 2015.

[188] S. P. Liu, N. Peyerimhoff, and A. Vdovina. Signatures, lifts, and eigenvalues of graphs. *Discrete and Continuous Models in the Theory of Networks*, pages 255–269, 2020.

[189] L. Lovász. Submodular functions and convexity. In A. Bachem, B. Korte, and M. Grötschel, editors, *Mathematical programming: The state of the art, Bonn 1982*, pages 235–257. Springer Berlin Heidelberg, 1983.

[190] L. Lovász. *Large networks and graph limits*, volume 60. American Mathematical Society, 2012.

[191] L. Lovász. Normal hypergraphs and the perfect graph conjecture. *Discrete Mathematics*, 2(3):253–267, 1972.

[192] A. Lubotzky. *Discrete groups, expanding graphs and invariant measures*, volume 125. Springer Science & Business Media, 1994.

[193] A. Lubotzky. Expander graphs in pure and applied mathematics. *Bulletin of the American Mathematical Society*, 49(1):113–162, 2012.

[194] A. Lubotzky. High dimensional expanders. In *Proceedings of the International Congress of Mathematicians: Rio de Janeiro 2018*, pages 705–730. World Scientific, 2018.

[195] A. Lubotzky, R. Phillips, and P. Sarnak. Ramanujan graphs. *Combinatorica*, 8(3):261–277, 1988.

[196] M. Lucia and F. Schuricht. Mountain pass solution for nonsmooth elliptic problems. *Minimax Theory and Its Applications*, 5(1):129–150, 2020.

[197] C. Ma, H. Liang, Q. Xie, and P. Wang. Some inequalities on the spectral radius of nonnegative tensors. *Open Mathematics*, 18(1):262–269, 2020.

[198] B. D. MacArthur, R. J. Sánchez-García, and J. W. Anderson. Symmetry in complex networks. *Discrete Applied Mathematics*, 156(18):3525–3531, 2008.

[199] G. A. Margulis. Explicit constructions of concentrators. *Problemy Peredachi Informatsii*, 9(4):71–80, 1973. English translation in: *Problems of Information Transmission* 9 (1975), 325–332.

[200] G. A. Margulis. Explicit constructions of graphs without short cycles and low density codes. *Combinatorica*, 2(1):71–78, 1982.

[201] G. A. Margulis. Explicit group-theoretical constructions of combinatorial schemes and their application to the design of expanders and concentrators. *Problemy Peredachi Informatsii*, 24(1):51–60, 1988.

[202] T. Martin, X. Zhang, and M. E. J. Newman. Localization and centrality in networks. *Physical Review E*, 90(5):052808, 2014.

[203] W. S. Massey. *A basic course in algebraic topology*. Graduate texts in mathematics. Springer-Verlag, 1991.

[204] V. G. Maz'ya. Classes of domains and imbedding theorems for function spaces. In *Doklady Akademii Nauk*, volume 133, pages 527–530. Russian Academy of Sciences, 1960.

[205] R. Meshulam and N. Wallach. Homological connectivity of random k-dimensional complexes. *Random Structures Algorithms*, 34(3):408–417, 2009.

[206] L. Miclo. On eigenfunctions of Markov processes on trees. *Probability Theory and Related Fields*, 142(3):561–594, 2008.

[207] Z. Milbers and F. Schuricht. Existence of a sequence of eigensolutions for the 1-Laplace operator. *Journal of the London Mathematical Society*, 82(1):74–88, 2010.

[208] Z. Milbers and F. Schuricht. Necessary condition for eigensolutions of the 1-Laplace operator by means of inner variations. *Mathematische Annalen*, 356(1):147–177, 2013.

[209] S. Minakshisundaram. Eigenfunctions on Riemannian manifolds. *Journal of the Indian Mathematical Society*, 27:158–165, 1953.

[210] A. Mohammadian. Graphs and their real eigenvectors. *Linear and Multilinear Algebra*, 64(2):136–142, 2016.

[211] B. Mohar. Isoperimetric numbers of graphs. *Journal of Combinatorial Theory, Series B*, 47(3):274–291, 1989.

[212] J. Moreno. *Who shall survive?: A new approach to the problem of human interrelations*. Nervous and Mental Disease Publishing Co., 1934.

[213] J. Moreno and H. Jennings. Statistics of social configurations. *Sociometry*, 1(3–4): 342–374, 1938.

[214] R. Mulas. A Cheeger cut for uniform hypergraphs. *Graphs and Combinatorics*, 37:2265–2286, 2021.

[215] R. Mulas. Spectral classes of hypergraphs. *Australasian Journal of Combinatorics*, 79:495–514, 2021.

[216] R. Mulas and M. J. Casey. Estimating cellular redundancy in networks of genetic expression. *Mathematical Biosciences*, 341:108713, 2021.

[217] R. Mulas, D. Horak, and J. Jost. Graphs, simplicial complexes and hypergraphs: Spectral theory and topology. In F. Battiston and G. Petri, editors, *Higher-order systems*, pages 1–58. Springer, 2022.

[218] R. Mulas, C. Kuehn, and J. Jost. Coupled dynamics on hypergraphs: Master stability of steady states and synchronization. *Physical Review E*, 101:062313, 2020.

[219] R. Mulas and D. Zhang. Spectral theory of Laplace operators on oriented hypergraphs. *Discrete Mathematics*, 344(6):112372, 2021.

[220] R. Mulas, D. Zhang, and G. Zucal. There is no going back: Properties of the non-backtracking Laplacian. *Linear Algebra and Its Applications*, 680:341–370, 2024.

[221] K. Murota. *Discrete convex analysis*. Society for Industrial and Applied Mathematics, 2003.

[222] X. Ouvrard, J.-M. Le Goff, and S. Marchand-Maillet. Adjacency and tensor representation in general hypergraphs part 1: e-adjacency tensor uniformisation using homogeneous polynomials. arXiv:1712.08189.

[223] X. Ouvrard, J.-M. Le Goff, and S. Marchand-Maillet. Adjacency and tensor representation in general hypergraphs. Part 2: Multisets, hb-graphs and related e-adjacency tensors. arXiv:1805.11952.

[224] E. Parini. An introduction to the Cheeger problem. *Surveys in Mathematics and its Applications*, 6:9–21, 2011.

[225] O. Parzanchevski, R. Rosenthal, and R. J. Tessler. Isoperimetric inequalities in simplicial complexes. *Combinatorica*, 36:195–227, 2015.

[226] R. Pastor-Satorras and C. Castellano. The localization of non-backtracking centrality in networks and its physical consequences. *Scientific Reports*, 10(1):1–12, 2020.

[227] Y. Peres, O. Schramm, S. Sheffield, and D. Wilson. Tug-of-war and the infinity Laplacian. *Journal of the American Mathematical Society*, 22(1):167–210, 2009.

[228] A. Pikovsky, M. Rosenblum, and J. Kurths. *Synchronization: A universal concept in nonlinear science*. Cambridge University Press, 2001.

[229] M. Pinsker. On the complexity of a concentrator. In *7th International Telegraffic Conference*, volume 4, pages 1–318. Citeseer, 1973.

[230] G. Pólya and S. Szegö. Isoperimetric inequalities in mathematical physics. *Annals of Mathematics Studies*, 27, 1951.

[231] L. Qi. Directed submodularity, ditroids and directed submodular flows. *Mathematical Programming*, 42:579–599, 1988.

[232] L. Qi. Eigenvalues of a real supersymmetric tensor. *Journal of Symbolic Computation*, 40(6):1302–1324, 2005.

[233] L. Qi and Z. Luo. *Tensor analysis*. Society for Industrial and Applied Mathematics, Philadelphia, PA, 2017. Spectral theory and special tensors.

[234] L. Qi and Y. Song. An even order symmetric B tensor is positive definite. *Linear Algebra and Its Applications*, 457:303–312, 2014.

[235] R. Read and D. Corneil. The graph isomorphism disease. *Journal of Graph Theory*, 1(4):339–363, 1977.

[236] B. Riemann. *Bernhard Riemann's gesammelte mathematische Werke und wissenschaftlicher Nachlass*. BG Teubner, 1892.

[237] B. Riemann. *Ueber die Hypothesen, welche der Geometrie zu Grunde liegen*. Edited with a commentary by J. Jost, Klassische Texte der Wissenschaft, Springer, Berlin etc., 2013.

[238] B. Riemann. *On the hypotheses which lie at the bases of geometry*. Translated by W. K. Clifford, edited with a commentary by J. Jost, Classic Texts in the Sciences, Birkhäuser, 2016.

[239] R. T. Rockafellar and R. Wets. *Variational analysis*, volume 317. Springer Science & Business Media, 2009.

[240] L. I. Rudin, S. Osher, and E. Fatemi. Nonlinear total variation based noise removal algorithms. *Physica D: Nonlinear Phenomena*, 60(1–4):259–268, 1992.

[241] O. Scherzer, M. Grasmair, H. Grossauer, M. Haltmeier, and F. Lenzen. *Variational methods in imaging*. Springer, 2009.

[242] U. Schöning. Graph isomorphism is in the low hierarchy. *Journal of Computer and System Sciences*, 37(3):312–323, 1988.

[243] F. Schuricht. An alternative derivation of the eigenvalue equation for the 1-Laplace operator. *Archiv der Mathematik*, 87(6):572–577, 2006.

[244] M. Schwarz. *Morse homology*. Birkhäuser, 1993.

[245] C. Semple and M. Steel. *Phylogenetics*, volume 24. Oxford University Press, 2003.

[246] C. Shannon. A mathematical theory of communication. *The Bell System Technical Journal*, 27(3):379–423, 1948.

[247] J.-Y. Shao. A general product of tensors with applications. *Linear Algebra and Its Applications*, 439(8):2350–2366, 2013.

[248] J.-Y. Shao, H.-Y. Shan, and B.-F. Wu. Some spectral properties and characterizations of connected odd-bipartite uniform hypergraphs. *Linear and Multilinear Algebra*, 63(12):2359–2372, 2015.

[249] J.-Y. Shao, H.-Y. Shan, and L. Zhang. On some properties of the determinants of tensors. *Linear Algebra and Its Applications*, 439(10):3057–3069, 2013.

[250] M. Shrestha, S. V. Scarpino, and C. Moore. Message-passing approach for recurrent-state epidemic models on networks. *Physical Review E*, 92(2):022821, 2015.

[251] M. A. Shubin. Discrete magnetic Laplacian. *Communications in Mathematical Physics*, 164(2):259–275, 1994.

[252] M. Sipser and D. Spielman. Expander codes. *IEEE Transactions on Information Theory*, 42(6):1710–1722, 1996.

[253] D. A. Spielman and S.-H. Teng. Spectral partitioning works: Planar graphs and finite element meshes. *Linear Algebra and Its Applications*, 421:284–305, 2007.

[254] J. Steenbergen, C. Klivans, and S. Mukherjee. A Cheeger-type inequality on simplicial complexes. *Advances in Applied Mathematics*, 56:56–77, 2014.

[255] T. Sunada. A discrete analogue of periodic magnetic schrödinger operators. *Contemporary Mathematics*, 173:283–283, 1994.

[256] A. Szlam and X. Bresson. Total variation and Cheeger cuts. In *Proceedings of the 27th International Conference on International Conference on Machine Learning*, ICML'10, pages 1039–1046, Madison, WI, USA, 2010. Omnipress.

[257] A. Terras. *Zeta functions of graphs: A stroll through the garden*, volume 128 of *Cambridge studies in advanced mathematics*. Cambridge University Press, 2010.

[258] M. Thüne. *Eigenvalues of matrices and graphs*. PhD thesis, Leipzig University, 2012.

[259] L. Torres. Non-backtracking Spectrum: Unitary Eigenvalues and Diagonalizability. *arXiv preprint* arXiv:2007.13611, 2020.

[260] L. Torres, K. S. Chan, H. Tong, and T. Eliassi-Rad. Node immunization with non-backtracking eigenvalues. *arXiv preprint* arXiv:2002.12309, 2020.

[261] L. Torres, K. S. Chan, H. Tong, and T. Eliassi-Rad. Nonbacktracking eigenvalues under node removal: X-centrality and targeted immunization. *SIAM Journal on Mathematics of Data Science*, 3(2):656–675, 2021.

[262] L. Torres, P. Suárez-Serrato, and T. Eliassi-Rad. Non-backtracking cycles: Length spectrum theory and graph mining applications. *Applied Network Science*, 4(1):1–35, 2019.

[263] L. Trevisan. Max cut and the smallest eigenvalue. *SIAM Journal on Computing*, 41(6):1769–1786, 2012.

[264] N. G. Trillos, R. Murray, and Thorpe. M. From graph cuts to isoperimetric inequalities: Convergence rates of Cheeger cuts on data clouds. *Archive for Rational Mechanics and Analysis*, 244(3):541–598, 2022.

[265] N. G. Trillos and D. Slepcev. Continuum limit of total variation on point clouds. *Archive for Rational Mechanics and Analysis*, 220:193–241, 2016.

[266] F. Tudisco and M. Hein. A nodal domain theorem and a higher-order Cheeger inequality for the graph p-Laplacian. *Journal of Spetral Theory*, 8(3):883–908, 2018.

[267] F. Tudisco and D. Zhang. Nonlinear spectral duality. *arXiv preprint* arXiv:2209.06241, 2022.

[268] D. Valtorta. *On the p-Laplace operator on Riemannian manifolds*. PhD thesis, Università degli Studi di Milano, 2014. arXiv:1212.3422v3.

[269] Edwin R. Van Dam and Willem H. Haemers. Which graphs are determined by their spectrum? *Linear Algebra and Its applications*, 373:241–272, 2003.

[270] U. von Luxburg. A tutorial on spectral clustering. *Statistics and Computing*, 17:395–416, 2007.

[271] U. von Luxburg, M. Belkin, and O. Bousquet. Consistency of spectral clustering. *Annals of Statistics*, 36(2):555–586, 04 2008.

[272] H. Weyl. Über die asymptotische Verteilung der Eigenwerte. *Nachrichten der Königlichen Gesellschaft der Wissenschaften zu Göttingen, Mathematisch-Physikalische Klasse*, pages 110–117, 1911.

[273] H. Weyl. Das asymptotische Verteilungsgesetz der Eigenwerte linearer partieller Differentialgleichungen. *Mathematische Annalen*, 71:441–469, 1912.

[274] S. T. Yau. Isoperimetric constants and the first eigenvalue of a compact Riemannian manifold. *Annales Scientifiques de l'École Normale Supérieure, (4)*, 8(4):487–507, 1975.

[275] T. Zaslavsky. Signed graphs. *Discrete Applied Mathematics*, 4:47–74, 1982.

[276] T. Zaslavsky. Matrices in the theory of signed simple graphs. In *Advances in discrete mathematics and applications: Mysore, 2008*, volume 13 of *Ramanujan Mathematical Society Lecture Notes Series*, pages 207–229. Ramanujan Mathematical Society, Mysore, 2010.

[277] E. Zeidler. *Nonlinear functional analysis and its applications: III: Variational methods and optimization*. Springer, 1985.

[278] D. Zhang. Homological eigenvalues of graph p-Laplacians. *Journal of Topology and Analysis*, 17: 555–606, 2025.

[279] D. Zhou, J. Huang, and B. Schölkopf. Learning with hypergraphs: Clustering, classification, and embedding. *Advances in Neural Information Processing Systems*, 19:1601–1608, 2006.

[280] G. Zucal. Action convergence of general hypergraphs and tensors. *arXiv preprint* arXiv:2308.00226, 2023.

Index

1-form, 60
1-Laplacian, 72

absolutely comonotonic, 188
adjacency matrix, 9, 81, 86, 100, 379
adjoint, 64
algebraic graph Laplacian, 80, 81, 152, 379
algebraic graph theory, 9
algebraic Laplacian, 118
ambiguous vertex, 151
antibalanced graph, 139
automorphism, 45
averaging property, 35

balanced, 151
balanced cycle, 139
balanced graph, 139
Banach space, 5
Betti number, 23
bipartite graph, 96, 97
bistar, 106, 117
book graph, 104, 137, 149, 319, 325
boundary of a vertex set, 379
boundary operator, 18, 36, 176
bounded variation, 70

category, 14
Cauchy interlacing theorem, 297
Cayley graph, 381
Čech complex, 15
chain, 92
Cheeger constant, 11, 66, 71, 116, 379
Cheeger constant on a chemical hypergraph,
    197
Cheeger cut, 11
Cheeger cut problem, 12
Cheeger estimate, 11, 67

Cheeger inequality, 198
Cheeger problem, 114
Cheeger set, 71
chemical hypergraph, 16, 197, 203
Chung Laplacian, 9
clique, 101
clique of hypergraph, 230
clustering coefficient, 110
coarea formula, 67
coboundary, 22
cochain, 31
code, 382
cohomology group, 23, 32
collapse, 304
comonotonic, 186
complementary graph, 83
complete bipartite graph, 131, 212, 319, 325
complete graph, 82, 117, 137, 212, 319, 325
complete graph with self-loops, 82
concentrator, 12
connection Laplacian, 138
contraction, 314
cotangent vector, 60
coupling of measures, 323
covering, 47, 148, 311
critical point, 44
cube graph, 321, 325
cycle, 92, 96, 139, 363
cycle graph, 100, 319

d'Alembert operator, 4
dead end vertex, 151
degenerate weighted simplicial complex, 33
degree, 32, 77
degree matrix, 81, 86
denoising, 58, 72
diameter, 111

diffusion through edge, 92
directed graph, 15, 151
directed hypergraph, 16
disjoint-pair Lovász extension, 187
distance, 111
divergence, 35
down Laplace operator, 29, 48
d-regular, 379
dual Cheeger constant, 123
dual signed Rayleigh quotient, 146
duplicated motif, 46
duplication of edge, 319
duplication of vertex, 99

Eckmann Laplacian, 8
edge, 76
edge boundary, 114
eigenfunction, 6, 8, 38, 52, 79
eigenpair, 188
eigenvalue, 6, 8, 52
eigenvalue equation, 6
eigenvalue of function pair, 188
eigenvalues, 78
eigenvalues of graph, 86
eigenvector centrality, 374
eigenvector of function pair, 188
Eigenwert, 8
Einstein summation convention, 60
elementary collapse, 304
elementary contraction, 314
error correcting code, 382
expander, 12
expander family, 381
exterior algebra, 7
exterior derivative, 62
exterior form, 61
exterior product, 61

(f, g)–Cheeger constant, 194
face, 21
facet, 21
fidelity term, 59, 72
Floer Theory, 176
Fourier series, 5
free face, 304
functor, 14

generalized gradient, 43
gradient flow, 175
graph, 15, 76
graph doubling, 320

Hamming distance, 383
Harary's characterization of balanced graphs, 139
harmonic form, 64
harmonic function, 65
Hashimoto graph, 351
heat equation, 7
Hilbert space, 5, 52
Hodge decomposition, 29
Hodge Laplacian, 8, 66
homology group, 178
homomorphism, 47
hyperedge, 15, 24
hyperflower, 170
hypergraph, 15
hypergraph expander, 384
hypervertex, 24

image denoising, 58, 72
independence number, 212, 229
independent set, 212
index of critical point, 175
induced subgraph, 101
integral equation, 8
interlacing theorem, 297
isoperimetric problem, 113
isospectral, 4, 94, 100, 370

Kazhdan's property (T), 381
Kirchhoff Laplacian, 9, 81
Krasnoselskii genus, 42, 211

Laplace operator, 6, 29, 48, 51, 52, 77
Laplace operator for hypergraphs, 18
Laplace operator of graph, 77, 85
Laplace operator of oriented hypergraph, 37
Laplace-Beltrami operator, 7, 64, 65
Laplacian, 6, 18
Laplacian of graph, 85
Laplacian of signed graph, 141
Lebesgue integral, 5
Lovász extension, 183, 184, 199
lower semi-continuous, 53
Lusternik-Schnirelman theory, 43

magical graph, 382
magnetic Laplacian, 138
Margulis expander, 381
metric tensor, 63
minimizing sequence, 53
min-max principle, 37

modular, 185
morphism, 14
Morse function, 174
motif, 46

negative basis, 62
neighbor, 77
neighborhood graph, 129
network, 3
nodal domain, 269
node, 76
non-backtracking graph, 351, 354, 356
non-backtracking Laplacian, 354
non-backtracking matrix, 353
non-local model, 73
non-local TV model, 73
normalized combinatorial Laplace operator,
    32, 85
normalized graph Laplacian, 85, 151, 379

opposite graph, 138
orientable manifold, 64
orientation, 16, 22, 61
oriented hypergraph, 16

page rank algorithm, 375
partial differential equation, 4
petal graph, 104, 105, 117, 137, 149, 319, 325
p-form, 61
p-Laplacian, 68
Poincaré inequality, 53
Pólya-Cheeger constant, 11, 116
positive basis, 62
proper difference, 34
property (T), 381

quantum mechanics, 8

Ramanujan graph, 380
random graph, 4, 13, 322
rate of a code, 383
Rayleigh quotient, 38, 41, 54, 69, 79, 80, 163
Rayleigh-Ritz quotient, 38
relative cohomology, 32
Rellich compactness theorem, 53
Riemannian geometry, 7
Riemannian metric, 63
Riemann's habilitation thesis, 4

self-loop, 24, 77
separation of variables, 5

signed Cheeger constant, 150
signed graph, 138
signed Rayleigh quotient, 146
simplicial complex, 21
simplicial complexes with loops, 33
simplicial map, 305
smoothness term, 59, 72
Sobolev space, 52
spectral class, 325
spectral distance, 324
spectral gap, 105, 360
spectral theory of linear operators, 4
spectrum, 30
    bipartite graph, 97
    bistar, 107
    book graph, 104, 150
    complete bipartite graph, 100
    complete graph, 82, 92
    complete graph with self-loops, 82, 92
    cube graph, 321
    petal graph, 104
    signed graph, 141
    windmill graph, 105
star, 47
star graph, 83, 101, 117, 212, 319
strong covering, 312
strongly connected, 151
subcomplex, 34
subdifferentiable, 44, 182
subgradient, 44, 183
submodular, 185
super concentrator, 382
switching, 140
switching equivalent, 140
switching function, 140
symmetric bistar, 106

tangent vector, 60
tensor, 328
topological expander, 384
tree, 92, 96, 102
triangle, 110
tripartite, 108
TV model, 72

unweighted graph, 85
up Laplace operator, 29, 48
up path connected, 313

vector field, 60
vertex, 76

vertex boundary, 114
vibration, 6
volume, 111

water wheel graph, 319, 325
wave equation, 6
wave operator, 4
weak convergence, 52

weakly connected, 151
weighted adjacency matrix, 151
weighted directed graph, 151
weighted graph, 85
weighted simplicial complex, 31
Weyl estimate, 58, 66
wheel graph, 101, 319
windmill graph, 105, 106

For EU product safety concerns, contact us at Calle de José Abascal, 56–1°, 28003 Madrid, Spain or eugpsr@cambridge.org.